Quantum Image Processing in Practice

Quantum Image Processing in Practice

A Mathematical Toolbox

Artyom M. Grigoryan

Sos S. Agaian

WILEY

Published by John Wiley & Sons, Inc., Hoboken, New Jersey.
Published simultaneously in Canada.

Library of Congress Cataloging-in-Publication Data is applied for:

Hardback ISBN 9781394265152

Cover Design: Wiley
Cover Image: © Jose A. Bernat Bacete/Getty Images

Set in 9.5/12.5pt STIXTwoText by Straive, Pondicherry, India

To Anoush
and
To Sarkis and Gayane

Contents

Preface

The modern world has witnessed remarkable applications in the dynamic field of image processing, where operations transform an image to enhance it or extract vital information. It is a vibrant and diverse field encompassing various applications, such as facial recognition, image segmentation and compression, noise reduction, and more. These applications require sophisticated techniques to transform, enhance, and extract image information. However, these techniques also demand substantial computational resources for image storage and processing, which pose significant challenges for scalability and efficiency. Therefore, there is a critical need for more advanced and innovative methods to handle visual information. On the other hand, quantum computing defines a probabilistic approach to represent classical information using methods from quantum theory. Quantum computing offers a probabilistic and parallel approach to computation, which differs fundamentally from the deterministic and sequential approach of classical computing. The basic unit of quantum information, the qubit, can exist in a superposition of two states until measured, which enables quantum parallelism and entanglement. These quantum phenomena can provide exponential speedups and enhanced security for specific computational tasks, such as factoring large numbers, searching unsorted databases, simulating quantum systems, and solving linear systems of equations.

Quantum image processing (QIP) is a research branch of quantum information and computing that aims to exploit the advantages of quantum computing for image processing. QIP studies how to encode and process images using various quantum image representations and operations in a quantum computer. QIP has the potential to outperform classical image processing in terms of computing speed, security, and minimum storage requirements. However, QIP also faces many challenges and open questions, such as quantum superiority, reading the classical data, measurement, noise and error correction, scalability and compatibility, and the practical implementation of QIP algorithms and circuits.

In this book, we provide a comprehensive introduction to QIP, covering the theoretical foundations, methodological developments, quaternion color imaging, and practical QIP applications. We describe the existing quantum image representations and their operations, such as geometric transformations, color transformations, filtering, and enhancement. We also explore the emerging topics and applications of QIP, such as quantum image filtration in the frequency domain, convolution, and fast unitary transforms. We discuss the current state of QIP research, addressing the controversies and opportunities, as well as the challenges and future directions of QIP. We illustrate the QIP algorithms and circuits with detailed examples, diagrams, and code snippets using the Qiskit framework. We also provide exercises and references for further learning and research.

Organization of the Book

This book is organized into 14 chapters, as follows:

Chapter 1: Introduction
This chapter provides an overview of the main concepts and motivations of quantum computing and image processing. It outlines the structure and objectives of the book.

Chapter 2: Basic Concepts of Qubits
This chapter delves into the core concepts and principles, such as computational qubit states, superposition, operations on qubits, permutations, elementary gates, and qubit measurement. It presents the operations and gates in matrix and graphical notations and illustrates them with examples; 3-D model of qubits is presented together with the known Bloch sphere.

Chapter 3: Understanding 2-Qubit Systems
This chapter focuses on 2-qubit systems, which are the building blocks of multi-qubit systems. It discusses the mathematical tools and techniques for manipulating 2-qubit systems, such as projection operators, Kronecker product and sum, qubit entanglement, orthogonality, and unitary transformations. It also describes the elementary operations and main gates for 2-qubit systems, such as CNOT, SWAP, local and controlled gates, and explores their properties and applications with practical examples.

Chapter 4: Multi-Qubit Superpositions and Operations
This chapter extends the concepts and methods of 2-qubit systems to multi-qubit systems, which are essential for quantum image processing. It examines multi-qubit superpositions of different types. Many 3-qubit gates with 1 and 2 control bits are described with matrices and circuit elements. It also highlights the key 3-qubit gates, such as Toffoli and Fredkin, bit SWAP, and Hadamard gates, and shows how they can be used to implement classical logic functions and reversible circuits.

Chapter 5: Fast Transforms in Quantum Computation
This chapter introduces the quantum analogs of the classical fast transforms, such as the discrete paired, Fourier, and Hadamard transforms which are widely used in image processing. It provides detailed descriptions of the algorithms and implementations of these quantum-fast transforms, supported by examples and circuit designs. It also compares the advantages and disadvantages of these quantum fast transforms concerning their classical counterparts. Examples and circuits of these transforms and their inverses on 2-, 3-, and 4-qubits are presented. The paired transform is the core of the Fourier and Hadamard transforms. Therefore, the quantum paired transform is described in detail. The 1-D quantum circular convolution for phase filters with circuits is also presented with examples.

Chapter 6: Quantum Signal-Induced Heap Transform
This chapter presents a novel concept of quantum fast transform, which refers to the so-called discrete signal-induced heap transform (DsiHT), which can generate a unique unitary and fast transform for any given signal. It explains the theory and algorithm of quantum signal-induced heap transform (QsiHT) and demonstrates its applications in quantum cosine and Hartley transforms with quantum circuits. It also shows how DsiHT can be used to factorize and decompose any transform in a set of rotated gates and permutation, as well to initiate any quantum superposition from the basis state 0.

Chapter 7: Quantum Image Representation with Examples
This chapter describes the various quantum image representations proposed in the literature and compares their features and limitations. It covers the models for both grayscale and color images, such as QLM, NEQR, FRQI, RKL, GQIP, QIRHSI, and MQFTR. It also presents the 2×2 model of color image respresentation as a single

grayscale image in quantum computation. It explores the quantum representation of different color models, such as RGB, CMY, XYZ, HSV, and HSI, and discusses the challenges and opportunities of quantum color image processing.

Chapter 8: *Image Representation on the Unit Circle and MQFTR*
This chapter focuses on a specific type of quantum image representation, called the multi-qubit Fourier transform representation (MQFTR), which encodes the image information on the unit circle using the Kronecker product of qubits. It explains the advantages and disadvantages of MQFTR. It presents some extensions and variations of MQFTR, such as MQFTR with phase shift and MQFTR with amplitude modulation. It also describes the quantum schemes for 2-D quantum Fourier transform with examples for 4×4 and 8×8 images.

Chapter 9: *New Arithmetic on Qubits*
This chapter introduces some novel concepts and methods of arithmetic operations on qubits, such as multiplication, division, conjugate, and inverse. It extends these operations to multi-qubit superpositions and discusses their properties and applications. It also shows how these operations can be implemented using quantum circuits and algorithms in image summation, linear convolution and filtration with examples.

Chapter 10: *Quaternion-Based Arithmetic in Quantum Image Processing*
This chapter explores the non-commutative quaternion arithmetic in quantum color image processing, which can offer some advantages over conventional complex arithmetic. It differentiates between the traditional and commutative quaternion algebras and discusses their properties and applications. It presents a new concept of the multiplicative group of 2-qubits and describes the main properties of the multiplication of 2-qubits and 2-qubit-based superpositions. It also describes the graphical representation of 2-qubits.

Chapter 11: *Quantum Schemes for Multiplication of 2-Qubits*
This chapter presents detailed quantum 2-qubit multiplication circuits that underlie many quantum arithmetic operations. It showcases a few circuits composed using the QsiHT concept and compares their performance and complexity. It shows how to design the quantum 4×4-gates of multiplication with 4, 5, and 6 rotations. It also describes 12 Hadamard matrices as multiplication gates.

Chapter 12: *Quaternion Qubit Image Representation (QQIR)*
This chapter introduces a new quantum image representation, called the quaternion qubit image representation (QQIR), which combines the features of 4-D quaternion arithmetic and MQFTR. It covers some basic operations in QQIR, such as square root, power, and exponentiation, and shows how they can be used for image processing. It explores some advanced operations in QQIR, such as the convolution and gradient calculation, and demonstrates their applications in image filtering, edge detection, and feature extraction. It also describes the concept of the quantum quaternion Fourier transform (QQFT) and ideal filtration by this transform.

Chapter 13: *Quantum Neural Networks (QNN)*
This chapter bridges quantum computing and machine learning and discusses the development and applications of quantum neural networks inspired by classical neural networks. It highlights the differences, synergies, and challenges of quantum and classical neural networks and reviews some existing models and architectures of quantum neural networks. It also explores some potential applications of quantum neural networks in image processing, such as image classification, recognition, segmentation, restoration, and reconstruction.

Chapter 14: *Conclusion and Opportunities and Challenges of Quantum Image Processing*
This chapter summarizes the main contributions and findings of the book and reflects on the current state and future directions of quantum image processing. It discusses the opportunities and challenges of quantum image processing, such as quantum superiority, noise, and error correction, scalability and compatibility, and practical implementation. It also provides some suggestions and recommendations for further research and development in quantum image processing.

Designed for a diverse audience, from students to professionals, this book is accessible to those with some background in linear algebra, quantum mechanics, and image processing. Many examples and references throughout the text encourage further exploration and deeper understanding. We are grateful to everyone who read this book. We hope the reader will enjoy learning about quantum imaging and its applications and gain some ideas and inspiration from the methodologies and concepts discussed in this book. This book explores the emerging interdisciplinary field of quantum imaging, introducing fundamental concepts, state-of-the-art techniques, and applications.

An Invitation to Innovation: We also hope that this book will arouse readers' interest in QIP and inspire them to contribute to its development. Join us on a journey where the classical and quantum worlds intertwine, unlocking a future brimming with unprecedented potential for image processing innovation.

We appreciate all who assisted in the preparation of this book. We are grateful to Meruzhan Grigoryan, Alexis Gomez, and the reviewers for many suggestions and recommendations.

May 2024 *Artyom Grigoryan and Sos S. Agaian*

Acknowledgments

Artyom Grigoryan: I would like to express my deepest gratitude to my daughter, Anoush Grigoryan, for her unwavering support. My heartfelt thanks go to my brother, Meruzhan Grigoryan, for his insightful comments and suggestions, which significantly improved the material in Chapter 6. I am also thankful to our student, Alexis Gomez at UTSA, for his valuable assistance in developing Python codes for the examples in this book.

Sos Agaian: I want to express my sincere appreciation and dedication to my wife, Gayane Abrahamian, and my son, Sarkis Agaian, for their steadfast support throughout the writing process. Their constant encouragement, motivation, and love were the foundation of this work. This book would not have come to fruition without their presence and assistance.

Additionally, we extend our gratitude to Kavipriya for her diligent support, the reviewers for their invaluable feedback, senior commissioning editor Sandra Grayson for her expert guidance, senior editorial assistant Becky Cowan for her meticulous attention to detail, and cover designer Jose Bacede for his creative contributions.

Finally, we are grateful for the opportunity to present "*Quantum Image Processing: A Mathematical Toolbox*," a work that has been a labor of love and dedication.

About the Companion Website

This book is accompanied by a companion website

www.wiley.com/go/grigoryan/quantumimageprocessing

This website includes:

- QuAlgorithms

Part I

Mathematical Foundation of Quantum Computation

1

Introduction

Image processing represents a critical use of artificial intelligence in various applications, including biomedicine, entertainment, economics, and industry. For example, image processing is extensively used in fast-growing markets like facial recognition and autonomous vehicles. Recently, the rapidly increased volume of image data has become the critical driving force for further improving image processing and analysis efficiency. Quantum computing offers great promise for speedy computation of problems in digital image processing (DIP), namely, in processing grayscale and color images. Quantum image processing (QIP) is a research branch of quantum information and quantum computing. It studies how to use quantum mechanics' properties to represent images in a quantum computer and then implement various image operations based on that image format. The parallel execution of multiple computations is a quantum computer's main advantage [1, 2], which can be used for many image processing applications. Therefore, developing new tools for calculating not only known procedures in DIP but also new ones is essential, using special rules for calculating qubits. The basic unit in quantum computation is a single qubit with only two states. The mathematical description of a qubit is probabilistic. The qubit can be measured only once, and at any moment of its measurement, it will be in only one of its states 0 or 1. If we consider qubits as spins, then these two states are "spin up" and "spin down." Before the measurement, the state of the qubit is described as a linear combination of the basis states 0 and 1. The amplitudes of such a combination are defined by probabilities of the qubit being in these states. Namely, these amplitudes are square roots of the corresponding probabilities. Thus, in quantum computation, there are no numbers greater than 1, such as 2, 3, 5, and 12. Such numbers can only be written in basis states of multi-qubit superpositions. It is possible only to work and get such state numbers through their amplitudes. In other words, through the probabilities of qubits to be in these states. This is the main difference between the existing classical and future quantum computers. For example, to add two 3-bit numbers 3 and 4 and get 7 is a trivial operation for the classical computer. Solving this equation $3 + 4 = 7$ in a quantum computer with simple 3-bit operations $\oplus$ is impossible. All existent algorithms for the addition of numbers use more bits, or qubits, and other operations. In general, the composition of algorithms for processing images is a complex task. Storage and processing of information about 2^r states in one superposition of r-qubits is the main factor and a big advantage in quantum computation (QC). Therefore, there is an opinion that a quantum computer may surpass the much real possibilities of existing computers in the near future. All this impeded the development of image representation, storage, and processing methods in QC. In quantum imaging, several methods are currently available, such as interference, correlation, and entanglement-based quantum mapping [3–5]. These are the known methods in use with applications in quantum computation. Unlike the classical computer, images in QC can be represented by different quantum superpositions of states. These models include the qubit lattice model (QLM) [6], flexible representation for quantum images (FRQI) [7], the novel enhanced quantum representation (NEQR) [8–10], novel representations of color images [11, 12], new models of the Fourier transform quantum representation (FTQR) [12, 13], and quaternion qubit image representation (QQIR) [14, 15].

Quantum Image Processing in Practice: A Mathematical Toolbox, First Edition. Artyom M. Grigoryan and Sos S. Agaian.

Companion website: www.wiley.com/go/grigoryan/quantumimageprocessing

This book aims to introduce the reader to the necessary part of QIP that plays an essential role in quantum information processing. It includes

- A review of some needed mathematical concepts of qubits and their operations, such as multiplication, division, inverse qubits, and their quantum circuits description.
- the quantum Fourier and Hadamard transform with their quantum circuits description.
- the processing images in the frequency domain, namely, applying quantum ideal low-pass and high-pass filters.
- the quantum grayscale and color image quaternion-based representations that allow efficient encoding of the classical data into a quantum state and future use in QIP applications.
- the basics of quantum neural networks.

The study shows that the number of researchers working in the QIP field is increasing, and some significant problems remain unsolved. We hope this book will accelerate the efforts to create more practical QIP-based technologies.

References

1 Nielsen, M.A. and Chuang, I.L. (2000). *Quantum Computation and Quantum Information*. Cambridge: Cambridge University Press.

2 Rieffel, E.G. and Polak, W.H. (2011). *Quantum Computing: A Gentle Introduction*. Cambridge: The MIT Press.

3 Latorre, J. (2005). Image compression and entanglement. *arXiv: quant-ph/0510031*.

4 Yongquan, C., Xiaowei, L., and Nan, J. (2018). A survey of quantum image representations. *Chinese Journal of Electronics* 27 (4): 9.

5 Latorre, J.L. (2005). Image compression and entanglement. 4. https://arxiv.org/abs/quant-ph/0510031.

6 Venegas-Andraca, S. and Bose, S. (2003). Storing, processing, and retrieving an image using quantum mechanics. *Proceedings of. SPIE 5105, Quantum Information and Computation* (4 August 2003). https://doi.org/10.1117/12.485960.

7 Le, P., Dong, F., and Hirota, K. (2011). A flexible representation of quantum images for polynomial preparation, image compression, and processing operations. *Quantum Information Processing* 10 (1): 63–84.

8 Zhang, Y., Lu, K., Gao, Y., and Wang, M. (2013). NEQR: a novel enhanced quantum representation of digital images. *Quantum Information Processing* 12 (8): 2833–2860.

9 Jiang, N. and Wang, L. (2015). Quantum image scaling using nearest neighbor interpolation. *Quantum Information Processing* 14 (5): 1559–1571.

10 Li, H.-S., Fan, P., Xia, H.-Y. et al. (2019). Quantum implementation circuits of quantum signal representation and type conversion. *IEEE Transactions on Circuits and Systems I: Regular Papers* 66 (1): 341–354.

11 Sang, J.Z., Wang, S., and Li, Q. (2017). A novel quantum representation of color digital images. *Quantum Information Processing* 16 (42): 42–56.

12 Liu, K., Zhang, Y., Lu, K., and Wang, X. (2018). An optimized quantum representation for color digital images. *International Journal of Theoretical Physics* 57 (10): 2938–2948.

13 Grigoryan, A.M. and Agaian, S.S. (2020). New look on quantum representation of images: Fourier transform representation. *Quantum Information Processing* 19 (148): 26. https://doi.org/10.1007/s11128-020-02643-3.

14 Grigoryan, A.M. and Agaian, S.S. (2020). Quaternion quantum image representation: new models. *Proceedings Volume 11399, Mobile Multimedia/Image Processing, Security, and Applications 2020*; 113990O. https://doi.org/10.1117/12.2557862.

15 Grigoryan, A.M. and Agaian, S.S. (2018). *Quaternion and Octonion Color Image Processing with MATLAB*. SPIE. https://doi.org/10.1117/3.2278810.

2

Basic Concepts of Qubits

In this section, the basic concept and operations, called gates, on qubits are described. We are interested in the mathematical side of this concept, namely the mathematical model of the qubit. The main gates are considered and then, in Section 9.1, we present our vision of a multiplicative qubit group using qubit multiplication and division operations. One can even solve quadratic equations with qubits.

In the theory of quantum computation, a quantum bit, which is called a qubit, may be in two states, $|0\rangle$ and $|1\rangle$. We can think of a qubit as a magical, fast-spinning 2D coin with two numbers 0 and 1 written on its sides. When we stop the rotation with our hand, we will see only one side of the coin in the palm of our hand, that is, only the number 0 or 1 (see Fig. 2.1). The events that the coin will show 0 or 1 may occur with different probabilities p_1 and p_2. The square roots of these probabilities, $a_0 = \sqrt{p_0}$ and $a_1 = \sqrt{p_1}$, are called the amplitudes of the qubit. These two events are called the basis states of the qubit. The state 0 is usually written with 2 bits as 10, and the state 1 as 01. Mathematically, a single qubit is defined as a linear combination of these basis states with given amplitudes a_0 and a_1. Any quantum system with only two states can be considered as a qubit. For example, (i) a nuclear spin with two energy levels and (ii) a photon with vertical and horizontal polarization.

For column and row vectors, Dirac's ket-bra notation is used. Ket-notation $|\boldsymbol{x}\rangle$ is the column vector for $\boldsymbol{x}$, and bra-notation $\langle \boldsymbol{y}|$ is the row vector for $\boldsymbol{y}$. The state 0 as the unit vector $[1, 0]'$ is the ket-0$|0\rangle$, and the basis state 1 as the vector $[0, 1]'$ is the ket-1 $|1\rangle$. Here, the symbol$'$ is used for the transpose vector. The operation $\langle \boldsymbol{y}|\boldsymbol{x}\rangle$ is the inner product, that is, $\langle \boldsymbol{y}||\boldsymbol{x}\rangle$. For examples, if $\boldsymbol{x} = [1, 2]$ and $\boldsymbol{y} = [3, -1]$, then

$$\langle \boldsymbol{y}|\boldsymbol{x}\rangle = \langle \boldsymbol{y}||\boldsymbol{x}\rangle = \boldsymbol{y}\boldsymbol{x}' = 1 \cdot 3 + 2 \cdot (-1) = 1.$$

If the vectors are complex, the inner product is defined as $\langle \boldsymbol{y}|\boldsymbol{x}\rangle = \langle \boldsymbol{y}|\overline{|\boldsymbol{x}\rangle}$. The bra vector $\langle \boldsymbol{x}|$ is complex conjugate of $|\boldsymbol{x}\rangle$.

In a real Euclidean space, the inner product is defined as a rule by means of which for each of the two elements x and y of the space a real number is associated (*inner product*), that is denoted as (x, y) and that satisfies the following four Axioms:

1) $(x, y) = (y, x)$;
2) $(x + z, y) = (x, y) + (z, y)$;
3) $(kx, y) = k(x, y),\ k \in R$;
4) $(x, x) = |x|^2 \geq 0$, and $(x, x) = 0$ if only $x = 0$.

Quantum Image Processing in Practice: A Mathematical Toolbox, First Edition. Artyom M. Grigoryan and Sos S. Agaian.

Companion website: www.wiley.com/go/grigoryan/quantumimageprocessing

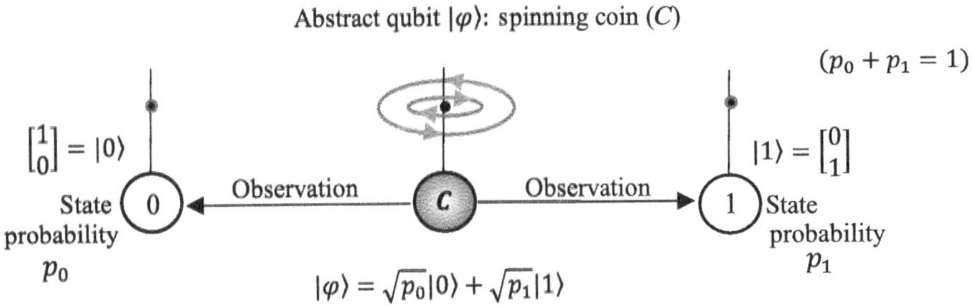

Fig. 2.1 Abstract model: 2D spinning coin with two numbers 0 and 1 written on the sides of the 2D coin.

Example 2.1 Vector Space of Dimension *n*
In the linear space of sets of n real numbers (or n-dimensional vectors) $\boldsymbol{x} = (x_0, x_1, \ldots, x_{n-1})$, $\boldsymbol{y} = (y_0, y_1, \ldots, y_{n-1})$, ..., the inner product is defined as

$$(\boldsymbol{x},\boldsymbol{y}) = x_0y_0 + x_1y_1 + \ldots + x_{n-1}y_{n-1}.$$

It is equal to the multiplication of vectors, $(\boldsymbol{x}, \boldsymbol{y}) = \boldsymbol{x}\boldsymbol{y}'$. In the ket-bra notation $(\boldsymbol{x}, \boldsymbol{y}) = \langle \boldsymbol{x}|\boldsymbol{y}\rangle$.

If the space of elements is complex, the inner product is the complex function. Therefore, the complex conjugate operation is considered in the first Axiom, $(\boldsymbol{x},\boldsymbol{y}) = \overline{(\boldsymbol{y},\boldsymbol{x})}$. Also, the constants k in the third Axiom is taken from the complex plane, that is, $k \in C$. It will be written as, $(k\boldsymbol{x},\boldsymbol{y}) = \overline{k}(\boldsymbol{x},\boldsymbol{y})$. For the above example, the inner product in the complex vector space of vectors $\boldsymbol{x}$ and $\boldsymbol{y}$ is defined as

$$(\boldsymbol{x},\boldsymbol{y}) = \overline{x}_0y_0 + \overline{x}_1y_1 + \ldots + \overline{x}_{n-1}y_{n-1} = \langle \boldsymbol{x}||\boldsymbol{y}\rangle = \langle \boldsymbol{x}|\boldsymbol{y}\rangle.$$

Other brackets can be used instead of parentheses, such as $\langle \boldsymbol{x}, \boldsymbol{y}\rangle$ and we can just write it like $\langle \boldsymbol{x}\boldsymbol{y}\rangle$. The inner product is linear in the second argument, $\langle \boldsymbol{x}, k\boldsymbol{y}\rangle = k\langle \boldsymbol{x}, \boldsymbol{y}\rangle$, and antilinear in the first argument, $\langle k\boldsymbol{x},\boldsymbol{y}\rangle = \overline{k}\langle \boldsymbol{x},\boldsymbol{y}\rangle$. Note that mathematicians usually define the inner product as $\langle \boldsymbol{x}|\boldsymbol{y}\rangle = \langle \boldsymbol{x},\boldsymbol{y}\rangle = x_0\overline{y_0} + x_1\overline{y_1} + \ldots + x_{n-1}\overline{y}_{n-1}$, which is linear in the first argument.

If A is a linear operator on vectors, the notation $\langle \boldsymbol{x}|A|\boldsymbol{y}\rangle$ denotes the inner product $\langle \boldsymbol{x}, A\boldsymbol{y}\rangle$, or $\langle \boldsymbol{x}|$ and $|A\boldsymbol{y}\rangle$. It is also the inner product of the vectors $\langle \boldsymbol{x}|A$ and $|\boldsymbol{y}\rangle$. Therefore, the conjugate operator A^* is defined by $\langle A^*\boldsymbol{x}| = \langle \boldsymbol{x}|A$. Then, $\langle \boldsymbol{x}, A\boldsymbol{y}\rangle = \langle A^*\boldsymbol{x}, \boldsymbol{y}\rangle$, or $\langle \boldsymbol{x}\,|\,A|\boldsymbol{y}\rangle = \langle A^*\boldsymbol{x}|\boldsymbol{y}\rangle$.

A single qubit is a superposition of two standard (computational) basis states with amplitudes a_0 and a_1,

$$|\varphi\rangle = a_0|0\rangle + a_1|1\rangle = a_0\begin{bmatrix}1\\0\end{bmatrix} + a_1\begin{bmatrix}0\\1\end{bmatrix} = \begin{bmatrix}a_0\\a_1\end{bmatrix}. \tag{2.1}$$

Figure 2.2 illustrates such a qubit for the above example with a spinning-coin. Equation 2.1 is the mathematical model of the single qubit. This model is probabilistic, because the amplitudes of the basis states define the probabilities of the qubit to be in state $|0\rangle$ and $|1\rangle$, after measuring the qubit.

In the general case, the coefficients, or amplitudes a_0 and a_1 may be complex numbers, such that $|a_0|^2 + |a_1|^2 = 1$. The case of complex amplitudes in the superposition $|\varphi\rangle$ is difficult to imagine and understand since the complex numbers are points in the plane or pairs of real numbers.

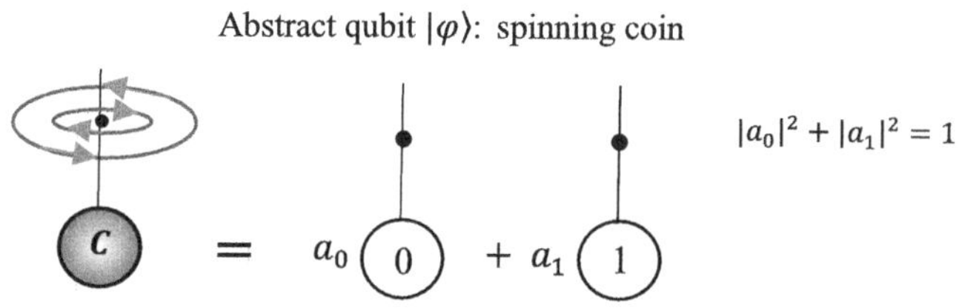

$$|\varphi\rangle = a_0|0\rangle + a_1|1\rangle \;=\; a_0\begin{bmatrix}1\\0\end{bmatrix} + a_1\begin{bmatrix}0\\1\end{bmatrix} = \begin{bmatrix}a_1\\a_2\end{bmatrix}$$

Fig. 2.2 The coin-qubit as a linear superposition of the states 0 and 1.

2.1 Measurement of the Qubit

A qubit $|\varphi\rangle = a_0|0\rangle + a_1|1\rangle$ can only be measured once, and only one of its states will be the result of the measurement. It is the state $|0\rangle$ or $|1\rangle$ with probability $|a_0|^2$ or $|a_1|^2$, respectively. The values of these two numbers are calculated after repeatedly measuring the qubit in its states state $|0\rangle$ and $|1\rangle$. When a qubit is measured, it enters the measurement state. Its original superposition cannot be restored. Since $\langle 0|0\rangle = 1$ and $\langle 0|1\rangle = \langle 1|0\rangle = 0$, the basis states $|0\rangle$ and $|1\rangle$ are orthogonal. Therefore, the amplitudes can be calculated as the projections of $|\varphi\rangle$ on these states; $\langle 0|\varphi\rangle = a_0$ and $\langle 1|\varphi\rangle = a_1$. The probability of the qubit $|\varphi\rangle$ to be in state $|0\rangle$ after measurement (M) can be calculated by the inner product as

$$\Pr(M|\varphi\rangle = |0\rangle) = |\langle 0|\varphi\rangle|^2 = |\langle 0|(a_0|0\rangle + a_1|1\rangle)\rangle|^2 = |a_0\langle 0|0\rangle + a_1\langle 0|1\rangle|^2 = |a_0|^2.$$

Similarly, the probability of the qubit $|\varphi\rangle$ to be in state $|1\rangle$ after measurement can be calculated as follows:

$$\Pr(M|\varphi\rangle = |1\rangle\,) = |\langle 1|\varphi\rangle|^2 = |\langle 1|(a_0|0\rangle + a_1|1\rangle)\rangle|^2 = |a_0\langle 1|0\rangle + a_1\langle 1|1\rangle|^2 = |a_1|^2.$$

Note that $\langle 0|\varphi\rangle|0\rangle = a_0|0\rangle$ and $\langle 1|\varphi\rangle|1\rangle = a_1|1\rangle$.

Let us consider the superposition $|\varphi\rangle$ with real amplitudes. Measurement of the qubit in states $|0\rangle$ and $|1\rangle$ can be described by orthogonal projections P_0 and P_1 which are described by the matrices

$$P_0 = \begin{bmatrix}1 & 0\\0 & 0\end{bmatrix} = |0\rangle\langle 0\,| \quad\text{and}\quad P_1 = \begin{bmatrix}0 & 0\\0 & 1\end{bmatrix} = |1\rangle\langle 1|.$$

Indeed,

$$M[|\varphi\rangle = |0\rangle] = \frac{P_0|\varphi\rangle}{\sqrt{\Pr(M|\varphi\rangle = |0\rangle)}} = \frac{P_0|\varphi\rangle}{\sqrt{\langle\overline{\varphi}|P_0|\varphi\rangle}} = \frac{1}{\sqrt{|\,a_0^2\,|}}\begin{bmatrix}1 & 0\\0 & 0\end{bmatrix}\begin{bmatrix}a_0\\a_1\end{bmatrix} = \frac{1}{|\,a_0\,|}\begin{bmatrix}a_0\\0\end{bmatrix} = \pm|0\rangle,$$

where

$$\langle\overline{\varphi}|P_0|\varphi\rangle = [\overline{a_0}\;\overline{a_1}]\begin{bmatrix}1 & 0\\0 & 0\end{bmatrix}\begin{bmatrix}a_0\\a_1\end{bmatrix} = [\overline{a_0}\quad\overline{a_1}]\begin{bmatrix}a_0\\0\end{bmatrix} = |a_0|^2.$$

Similarly,

$$\langle\overline{\varphi}|P_1|\varphi\rangle = [\overline{a_0}\;\;\overline{a_1}]\begin{bmatrix}0 & 0\\0 & 1\end{bmatrix}\begin{bmatrix}a_0\\a_1\end{bmatrix} = [\overline{a_0}\quad\overline{a_1}]\begin{bmatrix}0\\a_1\end{bmatrix} = |a_1|^2$$

and

$$M[|\varphi\rangle = |1\rangle] = \frac{P_1|\varphi\rangle}{\sqrt{\Pr(M|\varphi\rangle = |1\rangle)}} = \frac{P_1|\varphi\rangle}{\sqrt{\langle\overline{\varphi} \mid P_1|\varphi\rangle}} = \frac{1}{\sqrt{|\,a_1^2\,|}}\begin{bmatrix}0 & 0\\ 0 & 1\end{bmatrix}\begin{bmatrix}a_0\\ a_1\end{bmatrix} = \frac{1}{|\,a_1\,|}\begin{bmatrix}0\\ a_1\end{bmatrix} = \pm|1\rangle.$$

If the amplitudes are real numbers, then the qubit can be described by the 2D vector with length 1. Thus, each 2D vector after normalization can be considered as a qubit. For example, for the vector $\boldsymbol{x} = [3,\ 4]'$ with norm $\sqrt{3^2+4^2} = 5$, the corresponding qubit $|\boldsymbol{x}\rangle$ is defined as

$$\boldsymbol{x} = \begin{bmatrix}3\\ 4\end{bmatrix} = 3\begin{bmatrix}1\\ 0\end{bmatrix} + 4\begin{bmatrix}0\\ 1\end{bmatrix} = 3|0\rangle + 4|1\rangle \rightarrow |\boldsymbol{x}\rangle = \frac{3|0\rangle + 4|1\rangle}{5} = \frac{3}{5}|0\rangle + \frac{4}{5}|1\rangle.$$

Examples of well-known qubits are the Hadamard superpositions

$$|\varphi_1\rangle = \frac{|0\rangle + |1\rangle}{\sqrt{2}} \quad \text{and} \quad |\varphi_2\rangle = \frac{|0\rangle - |1\rangle}{\sqrt{2}}. \tag{2.2}$$

These two qubits are orthogonal, that is, the inner product of qubits is equal to zero. Indeed, we have the following:

$$\langle\varphi_1|\varphi_2\rangle = \langle\varphi_1\,||\varphi_2\rangle = \frac{1}{2}[1\ 1]\begin{bmatrix}1\\ -1\end{bmatrix} = 0.$$

Therefore, a single qubit $|\varphi\rangle$ can also be written in the basis states $\{|\varphi_1\rangle, |\varphi_2\rangle\}$, that is,

$$|\varphi\rangle = a_0|0\rangle + a_1|1\rangle = b_0\left(\frac{|0\rangle + |1\rangle}{\sqrt{2}}\right) + b_1\left(\frac{|0\rangle - |1\rangle}{\sqrt{2}}\right) = b_0|\varphi_1\rangle + b_1|\varphi_2\rangle, \tag{2.3}$$

where the amplitudes $b_0 = (a_0 + a_1)/\sqrt{2}$ and $b_1 = (a_0 - a_1)/\sqrt{2}$.

If the amplitudes of the qubit superposition are not real numbers, the qubit without global phase can be described as [1]

$$|\varphi\rangle = a_0|0\rangle + a_1|1\rangle = \cos\frac{\vartheta}{2}|0\rangle + e^{i\phi}\sin\frac{\vartheta}{2}|1\rangle, \quad \phi \in [0, 2\pi], \vartheta \in [0, \pi]. \tag{2.4}$$

This is the case when the amplitude a_0 is real, and the amplitude $a_1 = a_{1,r} + ia_{1,i}$ is a complex number. Therefore, the single qubit can be presented by the 3D point $q = q(\vartheta/2, \phi) = (x, y, z)$ on the unit sphere, where

$$z = a_1 = \cos\frac{\vartheta}{2}, \quad x = a_{1,r} = \sin\frac{\vartheta}{2}\cos\phi, \quad y = a_{1,i} = i\sin\frac{\vartheta}{2}\sin\phi. \tag{2.5}$$

Figure 2.3 shows the geometry of this point together with points that present the basis states $|0\rangle$ and $|1\rangle$ in part (a). Here, X–Y presents the complex plane. The state $|0\rangle$ corresponds to the points $q(0, 0) = (0, 0, 1)$ and the state $|1\rangle$ to the point $q(\pi, 0) = (1, 0, 0)$.

The qubits in states $(|0\rangle \pm |1\rangle)/\sqrt{2}$ correspond to the points $q(\pi/4, 0)$ and $q = q(-\pi/4, \pi)$, which are shown in part (b). Indeed, when the angles are $\vartheta/2 = \pm\pi/4$ and $\phi = 0$, the corresponding points $q_{1,2} = q(\pm\pi/4, 0)$ have the coordinates

$$z = \cos\frac{\vartheta}{2} = \frac{1}{\sqrt{2}}, \quad x = \cos\phi\sin\frac{\vartheta}{2} = \pm\frac{1}{\sqrt{2}}, \quad y = \sin\phi\sin\frac{\vartheta}{2} = 0.$$

The corresponding qubits are

$$|\varphi_1\rangle = \frac{1}{\sqrt{2}}(|0\rangle + |1\rangle) \quad \text{and} \quad |\varphi_2\rangle = \frac{1}{\sqrt{2}}(|0\rangle - |1\rangle).$$

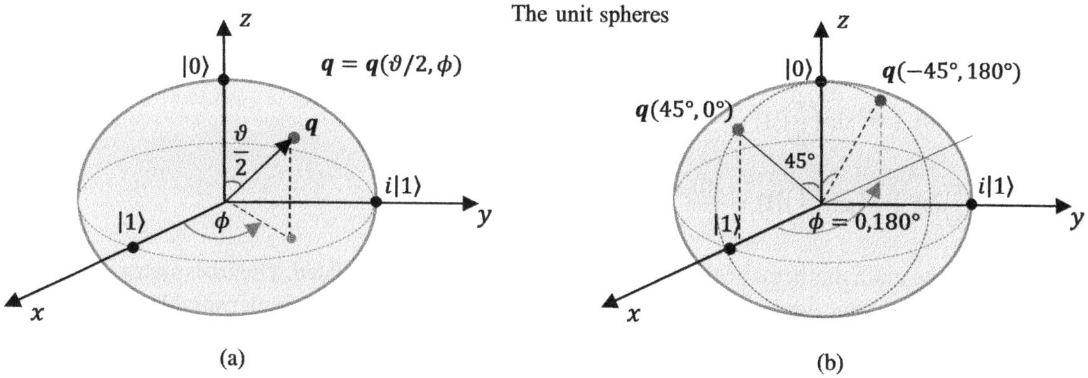

Fig. 2.3 (a) The 3D model of the qubit with (b) the points of the Hadamard states.

In the general $\phi = 0$ case, the qubit

$$|\varphi\rangle = a_0|0\rangle + a_1|1\rangle = \cos\frac{\vartheta}{2}|0\rangle + \sin\frac{\vartheta}{2}|1\rangle, \quad \vartheta \in [0, \pi],$$

and it is represented by a point lying on the unit circle in X–Z plane.

Here, we mention the well-known Bloch unit sphere, where the 3D points $b = b(\vartheta, \phi) = (x, y, z)$ for the same qubit $|\varphi\rangle$ are calculated as follows:

$$z = \cos\vartheta, \quad x = \cos\phi\sin\vartheta, \quad y = \sin\phi\sin\vartheta, \quad \vartheta \in [0, \pi], \ \phi \in [0, 2\pi]. \tag{2.6}$$

The traditional spherical coordinates are used and the sphere is shown in Fig. 2.4.

Thus, the coordinates of the amplitudes of the qubit $|\varphi\rangle$ in Eq. 2.5 are changed as

$$\boldsymbol{q} = \left(\cos\phi\sin\frac{\vartheta}{2}, \sin\phi\sin\frac{\vartheta}{2}, i\cos\frac{\vartheta}{2}\right) \to \boldsymbol{b} = (\cos\phi\sin\vartheta, \sin\phi\sin\vartheta, \cos\vartheta).$$

In this sphere, the points (0, 0, 1) and (0, 0, − 1) correspond to the angles $(\vartheta, \phi) = (0, \phi)$ and (π, ϕ), respectively. Considering $\phi = 0$, we obtain two basis states

$$|\varphi_{001}\rangle = \cos(0)|0\rangle + e^{i0}\sin(0)|1\rangle = |0\rangle \quad \text{and} \quad |\varphi_{00-1}\rangle = \cos\left(\frac{\pi}{2}\right)|0\rangle + e^{i0}\sin\left(\frac{\pi}{2}\right)|1\rangle = |1\rangle.$$

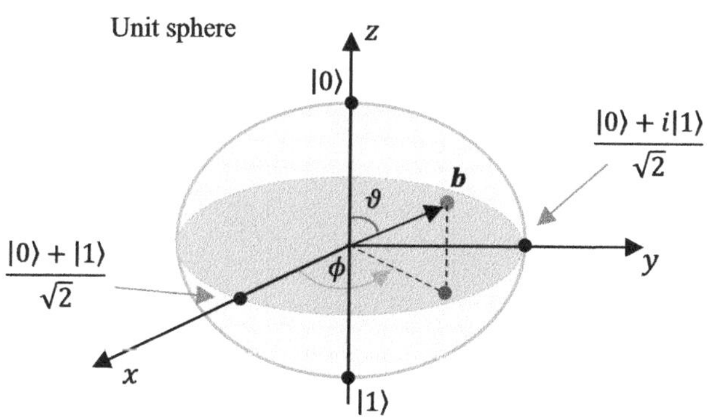

Fig. 2.4 The Bloch sphere with points presenting the qubits and basis states.

These points are shown in Fig. 2.4 together with other two points (1, 0, 0) and (0, 1, 0) corresponds to the angles $(\vartheta, \phi) = (\pi/2, 0)$ and $(\pi/2, \pi/2)$, respectively. The corresponding qubits are

$$|\varphi_{100}\rangle = \cos\left(\frac{\pi}{2}\right)|0\rangle + e^{i0}\sin\left(\frac{\pi}{2}\right)|1\rangle = \frac{|0\rangle + |1\rangle}{\sqrt{2}},$$

$$|\varphi_{010}\rangle = \cos\left(\frac{\pi}{2}\right)|0\rangle + e^{\frac{i\pi}{2}}\sin\left(\frac{\pi}{2}\right)|1\rangle = \frac{|0\rangle + i|1\rangle}{\sqrt{2}}.$$

In the Bloch sphere, the main mathematical description of the basis states is violated. The basis states $|0\rangle$ and $|1\rangle$ are on the same vertical and opposite in sign. Any vectors on one straight line are commensurable, or let us say linearly dependent. It is not possible to measure them at the same time in a qubit. The basis states $|0\rangle$ and $|1\rangle$ are opposite to each other, more precisely orthogonal if we use the concept of inner product, $\langle 0|1\rangle = \langle 1|0\rangle = 0$. Therefore, the model of a single qubit, represented in the sphere shown in Fig. 2.3 is considered more accurate.

2.1.1 Operations on Qubits

Superpositions can be added and subtracted with subsequent normalization. For instance, if we have two different qubits $|\varphi\rangle = a_0|0\rangle + a_1|1\rangle$ and $|\psi\rangle = b_0|0\rangle + b_1|1\rangle$, then the qubit of the sum is defined as the superposition

$$|\varphi + \psi\rangle = \frac{1}{K}[(a_0 + b_0)|0\rangle + (a_1 + b_1)|1\rangle]. \tag{2.7}$$

The normalization coefficient $K = \sqrt{(a_0 + b_0)^2 + (a_1 + b_1)^2}$. If the qubits are equal, this operation does not change the qubit, that is, $|\varphi + \varphi\rangle = |\varphi\rangle$. In other words, the superpositions $a_0|0\rangle + a_1|1\rangle$, $2a_0|0\rangle + 2a_1|1\rangle$, $3a_0|0\rangle + 3a_1|1\rangle$,... correspond to the same qubit.

2.1.2 Elementary Gates

Operators on qubits present themselves 2×2 unitary matrices U. The inverse U^{-1} of a unitary matrix is its conjugate transpose, $U^* = \overline{U'}$. In quantum mechanics, this matrix is referred to as Hermitian adjoint of U and denoted by $U^\dagger$. Note that any matrix (gate) can be written as product of vectors in ket-bra notations. Indeed, a matrix A with only one nonzero element at position (n, m) can be written as $|n\rangle\langle m|$. For example,

$$A = \begin{bmatrix} 0 & 1 \\ 0 & 0 \end{bmatrix} = \begin{bmatrix} 1 \\ 0 \end{bmatrix}[0 \quad 1] = |0\rangle\langle 1|.$$

The list of known unitary matrices, or gates, on qubits includes the following gates.

1) Elementary rotation, or Givens rotation

$$W = W(\vartheta) = \begin{bmatrix} \cos\vartheta & -\sin\vartheta \\ \sin\vartheta & \cos\vartheta \end{bmatrix}. \tag{2.8}$$

2) Y-rotation by angle ϑ, $R_y(\vartheta) = W(\vartheta/2)$.
3) X-rotation by angle ϑ,

$$R_x(\vartheta) = \begin{bmatrix} \cos\vartheta/2 & -i\sin\vartheta/2 \\ i\sin\vartheta/2 & \cos\vartheta/2 \end{bmatrix}.$$

4) Z-rotation by angle ϑ,

$$R_z(\vartheta) = \begin{bmatrix} e^{-i\vartheta/2} & 0 \\ 0 & e^{i\vartheta/2} \end{bmatrix} = e^{-i\vartheta/2}\begin{bmatrix} 1 & 0 \\ 0 & e^{i\vartheta} \end{bmatrix}.$$

5) Pauli NOT gate X and Pauli Z and Y-gates,

$$X = |0\rangle\langle 1| + |1\rangle\langle 0| = \begin{bmatrix} 0 & 1 \\ 1 & 0 \end{bmatrix}, \quad Z = |0\rangle\langle 0| - |1\rangle\langle 1| = \begin{bmatrix} 1 & 0 \\ 0 & -1 \end{bmatrix}, \tag{2.9}$$

$$Y = -i|0\rangle\langle 1| + i|1\rangle\langle 0| = i\begin{bmatrix} 0 & -1 \\ 1 & 0 \end{bmatrix} = \begin{bmatrix} 0 & -i \\ i & 0 \end{bmatrix}. \tag{2.10}$$

The results of these gates on a single qubit $|\varphi\rangle = a_0|0\rangle + b_0|1\rangle$ are

$$X: \; a_0|0\rangle + b_0|1\rangle \to b_0|0\rangle + a_0|1\rangle, \quad Z: \; a_0|0\rangle + b_0|1\rangle \to a_0|0\rangle - b_0|1\rangle,$$
$$Y: \; a_0|0\rangle + b_0|1\rangle \to -ib_0|0\rangle + ia_0|1\rangle.$$

6) V-gate, or $\sqrt{\text{NOT}}$ gate

$$V = \sqrt{X} = \frac{1}{2}\begin{bmatrix} 1+i & 1-i \\ 1-i & 1+i \end{bmatrix}. \tag{2.11}$$

The determinant of this matrix is i, and the Hermitian adjoint of matrix V is

$$V^* = \frac{1}{2}\begin{bmatrix} 1-i & 1+i \\ 1+i & 1-i \end{bmatrix}.$$

7) Hadamard gate

$$H = H_2 = \frac{1}{\sqrt{2}}(X+Z) = \frac{1}{\sqrt{2}}\begin{bmatrix} 1 & 1 \\ 1 & -1 \end{bmatrix}. \tag{2.12}$$

The Hadamard transform of the qubit $|\varphi\rangle = a_0|0\rangle + b_0|1\rangle$ is the qubit in the superposition

$$H|\varphi\rangle = \frac{a_0+b_0}{\sqrt{2}}|0\rangle + \frac{a_0-b_0}{\sqrt{2}}|1\rangle. \tag{2.13}$$

Note, that the determinant of this matrix is equal to -1, but $H^2 = HH$ is the identity 2×2 matrix

$$H^2 = I = \begin{bmatrix} 1 & 0 \\ 0 & 1 \end{bmatrix}.$$

8) Parameterized phase P-gate and its particular cases, the T- and S-gates

$$P(\phi) = \begin{bmatrix} 1 & 0 \\ 0 & e^{i\phi} \end{bmatrix}, \quad T = P(\pi/4) = \begin{bmatrix} 1 & 0 \\ 0 & e^{i\pi/4} \end{bmatrix}, \quad S = P(\pi/2) = \begin{bmatrix} 1 & 0 \\ 0 & i \end{bmatrix}. \tag{2.14}$$

The results of these gates on a single qubit $|\varphi\rangle = a_0|0\rangle + b_0|1\rangle$ are

$$P(\phi)|\varphi\rangle = a_0|0\rangle + e^{i\phi}b_0|1\rangle, \quad T|\varphi\rangle = a_0|0\rangle + e^{i\pi/4}b_0|1\rangle,$$
$$S|\varphi\rangle = a_0|0\rangle + ib_0|1\rangle.$$

Note that $S = \sqrt{Z}$ and $T = \sqrt{S}$.

9) Paired transform gate with the matrix

$$A_2 = W(\pi/4) = \frac{1}{\sqrt{2}}\begin{bmatrix} 1 & -1 \\ 1 & 1 \end{bmatrix}. \quad (2.15)$$

The determinant of this matrix is 1,

$$A_2 = XH_2 = \begin{bmatrix} 0 & 1 \\ 1 & 0 \end{bmatrix} \times \frac{1}{\sqrt{2}}\begin{bmatrix} 1 & 1 \\ 1 & -1 \end{bmatrix} \text{ (or } H_2 = XA_2),$$

and

$$A_2^2 = \begin{bmatrix} 0 & -1 \\ 1 & 0 \end{bmatrix} = XZ = \begin{bmatrix} 0 & 1 \\ 1 & 0 \end{bmatrix}\begin{bmatrix} 1 & 0 \\ 0 & -1 \end{bmatrix}.$$

The paired transform of the qubit $|\varphi\rangle = a_0|0\rangle + b_0|1\rangle$ is the qubit

$$A_2|\varphi\rangle = \frac{a_0 - b_0}{\sqrt{2}}|0\rangle + \frac{a_0 + b_0}{\sqrt{2}}|1\rangle. \quad (2.16)$$

This gate is also known as the inverse pseudo-Hadamard gate ($H^\dagger$).

10) For basis states $|0\rangle$ and $|1\rangle$, the modulo 2 addition operation $\oplus$ is used. $|0\rangle \oplus |1\rangle = |1\rangle \oplus |0\rangle = |1\rangle$ and $|0\rangle \oplus |0\rangle = |1\rangle \oplus |1\rangle = |0\rangle$.

Table 2.1 shows the above gates and circuit symbols. The horizontal lines are used for the input and output qubits. These gates can be combined into different networks [1–3]. For instance, with two Hadamard gates and phase shift gates, we can construct the circuit shown in Fig. 2.5 in part (a). The result of this circuit on the input basis state $|0\rangle$ is the general state of qubit, which is written in Eq. 2.4, with the global phase $\vartheta/2$. Therefore, the qubit can be written as

$$|\varphi\rangle = P(\phi + \pi/2)HP(\vartheta)H|0\rangle = e^{i\vartheta/2}\left(\cos\frac{\vartheta}{2}|0\rangle + e^{i\phi}\sin\frac{\vartheta}{2}|1\rangle\right). \quad (2.17)$$

Indeed,

$$\begin{bmatrix} 1 & 0 \\ 0 & ie^{i\phi} \end{bmatrix} \times \frac{1}{\sqrt{2}}\begin{bmatrix} 1 & 1 \\ 1 & -1 \end{bmatrix} \times \begin{bmatrix} 1 & 0 \\ 0 & e^{i\vartheta} \end{bmatrix} \times \frac{1}{\sqrt{2}}\begin{bmatrix} 1 & 1 \\ 1 & -1 \end{bmatrix}\begin{bmatrix} 1 \\ 0 \end{bmatrix} = e^{i\vartheta/2}\begin{bmatrix} \cos(\vartheta/2) \\ e^{i\phi}\sin(\vartheta/2) \end{bmatrix}.$$

Without the global phase, the qubit can be presented as follows:

$$|\varphi\rangle = \cos\frac{\vartheta}{2}|0\rangle + e^{i\phi}\sin\frac{\vartheta}{2}|1\rangle = P(\phi)W\left(\frac{\vartheta}{2}\right)|0\rangle = \begin{bmatrix} 1 & 0 \\ 0 & e^{i\phi} \end{bmatrix}\begin{bmatrix} \cos(\vartheta/2) & -\sin(\vartheta/2) \\ \sin(\vartheta/2) & \cos(\vartheta/2) \end{bmatrix}\begin{bmatrix} 1 \\ 0 \end{bmatrix} = \begin{bmatrix} \cos(\vartheta/2) \\ e^{i\phi}\sin(\vartheta/2) \end{bmatrix}.$$

Note that global phase can be added using the $G(\vartheta/2)$ gate if necessary, as shown in Fig. 2.5 in part (b). Here and further, we draw abstract horizontal lines in the circuit. Imagine the line in the circuit which represents a physical particle, for instance, a photon flowing in the space, from left to right through the gates that change the state of the particle.

It also known [4, p. 8] that every 2×2 matrix U can be expressed by the global phase matrix and 3 rotations,

$$U = G(\phi_0)R_z(-\vartheta_1)R_y(-\vartheta_2)R_z(-\vartheta_3), \quad \text{where } \phi_0, \vartheta_1, \vartheta_2, \vartheta_3 \in [0, 2\pi].$$

Table 2.1 The list of single qubit gates and their description.

Gate name	Matrix of the gate	Symbol in circuits	Brief description and properties
Givens rotation	$W(\vartheta) = \begin{bmatrix} \cos\vartheta & -\sin\vartheta \\ \sin\vartheta & \cos\vartheta \end{bmatrix}$	W	Rotates the vector of amplitudes by angle ϑ. It is also called the elementary rotation.
Y-rotation	$R_y(\vartheta) = W(\vartheta/2)$	R_y	Rotation by angle $\vartheta/2$ around y-axis $R_x(\vartheta) = \exp(-i\vartheta/2Y)$.
X-rotation	$R_x(\vartheta) = \begin{bmatrix} \cos\vartheta/2 & -i\sin\vartheta/2 \\ i\sin\vartheta/2 & \cos\vartheta/2 \end{bmatrix}$	R_x	Rotation by angle $\vartheta/2$ around x-axis $R_x(\vartheta) = \exp(-i\vartheta/2X)$.
Z-rotation	$R_z(\vartheta) = \begin{bmatrix} e^{-i\vartheta/2} & 0 \\ 0 & e^{+i\vartheta/2} \end{bmatrix}$	R_z	Adds a relative shift of ϑ. Rotates vector of amplitudes about z-axis.
Pauli NOT, X	$X = \begin{bmatrix} 0 & 1 \\ 1 & 0 \end{bmatrix}$	X, $\oplus$	Switches the amplitudes of the states $\|0\rangle$ and $\|1\rangle$.
Pauli Z	$Z = \begin{bmatrix} 1 & 0 \\ 0 & -1 \end{bmatrix}$ $Z = 2\|0\rangle\langle 0\| - I$	Z	Changes the sign of the amplitude of $\|1\rangle$. Adds a relative phase shift of π between states $\|0\rangle$ and $\|1\rangle$; $Z = P(\pi)$.
Pauli Y	$Y = \begin{bmatrix} 0 & -i \\ i & 0 \end{bmatrix}$	Y	$Y = iXZ$, $Y^2 = I$.
V, $\sqrt{\text{NOT}}$	$V = \frac{1}{2}\begin{bmatrix} 1+i & 1-i \\ 1-i & 1+i \end{bmatrix}$	V	$V^2 = X$, $\det V = i$ $V = (1+i)(I - iX)/2$
Walsh-Hadamard	$H = \frac{1}{\sqrt{2}}\begin{bmatrix} 1 & 1 \\ 1 & -1 \end{bmatrix}$	H	$H^2 = I$, presents the operation of butterfly in fast radix-2 Fourier transform.
Paired ($H^\dagger$)	$A_2 = \frac{1}{\sqrt{2}}\begin{bmatrix} 1 & -1 \\ 1 & 1 \end{bmatrix}$	A_2	$A_2 = XH$. The main 2×2 butterfly of the paired transform (PT). It is also called the inverse pseudo-Hadamard gate.
Phase shift, P and Phase shift, P_-	$P(\phi) = \begin{bmatrix} 1 & 0 \\ 0 & e^{i\phi} \end{bmatrix}$ $P_-(\phi) = \begin{bmatrix} e^{i\phi} & 0 \\ 0 & 1 \end{bmatrix}$	ϕ (filled dot), ϕ (open dot)	Adds a relative phase shift of ϕ between states $\|0\rangle$ and $\|1\rangle$. $P_-(\phi) = XP(\phi)X$
T, $\pi/8$	$T = \begin{bmatrix} 1 & 0 \\ 0 & e^{i\pi/4} \end{bmatrix}$	$\pi/4$	$T = e^{i\phi/8}\begin{bmatrix} e^{-i\phi/8} & 0 \\ 0 & e^{i\phi/8} \end{bmatrix} = e^{\frac{i\phi}{8}} R_z(\phi/4)$
S	$S = \begin{bmatrix} 1 & 0 \\ 0 & i \end{bmatrix} = P(\pi/2)$	$\pi/2$	Adds a relative phase shift of $\pi/2$ between states $\|0\rangle$ and $\|1\rangle$. $S = P(\pi/2) = T^2 = \sqrt{Z}$.
Measurement, Observation, M	M	M	Measurement (M) of the single qubit, 2 lines denote a wire for the classical bit; one line (input) denote a single qubit wire.
Global phase, G	$G(\phi_0) = e^{i\phi_0}\begin{bmatrix} 1 & 0 \\ 0 & 1 \end{bmatrix}$	$e^{i\phi_0}$	Add the global phase to the qubit, $e^{i\phi_0}\|\varphi\rangle$; $G = P(2\phi_0)R_z(-2\phi_0)$

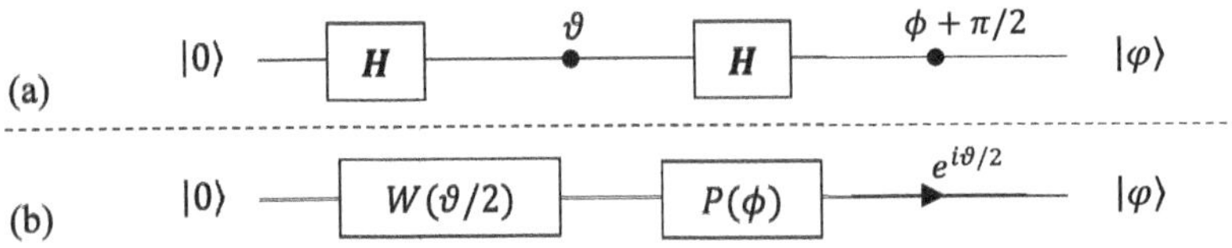

Fig. 2.5 Two circuits for preparation of the single qubit (a) without and (b) with a global phase.

References

1 Nielsen, M. and Chuang, I. (2001). *Quantum Computation and Quantum Information*, 2e. Cambridge University Press.

2 Strubell, E. (2011). *An Introduction to Quantum Algorithms*, vol. COS498. Springer.

3 Kaye, P., Laflamme, R., and Mosca, M. (2004). *An Introduction to Quantum Computing*. Oxford University Press.

4 Barenco, A., Bennett, C.H., Cleve, R. et al. (1995). Elementary gates for quantum computation. *Physical Review A* 52 (2): 3457–3467.

3

Understanding of Two Qubit Systems

In this section, we work with 2-qubit superposition with can be observed, or measured, in four different basis states. Many such superpositions can be obtained by combining two qubits, since each of them is in two states. In general, a 2-qubit superposition can be obtained without working with individual qubits. Here, we will look at the basic operations on them, and then in another section, we will describe the rich multiplicative structure of 2-cubes, which allows us to introduce analogs of the arithmetic operations on 2-cubes known to us. They include multiplication, division, inverse, square root, and power operations.

The importance of 2 qubits in color images relates to their ability to represent and manipulate the quantum states associated with the color channels in an image. The following must be considered:

- 2 qubits can be entangled, meaning their quantum states become correlated. In the context of color images, entangled qubits can represent complex relationships between the color channels. For instance, if two qubits are entangled, changes in one qubit's state may affect the color combination represented by the other qubit. This entanglement can enable more sophisticated image-processing operations involving color channel correlations.
- Using 2 qubits, it is possible to represent a subset of color information from an image. Selecting specific combinations of color channels through manipulating qubit states makes it possible to compress or represent images in a more compact form. This capability can benefit tasks like image storage, transmission, and analysis, where efficient representation and compression are desired.
- Quantum algorithms can be designed to process color images using 2 qubits. These algorithms can leverage the principles of quantum mechanics to perform operations such as image enhancement, noise reduction, edge detection, and feature extraction. By harnessing the power of quantum parallelism and entanglement, 2-qubit systems can enable the development of novel and potentially more efficient algorithms for image-processing tasks.
- As quantum computing advances, exploring color images using 2 qubits opens up possibilities for new image technologies. Quantum-inspired image processing algorithms, which may exploit certain quantum concepts even on classical hardware, could benefit from the insights gained by working with 2-qubit systems. Additionally, as quantum computers scale up and become more powerful, higher-dimensional quantum states can represent more complex color information, enabling the development of advanced quantum image processing techniques.

In summary, 2-qubits are critical for color imaging as they allow the representation, manipulation, and processing of quantum states associated with color channels. Their ability to leverage quantum superposition, entanglement, and advanced algorithms offers more efficient image representation, compression, and processing opportunities. Exploring the application of 2-qubit systems in color image tasks can contribute to the development of future quantum imaging technologies and algorithms.

2-qubit is defined as a quantum superposition with four basis states. If we could, we would imagine a magical, fast-spinning 4D coin with numbers 0, 1, 2, and 3 written on its four sides. As in the above case with a magic 2D coin, if we

Quantum Image Processing in Practice: A Mathematical Toolbox, First Edition. Artyom M. Grigoryan and Sos S. Agaian.

Companion website: www.wiley.com/go/grigoryan/quantumimageprocessing

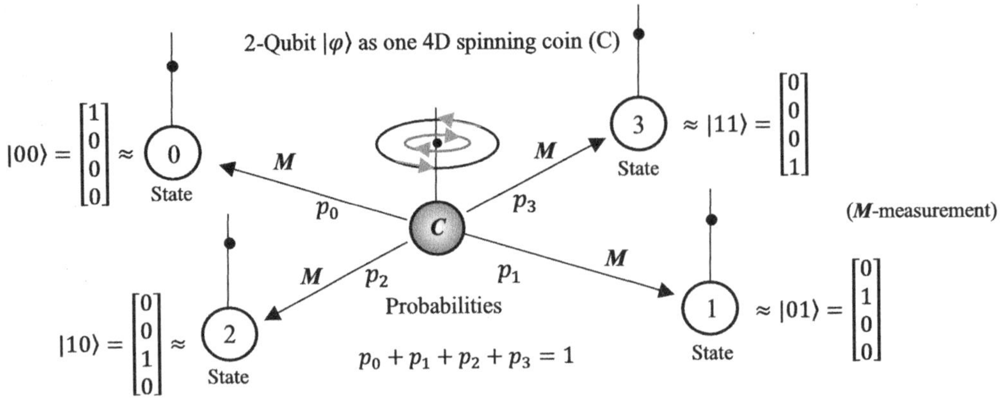

Fig. 3.1 Abstract model of 2-qubit: A 4-D spinning coin with numbers 0, 1, 2, and 3 written on four sides of the coin.

stop the spinning coin, we would see only one side of the coin in the palm of our hand, that is, only the number 0, 1, 2, or 3 (see Fig. 3.1). The event that the coin will show number k, where $k \in \{0, 1, 2, 3\}$ may occur with probability p_k. These four events are called the computational or standard basis states of the 2-qubit. To write each of these states, 2 bits are used in ket-bra notations.

In the computational basis states, a two-qubit quantum superposition (linear combination) is written as

$$|\varphi\rangle = a_0|0\rangle + a_1|1\rangle + a_2|2\rangle + a_3|3\rangle.$$

The amplitudes a_k, $k = 0:3$, define the probabilities $|a_k|^2$ of the 2-qubit to stay in the states $|k\rangle$. The basis states can be presented by the 4D unit vectors $|0\rangle = [1, 0, 0, 0]'$, $|1\rangle = [0, 1, 0, 0]'$, $|2\rangle = [0, 0, 1, 0]'$, and $|3\rangle = [0, 0, 0, 1]'$. These four vectors in binary representation can be written as $|00\rangle$, $|01\rangle$, $|10\rangle$, and $|11\rangle$, respectively. Therefore, we can write

$$|\varphi\rangle = a_0|00\rangle + a_1|01\rangle + a_2|10\rangle + a_3|11\rangle, \tag{3.1}$$

$$\text{or } |\varphi\rangle = a_{00}|00\rangle + a_{01}|01\rangle + a_{10}|10\rangle + a_{11}|11\rangle.$$

Many 2-qubit superpositions can also be prepared by two single qubits. Figure 3.2 illustrates such an abstract example with two 2D spinning-coins. The coins can spin independently, and it is also possible to imagine that there is some kind of interaction between the two coins when rotating.

3.1 Measurement of 2-Qubits

2-qubit can only be in one state $|00\rangle$, $|01\rangle$, $|10\rangle$, or $|11\rangle$ when measured or observed [1]. As for the single-qubit, the probability of the qubit $|\varphi\rangle$ to be in its single state $|k\rangle$ after measurement (M) can be described by the inner product. For instance, for the state $|1\rangle = |01\rangle$,

$$\Pr(M|\varphi\rangle = |01\rangle) = |\langle 01\,|\varphi\rangle|^2 = |\langle 01\,|(a_0|00\rangle + a_1|01\rangle + a_2|10\rangle + a_3|11\rangle)\rangle|^2 = |\,a_1\langle 01\,|01\rangle|^2 = |a_1|^2.$$

Here, $\langle 01|01\rangle = 1$ and $\langle 01|00\rangle = \langle 01|10\rangle = \langle 01|11\rangle = 0$.

If only one qubit is measured in $|\varphi\rangle$, for instance, the second qubit, and it will be in state $|1\rangle$, then the measured 2-qubit will be in superposition with the basis states $|01\rangle$ and $|11\rangle$. The probability of measured 2-qubit equals

$$\Pr\ = |\langle 01\,|\varphi\rangle|^2 + |\langle 11\,|\varphi\rangle|^2 = |\,a_1|^2 + |\,a_3|^2.$$

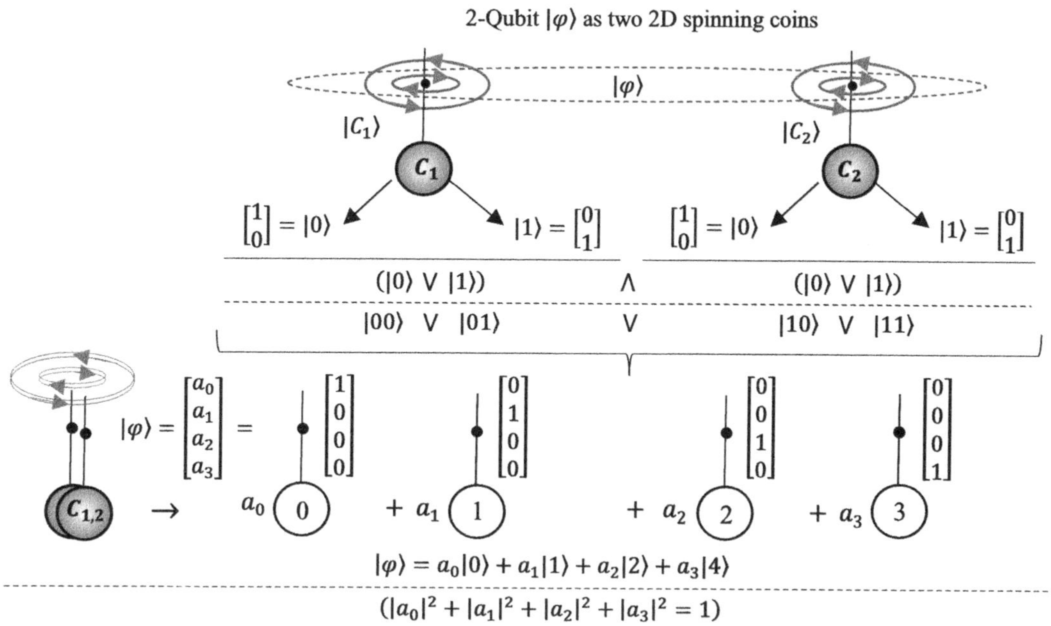

Fig. 3.2 Abstract model: 2-qubit as a linear superposition of four computational basis states 0, 1, 2, and 3.

Then, the measured 2-qubit will be in the superposition calculated as

$$|\varphi_1\rangle = M|\varphi\rangle = \frac{\langle 01\ |\varphi\rangle|01\rangle + \langle 11\ |\varphi\rangle|11\rangle}{\sqrt{|\ a_1|^2 + \ |\ a_3|^2}} = \frac{a_1|01\rangle + a_3|11\rangle}{\sqrt{|\ a_1|^2 + \ |\ a_3|^2}} = \frac{a_1|0\rangle + a_3|1\rangle}{\sqrt{|\ a_1|^2 + \ |\ a_3|^2}}|1\rangle. \tag{3.2}$$

Similarly, we can consider the subspace of superpositions with basis states $|00\rangle$ and $|10\rangle$. The probability that the first qubit of $M|\varphi\rangle$ is in $|0\rangle$ equals to $|a_0|^2 + \ |\ a_2|^2$ and the measured 2-qubit will be in the superposition

$$|\varphi_0\rangle = M|\varphi\rangle = \frac{\langle 00\ |\varphi\rangle|00\rangle + \langle 10|\varphi\rangle|10\rangle}{\sqrt{|\ a_0|^2 + \ |\ a_2|^2}} = \frac{a_0|00\rangle + a_2|10\rangle}{\sqrt{|\ a_0|^2 + \ |\ a_2|^2}} = \frac{a_0|0\rangle + a_2|1\rangle}{\sqrt{|\ a_0|^2 + \ |\ a_2|^2}}|0\rangle. \tag{3.3}$$

Note that the above measurement could be better defined as an observation. This does not apply to a register in a classical computer, where one can simply take and change any bit in a memory register at any time. In a quantum system, the measurement or observation results in a change in the true superposition; it will be lost. The quantum system is closed, it does not allow itself to be seen completely when observed or interfered with from the outside. With any interference in its world, the invisible state immediately passes into one of its substates. This is the so-called phenomenon of destruction, collapse.

3.1.1 Projection Operators

The above measurements in Eq. 3.2 can be simplified to better understand what we can observe in a 2-qubit. Let us assume that a 2-qubit $|\varphi\rangle$ represents a pair of qubits $|\psi_0\rangle = a_0|0\rangle + b_0|1\rangle$ and $|\psi_1\rangle = a_1|0\rangle + b_1|1\rangle$, that is,

$$|\varphi\rangle = |\psi_0\rangle \otimes |\psi_1\rangle = a_0a_1|00\rangle + a_0b_1|01\rangle + b_0a_1|10\rangle + b_0b_1|11\rangle.$$

Then, the observation of the first qubit in k, where $k = 0$ or 1, will show the new superposition

$$M|\varphi\rangle_{|k} = a_1|k0\rangle + b_1|k1\rangle = |k\rangle(a_1|0\rangle + b_1|1\rangle) = |k\rangle|\psi_1\rangle.$$

Here, one sees the second qubit $|\psi_1\rangle$. Measuring the first qubit, the second qubit is observed. Similarly, measuring the second qubit, the first qubit is observed in the post-measurement state $|\psi_0\rangle|k\rangle$.

The measurement of 2-qubit can be described by the projection operator [2]. Measurement of the first qubit in the 2-qubit in the computational basis can be described by the following orthogonal projections with the matrices:

$$T_0 = T_{00,01} = \begin{bmatrix} 1 & 0 & 0 & 0 \\ 0 & 1 & 0 & 0 \\ 0 & 0 & 0 & 0 \\ 0 & 0 & 0 & 0 \end{bmatrix} \quad \text{and} \quad T_1 = T_{10,11} = \begin{bmatrix} 0 & 0 & 0 & 0 \\ 0 & 0 & 0 & 0 \\ 0 & 0 & 1 & 0 \\ 0 & 0 & 0 & 1 \end{bmatrix}. \tag{3.4}$$

Indeed,

$$M|\varphi\rangle_{|00,01} = \frac{T_0|\varphi\rangle}{\sqrt{\Pr\left(M|\varphi\rangle_{|00,01}\right)}} = \frac{T_0|\varphi\rangle}{\sqrt{\langle\overline{\varphi} \mid T_0|\varphi\rangle}} = \frac{1}{\sqrt{|a_0|^2}} \begin{bmatrix} 1 & 0 & 0 & 0 \\ 0 & 1 & 0 & 0 \\ 0 & 0 & 0 & 0 \\ 0 & 0 & 0 & 0 \end{bmatrix} \begin{bmatrix} a_0a_1 \\ a_0b_1 \\ b_0a_1 \\ b_0b_1 \end{bmatrix} = \frac{1}{|a_0|} \begin{bmatrix} a_0a_1 \\ a_0b_1 \\ 0 \\ 0 \end{bmatrix}$$

$$= \pm(a_1|00\rangle + b_1|01\rangle) = \pm|0\rangle|\psi_1\rangle, \tag{3.5}$$

where

$$\langle\overline{\varphi} \mid T_0|\varphi\rangle = \begin{bmatrix} \overline{a_0a_1} & \overline{a_0b_1} & \overline{b_0a_1} & \overline{b_0b_1} \end{bmatrix} \begin{bmatrix} 1 & 0 & 0 & 0 \\ 0 & 1 & 0 & 0 \\ 0 & 0 & 0 & 0 \\ 0 & 0 & 0 & 0 \end{bmatrix} \begin{bmatrix} a_0a_1 \\ a_0b_1 \\ b_0a_1 \\ b_0b_1 \end{bmatrix} = \begin{bmatrix} \overline{a_0a_1} & \overline{a_0b_1} & \overline{b_0a_1} & \overline{b_0b_1} \end{bmatrix} \begin{bmatrix} a_0a_1 \\ a_0b_1 \\ 0 \\ 0 \end{bmatrix}$$

$$= |a_0a_1|^2 + |a_0b_1|^2 = |a_0|^2(|a_1|^2 + |b_1|^2) = |a_0|^2.$$

Similarly, $\langle\overline{\varphi} \mid T_1|\varphi\rangle = |b_0|^2$ and

$$M|\varphi\rangle_{|10,11} = \frac{T_1|\varphi\rangle}{\sqrt{\Pr\left(M|\varphi\rangle_{|10,11}\right)}} = \frac{T_1|\varphi\rangle}{\sqrt{\langle\overline{\varphi} \mid T_1|\varphi\rangle}} = \frac{1}{\sqrt{|b_0|^2}} \begin{bmatrix} 0 & 0 & 0 & 0 \\ 0 & 0 & 0 & 0 \\ 0 & 0 & 1 & 0 \\ 0 & 0 & 0 & 1 \end{bmatrix} \begin{bmatrix} a_0a_1 \\ a_0b_1 \\ b_0a_1 \\ b_0b_1 \end{bmatrix} = \frac{1}{|b_0|} \begin{bmatrix} 0 \\ 0 \\ b_0a_1 \\ b_0b_1 \end{bmatrix}$$

$$= \pm(a_1|10\rangle + b_1|11\rangle) = \pm|1\rangle|\psi_1\rangle.$$

After measuring the first qubit, the state will be in the second qubit $|\psi_1\rangle$. Everything is logical here. The measurement of the first qubit will leave the 2-qubit in the second qubit.

Measurement of the second qubit in the 2-qubit can be described, for instance, by the orthogonal projection

$$T = T_{00,10} = \begin{bmatrix} 1 & 0 & 0 & 0 \\ 0 & 0 & 0 & 0 \\ 0 & 0 & 1 & 0 \\ 0 & 0 & 0 & 0 \end{bmatrix} = \begin{bmatrix} 1 \\ 0 \\ 0 \\ 0 \end{bmatrix} \begin{bmatrix} 1 & 0 & 0 & 0 \end{bmatrix} + \begin{bmatrix} 0 \\ 0 \\ 1 \\ 0 \end{bmatrix} \begin{bmatrix} 0 & 0 & 1 & 0 \end{bmatrix} = |00\rangle\langle 00| + |10\rangle\langle 10|.$$

Indeed,

$$M|\varphi\rangle_{|00,10} = \frac{T|\varphi\rangle}{\sqrt{\Pr\left(M|\varphi\rangle_{|00,01}\right)}} = \frac{T|\varphi\rangle}{\sqrt{\langle\overline{\varphi}\mid T|\varphi\rangle}} = \frac{1}{\sqrt{|a_1|^2}}\begin{bmatrix}1&0&0&0\\0&0&0&0\\0&0&1&0\\0&0&0&0\end{bmatrix}\begin{bmatrix}a_0a_1\\a_0b_1\\b_0a_1\\b_0b_1\end{bmatrix} = \frac{1}{|a_1|}\begin{bmatrix}a_0a_1\\0\\b_0a_1\\0\end{bmatrix}$$

$$= \pm(a_0|00\rangle + b_0|10\rangle) = \pm|\psi_0\rangle|0\rangle,$$

where

$$\langle\overline{\varphi}\mid T|\varphi\rangle = \begin{bmatrix}\overline{a_0a_1} & \overline{a_0b_1} & \overline{b_0a_1} & \overline{b_0b_1}\end{bmatrix}\begin{bmatrix}1&0&0&0\\0&0&0&0\\0&0&1&0\\0&0&0&0\end{bmatrix}\begin{bmatrix}a_0a_1\\a_0b_1\\b_0a_1\\b_0b_1\end{bmatrix} = \begin{bmatrix}\overline{a_0a_1} & \overline{a_0b_1} & \overline{b_0a_1} & \overline{b_0b_1}\end{bmatrix}\begin{bmatrix}a_0a_1\\0\\b_0a_1\\0\end{bmatrix}$$

$$= |a_0a_1|^2 + |b_0a_1|^2 = \left(|a_0|^2 + |b_0|^2\right)|a_1|^2 = |a_1|^2.$$

After measuring the second qubit, the state will be in the first qubit $|\psi_0\rangle$.

In general, there are four basis states, and a 2-qubit superposition $|\varphi\rangle = a_0|00\rangle + a_1|01\rangle + a_2|10\rangle + a_3|11\rangle$ is represented by a liner combination of these states. This superposition may not be the Kronecker product of two qubits, that is, it is entangled. We denote this space by V and let V' be a subspace of V, which is defined by one or few of these states. A projection operator T can then be defined that maps V onto V' in a manner similar to the projections described above. The operator T is called *the orthogonal projection* because the basis states of the space V (as well as the subspace ′) are orthogonal. The construction of such projections is described below in the examples.

Example 3.1 Projection on {|01⟩, | 11⟩}-Subspace

Let V' is the subspace with basis states $|01\rangle$ and $|11\rangle$. We denote this subspace V' by $V_1 + V_3$. The projection operator T is described by the matrix which is the sum of ket-bra operations of basis states $|01\rangle$ and $|11\rangle$. In other words,

$$T = T_{01,11} = \begin{bmatrix}0&0&0&0\\0&1&0&0\\0&0&0&0\\0&0&0&1\end{bmatrix} = \begin{bmatrix}0\\1\\0\\0\end{bmatrix}\begin{bmatrix}0&1&0&0\end{bmatrix} + \begin{bmatrix}0\\0\\0\\1\end{bmatrix}\begin{bmatrix}0&0&0&1\end{bmatrix} = |01\rangle\langle 01| + |11\rangle\langle 11|. \tag{3.6}$$

Indeed, $T|\varphi\rangle = T[a_0, a_1, a_2, a_3]' = [0, a_1, 0, a_3]' = a_1|01\rangle + a_3|11\rangle$. We can also write that

$$T|\varphi\rangle = |01\rangle\langle 01||\varphi\rangle + |11\rangle\langle 11||\varphi\rangle, \quad \text{or} \quad T|\varphi\rangle = \langle 01|\varphi\rangle|01\rangle + \langle 11|\varphi\rangle|11\rangle = a_1|01\rangle + a_3|11\rangle.$$

After normalization, the measured 2-qubit superposition is

$$M|\varphi\rangle = \frac{T|\varphi\rangle}{|T|\varphi\rangle|} = \frac{\langle 01|\varphi\rangle|01\rangle + \langle 11|\varphi\rangle|11\rangle}{\sqrt{|\langle 01|\varphi\rangle|^2 + |\langle 11|\varphi\rangle|^2}} = \frac{a_1|01\rangle + a_3|11\rangle}{\sqrt{|a_1|^2 + |a_3|^2}}. \tag{3.7}$$

Example 3.2 Projection on {|01⟩, | 10⟩}-Subspace

Consider the subspace $V' = V_1 + V_2$ with basis states $|01\rangle$, and $|10\rangle$. The projection operator T is described by the matrix which is the sum of ket-bra operations of these basis states, In other words,

$$T = |01\rangle\langle 01| + |10\rangle\langle 10| = \begin{bmatrix}0&0&0&0\\0&1&0&0\\0&0&1&0\\0&0&0&0\end{bmatrix}.$$

In this case, the projector operator on the 2-qubit superposition $|\varphi\rangle$ results in the following superposition:

$$T|\varphi\rangle = \langle 01|\varphi\rangle|01\rangle + \langle 10|\varphi\rangle|10\rangle = a_1|01\rangle + a_2|10\rangle.$$

The measured superposition will be in the state

$$M|\varphi\rangle = \frac{T|\varphi\rangle}{|T|\varphi\rangle|} = \frac{\langle 01|\varphi\rangle|01\rangle + \langle 10|\varphi\rangle|10\rangle}{\sqrt{|\langle 01|\varphi\rangle|^2 + |\langle 10|\varphi\rangle|^2}} = \frac{a_1|01\rangle + a_2|10\rangle}{\sqrt{|a_1|^2 + |a_2|^2}}. \tag{3.8}$$

Example 3.3 Projection on $\{|00\rangle, |01\rangle, |10\rangle\}$-Subspace
Consider the subspace $V' = V_0 + V_1 + V_2$ with basis states $|00\rangle$, $|01\rangle$, and $|10\rangle$. The projection operator T is described by the matrix which is the sum of ket-bra operations of these three basis states, In other words,

$$T = |00\rangle\langle 00| + |01\rangle\langle 01| + |10\rangle\langle 10| = \begin{bmatrix} 1 & 0 & 0 & 0 \\ 0 & 1 & 0 & 0 \\ 0 & 0 & 1 & 0 \\ 0 & 0 & 0 & 0 \end{bmatrix}.$$

In this case, the projector operator on $|\varphi\rangle$ results in the following superposition:

$$T|\varphi\rangle = \langle 00|\varphi\rangle|00\rangle + \langle 01|\varphi\rangle|01\rangle + \langle 10|\varphi\rangle|10\rangle = a_0|00\rangle + a_1|01\rangle + a_2|10\rangle.$$

After normalization, the superposition of measured 2-qubit is

$$M|\varphi\rangle = \frac{T|\varphi\rangle}{|T|\varphi\rangle|} = \frac{\langle 00|\varphi\rangle|00\rangle + \langle 01|\varphi\rangle|01\rangle + \langle 10|\varphi\rangle|10\rangle}{\sqrt{|\langle 00|\varphi\rangle|^2 + |\langle 01|\varphi\rangle|^2 + |\langle 10|\varphi\rangle|^2}}. \tag{3.9}$$

In other words, $M|\varphi\rangle = (a_0|00\rangle + a_1|01\rangle + a_2|10\rangle)/\sqrt{|a_0|^2 + |a_1|^2 + |a_2|^2}$.

3.2 Operation of Kronecker Product

The operation $\otimes$ of the Kronecker product, or the tensor product, from the right, on matrices and vectors, is used widely in quantum calculations [3]. To illustrate this operation, we consider a 2×2 matrix A. The tensor product of this matrix with another matrix B is defined as the block matrix of twice size,

$$A \otimes B = \begin{bmatrix} a_{0,0} & a_{0,1} \\ a_{1,0} & a_{1,1} \end{bmatrix} \otimes B = \begin{bmatrix} a_{0,0}B & a_{0,1}B \\ a_{1,0}B & a_{1,1}B \end{bmatrix}. \tag{3.10}$$

Consider the following example:

$$\begin{bmatrix} 1 & 0 \\ 0 & -1 \end{bmatrix} \otimes \begin{bmatrix} 1 & 1 \\ 1 & -1 \end{bmatrix} = \begin{bmatrix} 1 & 1 & 0 & 0 \\ 1 & -1 & 0 & 0 \\ 0 & 0 & -1 & -1 \\ 0 & 0 & -1 & 1 \end{bmatrix}.$$

In the general case, the Kronecker product of two matrices $A = [a_{i,j}]$ and B of sizes $N \times M$ and $p \times q$, respectively, is defined as the following $Np \times Mq$ matrix:

$$A \otimes B = \left[a_{i,j}B\right]_{i=0:N-1, j=0:(M-1)}.$$

3.2.1 Tensor Product of Single Qubits

The tensor product, or Kronecker product, of two single qubits is a 2-qubit superposition. For qubits $|\varphi_1\rangle = a_0|0\rangle + a_1|1\rangle$ and $|\varphi_2\rangle = b_0|0\rangle + b_1|1\rangle$, the tensor product is defined as

$$|\varphi_1,\varphi_2\rangle \triangleq |\varphi_1\rangle \otimes |\varphi_2\rangle = \begin{bmatrix} a_0 \\ a_1 \end{bmatrix} \otimes \begin{bmatrix} b_0 \\ b_1 \end{bmatrix} = \begin{bmatrix} a_0b_0 \\ a_0b_1 \\ a_1b_0 \\ a_1b_1 \end{bmatrix}, \tag{3.11}$$

or $|\varphi_1,\varphi_2\rangle = a_0b_0|00\rangle + a_0b_1|01\rangle + a_1b_0|10\rangle + a_1b_1|11\rangle$.

Consider the following examples:

$$|01\rangle = |0\rangle \otimes |1\rangle = \begin{bmatrix} 1 \\ 0 \end{bmatrix} \otimes \begin{bmatrix} 0 \\ 1 \end{bmatrix} = \begin{bmatrix} 0 \\ 1 \\ 0 \\ 0 \end{bmatrix}, \quad |10\rangle = |1\rangle \otimes |0\rangle = \begin{bmatrix} 0 \\ 1 \end{bmatrix} \otimes \begin{bmatrix} 1 \\ 0 \end{bmatrix} = \begin{bmatrix} 0 \\ 0 \\ 1 \\ 0 \end{bmatrix}.$$

In the computational basis states $\{|00\rangle, |01\rangle, |10\rangle, |11\rangle\} = \{|0\rangle, |1\rangle, |2\rangle, |3\rangle\}$, the above two ket-vectors can be written as $|1\rangle$ and $|2\rangle$, respectively.

Consider the following examples of tensor product:

If $|n\rangle$ is the computational basis state in the space of r qubits, where $N = 2^r$, then $|1\rangle \otimes |n\rangle = |n + N\rangle$ in the space of $(r + 1)$ qubits.

1) The tensor products of the basis state $|n\rangle$ with states of 2-qubits are

$$|n,00\rangle = |4n\rangle, |n,01\rangle = |4n+1\rangle, |n,10\rangle = |4n+2\rangle, \text{and } |n,11\rangle = |4n+3\rangle.$$

2) In the general case, when $|n\rangle$ is the basis state in the space of r qubits and $|m\rangle$ is the basis state in the space of s qubits, then

$$|n\rangle \otimes |m\rangle = |2^s n + m\rangle, \quad m = 0, 1, \ldots, 2^s - 1. \tag{3.12}$$

Note that

1) $|0\rangle \otimes |1\rangle \neq |1\rangle \otimes |0\rangle$, that is, the operation of the tensor product is not commutative.
2) There are many 2-qubits that cannot be written as a tensor product of single qubits. Such quantum superpositions are called *entangled*. This is a phenomenon of interdependence that is observed in quantum systems. For example, in a pair of photons, the helicity of one photon will be the opposite of the helicity of the other photon when measured.

   For instance, the following four superpositions (known as Bell states) are entangled

$$|\beta_{1,2}\rangle = \frac{1}{\sqrt{2}}(|00\rangle \pm |11\rangle) \quad \text{and} \quad |\beta_{3,4}\rangle = \frac{1}{\sqrt{2}}(|01\rangle \pm |10\rangle). \tag{3.13}$$

In general, let us assume that a 2-qubit $|\varphi\rangle = a|0\rangle + b|1\rangle + c|0\rangle + d|1\rangle$ is the tensor product of qubits $|\varphi_1\rangle = a_0|0\rangle + a_1|1\rangle$ and $|\varphi_2\rangle = b_0|0\rangle + b_1|1\rangle$. Thus, in the matrix form, we have

$$|\varphi\rangle = \begin{bmatrix} a \\ b \\ c \\ d \end{bmatrix} = |\varphi_1,\varphi_2\rangle = \begin{bmatrix} a_0 \\ a_1 \end{bmatrix} \otimes \begin{bmatrix} b_0 \\ b_1 \end{bmatrix} = \begin{bmatrix} a_0b_0 \\ a_0b_1 \\ a_1b_0 \\ a_1b_1 \end{bmatrix}.$$

Then, $a = a_0b_0$, $b = a_0b_1$, $c = a_1b_0$, and $d = a_1b_1$. The following ratios are correct: $c/a = d/b$, or $bc = ad$. For instance, for the 2-qubit with amplitudes $[1, 2, 3, 6]/\sqrt{50}$, this condition holds, $2\times 3 = 1\times 6$ and therefore

$$|\varphi\rangle = \frac{|0\rangle + 2|1\rangle + 3|2\rangle + 6|3\rangle}{\sqrt{50}} = \frac{|0\rangle + 3|1\rangle}{\sqrt{10}} \otimes \frac{|0\rangle + 2|1\rangle}{\sqrt{5}}.$$

In vector form, this operation can be written as

$$\left(\frac{1}{\sqrt{10}}\begin{bmatrix}1\\3\end{bmatrix}\right)\otimes\left(\frac{1}{\sqrt{5}}\begin{bmatrix}1\\2\end{bmatrix}\right) = \frac{1}{\sqrt{50}}\begin{bmatrix}1\\2\\3\\6\end{bmatrix}.$$

For the 2-qubit with amplitudes $[1, 2, 3, 4]/\sqrt{30}$, the required condition does not hold, $2\times 3 \neq 1\times 4$. Therefore, the 2-qubit $|\varphi\rangle = (|0\rangle + 2|1\rangle + 3|2\rangle + 4|1\rangle)/\sqrt{30}$ is entangled. The same is for the above Bell states. For instance, for the superposition $|\beta_1\rangle = (|0\rangle + 0|1\rangle + 0|2\rangle + |3\rangle)/\sqrt{2}$, we have $0\times 0 \neq 1\times 1$. For the Hadamard state of 2-qubit $|\varphi_H\rangle = (|0\rangle + |1\rangle + |2\rangle + |1\rangle)/2$, the required condition holds, $1\times 1 = 1\times 1$ and this 2-qubit is not entangled,

$$|\varphi_H\rangle = \frac{|0\rangle + |1\rangle + |2\rangle + |3\rangle}{2} = \frac{|0\rangle + |1\rangle}{\sqrt{2}} \otimes \frac{|0\rangle + |1\rangle}{\sqrt{2}}.$$

The 2-qubits $(|0\rangle + |1\rangle - |2\rangle + |1\rangle)/2$ and $(|0\rangle - |1\rangle + |2\rangle + |1\rangle)/2$ are entangled.

If the condition $bc = ad$ holds for the 2-qubit with positive amplitudes, then the single qubits $|\varphi_1\rangle$ and $|\varphi_2\rangle$ can be found as follows:

$$|\varphi_1\rangle = \cos\vartheta_1|0\rangle + \sin\vartheta_1|1\rangle \text{ and } |\varphi_2\rangle = \cos\vartheta_2|0\rangle + \sin\vartheta_2|1\rangle$$

where $\vartheta_1 = \text{atan}\, c/a$ and $\vartheta_2 = \text{atan}\, b/a$, if $a \neq 0$. If $a = 0$, then b or c equals zero and this case should be considered separately. In the case with negative amplitudes in 2-qubit $|\varphi\rangle$, for instance, $a < 0$, the angle of the first qubit is considered as $\vartheta_1 + \pi$. For example, the 2-qubit

$$|\varphi\rangle = \frac{-|0\rangle - 3|1\rangle + 2|2\rangle + 6|3\rangle}{\sqrt{50}} = \frac{|0\rangle - 2|1\rangle}{\sqrt{5}} \otimes \frac{-|0\rangle - 3|1\rangle}{\sqrt{10}} = \frac{-|0\rangle + 2|1\rangle}{\sqrt{5}} \otimes \frac{|0\rangle + 3|1\rangle}{\sqrt{10}}.$$

3) The tensor product is not a multiplication of qubits, or 2-qubits, or quantum superpositions. It extends the dimension of superpositions. For instance, as follows from Eq. 3.11, the tensor product of two single qubits is a 2-qubit. The tensor product is not a quantum gate on qubits. In other words, this operation cannot be described by a matrix, not to mention the unitary matrix.

3.3 Operation of Kronecker Sum

Kronecker sum (or direct sum) of matrices A and B is defined as the block diagonal matrix

$$A \oplus B = \begin{bmatrix}A & 0\\0 & B\end{bmatrix}. \tag{3.14}$$

For example, we consider the controlled-phase gate with matrix representation

$$C = I \oplus S = \begin{bmatrix}I & 0\\0 & S\end{bmatrix} = \begin{bmatrix}1&0&0&0\\0&1&0&0\\0&0&1&0\\0&0&0&i\end{bmatrix}, \quad S = \begin{bmatrix}1 & 0\\0 & i\end{bmatrix}. \tag{3.15}$$

The symbol $\bigoplus$ will also be used as XOR for addition modulo 2 of numbers, $x \bigoplus y$, when $x, y = 0$, or 1.

3.3.1 Properties on Matrices

1) $(AB)' = B'A'$, where X', or X^T, denotes the transpose of a matrix X.
2) $A \oplus B \neq B \oplus A$, if $A \neq B$.
3) $A \otimes B \neq B \otimes A$, if $A \neq B$,

$$\begin{aligned}
&(A + C) \oplus B = A \oplus B + C \oplus B,\\
&k(A \otimes B) = kA \oplus B = A \oplus kB, \quad k \text{ is a constrant} \neq 0,\\
&A \otimes (B \oplus C) = (A \otimes B) \oplus (A \otimes C),\\
&(A \otimes B) \otimes C = A \otimes (B \otimes C)\\
&(B \oplus C) \otimes A = (B \otimes A) \oplus (C \otimes A),\\
&(A \oplus B)(C \oplus D) = (AC) \oplus (BD) = \begin{bmatrix} AC & 0 \\ 0 & BD \end{bmatrix},\\
&(A \otimes B)(C \otimes D) = (AC) \otimes (BD).
\end{aligned}$$

4) $(A \otimes B)^{-1} = A^{-1} \otimes B^{-1}$ and $(A \otimes B)' = A' \otimes B'$,

$$(A \oplus B)^{-1} = A^{-1} \oplus B^{-1},$$

$$(AB)^{-1} = B^{-1}A^{-1}.$$

5) For vectors v_n, $n = 0:N-1$, and w_m, $m = 0:N-1$,

$$\sum_{n=0}^{N-1}\sum_{m=0}^{M-1} a_{n,m}(v_n \otimes w_m) = \sum_{n=0}^{N-1} v_n \otimes \sum_{m=0}^{M-1} a_{n,m} w_m$$

and

$$\left(\sum_{n=0}^{N-1} a_n v_n\right) \otimes \left(\sum_{m=0}^{M-1} b_m w_m\right) = \sum_{n=0}^{N-1}\sum_{m=0}^{M-1} a_n b_m (v_n \otimes w_m). \tag{3.16}$$

For tensor product of two superpositions

$$|\varphi\rangle_1 \otimes |\varphi\rangle_2 = \sum_{n=0}^{N-1} a_n |n\rangle \otimes \sum_{m=0}^{M-1} b_n |m\rangle = \sum_{n=0}^{N-1}\sum_{m=0}^{M-1} a_n b_m |n\rangle |m\rangle,$$

and for two operations with $N \times N$ and $M \times M$ matrices U_1 and U_2, respectively,

$$(U_1 \otimes U_2)\left(|\varphi\rangle_1 \otimes |\varphi\rangle_2\right) = \left(U_1|\varphi\rangle_1\right) \otimes \left(U_2|\varphi\rangle_2\right) = \sum_{n=0}^{N-1}\sum_{m=0}^{M-1} a_n b_m (U_1|n\rangle)(U_2|m\rangle). \tag{3.17}$$

3.3.2 Orthogonality of Matrices

Let A be the $N \times N$ matrix, $A = [a_{n,m}]_{n,m\,=\,0:N-1}$.

1) The conjugate matrix $A^* = [\bar{a}_{m,n}]_{n,m\,=\,0:N-1}$ is a complex conjugate transpose matrix of A. The matrix A^* is also called the Hermitian adjoint of A. If the A matrix is real, then A^* is transpose matrix, $A^* = A^T = [a_{m,n}]_{n,\,m\,=\,0:N-1}$.

2) The matrix is unitary, if $AA^* = I_N$, where I_N is the identity $N \times N$ matrix.
   The inverse matrix A^{-1} coincides with the conjugate A^*.
   Orthogonality of rows and columns takes place [3]:

$$\sum_{k=0}^{N-1} a_{n,k}\bar{a}_{m,k} = \sum_{k=0}^{N-1} a_{k,m}\bar{a}_{k,m} = \delta_{n,m} = \begin{cases} 1, & \text{if } n = m, \\ 0, & \text{otherwise,} \end{cases} \qquad n, m = 0 : (N-1).$$

Example 3.4 4 × 4 Matrix
The unitary matrix of the 4-point discrete Fourier transform is

$$A = \frac{1}{2}\begin{bmatrix} 1 & 1 & 1 & 1 \\ 1 & -i & -1 & i \\ 1 & -1 & 1 & -1 \\ 1 & i & -1 & -i \end{bmatrix}, \qquad \det(A) = -i.$$

If we consider the second and fourth rows, then

$$\sum_{k=0}^{3} a_{2,k}\bar{a}_{4,k} = \frac{1}{4}[1(1) - i(-i) - 1(-1) + i(i)] = 0.$$

3) The real unitary matrix A is called orthogonal. The inverse A^{-1} coincides with the transpose matrix A^T.

3.4 Permutations

Permutations are important elements (gates) in quantum computation. A permutation (P) of a set of elements is a rearrangement of elements. For example, the order of six elements can be changed as $P : (0, 1, 2, 3, 4, 5) \rightarrow (2, 4, 3, 0, 1, 5)$. We will use Cauchy's 2-line notations and cycle notations. For the permutation in this example, we can write

$$P = \begin{pmatrix} 012345 \\ 243015 \end{pmatrix} \quad \text{or} \quad P = (0, 2, 3)(1, 4).$$

Each permutation is described by an orthogonal matrix with only one coefficient of 1 on each column and row. For the above permutation, this matrix is

$$P = P_{(023,14)} = P_{(023)}\,P_{(14)} = \begin{bmatrix} 0 & 0 & 0 & 1 & 0 & 0 \\ 0 & 1 & 0 & 0 & 0 & 0 \\ 1 & 0 & 0 & 0 & 0 & 0 \\ 0 & 0 & 1 & 0 & 0 & 0 \\ 0 & 0 & 0 & 0 & 1 & 0 \\ 0 & 0 & 0 & 0 & 0 & 1 \end{bmatrix} \begin{bmatrix} 1 & 0 & 0 & 0 & 0 & 0 \\ 0 & 0 & 0 & 0 & 1 & 0 \\ 0 & 0 & 1 & 0 & 0 & 0 \\ 0 & 0 & 0 & 1 & 0 & 0 \\ 0 & 1 & 0 & 0 & 0 & 0 \\ 0 & 0 & 0 & 0 & 0 & 1 \end{bmatrix} = \begin{bmatrix} 0 & 0 & 0 & 1 & 0 & 0 \\ 0 & 0 & 0 & 0 & 1 & 0 \\ 1 & 0 & 0 & 0 & 0 & 0 \\ 0 & 0 & 1 & 0 & 0 & 0 \\ 0 & 1 & 0 & 0 & 0 & 0 \\ 0 & 0 & 0 & 0 & 0 & 1 \end{bmatrix}.$$

and $\det P = -1$. The permutation is orthogonal, $PP' = I_6$.

Many quantum circuits involve permutations. Permutations are usually accompanied by many ±90° degree rotations along with CNOT operations.

3.4.1 Elementary Operations on 2-Qubits

With each additional qubit, the number of operations and gates on the quantum superposition increases, as well as their complexity [2]. In this section, we will try to describe elementary operations of 2-quit superpositions with examples. We need to understand them well, since it is difficult to imagine a quantum circuit in signal and image processing without such similar operations.

Operators on a 2-qubit superposition $|\varphi\rangle$ present themselves 4×4 unitary matrices $\boldsymbol{U}$ applied to column vector $q = (a_0, a_1, a_2, a_3)'$ composed by the amplitudes of this superposition. The list of known unitary matrices on 2-qubits includes the following:

1) Permutation of the last two amplitudes a_2 and a_3, which is called Controlled-NOT, or CNOT,

$$C_{NOT} = I \oplus X = \begin{bmatrix} 1 & 0 & 0 & 0 \\ 0 & 1 & 0 & 0 \\ 0 & 0 & 0 & 1 \\ 0 & 0 & 1 & 0 \end{bmatrix}. \tag{3.18}$$

The first qubit (on the line with ullet) is called the control qubit and the second qubit (lower) is the target qubit. This gate changes 2-qubit states as (00, 01, 10, 11) $\rightarrow$ (00, 01, 11, 10), that is, it flips the target qubit. On basis states $|\varphi_1\rangle$ and $|\varphi_2\rangle \in \{0, 1\}$, the mapping of this gate can be written as $|\varphi_1, \varphi_2\rangle \rightarrow |\varphi_1, \varphi_1 \oplus \varphi_2\rangle$.

Comments on Line Drawing in Quantum Circuits: Consider the example with two lines and 4×4 gate U_4 which is shown in Fig. 3.3 in part (a). The input to this gate is the 2-qubit superposition equal to the tensor product $|\varphi\rangle = |\varphi_1\rangle \otimes |\varphi_2\rangle$ of qubits $|\varphi_1\rangle = a_1|0\rangle + b_1|1\rangle$ and $|\varphi_2\rangle = a_2|1\rangle + b_2|1\rangle$. The general case of the input being 2-qubit superposition $|\varphi\rangle$ is shown in part (b). In this case, $|\varphi\rangle$ need not be the tensor product of two qubits. The same is for the output $|\phi\rangle$. When qubits $|\varphi_1\rangle$ and $|\varphi_2\rangle$ are not specified in the first diagram, we consider these two drawings to be equal.

As an example, we consider the C-NOT gate as U_4 on the tensor product of two qubits. The input is

$$|\varphi\rangle = |\varphi_1\rangle \otimes |\varphi_2\rangle = (a_1|0\rangle + b_1|1\rangle) \otimes (a_2|0\rangle + b_2|1\rangle) = a_1a_2|00\rangle + a_1b_2|01\rangle + b_1a_2|10\rangle + b_1b_2|11\rangle.$$

The output of the gate can be written in the matrix form as

$$|\phi\rangle = \begin{bmatrix} 1 & 0 & 0 & 0 \\ 0 & 1 & 0 & 0 \\ 0 & 0 & 0 & 1 \\ 0 & 0 & 1 & 0 \end{bmatrix} \begin{bmatrix} a_1a_2 \\ a_1b_2 \\ b_1a_2 \\ b_1b_2 \end{bmatrix} = \begin{bmatrix} a_1a_2 \\ a_1b_2 \\ b_1b_2 \\ b_1a_2 \end{bmatrix} = a_1a_2|00\rangle + a_1b_2|01\rangle + b_1b_2|10\rangle + b_1a_2|11\rangle$$

$$= a_1|0\rangle\underbrace{(a_2|0\rangle + b_2|1\rangle)}_{|\varphi_2\rangle} + b_1|1\rangle\underbrace{(b_2|0\rangle + a_2|1\rangle)}_{|\varphi_2\rangle \text{ or} \neq |\varphi_2\rangle}.$$

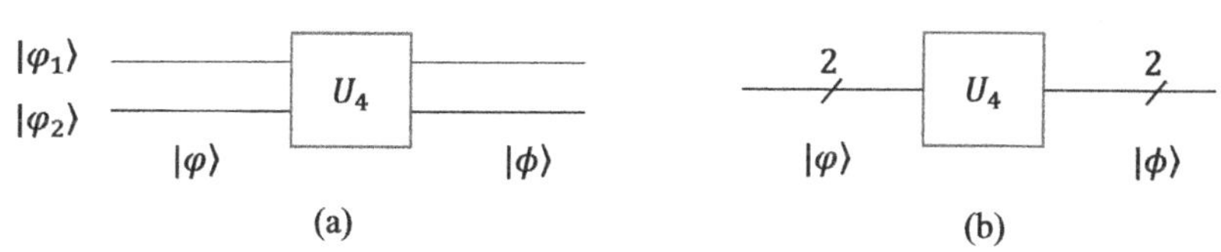

Fig. 3.3 The circuit (a) with two qubits and (b) with a 2-qubit superposition.

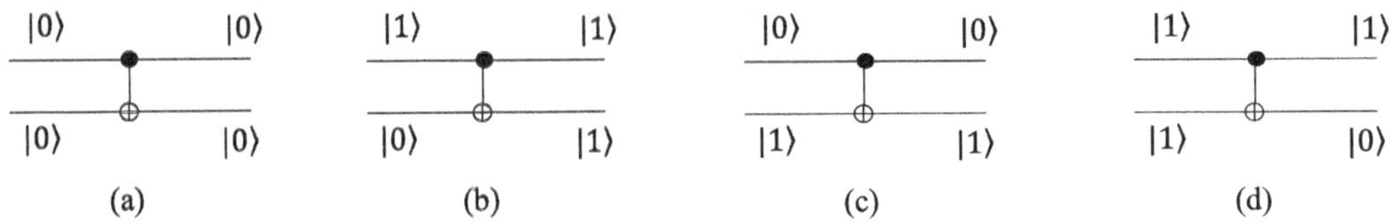

Fig. 3.4 The CNOT gate with basis state inputs.

In the case, when $b_2 = \pm a_2$, that is, when the second qubit is in one of the Hadamard states, then the output is the tensor product of two qubits,

$$|\phi\rangle = (a_1|0\rangle + b_1|1\rangle) \otimes \left(\frac{|0\rangle + |1\rangle}{\sqrt{2}}\right) = |\varphi_1\rangle \otimes |\varphi_2\rangle, \quad (\text{if } b_2 = a_2),$$

and

$$|\phi\rangle = (a_1|0\rangle - b_1|1\rangle) \otimes \left(\frac{|0\rangle - |1\rangle}{\sqrt{2}}\right) = [Z\,|\varphi_1\rangle] \otimes |\varphi_2\rangle, \quad (\text{if } b_2 = -a_2).$$

Note that for the second qubit being the basis states, $|\varphi_2\rangle = |0\rangle$ and $|1\rangle$, we obtain two entangled superpositions

$$C_{NOT}(|\varphi_1\rangle|0\rangle) = a_1|00\rangle + b_1|11\rangle, \qquad C_{NOT}(|\varphi_1\rangle|1\rangle) = a_1|01\rangle + b_1|10\rangle.$$

When $|\varphi_1\rangle = |0\rangle$ and $|1\rangle$, the outputs are shown in Fig. 3.4. Parts (a) and (b) illustrate the copying of both basis states $|0\rangle$ and $|1\rangle$, when the second input is $|0\rangle$.

2) SWAP, or the permutation of amplitudes a_1 and a_2 in the vector $(a_0, a_1, a_2, a_3)'$,

$$P_X = P_{(1,2)} = \begin{bmatrix} 1 & 0 & 0 & 0 \\ 0 & 0 & 1 & 0 \\ 0 & 1 & 0 & 0 \\ 0 & 0 & 0 & 1 \end{bmatrix}. \tag{3.19}$$

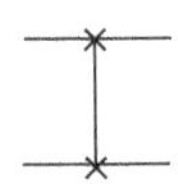

3) The controlled-phase gate

$$CS = I \oplus S = \begin{bmatrix} I & 0 \\ 0 & S \end{bmatrix} = \begin{bmatrix} 1 & 0 & 0 & 0 \\ 0 & 1 & 0 & 0 \\ 0 & 0 & 1 & 0 \\ 0 & 0 & 0 & i \end{bmatrix}. \tag{3.20}$$

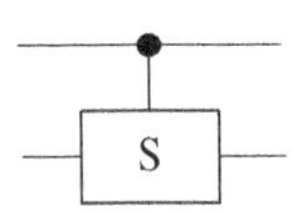

4) The controlled-phase shift gate

$$CP = I \oplus P(\phi) = \begin{bmatrix} I & 0 \\ 0 & P(\phi) \end{bmatrix} = \begin{bmatrix} 1 & 0 & 0 & 0 \\ 0 & 1 & 0 & 0 \\ 0 & 0 & 1 & 0 \\ 0 & 0 & 0 & e^{i\phi} \end{bmatrix}. \tag{3.21}$$

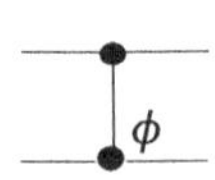

Also, $CP|11\rangle = e^{i\phi}|11\rangle$. For other three states $|0\rangle$, $|1\rangle$, and $|2\rangle$, the phase shift gates on the basis states $|n\rangle \rightarrow e^{i\phi}|n\rangle$ can be described as follows:

$$e^{i\phi}|0\rangle = \left(P_{(\phi)} \oplus I\right)|0\rangle = \begin{bmatrix} e^{i\phi} & 0 & 0 & 0 \\ 0 & 1 & 0 & 0 \\ 0 & 0 & 1 & 0 \\ 0 & 0 & 0 & 1 \end{bmatrix} \begin{bmatrix} 1 \\ 0 \\ 0 \\ 0 \end{bmatrix}, \quad e^{i\phi}|1\rangle = (P(\phi) \oplus I)|1\rangle = \begin{bmatrix} 1 & 0 & 0 & 0 \\ 0 & e^{i\phi} & 0 & 0 \\ 0 & 0 & 1 & 0 \\ 0 & 0 & 0 & 1 \end{bmatrix} \begin{bmatrix} 0 \\ 1 \\ 0 \\ 0 \end{bmatrix},$$

$$e^{i\phi}|2\rangle = \left(I \oplus P_{(\phi)}\right)|2\rangle = \begin{bmatrix} 1 & 0 & 0 & 0 \\ 0 & 1 & 0 & 0 \\ 0 & 0 & e^{i\phi} & 0 \\ 0 & 0 & 0 & 1 \end{bmatrix} \begin{bmatrix} 0 \\ 0 \\ 1 \\ 0 \end{bmatrix}.$$

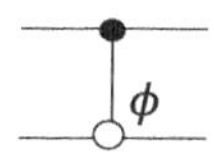

5) The controlled-Z gate (CSIGN)

$$CZ = I \oplus Z = \begin{bmatrix} I & 0 \\ 0 & Z \end{bmatrix} = \begin{bmatrix} 1 & 0 & 0 & 0 \\ 0 & 1 & 0 & 0 \\ 0 & 0 & 1 & 0 \\ 0 & 0 & 0 & -1 \end{bmatrix}. \tag{3.22}$$

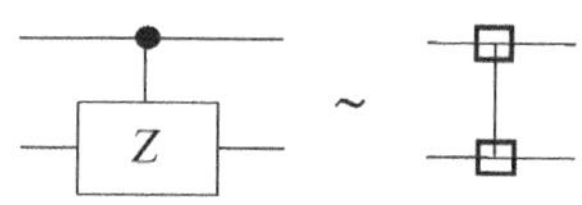

This gate can be used to obtain a CNOT gate. Indeed, considering that

$$ZR_y(-\pi) = \begin{bmatrix} 1 & 0 \\ 0 & -1 \end{bmatrix} \begin{bmatrix} 0 & 1 \\ -1 & 0 \end{bmatrix} = \begin{bmatrix} 0 & 1 \\ 1 & 0 \end{bmatrix} = X,$$

we obtain

$$\begin{bmatrix} I & 0 \\ 0 & ZR_y(-\pi) \end{bmatrix} = \begin{bmatrix} I & 0 \\ 0 & X \end{bmatrix}.$$

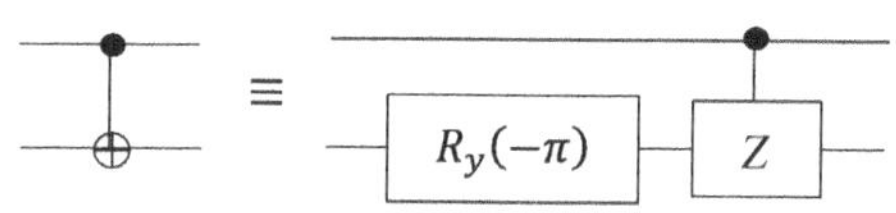

6) The general controlled-U gate for a given 2×2 unitary matrix U

$$CU = I \oplus U = \begin{bmatrix} I & 0 \\ 0 & U \end{bmatrix} = \begin{bmatrix} 1 & & \\ & 1 & \\ & & U \end{bmatrix}.$$

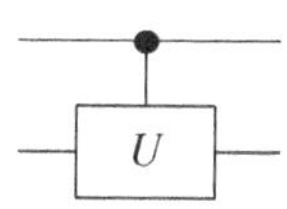

This gate applies the unitary transform U on the second (target) qubit if only the first (control) qubit is $|1\rangle$. Therefore, the gate can be called the 1-controlled-U gate and denoted as C_1U. Considering the transform matrix

$$U = U_2 = \begin{bmatrix} u_{0,0} & u_{0,1} \\ u_{1,0} & u_{1,1} \end{bmatrix},$$

the controlled gate C_1U can be described as shown in Table 3.1. The numbering of qubits, or bit planes, from the left. The first qubit is the control qubit and the second one is the target qubit.

Table 3.1 Controlled gate, or 1-controlled gate, with 2 × 2 transform U_2.

k	$\lvert\varphi_k\rangle$	1st bit	2nd bit	$CU_2 \mid \varphi_k\rangle$	Gate
0	$\lvert 0, 0\rangle$	0	0	$\lvert 0, 0\rangle$	1 0 0 0
1	$\lvert 0, 1\rangle$	0	1	$\lvert 0, 1\rangle$	0 1 0 0
2	$\lvert 1, 0\rangle$	1	0	$\lvert 1, U_2(0)\rangle = u_{0,0}\lvert 10\rangle + u_{1,0} \mid 11\rangle$	0 0 $u_{0,0}$ $u_{0,1}$
3	$\lvert 1, 1\rangle$	1	1	$\lvert 1, U_2(1)\rangle = u_{0,1}\lvert 10\rangle + u_{1,1} \mid 11\rangle$	0 0 $u_{1,0}$ $u_{1,1}$
	Basis state	Control	Target	Result	4×4 matrix

Table 3.2 The 0-Controlled gate with the 2 × 2 transform U_2.

k	$\lvert\varphi_k\rangle$	1st bit	2nd bit	$C_0U_2 \mid \varphi_k\rangle$	Gate
0	$\lvert 0, 0\rangle$	0	0	$\lvert 0, U_2(0)\rangle = u_{0,0}\lvert 00\rangle + u_{1,0} \mid 01\rangle$	$u_{0,0}$ $u_{0,1}$ 0 0
1	$\lvert 0, 1\rangle$	0	1	$\lvert 0, U_2(1)\rangle = u_{0,1}\lvert 00\rangle + u_{1,1} \mid 01\rangle$	$u_{1,0}$ $u_{1,1}$ 0 0
2	$\lvert 1, 0\rangle$	1	0	$\lvert 1, 0\rangle$	0 0 1 0
3	$\lvert 1, 1\rangle$	1	1	$\lvert 1, 1\rangle$	0 0 0 1
	Basis state	Control	Target	Result	4×4 matrix

Usually the controlled gates are drawn in small circles on the diagrams. Gates with 1-control bits are indicated by one or more filled circles. For 0-controlled gates, unfilled circles are used.

7) The general 0-controlled-U gate for a given 2×2 unitary matrix U

$$C_0U = U \oplus I = \begin{bmatrix} U & 0 \\ 0 & I \end{bmatrix} = \begin{bmatrix} U & & \\ & 1 & \\ & & 1 \end{bmatrix}.$$

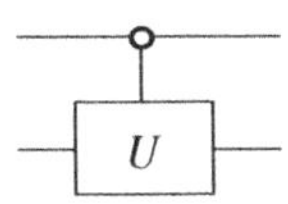

Table 3.2 illustrates the calculation of this gate.
The gate applies the unitary transform U on the first qubit if only the first qubit is $\lvert 0\rangle$. For example,

- The 0-controlled-Z gate

$$C_0Z = Z \oplus I = \begin{bmatrix} Z & 0 \\ 0 & I \end{bmatrix} = \begin{bmatrix} 1 & 0 & 0 & 0 \\ 0 & -1 & 0 & 0 \\ 0 & 0 & 1 & 0 \\ 0 & 0 & 0 & 1 \end{bmatrix}.$$

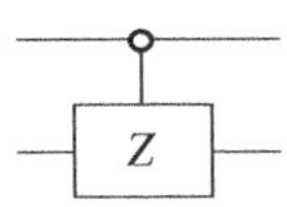

- The 0-controlled-X gate

$$C_0X = X \oplus I = \begin{bmatrix} X & 0 \\ 0 & I \end{bmatrix} = \begin{bmatrix} 0 & 1 & 0 & 0 \\ 1 & 0 & 0 & 0 \\ 0 & 0 & 1 & 0 \\ 0 & 0 & 0 & 1 \end{bmatrix}.$$

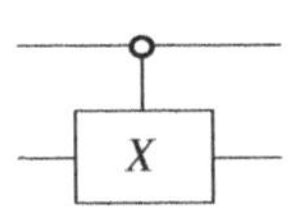

- The 0-controlled-W gate, or 0-controlled Givens rotation gate,

$$C_0W = W \oplus I = \begin{bmatrix} W & 0 \\ 0 & I \end{bmatrix} = \begin{bmatrix} \cos\vartheta & -\sin\vartheta & 0 & 0 \\ \sin\vartheta & \cos\vartheta & 0 & 0 \\ 0 & 0 & 1 & 0 \\ 0 & 0 & 0 & 1 \end{bmatrix}.$$

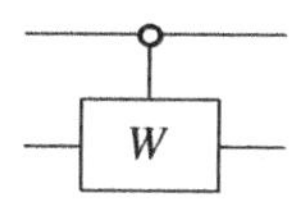

It is clear that the 1-controlled-U gate can be obtained from the 0-controlled-U, and vice versa. Indeed, let us consider the permutation (02)(13) with the following symmetric matrix and symbol (of sort):

$$P = P_1 = \begin{bmatrix} 0 & 0 & 1 & 0 \\ 0 & 0 & 0 & 1 \\ 1 & 0 & 0 & 0 \\ 0 & 1 & 0 & 0 \end{bmatrix} = X \otimes I_2.$$

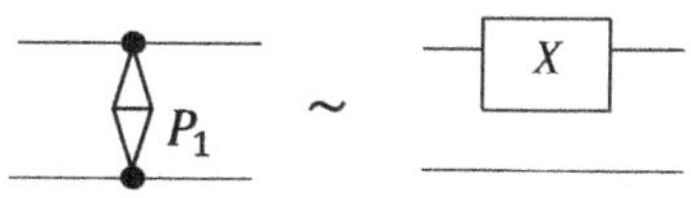

Then, for a 2×2 unitary matrix $U = [a, b; c, d]$, we obtain the following:

$$\begin{bmatrix} 0 & 0 & 1 & 0 \\ 0 & 0 & 0 & 1 \\ 1 & 0 & 0 & 0 \\ 0 & 1 & 0 & 0 \end{bmatrix} \begin{bmatrix} 1 & 0 & 0 & 0 \\ 0 & 1 & 0 & 0 \\ 0 & 0 & a & b \\ 0 & 0 & c & d \end{bmatrix} \begin{bmatrix} 0 & 0 & 1 & 0 \\ 0 & 0 & 0 & 1 \\ 1 & 0 & 0 & 0 \\ 0 & 1 & 0 & 0 \end{bmatrix} = \begin{bmatrix} a & b & 0 & 0 \\ c & d & 0 & 0 \\ 0 & 0 & 1 & 0 \\ 0 & 0 & 0 & 1 \end{bmatrix}$$

and since $P^2 = I_4$,

$$\begin{bmatrix} 0 & 0 & 1 & 0 \\ 0 & 0 & 0 & 1 \\ 1 & 0 & 0 & 0 \\ 0 & 1 & 0 & 0 \end{bmatrix} \begin{matrix} a & b & 0 & 0 \\ c & d & 0 & 0 \\ 0 & 0 & 1 & 0 \\ 0 & 0 & 0 & 1 \end{matrix} \begin{bmatrix} 0 & 0 & 1 & 0 \\ 0 & 0 & 0 & 1 \\ 1 & 0 & 0 & 0 \\ 0 & 1 & 0 & 0 \end{bmatrix} = \begin{bmatrix} 1 & 0 & 0 & 0 \\ 0 & 1 & 0 & 0 \\ 0 & 0 & a & b \\ 0 & 0 & c & d \end{bmatrix}.$$

Thus,

$$C_1U = I \oplus U = P(U \oplus I)P = P(C_0U)P \quad \text{and} \quad C_0U = P(C_1U)P.$$

The circuit elements of these controlled gates are shown in Fig. 3.5.
For instance, for the rotation gates of $W(\vartheta)$,

$$\begin{bmatrix} \cos\vartheta & -\sin\vartheta & 0 & 0 \\ \sin\vartheta & \cos\vartheta & 0 & 0 \\ 0 & 0 & 1 & 0 \\ 0 & 0 & 0 & 1 \end{bmatrix} = \begin{bmatrix} 0 & 0 & 1 & 0 \\ 0 & 0 & 0 & 1 \\ 1 & 0 & 0 & 0 \\ 0 & 1 & 0 & 0 \end{bmatrix} \begin{bmatrix} 1 & 0 & 0 & 0 \\ 0 & 1 & 0 & 0 \\ 0 & 0 & \cos\vartheta & -\sin\vartheta \\ 0 & 0 & \sin\vartheta & \cos\vartheta \end{bmatrix} \begin{bmatrix} 0 & 0 & 1 & 0 \\ 0 & 0 & 0 & 1 \\ 1 & 0 & 0 & 0 \\ 0 & 1 & 0 & 0 \end{bmatrix}.$$

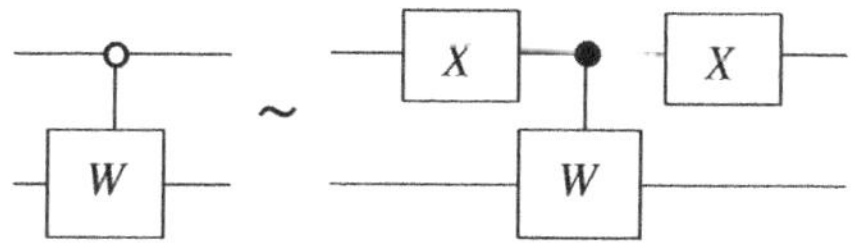

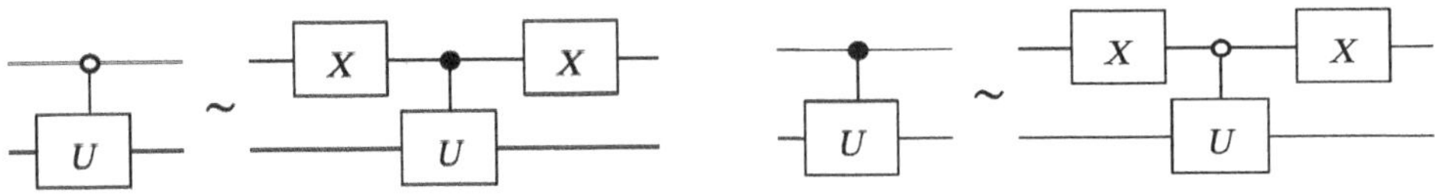

Fig. 3.5 Circuits for the controlled gates.

Example 3.5 Gate with Two Rotations
Consider the following 2-qubit gate with two rotations by angle -ϑ and ϑ, when $\vartheta \in (0, \pi)$:

$$B = W(-\vartheta) \oplus W(\vartheta) = \begin{bmatrix} \cos\vartheta & \sin\vartheta & 0 & 0 \\ -\sin\vartheta & \cos\vartheta & 0 & 0 \\ 0 & 0 & \cos\vartheta & -\sin\vartheta \\ 0 & 0 & \sin\vartheta & \cos\vartheta \end{bmatrix}.$$

The matrix $W(-\vartheta)$ can be calculated by $W(\vartheta)$ as

$$W(-\vartheta) = \begin{bmatrix} \cos\vartheta & \sin\vartheta \\ -\sin\vartheta & \cos\vartheta \end{bmatrix} = \begin{bmatrix} 1 & 0 \\ 0 & -1 \end{bmatrix} \begin{bmatrix} \cos\vartheta & -\sin\vartheta \\ \sin\vartheta & \cos\vartheta \end{bmatrix} \begin{bmatrix} 1 & 0 \\ 0 & -1 \end{bmatrix} = ZW(\vartheta)Z. \tag{3.23}$$

Therefore, we can write

$$B = \begin{bmatrix} ZW(\vartheta)Z & 0 \\ 0 & W(\vartheta) \end{bmatrix} = \begin{bmatrix} Z & 0 \\ 0 & I_2 \end{bmatrix} \begin{bmatrix} W(\vartheta) & 0 \\ 0 & W(\vartheta) \end{bmatrix} \begin{bmatrix} Z & 0 \\ 0 & I_2 \end{bmatrix} = C_0(Z)(I_2 \otimes W(\vartheta))C_0(Z). \tag{3.24}$$

The corresponding circuit for the gate B is shown in Fig. 3.6. It includes one local gate of rotation and two 0-controlled Pauli Z gates. This gate can also be written as $B = C_0(W(-\vartheta))C(W(\vartheta))$.

8) The permutation of two qubits $\{|\varphi_1\rangle, |\varphi_2\rangle\} \rightarrow \{|\varphi_2\rangle, |\varphi_1\rangle\}$ is described by the above permutation P = (0, 2)(1, 3): {0123} → {2301}. The following composition holds for the permutation matrix:

$$P_{1\leftrightarrow 2} = P = \begin{bmatrix} 0 & 0 & 1 & 0 \\ 0 & 1 & 0 & 0 \\ 1 & 0 & 0 & 0 \\ 0 & 0 & 0 & 1 \end{bmatrix} \begin{bmatrix} 1 & 0 & 0 & 0 \\ 0 & 0 & 0 & 1 \\ 0 & 0 & 1 & 0 \\ 0 & 1 & 0 & 0 \end{bmatrix} = \begin{bmatrix} 0 & 0 & 1 & 0 \\ 0 & 0 & 0 & 1 \\ 1 & 0 & 0 & 0 \\ 0 & 1 & 0 & 0 \end{bmatrix}, \quad \det P_{1\rightarrow 2} = 1. \tag{3.25}$$

9) The 4-point Walsh–Hadamard transform

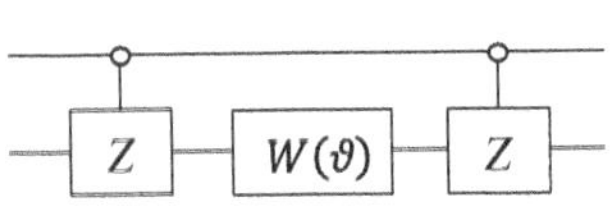

Fig. 3.6 The circuit of the B gate.

$$H_4 = H^{\otimes 2} = H \otimes H = \begin{bmatrix} H & H \\ H & -H \end{bmatrix} = \frac{1}{2} \begin{bmatrix} 1 & 1 & 1 & 1 \\ 1 & -1 & 1 & -1 \\ 1 & 1 & -1 & -1 \\ 1 & -1 & -1 & 1 \end{bmatrix}, \quad \det H_4 = 1. \tag{3.26}$$

Example 3.6 2-Qubit State Preparation
Consider the transform of the state $|00\rangle$,

$$\frac{1}{2}\begin{bmatrix}1 & 1 & 1 & 1\\ 1 & -1 & 1 & -1\\ 1 & 1 & -1 & -1\\ 1 & -1 & -1 & 1\end{bmatrix}\begin{bmatrix}1\\0\\0\\0\end{bmatrix}=\frac{1}{2}\begin{bmatrix}1\\1\\1\\1\end{bmatrix}=\frac{|00\rangle+|01\rangle+|10\rangle+|11\rangle}{2}. \tag{3.27}$$

Thus, $H_4|00\rangle=\frac{1}{2}(|0\rangle+|1\rangle+|2\rangle+|3\rangle)$.

10) The 4-point A_2-based Hadamard gate

$$A_4=A_2^{\otimes 2}=A_2\otimes A_2=\begin{bmatrix}A_2 & -A_2\\ A_2 & A_2\end{bmatrix}=\frac{1}{2}\begin{bmatrix}1 & -1 & -1 & -1\\ 1 & 1 & -1 & 1\\ 1 & -1 & 1 & -1\\ 1 & 1 & 1 & 1\end{bmatrix},\quad \det A_4=1.$$

11) The 4-point paired transform [4] with the unitary matrix

$$\chi_4'=\frac{1}{2}\begin{bmatrix}\sqrt{2} & 0 & -\sqrt{2} & 0\\ 0 & \sqrt{2} & 0 & -\sqrt{2}\\ 1 & -1 & 1 & -1\\ 1 & 1 & 1 & 1\end{bmatrix},\quad \det\chi_4'=1. \tag{3.28}$$

One can note that

$$\chi_4'=\begin{bmatrix}1 & 0 & 0 & 0\\ 0 & 1 & 0 & 0\\ 0 & 0 & \frac{1}{\sqrt{2}} & -\frac{1}{\sqrt{2}}\\ 0 & 0 & \frac{1}{\sqrt{2}} & \frac{1}{\sqrt{2}}\end{bmatrix}\times\frac{1}{\sqrt{2}}\begin{bmatrix}1 & 0 & -1 & 0\\ 0 & 1 & 0 & -1\\ 1 & 0 & 1 & 0\\ 0 & 1 & 0 & 1\end{bmatrix}$$

and

$$\frac{1}{\sqrt{2}}\begin{bmatrix}1 & 0 & -1 & 0\\ 0 & 1 & 0 & -1\\ 1 & 0 & 1 & 0\\ 0 & 1 & 0 & 1\end{bmatrix}=\frac{1}{\sqrt{2}}\begin{bmatrix}1 & -1\\ 1 & 1\end{bmatrix}\otimes\begin{bmatrix}1 & 0\\ 0 & 1\end{bmatrix}=A_2\otimes I_2.$$

Therefore, we can write the 4-point paired transform as

$$\chi_4'=(I_2\oplus A_2)(A_2\otimes I_2) \tag{3.29}$$

and consider the quantum circuit shown in Fig. 3.7.

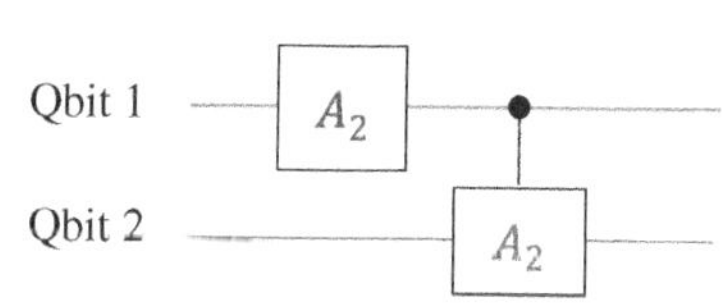

Fig. 3.7 The 4-point paired transform gate circuit.

Example 3.7 Transformation of Two Basis States
The paired transforms of the states $|00\rangle$ and $|01\rangle$ are calculated as follows:

$$\chi'_4|00\rangle = \frac{1}{2}\begin{bmatrix} \sqrt{2} & 0 & -\sqrt{2} & 0 \\ 0 & \sqrt{2} & 0 & -\sqrt{2} \\ 1 & -1 & 1 & -1 \\ 1 & 1 & 1 & 1 \end{bmatrix}\begin{bmatrix} 1 \\ 0 \\ 0 \\ 0 \end{bmatrix} = \frac{1}{2}\begin{bmatrix} \sqrt{2} \\ 0 \\ 1 \\ 1 \end{bmatrix} = \frac{1}{\sqrt{2}}|00\rangle + \frac{|10\rangle + |11\rangle}{2}, \tag{3.30}$$

$$\chi'_4|01\rangle = \frac{1}{2}\begin{bmatrix} \sqrt{2} & 0 & -\sqrt{2} & 0 \\ 0 & \sqrt{2} & 0 & -\sqrt{2} \\ 1 & -1 & 1 & -1 \\ 1 & 1 & 1 & 1 \end{bmatrix}\begin{bmatrix} 0 \\ 1 \\ 0 \\ 0 \end{bmatrix} = \frac{1}{2}\begin{bmatrix} 0 \\ \sqrt{2} \\ -1 \\ 1 \end{bmatrix} = \frac{1}{\sqrt{2}}|01\rangle - \frac{|10\rangle - |11\rangle}{2}. \tag{3.31}$$

In Fig. 3.7, an example of a local unitary operation on the first qubit is shown. Let us consider three such local gates with 2-point unitary transforms U_2 and V_2, which are shown in Fig. 3.8.

1) The 2-qubit operation shown in part (a) of Fig. 3.8 describes the unitary operation with the matrix

$$A = U_2 \otimes I_2 = \begin{bmatrix} a & & b & \\ & a & & b \\ c & & d & \\ & c & & d \end{bmatrix}, \qquad U_2 = \begin{bmatrix} a & b \\ c & d \end{bmatrix}. \tag{3.32}$$

2) The 2-qubit operation shown in part (b) describes the unitary operation with the matrix

$$A = I_2 \otimes V_2 = \begin{bmatrix} v_1 & v_2 & & \\ v_3 & v_4 & & \\ & & v_1 & v_2 \\ & & v_3 & v_4 \end{bmatrix} = V_2 \oplus V_2, \qquad V_2 = \begin{bmatrix} v_1 & v_2 \\ v_3 & v_4 \end{bmatrix}. \tag{3.33}$$

3) The 2-qubit operation shown in part (c) describes the sequential execution of U_2 and V_2 operations. As a result, we obtain the unitary operation with the matrix

$$A = (U_2 \otimes I_2)(I_2 \otimes V_2) = \begin{bmatrix} a & & b & \\ & a & & b \\ c & & d & \\ & c & & d \end{bmatrix}\begin{bmatrix} v_1 & v_2 & & \\ v_3 & v_4 & & \\ & & v_1 & v_2 \\ & & v_3 & v_4 \end{bmatrix} = \begin{bmatrix} av_1 & av_2 & bv_1 & bv_2 \\ av_3 & av_4 & bv_3 & bv_4 \\ cv_1 & cv_2 & dv_1 & dv_2 \\ cv_3 & cv_4 & dv_3 & dv_4 \end{bmatrix}. \tag{3.34}$$

Due to the property of the Kronecker product, this matrix is equal to $A = (U_2I_2 \otimes I_2V_2) = U_2 \otimes V_2$.

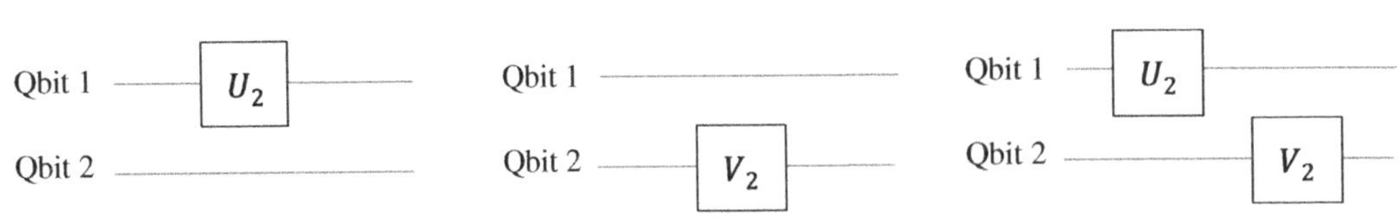

(a) $U_2 \otimes I_2$ (b) $I_2 \otimes V_2$ (c) $U_2 \otimes V_2$

Fig. 3.8 Local gates that operates on (a) the first qubit, (b) the second qubit, and (c) on two qubits.

However, the Kronecker product is not commutative operation, the order of U_2 and V_2 operations is important, when $V_2 \neq U_2$. In the figure, the gate U_2 is shown to the left of gate V_2, as if it were the first to be executed. Therefore, we assume that the gate on the first qubit is executed first, and then the gate on the second qubit is executed.

Example 3.8 Hadamard Transform Local Gates
Consider three circuits shown in Fig. 3.9 with local gates H_2 and A_2.

1) The parallel operation of the Hadamard gates H_2 has the matrix of the 4-point discrete Walsh–Hadamard transform H_4. Indeed, this matrix is calculated by

$$H_2 \otimes H_2 = \frac{1}{\sqrt{2}}\begin{bmatrix} 1 & 1 \\ 1 & -1 \end{bmatrix} \otimes \frac{1}{\sqrt{2}}\begin{bmatrix} 1 & 1 \\ 1 & -1 \end{bmatrix} = \frac{1}{2}\begin{bmatrix} 1 & 1 & 1 & 1 \\ 1 & -1 & 1 & -1 \\ 1 & 1 & -1 & -1 \\ 1 & -1 & -1 & 1 \end{bmatrix}.$$

When the input is $|01\rangle$, that is, two qubits in circuit in part (a) are $|0\rangle$ and $|1\rangle$, the output of the gate H_4 equals to

$$H_2(|01\rangle) = \frac{1}{2}(|00\rangle - |01\rangle + |10\rangle - |11\rangle) = \frac{1}{2}\sum_{k=0}^{3}(-1)^k|k\rangle. \tag{3.35}$$

2) The parallel operation of the gates A_2 and H_2 has the Hadamard matrix calculated by

$$A_2 \otimes H_2 = \frac{1}{\sqrt{2}}\begin{bmatrix} 1 & -1 \\ 1 & 1 \end{bmatrix} \otimes \frac{1}{\sqrt{2}}\begin{bmatrix} 1 & 1 \\ 1 & -1 \end{bmatrix} = \frac{1}{2}\begin{bmatrix} 1 & 1 & -1 & -1 \\ 1 & -1 & -1 & 1 \\ 1 & 1 & 1 & 1 \\ 1 & -1 & 1 & -1 \end{bmatrix}.$$

When the input is $|00\rangle$, the output of this gate equals to

$$(A_2 \otimes H_2)(|00\rangle) = \frac{1}{2}(|00\rangle + |01\rangle - |10\rangle - |11\rangle).$$

3) The parallel operation of the gates A_2 has another Hadamard matrix calculated by

$$A_2 \otimes A_2 = \frac{1}{\sqrt{2}}\begin{bmatrix} 1 & -1 \\ 1 & 1 \end{bmatrix} \otimes \frac{1}{\sqrt{2}}\begin{bmatrix} 1 & -1 \\ 1 & 1 \end{bmatrix} = \frac{1}{2}\begin{bmatrix} 1 & -1 & -1 & 1 \\ 1 & 1 & -1 & -1 \\ 1 & -1 & 1 & -1 \\ 1 & 1 & 1 & 1 \end{bmatrix}.$$

Here, the inputs $|00\rangle$ and $|11\rangle$ result in the following superpositions:

$$(A_2 \otimes A_2)(|00\rangle) = \frac{1}{2}(|00\rangle - |01\rangle - |10\rangle + |11\rangle),$$

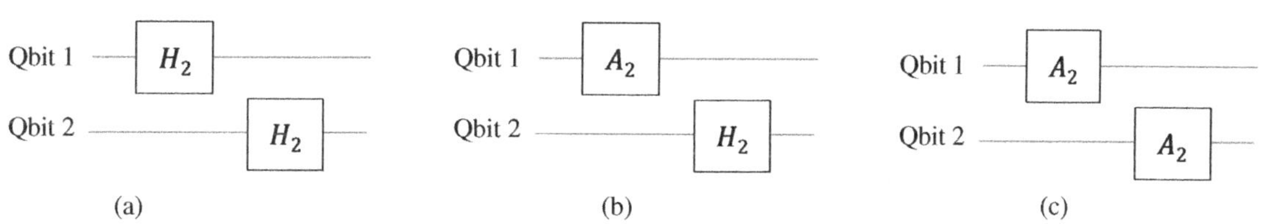

Fig. 3.9 Three local gates on the second qubit.

Fig. 3.10 Two circuits with one local gate and one controlled gate.

$$(A_2 \otimes A_2)(|11\rangle) = \frac{1}{2}(|00\rangle + |01\rangle + |10\rangle + |11\rangle) = \frac{1}{2}\sum_{k=0}^{3}|k\rangle.$$

It can be seen that these three Hadamard matrices are equivalent up to row permutations.

Now we consider two quantum circuits shown in Fig. 3.10.

The 2×2 unitary matrices in the circuit in part (a) are

$$U_2 = \begin{bmatrix} u_0 & u_1 \\ u_2 & u_3 \end{bmatrix} \quad \text{and} \quad V_2 = \begin{bmatrix} v_0 & v_1 \\ v_2 & v_3 \end{bmatrix}.$$

The 4×4 unitary matrix with local gate U_2 and controlled gate V_2 is calculated by

$$A = \begin{bmatrix} 1 & & & \\ & 1 & & \\ & & v_0 & v_1 \\ & & v_2 & v_3 \end{bmatrix} \begin{bmatrix} u_0 & 0 & u_1 & 0 \\ 0 & u_0 & 0 & u_1 \\ u_2 & 0 & u_3 & 0 \\ 0 & u_2 & 0 & u_3 \end{bmatrix} = \begin{bmatrix} u_0 & 0 & u_1 & 0 \\ 0 & u_0 & 0 & u_1 \\ v_0u_2 & v_1u_2 & v_0u_3 & v_1u_3 \\ v_2u_2 & v_3u_2 & v_2u_3 & v_3u_3 \end{bmatrix}. \tag{3.36}$$

In the case, when $V_2 = X$, we obtain

$$A = \begin{bmatrix} 1 & & & \\ & 1 & & \\ & & 0 & 1 \\ & & 1 & 0 \end{bmatrix} \begin{bmatrix} u_0 & 0 & u_1 & \\ 0 & u_0 & 0 & u_1 \\ u_2 & 0 & u_3 & 0 \\ 0 & u_2 & 0 & u_3 \end{bmatrix} = \begin{bmatrix} u_0 & 0 & u_1 & 0 \\ 0 & u_0 & 0 & u_1 \\ 0 & u_2 & 0 & u_3 \\ u_2 & 0 & u_3 & 0 \end{bmatrix}. \tag{3.37}$$

The application of this matrix on 4 basis states $|00\rangle, |01\rangle, |10\rangle$, and $|11\rangle$ result in the following 2-qubits:

$$|\varphi_{00}\rangle = A\begin{bmatrix} 1 \\ 0 \\ 0 \\ 0 \end{bmatrix} = \begin{bmatrix} u_0 \\ 0 \\ 0 \\ u_2 \end{bmatrix}, \quad |\varphi_{01}\rangle = A\begin{bmatrix} 0 \\ 1 \\ 0 \\ 0 \end{bmatrix} = \begin{bmatrix} 0 \\ u_0 \\ u_2 \\ 0 \end{bmatrix}, \quad |\varphi_{10}\rangle = A\begin{bmatrix} 0 \\ 0 \\ 1 \\ 0 \end{bmatrix} = \begin{bmatrix} u_1 \\ 0 \\ 0 \\ u_3 \end{bmatrix}, \quad |\varphi_{11}\rangle = A\begin{bmatrix} 0 \\ 0 \\ 0 \\ 1 \end{bmatrix} = \begin{bmatrix} 0 \\ u_1 \\ u_3 \\ 0 \end{bmatrix},$$

which are four untangled 2-qubits

$$|\varphi_{00}\rangle = u_0|00\rangle + u_2|11\rangle, \quad |\varphi_{01}\rangle = u_0|01\rangle + u_2|10\rangle, \tag{3.38}$$

$$|\varphi_{10}\rangle = u_1|00\rangle + u_3|11\rangle, \quad |\varphi_{11}\rangle = u_1|01\rangle + u_3|10\rangle. \tag{3.39}$$

The circuit in part (b) in Fig. 3.10 is a particular case of the first circuit, when U_2 is the Hadamard matrix H_2 and V_2 is the Pauli NOT matrix X. The corresponding matrix A equals to

$$A = \begin{bmatrix} 1 & & & \\ & 1 & & \\ & & 0 & 1 \\ & & 1 & 0 \end{bmatrix} \times \frac{1}{\sqrt{2}}\begin{bmatrix} 1 & 0 & 1 & 0 \\ 0 & 1 & 0 & 1 \\ 1 & 0 & -1 & 0 \\ 0 & 1 & 0 & -1 \end{bmatrix} = \frac{1}{\sqrt{2}}\begin{bmatrix} 1 & 0 & 1 & 0 \\ 0 & 1 & 0 & 1 \\ 0 & 1 & 0 & -1 \\ 1 & 0 & -1 & 0 \end{bmatrix}, \quad \det A = -1.$$

The corresponding four untangled 2-qubits are the following:

$$|\varphi_{00}\rangle = \frac{1}{\sqrt{2}}(|00\rangle + |11\rangle), \qquad |\varphi_{01}\rangle = \frac{1}{\sqrt{2}}(|01\rangle + |10\rangle),$$

$$|\varphi_{10}\rangle = \frac{1}{\sqrt{2}}(|00\rangle - |11\rangle), \qquad |\varphi_{11}\rangle = \frac{1}{\sqrt{2}}(|01\rangle - |10\rangle).$$

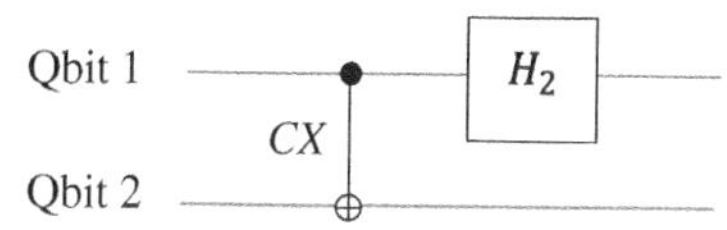

Fig. 3.11 The circuit reversed to the circuit in Fig. 3.10b.

These are the known Bell states.

Note that the transpose matrix is calculated by

$$A' = \frac{1}{\sqrt{2}}\begin{bmatrix} 1 & 0 & 1 & 0 \\ 0 & 1 & 0 & 1 \\ 1 & 0 & -1 & 0 \\ 0 & 1 & 0 & -1 \end{bmatrix}\begin{bmatrix} 1 & & & \\ & 1 & & \\ & & 0 & 1 \\ & & 1 & 0 \end{bmatrix} = \frac{1}{\sqrt{2}}\begin{bmatrix} 1 & 0 & 0 & 1 \\ 0 & 1 & 1 & 0 \\ 1 & 0 & 0 & -1 \\ 0 & 1 & -1 & 0 \end{bmatrix},$$

or $A' = (H_2 \otimes I_2)(CX)$. This matrix describes the circuit shown in Fig. 3.11.

We also consider two quantum circuits shown in Fig. 3.12. The 2-qubit operation shown in part (a) of this figure describes the unitary operation with the matrix

$$A(U_2) = \begin{bmatrix} 1 & 0 & 0 & 0 \\ 0 & a & 0 & b \\ 0 & 0 & 1 & 0 \\ 0 & c & 0 & d \end{bmatrix}, \qquad U_2 = \begin{bmatrix} a & b \\ c & d \end{bmatrix}. \tag{3.40}$$

In particular case, for the NOT gate $U_2 = X$, we obtain the matrix of the permutation (1, 3),

$$A(X) = P_{(1,3)} = \begin{bmatrix} 1 & 0 & 0 & 0 \\ 0 & 0 & 0 & 1 \\ 0 & 0 & 1 & 0 \\ 0 & 1 & 0 & 0 \end{bmatrix}. \tag{3.41}$$

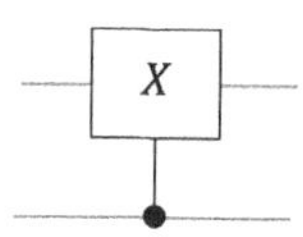

The 2-qubit operation shown in part (b) describes the product of the matrices of permutations (2, 3) and (1, 3),

$$C_{NOT}P_{(1,3)}C_{NOT} = \begin{bmatrix} 1 & 0 & 0 & 0 \\ 0 & 1 & 0 & 0 \\ 0 & 0 & 0 & 1 \\ 0 & 0 & 1 & 0 \end{bmatrix}\begin{bmatrix} 1 & 0 & 0 & 0 \\ 0 & 0 & 0 & 1 \\ 0 & 0 & 1 & 0 \\ 0 & 1 & 0 & 0 \end{bmatrix}\begin{bmatrix} 1 & 0 & 0 & 0 \\ 0 & 1 & 0 & 0 \\ 0 & 0 & 0 & 1 \\ 0 & 0 & 1 & 0 \end{bmatrix} = \begin{bmatrix} 1 & 0 & 0 & 0 \\ 0 & 0 & 1 & 0 \\ 0 & 1 & 0 & 0 \\ 0 & 0 & 0 & 1 \end{bmatrix},$$

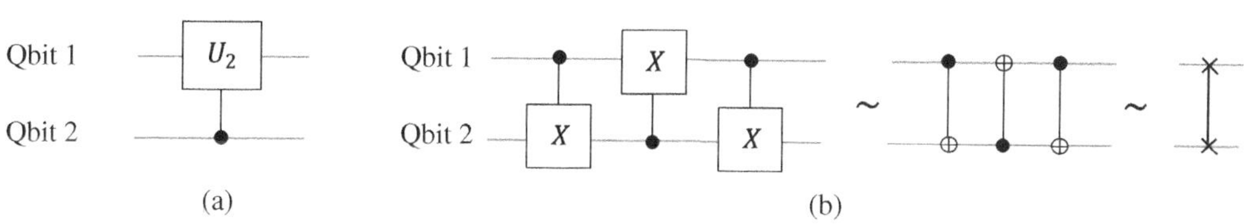

Fig. 3.12 Two circuits.

that is, it is the matrix of the SWAP gate, or the permutation (1, 2). If a 2-qubit superposition $|\varphi\rangle$ is not entangled, this operation swaps two qubits. Indeed, for qubits $|\varphi_0\rangle = a_0|0\rangle + b_0\,|\,1\rangle$ and $|\varphi_1\rangle = a_1|0\rangle + b_1\,|\,1\rangle$, the tensor product $|\varphi\rangle = |\,\varphi_0, \varphi_1\rangle$ is mapping to $|\varphi_0, \varphi_1\rangle$,

$$|\,\varphi\rangle = |\,\varphi_0\rangle \otimes |\,\varphi_1\rangle = \begin{bmatrix} a_0a_1 \\ a_0b_1 \\ b_0a_1 \\ b_0b_1 \end{bmatrix} \rightarrow \begin{bmatrix} 1 & 0 & 0 & 0 \\ 0 & 0 & 1 & 0 \\ 0 & 1 & 0 & 0 \\ 0 & 0 & 0 & 1 \end{bmatrix} \begin{bmatrix} a_0a_1 \\ a_0b_1 \\ b_0a_1 \\ b_0b_1 \end{bmatrix} = \begin{bmatrix} a_0a_1 \\ b_0a_1 \\ a_0b_1 \\ b_0b_1 \end{bmatrix} = |\,\varphi_1\rangle \otimes |\,\varphi_0\rangle.$$

References

1 Nielsen, M. and Chuang, I. (2001). *Quantum Computation and Quantum Information*, 2e. Cambridge UP.

2 Rieffel, E. and Polak, W. (2011). *Quantum Computing: A Gentle Introduction*. Cambridge, Massachusetts, London, England: The MIT Press.

3 Gel'fand, I.M. (1989). *Lectures on Linear Algebra*. Dover Publication.

4 Grigoryan, A.M. and Grigoryan, M.M. (2009). *Brief Notes in Advanced DSP: Fourier Analysis with MATLAB*. Taylor and Francis Group: CRC Press.

4

Multi-qubit Superpositions and Operations

This section describes the general case of working with qubits, the number of which can exceed two. Namely, we consider the multi-qubit superpositions. Each such superposition possibly describes some physical system of particles that can only be observed in a fixed number of states [1, 2]. Similar to 2-qubit superpositions, when the number of such independent states is 4. The theory and application of multi-qubit superpositions are key to the development of quantum computing and image processing. This section also introduces important qubit gates such as Toffoli and Fredkin, demonstrating how classical logic functions can be implemented in quantum circuits. The section details elementary operations in multi-qubit systems, starting with three-qubit operations using local gates and control bits.

Consider integer $N = 2^r$, $r \geq 2$, and the computational basis states $\{|0\rangle, |1\rangle, |2\rangle, \ldots, |N-1\rangle\}$. The numbers in states are considered in binary representation. For example, if $N = 16$, the basis state $|7\rangle$ is written as $|0111\rangle$. The multiple or r-qubit quantum superposition is written as

$$|\varphi\rangle = a_0|0\rangle + a_1|1\rangle + a_2|2\rangle + a_3|3\rangle + \cdots + a_{N-1}|N-1\rangle. \tag{4.1}$$

Each amplitude a_k, $k \in [0, N-1]$, defines the probability $p_k = |a_k|^2$ of measurement (or observing) the r-qubit to be in the state $|k\rangle$. Therefore, $|a_0|^2 + |a_1|^2 + \ldots + |a_{N-1}|^2 = 1$. The superposition with additional global phase $e^{i\theta}|\varphi\rangle$, where θ is an angle, is considered physically indistinguishable.

Several qubits make up a quantum system register, just like in a classical computer [3]. The information is stored in binary form. In such a register, each qubit is in a superposition of $|0\rangle$ and $|1\rangle$. The quantum registers for r-qubits contain all possible combinations of states of qubits, or bits 0 and 1. The state $|\varphi\rangle$ of the register is the vector presenting the tensor product of its constituent qubits. For instance, the above basis state $|7\rangle$, or $|0111\rangle$, will be written in the quantum register as the following string of quantum bits:

$$|0111\rangle = |0\rangle|1\rangle|1\rangle|1\rangle = \begin{bmatrix}1\\0\end{bmatrix} \otimes \begin{bmatrix}0\\1\end{bmatrix} \otimes \begin{bmatrix}0\\1\end{bmatrix} \otimes \begin{bmatrix}0\\1\end{bmatrix} = [0000\,0001\,0000\,0000]'.$$

The example of the quantum register for 4 qubits is shown in Table 4.1. This is the logical array of 4 qubits with all possible 16 states $|\varphi_k\rangle$ written in the natural order, as $|\varphi_k\rangle = |k\rangle$. Each state $|\varphi_k\rangle$ is a bit-string state, which is called the conventional state. All 16 states compose the basis $B_4 = \{|k\rangle, k = 0:15\}$ for 4 qubits to be in the superposition in Eq. 4.1 (for $N = 2^4 = 16$).

When reading the register, its state $|\varphi\rangle$ is considered to be the tensor product of 4 qubits, that is,

$$|\varphi\rangle = a_0|0\rangle + a_1|1\rangle + a_2|2\rangle + a_3|3\rangle + \cdots + a_{15}|15\rangle = |q_0\rangle \otimes |q_1\rangle \otimes |q_4\rangle \otimes |q_3\rangle.$$

Here, the amplitudes a_k, $k = 0:15$, are defined by the amplitudes of the 4 qubits. Thus, the readouts of the register are only in entangled superpositions. A quantum operation is performed on all 16 basis states simultaneously, unlike in a classical computer where only one state or bit-string, for instance (0010), is read from its register.

Quantum Image Processing in Practice: A Mathematical Toolbox, First Edition. Artyom M. Grigoryan and Sos S. Agaian.

Companion website: www.wiley.com/go/grigoryan/quantumimageprocessing

Table 4.1 Model of the 4-qubit quantum register.

Qubits	States $\|\varphi_k\rangle$, $k = 0{:}15$, in the register															
	$\|0\rangle$	$\|1\rangle$	$\|2\rangle$	$\|3\rangle$	$\|4\rangle$	$\|5\rangle$	$\|6\rangle$	$\|7\rangle$	$\|8\rangle$	$\|9\rangle$	$\|10\rangle$	$\|11\rangle$	$\|12\rangle$	$\|13\rangle$	$\|14\rangle$	$\|15\rangle$
$\|q_0\rangle$	0	0	0	0	0	0	0	0	1	1	1	1	1	1	1	1
$\|q_1\rangle$	0	0	0	0	1	1	1	1	0	0	0	0	1	1	1	1
$\|q_2\rangle$	0	0	1	1	0	0	1	1	0	0	1	1	0	0	1	1
$\|q_3\rangle$	0	1	0	1	0	1	0	1	0	1	0	1	0	1	0	1
	Amplitudes a_k, $k = 0{:}15$, of the states															
	a_0	a_1	a_2	a_3	a_4	a_5	a_6	a_7	a_8	a_9	a_{10}	a_{11}	a_{12}	a_{13}	a_{14}	a_{15}

It should be also noted that a common initial register state for many circuits is the zero state $|0\rangle^{\otimes r}$, or the state $|0\rangle^{\otimes 4} = |0000\rangle$ for this register. It will be shown in Section 6.1.2 that any r-qubit superposition can be prepared from this zero state.

4.1 Elementary Operations on Multi-qubits

Let us describe the main operations on multi-qubit superpositions. The operations of the sum and tensor product of r-qubits are similar to the operations for 2-qubits. Many other operators are described by $N \times N$ unitary matrices U on amplitudes of the r-qubit superpositions. These amplitudes together are considered as a unit norm vector $(a_0, a_1, \ldots, a_{N-1})'$ in the real or complex space, which is called a Hilbert space. The Hilbert space is liner, as the Euclidean vector space. In such a space, the sum of any two elements is also an element of the space. Since we work only with unit vectors, the set of such vectors does not describe a linear space. Therefore, in our work, we will not use the term and concept of Hilbert space. In this section, we describe several local and control gates on 3-qubits and then we consider the known permutations, Toffoli and Fredkin gates, which are universal logic gates for classical computation.

4.2 3-Qubit Operations with Local Gates

First, we consider a few local gates on 3-qubits, which are shown in Fig. 4.1.

The gates in this figure are described as follows.

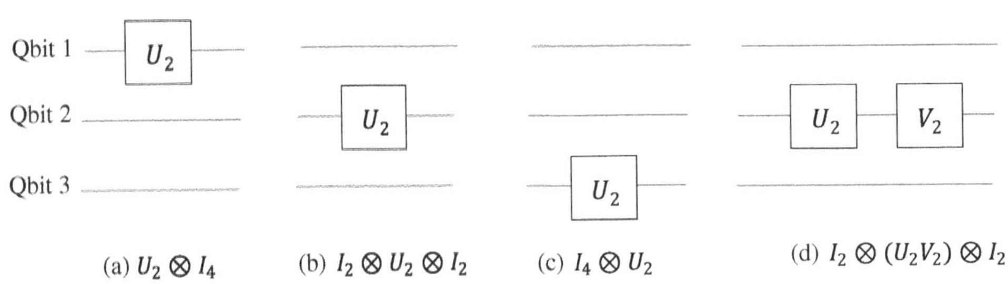

Fig. 4.1 3-Qubit local gates of operations applied (a) on the first qubit, (b) on the second qubit, (c) on the third qubit, and (d) the gate of two operations applied on the second qubit.

a) The 3-qubit operation in part (a) has the unitary matrix

$$A = U_2 \otimes I_4 = \begin{bmatrix} aI_4 & bI_4 \\ cI_4 & dI_4 \end{bmatrix} = \begin{bmatrix} a & & & & b & & & \\ & a & & & & b & & \\ & & a & & & & b & \\ & & & a & & & & b \\ c & & & & d & & & \\ & c & & & & d & & \\ & & c & & & & d & \\ & & & c & & & & d \end{bmatrix}, U_2 = \begin{bmatrix} a & b \\ c & d \end{bmatrix}. \tag{4.2}$$

b) The 3-qubit operation in part (b) has the unitary matrix

$$A = I_2 \otimes U_2 \otimes I_2 = (U_2 \oplus U_2) \otimes I_2 = \begin{bmatrix} a & & b & & & & & \\ & a & & b & & & & \\ c & & d & & & & & \\ & c & & d & & & & \\ & & & & a & & b & \\ & & & & & a & & b \\ & & & & c & & d & \\ & & & & & c & & d \end{bmatrix}. \tag{4.3}$$

c) The 3-qubit operation in part (c) has the unitary matrix

$$A = I_4 \otimes U_2 = U_2 \oplus U_2 \oplus U_2 \oplus U_2 = \begin{bmatrix} a & b & & & & & & \\ c & d & & & & & & \\ & & a & b & & & & \\ & & c & d & & & & \\ & & & & a & b & & \\ & & & & c & d & & \\ & & & & & & a & b \\ & & & & & & c & d \end{bmatrix}. \tag{4.4}$$

d) The 3-qubit operation in part (d) has the unitary matrix

$$A = I_2 \otimes (U_2V_2) \otimes I_2 = (U_2V_2 \oplus U_2V_2) \otimes I_2 = \begin{bmatrix} s_1 & & s_2 & & & & & \\ & s_1 & & s_2 & & & & \\ s_3 & & s_4 & & & & & \\ & s_3 & & s_4 & & & & \\ & & & & s_1 & & s_2 & \\ & & & & & s_1 & & s_2 \\ & & & & s_3 & & s_4 & \\ & & & & & s_3 & & s_4 \end{bmatrix}, U_2V_2 = \begin{bmatrix} s_1 & s_2 \\ s_3 & s_4 \end{bmatrix}. \tag{4.5}$$

Figures 4.2 and 4.3 show a few gates that are composed of two and three local gates on different qubits. As examples, we consider two operations shown in Fig. 4.4.
The gates in this figure are described as follows.

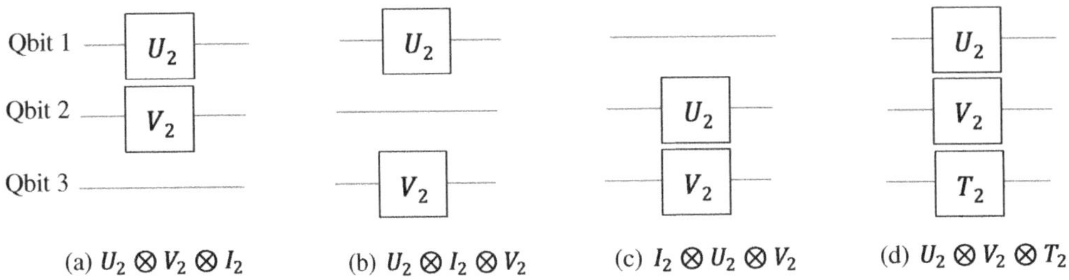

Fig. 4.2 Qubit local gates of two operations applied (a) on the first and second qubits, (b) on the second and third qubits, (c) on the second and third qubits, and (d) 3-qubit gates of three operations applied to different qubits.

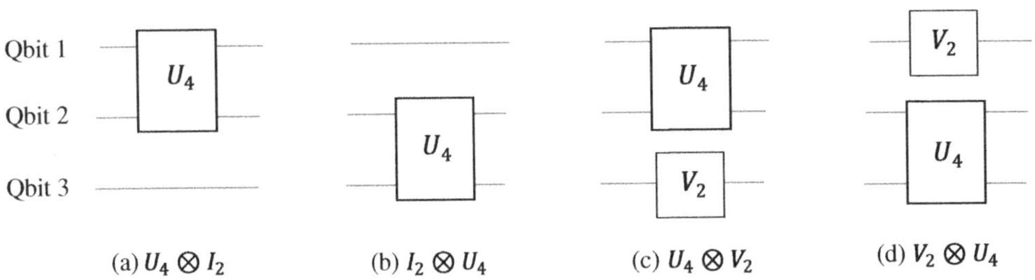

Fig. 4.3 (a) 3-Qubit operations with local 2-qubit and 1-qubit gates, U_4 and V_2, when (a) U_4 applied to the first and second qubits, (b) U_4 applied to the second and third qubits, and U_4 and V_2 applied respectively (c) to the first two qubits and last qubit and (d) to the last two qubits and first qubit.

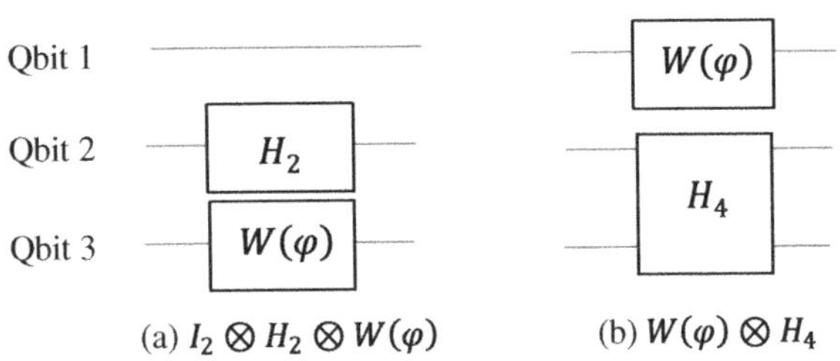

Fig. 4.4 3-Qubit operations with (a) Hadamard gate H_2 applied to the second qubit and rotation gate $W(\varphi)$ applied to the third qubit and (b) rotation gate $W(\varphi)$ applied to the first qubit and Hadamard gate H_4 applied to the last two qubits.

a) Denote $c = \cos(\varphi)$ and $s = \sin(\varphi)$. Then, the 3-qubit operation in part (a) has the unitary matrix

$$A = (I_2 \otimes H_2) \otimes W(\varphi) = (H_2 \oplus H_2) \otimes \begin{bmatrix} c & -s \\ s & c \end{bmatrix} = \frac{1}{\sqrt{2}} \begin{bmatrix} c & -s & c & -s & & & & \\ s & c & s & c & & & & \\ c & -s & -c & s & & & & \\ s & c & -s & -c & & & & \\ & & & & c & -s & c & -s \\ & & & & s & c & s & c \\ & & & & c & -s & -c & s \\ & & & & s & c & -s & c \end{bmatrix}.$$

b) The 3-qubit operation in part (b) has the unitary matrix

$$A = W(\varphi) \otimes H_4 = \begin{bmatrix} c & -s \\ s & c \end{bmatrix} \otimes H_4 = \frac{1}{2} \begin{bmatrix} c & c & c & c & -s & -s & -s & -s \\ c & -c & c & -c & -s & s & -s & s \\ c & c & -c & -c & -s & -s & s & s \\ c & -c & -c & c & -s & s & s & -s \\ s & s & s & s & c & c & c & c \\ s & -s & s & -s & c & -c & c & -c \\ s & s & -s & -s & c & c & -c & -c \\ s & -s & -s & s & c & -c & -c & c \end{bmatrix}.$$

4.3 3-Qubit Operations with Control Bits

A few 3-qubit gates with one control bits are shown in Fig. 4.5 and also given the symbolic designation of the gates. We will consider all these operations.

To describe the gates in this figure, we consider the unitary matrix

$$U_2 = \begin{bmatrix} a & b \\ c & d \end{bmatrix}$$

and all 8-bit planes (BP), which will be shown with control bits.

In part (a), we have the gate with the following 8×8 matrix:

$$(1, -, U_2) = I_4 \oplus U_2 \oplus U_2 = I_4 \oplus (I_2 \otimes U_2) = \begin{bmatrix} 1 & & & & & & & \\ & 1 & & & & & & \\ & & 1 & & & & & \\ & & & 1 & & & & \\ & & & & a & b & & \\ & & & & c & d & & \\ & & & & & & a & b \\ & & & & & & c & d \end{bmatrix}, BP = \begin{bmatrix} 0 & 0 & 0 \\ 0 & 0 & 1 \\ 0 & 1 & 0 \\ 0 & 1 & 1 \\ 1 & 0 & 0 \\ 1 & 0 & 1 \\ 1 & 1 & 0 \\ 1 & 1 & 1 \end{bmatrix}$$

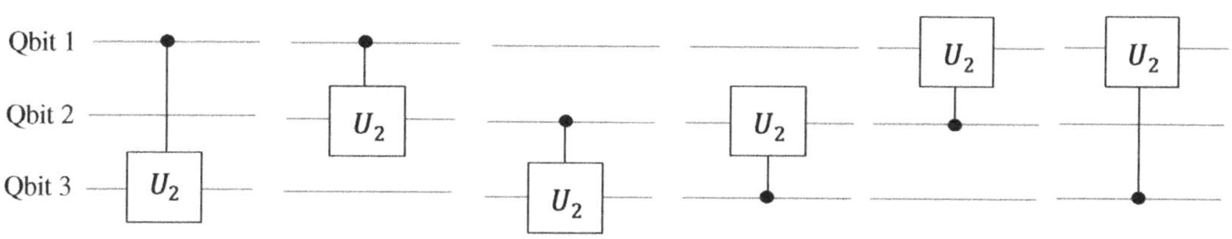

(a) $(1, -, U_2)$ (b) $(1, U_2, -)$ (c) $(-, 2, U_2)$ (d) $(-, U_2, 3)$ (e) $(U_2, 2, -)$ (f) $(U_2, -, 3)$

Fig. 4.5 Six 3-qubit operations with control qubits. The operations U_2 with the first control qubit is applied to (a) the third qubit and (b) the second qubit. U_2 with the second control qubit is applied to (c) the third qubit and (e) the first qubit. U_2 with the last control qubit is applied to (d) the second qubit and (f) the first qubit.

In part (b), we have the gate with the following 8×8 matrix:

$$(1, U_2, -) = I_4 \oplus (U_2 \otimes I_2) = \begin{bmatrix} 1 & & & & & & & \\ & 1 & & & & & & \\ & & 1 & & & & & \\ & & & 1 & & & & \\ & & & & a & & b & \\ & & & & & a & & b \\ & & & & c & & d & \\ & & & & & c & & d \end{bmatrix}, BP = \begin{bmatrix} 0 & 0 & 0 \\ 0 & 0 & 1 \\ 0 & 1 & 0 \\ 0 & 1 & 1 \\ 1 & 0 & 0 \\ 1 & 0 & 1 \\ 1 & 1 & 0 \\ 1 & 1 & 1 \end{bmatrix}$$

In part (c), we have the gate with the following 8×8 matrix:

$$(-, 2, U_2) = I_2 \otimes (I_2 \oplus U_2) = I_2 \oplus U_2 \oplus I_2 \oplus U_2 = \begin{bmatrix} 1 & & & & & & & \\ & 1 & & & & & & \\ & & a & b & & & & \\ & & c & d & & & & \\ & & & & 1 & & & \\ & & & & & 1 & & \\ & & & & & & a & b \\ & & & & & & c & d \end{bmatrix}, BP = \begin{bmatrix} 0 & 0 & 0 \\ 0 & 0 & 1 \\ 0 & 1 & 0 \\ 0 & 1 & 1 \\ 1 & 0 & 0 \\ 1 & 0 & 1 \\ 1 & 1 & 0 \\ 1 & 1 & 1 \end{bmatrix}$$

In part (d), we have the gate with the following 8×8 matrix:

$$(-, U_2, 3) = U_{2;1,3} \oplus U_{2;1,3} = I_2 \otimes U_{2;1,3} = \begin{bmatrix} 1 & & & & & & & \\ & a & & b & & & & \\ & & 1 & & & & & \\ & c & & d & & & & \\ & & & & 1 & & & \\ & & & & & a & & b \\ & & & & & & 1 & \\ & & & & & c & & d \end{bmatrix}, BP = \begin{bmatrix} 0 & 0 & 0 \\ 0 & 0 & 1 \\ 0 & 1 & 0 \\ 0 & 1 & 1 \\ 1 & 0 & 0 \\ 1 & 0 & 1 \\ 1 & 1 & 0 \\ 1 & 1 & 1 \end{bmatrix}$$

In part (e), we have the gate with the following 8×8 matrix:

$$(U_2, 2, -) = \begin{bmatrix} 1 & & & & & & & \\ & 1 & & & & & & \\ & & a & & & & b & \\ & & & a & & & & b \\ & & & & 1 & & & \\ & & & & & 1 & & \\ & & c & & & & d & \\ & & & c & & & & d \end{bmatrix}, BP = \begin{bmatrix} 0 & 0 & 0 \\ 0 & 0 & 1 \\ 0 & 1 & 0 \\ 0 & 1 & 1 \\ 1 & 0 & 0 \\ 1 & 0 & 1 \\ 1 & 1 & 0 \\ 1 & 1 & 1 \end{bmatrix}$$

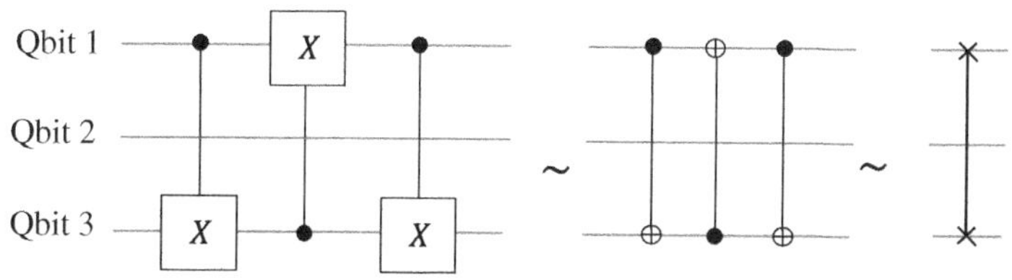

Fig. 4.6 The circuit of the 3-qubit SWAP of bits.

In part (f), we have the gate with the following 8×8 matrix:

$$(U_2, -, 3) = \begin{bmatrix} 1 & & & & & & & \\ & a & & & & b & & \\ & & 1 & & & & & \\ & & & a & & & & b \\ & & & & 1 & & & \\ & c & & & & d & & \\ & & & & & & 1 & \\ & & & c & & & & d \end{bmatrix}, BP = \begin{bmatrix} 0 & 0 & 0 \\ 0 & 0 & 1 \\ 0 & 1 & 0 \\ 0 & 1 & 1 \\ 1 & 0 & 0 \\ 1 & 0 & 1 \\ 1 & 1 & 0 \\ 1 & 1 & 1 \end{bmatrix}.$$

Example 4.1 Bit SWAP Operation on 3 Qubits
Consider the circuit shown in Fig. 4.6 in different forms. The unitary matrix of this circuit is calculated by

$$A = (I_4 \oplus X \oplus X) \begin{bmatrix} 1 & 0 & 0 & 0 & 0 & 0 & 0 & 0 \\ 0 & 0 & 0 & 0 & 0 & 1 & 0 & 0 \\ 0 & 0 & 1 & 0 & 0 & 0 & 0 & 0 \\ 0 & 0 & 0 & 0 & 0 & 0 & 0 & 1 \\ 0 & 0 & 0 & 0 & 1 & 0 & 0 & 0 \\ 0 & 1 & 0 & 0 & 0 & 0 & 0 & 0 \\ 0 & 0 & 0 & 0 & 0 & 0 & 1 & 0 \\ 0 & 0 & 0 & 1 & 0 & 0 & 0 & 0 \end{bmatrix} (I_4 \oplus X \oplus X) = \begin{bmatrix} 1 & 0 & 0 & 0 & 0 & 0 & 0 & 0 \\ 0 & 0 & 0 & 0 & 1 & 0 & 0 & 0 \\ 0 & 0 & 1 & 0 & 0 & 0 & 0 & 0 \\ 0 & 0 & 0 & 0 & 0 & 0 & 1 & 0 \\ 0 & 1 & 0 & 0 & 0 & 0 & 0 & 0 \\ 0 & 0 & 0 & 0 & 0 & 1 & 0 & 0 \\ 0 & 0 & 0 & 1 & 0 & 0 & 0 & 0 \\ 0 & 0 & 0 & 0 & 0 & 0 & 0 & 1 \end{bmatrix}.$$

It is the matrix of the permutation $P = (1, 4)(3, 6)$.

4.4 3-Qubit Operations with 2 Control Bits

The 3-qubit gates with 2 control bits are shown in Fig. 4.7. The gates are numbered by two digits and the symbol of the transform U_2. The numbers show the control qubits.

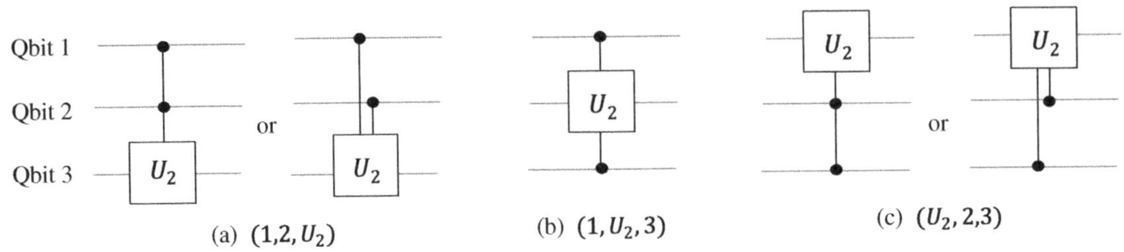

Fig. 4.7 Three 3-qubit gates with two 1-control qubit, when operation U_2 is applied to (a) the third qubit, (b) to the second qubit, and (c) to the first qubit.

In part (a), the 8×8 matrix of the gate can be written as

$$(1,2,U_2) = I_4 \oplus (I_2 \oplus U_2) = \begin{bmatrix} 1 &&&&&&& \\ & 1 &&&&&& \\ && 1 &&&&& \\ &&& 1 &&&& \\ &&&& 1 &&& \\ &&&&& 1 && \\ &&&&&& a & b \\ &&&&&& c & d \end{bmatrix}, BP = \begin{bmatrix} 0 & 0 & 0 \\ 0 & 0 & 1 \\ 0 & 1 & 0 \\ 0 & 1 & 1 \\ 1 & 0 & 0 \\ 1 & 0 & 1 \\ 1 & 1 & 0 \\ 1 & 1 & 1 \end{bmatrix}$$

In part (b), the 8×8 matrix of the gate can be written as

$$(1,U_2,3) = \begin{bmatrix} 1 &&&&&&& \\ & 1 &&&&&& \\ && 1 &&&&& \\ &&& 1 &&&& \\ &&&& 1 &&& \\ &&&&& a && b \\ &&&&&& 1 & \\ &&&&& c && d \end{bmatrix}, BP = \begin{bmatrix} 0 & 0 & 0 \\ 0 & 0 & 1 \\ 0 & 1 & 0 \\ 0 & 1 & 1 \\ 1 & 0 & 0 \\ 1 & 0 & 1 \\ 1 & 1 & 0 \\ 1 & 1 & 1 \end{bmatrix}$$

In part (c), the 8×8 matrix of the gate can be written as

$$(U_2,2,3) = \begin{bmatrix} 1 &&&&&&& \\ & 1 &&&&&& \\ && 1 &&&&& \\ &&& a &&&& b \\ &&&& 1 &&& \\ &&&&& 1 && \\ &&&&&& 1 & \\ &&& c &&&& d \end{bmatrix}, BP = \begin{bmatrix} 0 & 0 & 0 \\ 0 & 0 & 1 \\ 0 & 1 & 0 \\ 0 & 1 & 1 \\ 1 & 0 & 0 \\ 1 & 0 & 1 \\ 1 & 1 & 0 \\ 1 & 1 & 1 \end{bmatrix}.$$

We use the notation $U_{2;i,j}$ for the transform U_2 applied on i and j planes, where $i, j \in \{1, 2, ..., 7\}$. The above transforms are $U_{2;6,7}$, $U_{2;5,7}$, and $U_{2;3,7}$.

In a similar way, we can describe 3-qubit gates with 0-control bits. Figure 4.8 illustrates three such gates.

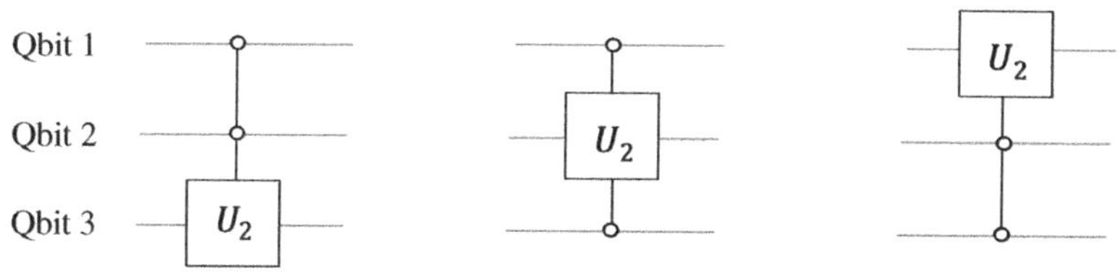

(a) $(1',2',U_2)$ (b) $(1',U_2,3')$ (c) $(U_2,2',3')$

Fig. 4.8 Three 0-controlled 3-qubit gates, when operation U_2 is applied to (a) the third qubit, (b) to the second qubit, and (c) to the first qubit.

In part (a), the gate is denoted as $(1', 2', U_2)$, where the numbers show the control qubits. Prime symbol "$'$" is used only for 0-control qubits. The 8×8 matrix of this gate can be written as

$$C_0(C_0U_2) = (C_0U_2)\oplus I_4 = U_2\oplus I_6 = \begin{bmatrix} a & b & & & & & & \\ c & b & & & & & & \\ & & 1 & & & & & \\ & & & 1 & & & & \\ & & & & 1 & & & \\ & & & & & 1 & & \\ & & & & & & 1 & \\ & & & & & & & 1 \end{bmatrix}, BP = \begin{bmatrix} 0 & 0 & 0 \\ 0 & 0 & 1 \\ 0 & 1 & 0 \\ 0 & 1 & 1 \\ 1 & 0 & 0 \\ 1 & 0 & 1 \\ 1 & 1 & 0 \\ 1 & 1 & 1 \end{bmatrix}.$$

The 8×8 matrix of the gate in part (b) can be written as

$$U_{2;0,2} = \begin{bmatrix} a & & b & & & & & \\ & 1 & & & & & & \\ c & & d & & & & & \\ & & & 1 & & & & \\ & & & & 1 & & & \\ & & & & & 1 & & \\ & & & & & & 1 & \\ & & & & & & & 1 \end{bmatrix}, BP = \begin{bmatrix} 0 & 0 & 0 \\ 0 & 0 & 1 \\ 0 & 1 & 0 \\ 0 & 1 & 1 \\ 1 & 0 & 0 \\ 1 & 0 & 1 \\ 1 & 1 & 0 \\ 1 & 1 & 1 \end{bmatrix}.$$

The 8×8 matrix of the gate with two 0-control bits in part (c) can be written as

$$U_{2;0,4} = \begin{bmatrix} a & & & & b & & & \\ & 1 & & & & & & \\ & & 1 & & & & & \\ & & & 1 & & & & \\ c & & & & d & & & \\ & & & & & 1 & & \\ & & & & & & 1 & \\ & & & & & & & 1 \end{bmatrix}, BP = \begin{bmatrix} 0 & 0 & 0 \\ 0 & 0 & 1 \\ 0 & 1 & 0 \\ 0 & 1 & 1 \\ 1 & 0 & 0 \\ 1 & 0 & 1 \\ 1 & 1 & 0 \\ 1 & 1 & 1 \end{bmatrix}$$

Also, we consider a few 0-controlled gates shown in Fig. 4.9.

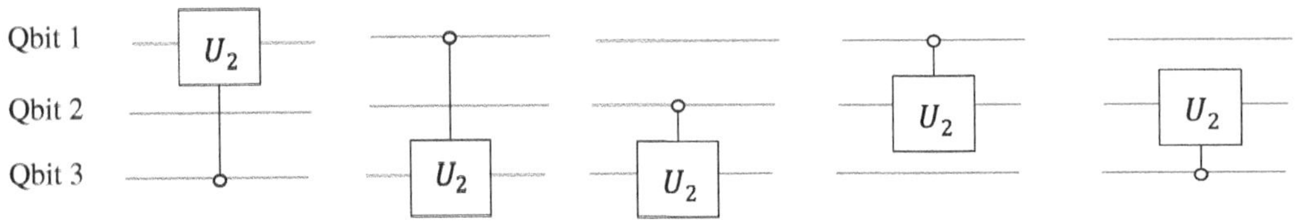

(a) $(U_2, -, 3',)$ (b) $(1', -, U_2)$ (c) $(-, 2', U_2)$ (d) $(1', U_2, -)$ (e) $(-, U_2, 3')$

Fig. 4.9 Five 0-controlled 3-qubit gates, when operation U_2 is applied (a) to the first qubit with the third control qubit, to the third qubit with (b) the first and (c) second control qubits, and to the second qubit with (d) the first and (e) the third control qubit.

The 8×8 matrix of the gate with two 0-control bits in part (a) can be written as

$$U_{2;04,26} = \begin{bmatrix} a & & & & b & & & \\ & 1 & & & & & & \\ & & a & & & & b & \\ & & & 1 & & & & \\ c & & & & d & & & \\ & & & & & 1 & & \\ & & c & & & & d & \\ & & & & & & & 1 \end{bmatrix}, BP = \begin{bmatrix} 0 & 0 & 0 \\ 0 & 0 & 1 \\ 0 & 1 & 0 \\ 0 & 1 & 1 \\ 1 & 0 & 0 \\ 1 & 0 & 1 \\ 1 & 1 & 0 \\ 1 & 1 & 1 \end{bmatrix}.$$

The 8×8 matrix of the gate with two 0-control bits in part (b) can be written as

$$U_{2;01,23} = (I_2 \otimes U_2) \oplus I_4 = \begin{bmatrix} a & b & & & & & & \\ c & d & & & & & & \\ & & a & b & & & & \\ & & c & d & & & & \\ & & & & 1 & & & \\ & & & & & 1 & & \\ & & & & & & 1 & \\ & & & & & & & 1 \end{bmatrix}, BP = \begin{bmatrix} 0 & 0 & 0 \\ 0 & 0 & 1 \\ 0 & 1 & 0 \\ 0 & 1 & 1 \\ 1 & 0 & 0 \\ 1 & 0 & 1 \\ 1 & 1 & 0 \\ 1 & 1 & 1 \end{bmatrix}.$$

In part (c), the corresponding 8×8 matrix of the gate with two 0-control bits can be written as

$$U_{2;01,45} = I_2 \otimes (U_2 \oplus I_2) = \begin{bmatrix} a & b & & & & & & \\ c & d & & & & & & \\ & & 1 & & & & & \\ & & & 1 & & & & \\ & & & & a & b & & \\ & & & & c & d & & \\ & & & & & & 1 & \\ & & & & & & & 1 \end{bmatrix}, BP = \begin{bmatrix} 0 & 0 & 0 \\ 0 & 0 & 1 \\ 0 & 1 & 0 \\ 0 & 1 & 1 \\ 1 & 0 & 0 \\ 1 & 0 & 1 \\ 1 & 1 & 0 \\ 1 & 1 & 1 \end{bmatrix}.$$

In part (d), the corresponding 8×8 matrix of the gate with 0-control bit can be written as

$$U_{2;02,13} = (U_2 \otimes I_2) \oplus I_4 = \begin{bmatrix} a & & b & & & & & \\ & a & & b & & & & \\ c & & d & & & & & \\ & c & & d & & & & \\ & & & & 1 & & & \\ & & & & & 1 & & \\ & & & & & & 1 & \\ & & & & & & & 1 \end{bmatrix}, BP = \begin{bmatrix} 0 & 0 & 0 \\ 0 & 0 & 1 \\ 0 & 1 & 0 \\ 0 & 1 & 1 \\ 1 & 0 & 0 \\ 1 & 0 & 1 \\ 1 & 1 & 0 \\ 1 & 1 & 1 \end{bmatrix}.$$

In part (e), the corresponding 8×8 matrix of the gate with 0-control bit can be written as

$$U_{2;02,46} = \begin{bmatrix} a & & b & & & & & \\ & 1 & & & & & & \\ c & & d & & & & & \\ & & & 1 & & & & \\ & & & & a & & b & \\ & & & & & 1 & & \\ & & & & c & & d & \\ & & & & & & & 1 \end{bmatrix}, BP = \begin{bmatrix} 0 & 0 & 0 \\ 0 & 0 & 1 \\ 0 & 1 & 0 \\ 0 & 1 & 1 \\ 1 & 0 & 0 \\ 1 & 0 & 1 \\ 1 & 1 & 0 \\ 1 & 1 & 1 \end{bmatrix}.$$

Figure 4.10 shows several mixed-type control gates, namely, 1- and 0-controled gates. The 8×8 matrix of the gate with 1- and 0-control bits in parts (a) to (f) can be written as follows:

$$U_{2;23} = (I_2 \oplus U_2) \oplus I_4 = \begin{bmatrix} 1 & & & & & & & \\ & 1 & & & & & & \\ & & a & b & & & & \\ & & c & d & & & & \\ & & & & 1 & & & \\ & & & & & 1 & & \\ & & & & & & 1 & \\ & & & & & & & 1 \end{bmatrix}, BP = \begin{bmatrix} 0 & 0 & 0 \\ 0 & 0 & 1 \\ 0 & 1 & 0 \\ 0 & 1 & 1 \\ 1 & 0 & 0 \\ 1 & 0 & 1 \\ 1 & 1 & 0 \\ 1 & 1 & 1 \end{bmatrix},$$

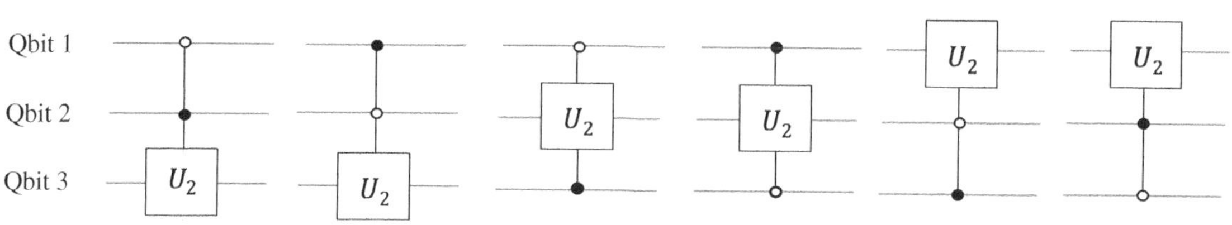

(a) $(1', 2, U_2)$ (b) $(1,2', U_2)$ (c) $(1', U_2, 3)$ (d) $(1, U_2, 3')$ (e) $(U_2, 2', 3)$ (f) $(U_2, 2,3')$

Fig. 4.10 Six 3-qubit gates with 2 different type control bits, when operation U_2 is applied to (a) the first 0 and second 1-control bits, (b) the first 1 and second 0-control bits, (c) first 0 and third 1-control bits, (d) first 1 and third 0-control bits, (e) second 0 and third 1-control bits, and (f) second 1 and third 0-control bits.

$$U_{2;45} = I_4 \oplus (U_2 \oplus I_2) = \begin{bmatrix} 1 & & & & & & & \\ & 1 & & & & & & \\ & & 1 & & & & & \\ & & & 1 & & & & \\ & & & & a & b & & \\ & & & & c & d & & \\ & & & & & & 1 & \\ & & & & & & & 1 \end{bmatrix}, BP = \begin{bmatrix} 0 & 0 & 0 \\ 0 & 0 & 1 \\ 0 & 1 & 0 \\ 0 & 1 & 1 \\ 1 & 0 & 0 \\ 1 & 0 & 1 \\ 1 & 1 & 0 \\ 1 & 1 & 1 \end{bmatrix},$$

$$U_{2;13} = \begin{bmatrix} 1 & & & & & & & \\ & a & & b & & & & \\ & & 1 & & & & & \\ & c & & d & & & & \\ & & & & 1 & & & \\ & & & & & 1 & & \\ & & & & & & 1 & \\ & & & & & & & 1 \end{bmatrix}, BP = \begin{bmatrix} 0 & 0 & 0 \\ 0 & 0 & 1 \\ 0 & 1 & 0 \\ 0 & 1 & 1 \\ 1 & 0 & 0 \\ 1 & 0 & 1 \\ 1 & 1 & 0 \\ 1 & 1 & 1 \end{bmatrix},$$

$$U_{2;46} = \begin{bmatrix} 1 & & & & & & & \\ & 1 & & & & & & \\ & & 1 & & & & & \\ & & & 1 & & & & \\ & & & & a & & b & \\ & & & & & 1 & & \\ & & & & c & & d & \\ & & & & & & & 1 \end{bmatrix}, BP = \begin{bmatrix} 0 & 0 & 0 \\ 0 & 0 & 1 \\ 0 & 1 & 0 \\ 0 & 1 & 1 \\ 1 & 0 & 0 \\ 1 & 0 & 1 \\ 1 & 1 & 0 \\ 1 & 1 & 1 \end{bmatrix},$$

$$U_{2;15} = \begin{bmatrix} 1 & & & & & & & \\ & a & & & & b & & \\ & & 1 & & & & & \\ & & & 1 & & & & \\ & & & & 1 & & & \\ & c & & & & d & & \\ & & & & & & 1 & \\ & & & & & & & 1 \end{bmatrix}, BP = \begin{bmatrix} 0 & 0 & 0 \\ 0 & 0 & 1 \\ 0 & 1 & 0 \\ 0 & 1 & 1 \\ 1 & 0 & 0 \\ 1 & 0 & 1 \\ 1 & 1 & 0 \\ 1 & 1 & 1 \end{bmatrix},$$

$$U_{2;26} = \begin{bmatrix} 1 & & & & & & & \\ & 1 & & & & & & \\ & & a & & & & b & \\ & & & 1 & & & & \\ & & & & 1 & & & \\ & & & & & 1 & & \\ & & c & & & & d & \\ & & & & & & & 1 \end{bmatrix}, BP = \begin{bmatrix} 0 & 0 & 0 \\ 0 & 0 & 1 \\ 0 & 1 & 0 \\ 0 & 1 & 1 \\ 1 & 0 & 0 \\ 1 & 0 & 1 \\ 1 & 1 & 0 \\ 1 & 1 & 1 \end{bmatrix}.$$

4.5 Known 3-Qubit Gates

Now, we mention the following three operations on qubits. The first two correspond to permutations and the last one to the 8-point discrete Hadamard transform (DHdT). Usually, among permutations on N inputs, one selects those that are symmetrical. That is, such permutations that coincide with their inverses.

1) The controlled-controlled-NOT gate, or Toffoli gate, with matrix representation

$$CC_{NOT} = I_4 \oplus C_{NOT} = \begin{bmatrix} \mathbf{1} & 0 & 0 & 0 & 0 & 0 & 0 & 0 \\ 0 & \mathbf{1} & 0 & 0 & 0 & 0 & 0 & 0 \\ 0 & 0 & \mathbf{1} & 0 & 0 & 0 & 0 & 0 \\ 0 & 0 & 0 & \mathbf{1} & 0 & 0 & 0 & 0 \\ 0 & 0 & 0 & 0 & \mathbf{1} & 0 & 0 & 0 \\ 0 & 0 & 0 & 0 & 0 & \mathbf{1} & 0 & 0 \\ 0 & 0 & 0 & 0 & 0 & 0 & 0 & \mathbf{1} \\ 0 & 0 & 0 & 0 & 0 & 0 & \mathbf{1} & 0 \end{bmatrix}. \qquad (4.6)$$

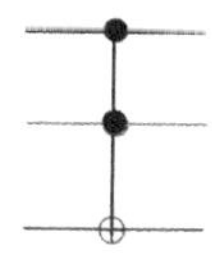

This is the Pauli NOT gate on planes 6 and 7, which we denote as $P_{(6,7)}$ gate, or $(1, 2, X)$-gate as shown in Fig. 4.7(a), for the $U_2 = X$ case. The determinant of this matrix equals to -1. This operator, or the permutation $(6, 7)$, swaps the last two amplitudes of a 3-qubit superposition. On basis states $|\varphi_1\rangle$, $|\varphi_2\rangle$, and $|\varphi_3\rangle \in \{0, 1\}$, the mapping of the Toffoli gate can be written as $|\varphi_1, \varphi_2, \varphi_3\rangle \to |\varphi_1, \varphi_2, \varphi_3 \oplus (\varphi_1 \wedge \varphi_2)\rangle$, where $\wedge$ denotes the logical AND operation. As any other permutation, the Toffoli operator is reversible; note that $(C_{NOT})^2 = I$. Toffoli gate is a universal logic gate for classical computation. For quantum computation, it can be used with other single qubit gates, to construct any reversible circuit [2].

The quantum circuit for the realization of the Toffoli gate is given in Fig. 4.11 [4]. The complex gate V is $\sqrt{\text{NOT}}$ which is given in Table 1, $V^2 = X$. The operator $V^\dagger$ is the Hermitian adjoint of V, which is the conjugate transpose (or Hermitian transpose $\overline{V}'$).

In general, such a representation holds for any matrix $(I_4 \oplus U_2)$, where U_2 is a unitary 2×2 matrix [2, Lemma 6.1, p. 3461].

It is not difficult to see that the gate CNOT requires one rotation by $-90°$ degree,

$$C_{NOT} = \begin{bmatrix} 1 & 0 & 0 & 0 \\ 0 & 1 & 0 & 0 \\ 0 & 0 & 0 & 1 \\ 0 & 0 & 1 & 0 \end{bmatrix} = \begin{bmatrix} 1 & 0 & 0 & 0 \\ 0 & 1 & 0 & 0 \\ 0 & 0 & 0 & 1 \\ 0 & 0 & -1 & 0 \end{bmatrix} \begin{bmatrix} 1 & 0 & 0 & 0 \\ 0 & 1 & 0 & 0 \\ 0 & 0 & -1 & 0 \\ 0 & 0 & 0 & 1 \end{bmatrix} = \left(I_2 \oplus W\left(-\frac{\pi}{2}\right)\right) \times (I_2 \oplus (-Z)).$$

Therefore, $CC_{NOT} = I_4 \oplus \left(I_2 \oplus W\left(-\frac{\pi}{2}\right)\right) \times I_4 \oplus (I_2 \oplus (-Z))$.

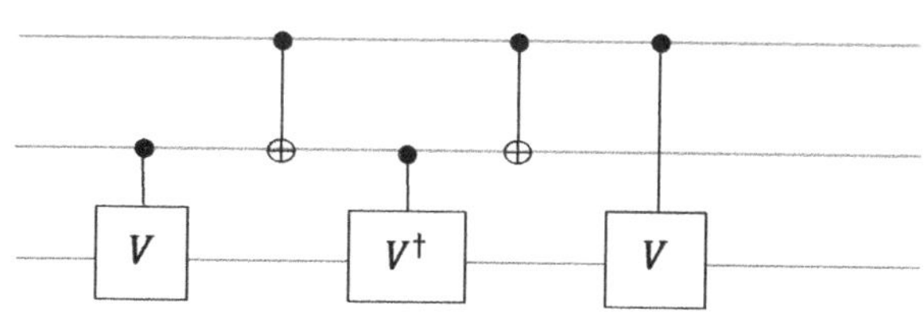

Fig. 4.11 The Toffoli gate circuit.

$V = (1 + i)(I - iX)/2$

2) The controlled-SWAP, or Fredkin gate, is the Kronecker sum of the identical matrix and the SWAP gate matrix,

$$F = I_4 \oplus P_{1,2} = \begin{bmatrix} \mathbf{1} & 0 & 0 & 0 & 0 & 0 & 0 & 0 \\ 0 & \mathbf{1} & 0 & 0 & 0 & 0 & 0 & 0 \\ 0 & 0 & \mathbf{1} & 0 & 0 & 0 & 0 & 0 \\ 0 & 0 & 0 & \mathbf{1} & 0 & 0 & 0 & 0 \\ 0 & 0 & 0 & 0 & \mathbf{1} & 0 & 0 & 0 \\ 0 & 0 & 0 & 0 & 0 & 0 & \mathbf{1} & 0 \\ 0 & 0 & 0 & 0 & 0 & \mathbf{1} & 0 & 0 \\ 0 & 0 & 0 & 0 & 0 & 0 & 0 & \mathbf{1} \end{bmatrix}. \tag{4.7}$$

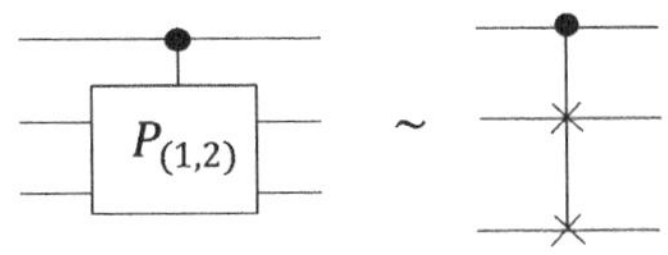

Here, $P_{1,2}$ is the permutation (1,2) of 4 numbers with the matrix

$$P_{(1,2)} = \begin{bmatrix} 1 & 0 & 0 & 0 \\ 0 & 0 & 1 & 0 \\ 0 & 1 & 0 & 0 \\ 0 & 0 & 0 & 1 \end{bmatrix}, \det(P_{1,2}) = -1.$$

Fredkin gate is the $X_{5,6}$ gate, or Pauli NOT gate X on planes 5 and 6. The determinant of the matrix F equals to -1. This operator, or the permutation (5,6), swaps the last 2-qubits if the first qubit is in state $|1\rangle$. Fredkin gate is another reversible universal logic gate for classical computation. The known circuit for the realization of the Fredkin gate is given in Fig. 4.12.

3) The 8-point Hadamard transform matrix

$$H_8 = H^{\oplus 3} = H \otimes H_4 = \begin{bmatrix} H_4 & H_4 \\ H_4 & -H_4 \end{bmatrix} = \frac{1}{\sqrt{8}} \begin{bmatrix} 1 & 1 & 1 & 1 & 1 & 1 & 1 & 1 \\ 1 & -1 & 1 & -1 & 1 & -1 & 1 & -1 \\ 1 & 1 & -1 & -1 & 1 & 1 & -1 & -1 \\ 1 & -1 & -1 & 1 & 1 & -1 & -1 & 1 \\ 1 & 1 & 1 & 1 & -1 & -1 & -1 & -1 \\ 1 & -1 & 1 & -1 & -1 & 1 & -1 & 1 \\ 1 & 1 & -1 & -1 & -1 & -1 & 1 & 1 \\ 1 & -1 & -1 & 1 & -1 & 1 & 1 & -1 \end{bmatrix}. \tag{4.8}$$

Example 4.2 Preparing Two 3-Qubit Superpositions
Consider the transforms of the states $|000\rangle$ and $|001\rangle$,

$$H_8|000\rangle = \frac{1}{\sqrt{8}} \sum_{n=0}^{7} |n\rangle \text{ and } H_8|001\rangle = \frac{1}{\sqrt{8}} \sum_{n=0}^{7} (-1)^n |n\rangle. \tag{4.9}$$

Gates can be used in series and parallel.

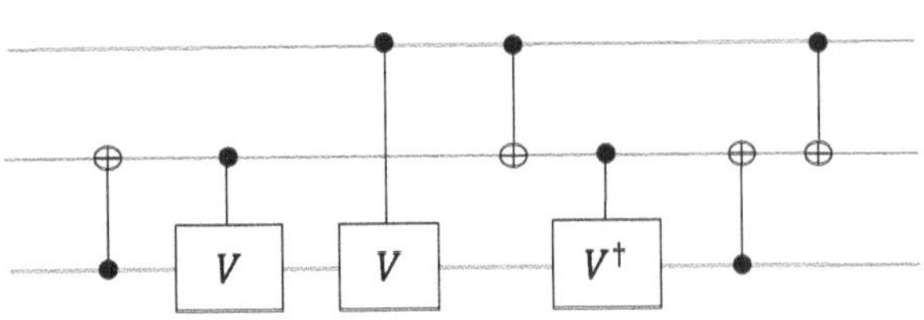

Fig. 4.12 The Fredkin gate circuit.

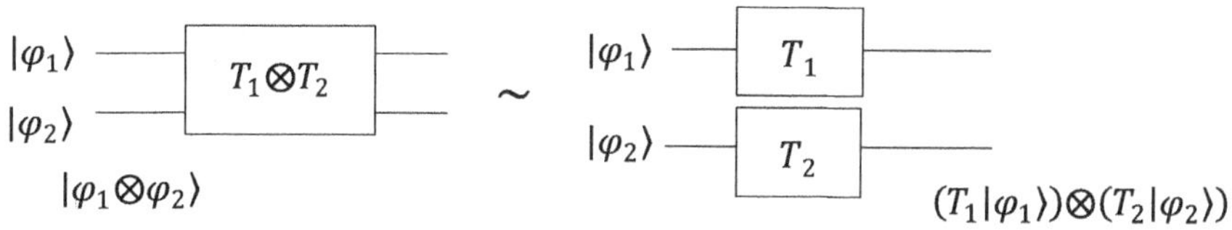

Fig. 4.13 Two equal circuits for two gates in parallel.

1) Two gates T_1 and T_2 in series are defined as $T_1[T_2|\varphi\rangle]$ and it is equal to the gate $[T_1 \cdot T_2]|\varphi\rangle$. The matrix of this gate is the multiplication of matrices of these two gates.
2) Two gates T_1 and T_2 in parallel (see Fig. 4.13) is defined by the tensor product of their matrices (gates),

$$(T_1 \otimes T_2)|\varphi_1 \otimes \varphi_2\rangle \triangleq (T_1|\varphi_1\rangle) \otimes (T_2|\varphi_2\rangle). \tag{4.10}$$

For example, if

$$T_1 = \frac{1}{\sqrt{2}}\begin{bmatrix} 1 & -1 \\ 1 & 1 \end{bmatrix} \text{ and } T_2 = \frac{1}{\sqrt{2}}\begin{bmatrix} 1 & 1 \\ -1 & 1 \end{bmatrix},$$

then

$$T_1 \otimes T_2 = \frac{1}{2}\begin{bmatrix} 1 & 1 & -1 & -1 \\ -1 & 1 & 1 & -1 \\ 1 & 1 & 1 & 1 \\ -1 & 1 & -1 & 1 \end{bmatrix}$$

and it is easy to check that the property of Eq. 4.10 holds for any two qubits $|\varphi_1\rangle$ and $|\varphi_2\rangle$, that is,

$$\frac{1}{2}\begin{bmatrix} 1 & 1 & -1 & -1 \\ -1 & 1 & 1 & -1 \\ 1 & 1 & 1 & 1 \\ -1 & 1 & -1 & 1 \end{bmatrix}|\varphi_1 \otimes \varphi_2\rangle = \left(\frac{1}{\sqrt{2}}\begin{bmatrix} 1 & -1 \\ 1 & 1 \end{bmatrix}|\varphi_1\rangle\right) \otimes \left(\frac{1}{\sqrt{2}}\begin{bmatrix} 1 & 1 \\ -1 & 1 \end{bmatrix}|\varphi_2\rangle\right).$$

4.6 Projection Operators

The measurement of r-qubits can be described by the projection operator, as for 2-qubits. Consider the space V of all superpositions in the basis states $\{|0\rangle, |1\rangle, |2\rangle, \ldots, |N-1\rangle\}$,

$$V = V_0 + V_1 + V_2 + \cdots + V_{N-1}. \tag{4.11}$$

Here, $N = 2^r$, V_k is the set of all states proportional to $|k\rangle$, that is, $V_k = \{a|k\rangle; a \in R \text{ or } C\}$. R and C denote the sets of real and complex numbers, respectively. Let V' be the subset $V' = V_{i_1} + V_{i_2} + \cdots + V_{i_n}$, where $i_1 \neq i_2 \neq \ldots \neq i_n \in [0, N-1]$, and $n < (N-1)$. The projector operator of the space V onto the subset V' is defined by the $N \times N$ matrix

$$T = |i_1\rangle\langle i_1| + |i_2\rangle\langle i_2| + \cdots + |i_n\rangle\langle i_n|.$$

This is the diagonal matrix with coefficients of 1 only in rows numbered $i_1, i_2, \ldots, i_n$, and 0 in all other rows. For example, we consider the $N = 4$ case and the subset $V' = V_1 + V_3$. Then, the corresponding projector operators are defined by

$$T = |1\rangle\langle 1| + |3\rangle\langle 3| = \begin{bmatrix} 0 & & & \\ & 1 & & \\ & & 0 & \\ & & & 0 \end{bmatrix} + \begin{bmatrix} 0 & & & \\ & 0 & & \\ & & 0 & \\ & & & 1 \end{bmatrix} = \begin{bmatrix} 0 & & & \\ & 1 & & \\ & & 0 & \\ & & & 1 \end{bmatrix}$$

and the projection on $|\varphi\rangle = a_0|0\rangle + a_1|1\rangle + a_2|2\rangle + a_3|3\rangle$ is calculated by

$$T|\varphi\rangle = \langle 1\ |\varphi\rangle|1\rangle + \langle 3\ |\varphi\rangle|3\rangle = \begin{bmatrix} 0 & & & \\ & 1 & & \\ & & 0 & \\ & & & 1 \end{bmatrix} \begin{bmatrix} a_0 \\ a_1 \\ a_2 \\ a_3 \end{bmatrix} = \begin{bmatrix} 0 \\ a_1 \\ 0 \\ a_3 \end{bmatrix} = a_1|1\rangle + a_3|3\rangle.$$

In the general case of the superposition $|\varphi\rangle = a_0|0\rangle + a_1|1\rangle + \ldots + a_{N-1}|N-1\rangle$, the projection operator T on the r-qubit quantum superposition $|\varphi\rangle$ can be written as

$$T|\varphi\rangle = \langle i_1\ |\varphi\rangle|i_1\rangle + \langle i_2\ |\varphi\rangle|i_2\rangle + \cdots + \langle i_n\ |\varphi\rangle|i_n\rangle = a_{i_1}|i_1\rangle + a_{i_2}|i_2\rangle + \cdots + a_{i_n}|i_n\rangle. \tag{4.12}$$

The superposition of measurement equals

$$M_T|\varphi\rangle = \frac{T|\varphi\rangle}{|\ T|\varphi\rangle\ |} = \frac{\langle i_1\ |\varphi\rangle|i_1\rangle + \langle i_2\ |\varphi\rangle|i_2\rangle + \cdots + \langle i_n\ |\varphi\rangle|i_n\rangle}{\sqrt{|\ \langle i_1\ |\varphi\rangle|^2 + |\ \langle i_2\ |\varphi\rangle|^2 + \cdots + |\ \langle i_n\ |\varphi\rangle|^2}}. \tag{4.13}$$

Thus,

$$M_T|\varphi\rangle = \frac{a_{i_1}|i_1\rangle + a_{i_2}|i_2\rangle + \cdots + a_{i_n}|i_n\rangle}{\sqrt{|\ a_{i_1}|^2 + |\ a_{i_2}|^2 + \cdots + |\ a_{i_n}|^2}}.$$

For the above example, $M_T|\varphi\rangle = (a_1|1\rangle + a_3|3\rangle)/\sqrt{|\ a_1|^2 + |\ a_3|^2}$.

References

1 Nielsen, M. and Chuang, I. (2001). *Quantum Computation and Quantum Information*, 2e. Cambridge University Press.

2 Barenco, A., Bennett, C., Cleve, R. et al. (1995). Elementary gates for quantum computation. *Physical Review A* 52: 3457–3467.

3 Rieffel, E. and Polak, W. (2011). *Quantum Computing: A Gentle Introduction*. Cambridge, MA/London, England: The MIT Press.

4 Williams, C.P. (2011). Texts in Computer Science, (Chapter 2). In: *Explorations in Quantum Computing*, 2ee (ed. D. Gries and F.B. Schneider). London: Springer-Verlag.

5

Fast Transforms in Quantum Computation

This section discusses quantum fast transforms that play a key role in the development of image processing techniques. Traditional methods such as the discrete Fourier transform, Haar transform, discrete cosine transform, Hadamard transform, and Hartley transform have been effective since the 1970s for image compression, restoration, filtration, and geometric transformations. These methods transform images from the spatial domain to the frequency domain for easier analysis and processing. Fast transforms are needed to efficiently switch between these domains, offering energy compaction and fast computation, making them indispensable for a variety of image processing tasks.

We describe concepts of the discrete paired, Hadamard, and Fourier transforms and their circuits in quantum computation. Fast discrete Fourier transform algorithms are well developed based on the first Cooley and Tukey algorithm [1–3]. Here, we focus on the discrete paired transform, which is a core of the N-point discrete Hadamard and Fourier transforms. For the most interesting case when N is a power of 2, the Hadamard and Fourier transform can be calculated by one fast paired transform-based algorithm [4–6].

5.1 Fast Discrete Paired Transform

The discrete paired transform (DPT) is the frequency–time representation of the signal that corresponds to a special splitting of both discrete Fourier and Hadamard transforms [7, 8]. In the case when $N = 2^r$, $r > 1$, the N-point DPT requires only $2(N-1)$ operations of addition and subtraction.

The transform has a matrix with only coefficients equal to 0 and ± 1. For the $N = 2$, 4, and 8, the transform has the following matrix representations:

$$[\chi'_2] = \begin{bmatrix} 1 & -1 \\ 1 & 1 \end{bmatrix}, \quad [\chi'_4] = \begin{bmatrix} 1 & 0 & -1 & 0 \\ 0 & 1 & 0 & -1 \\ 1 & -1 & 1 & -1 \\ 1 & 1 & 1 & 1 \end{bmatrix} \quad [\chi'_8] = \begin{bmatrix} 1 & 0 & 0 & 0 & -1 & 0 & 0 & 0 \\ 0 & 1 & 0 & 0 & 0 & -1 & 0 & 0 \\ 0 & 0 & 1 & 0 & 0 & 0 & -1 & 0 \\ 0 & 0 & 0 & 1 & 0 & 0 & 0 & -1 \\ 1 & 0 & -1 & 0 & 1 & 0 & -1 & 0 \\ 0 & 1 & 0 & -1 & 0 & 1 & 0 & -1 \\ 1 & -1 & 1 & -1 & 1 & -1 & 1 & -1 \\ 1 & 1 & 1 & 1 & 1 & 1 & 1 & 1 \end{bmatrix}. \tag{5.1}$$

Quantum Image Processing in Practice: A Mathematical Toolbox, First Edition. Artyom M. Grigoryan and Sos S. Agaian.

Companion website: www.wiley.com/go/grigoryan/quantumimageprocessing

The rows of these matrices are the non-normalized basis functions of the 2, 4, and 8-point DPTs. The paired transform is the frequency–time representation of the signal. The basis paired functions are defined as [9]

$$\chi'_{p,t}(n) = \begin{cases} 1, & \text{if } np = t \bmod N, \\ -1, & \text{if } np = \left(t + \dfrac{N}{2}\right) \bmod N, \\ 0, & \text{otherwise}, \end{cases} \qquad n = 0{:}\ (N-1). \tag{5.2}$$

These periodic functions are numbered by the frequency points $p = 1, 2, 4, 8, \ldots, N/2, 0$ and the time-points $t = 0, 1, 2, \ldots, N/(2p) - 1$, repectively. The first subset of functions $\chi'_{1,0}, \chi'_{1,1}, \chi'_{1,2}, \ldots, \chi'_{1,N/2-1}$ is generated by the frequency-point 1. The second subset of functions $\chi'_{2,0}, \chi'_{2,2}, \chi'_{2,4}, \ldots, \chi'_{N/4-2}$ is generated by the frequency-point 2, and so on. The last basis function is $\chi'_{0,0} \equiv 1$. In the matrix of the 8-point DPT in Eq. 5.1, the first four rows present the basis functions generated by the frequency-point $p = 1$, the next two rows are basis functions generated by $p = 2$, and the last two rows are the functions generated by the frequency-points 4 and 0, respectively.

For $N = 2^r$, $r > 1$, the complete set of N pared functions is [8, 9]

$$\chi' = \left\{\left\{\chi'_{2^k,2^k t}(n); k = 0{:}\ (r-1), t = 0{:}\ (2^{r-k-1} - 1\right\}, 1\right\}. \tag{5.3}$$

It is important to note that the basis paired functions can also be written as the following binarization of the cosine waves with frequencies $\omega_k = 2\pi/(2^{r-k})$:

$$\chi'_{2^k,2^k t}(x) = M\left(\cos\frac{2\pi(x-t)}{2^{r-k}}\right), \qquad x \in [0{:}\ N-1], \tag{5.4}$$

and $\chi'_{0,0}(x) \equiv 1$. Here, $M(x)$ is the function on the interval $[-1,1]$ with values $M(\pm 1) = \pm 1$ and $M(x) = 0$, otherwise.

Example 5.1 8-Point DPT

Consider the $N = 8$ case. Figure 5.1 shows eight cosine waves with frequencies $\pi/4$, $\pi/2$, and π in part (a) and the corresponding basis paired functions in part (b).

The N-point discrete DPT of the signal f_n, $n = 0 : (N-1)$ is calculated by

$$f'_{2^k,2^k t} = \sum_{n=0}^{N-1} \chi'_{2^k,2^k t}(n) f_n. \tag{5.5}$$

This transform is the set of $(r+1)$ short signals, which are called *the splitting-signals*,

$$f'_1 = \left(f'_{1,0}, f'_{1,1}, f'_{1,2}, \ldots, f'_{1,N/2-1}\right)\ \left(\text{of length } \frac{N}{2}\right), \tag{5.6}$$

$f'_2 = \left(f'_{2,0}, f'_{2,2}, f'_{2,4}, \ldots, f'_{2,N/2-2}\right)$ (of length $N/4$), $f'_4 = \left(f'_{4,0}, f'_{4,4}, f'_{4,8}, \ldots, f'_{4,N/2-4}\right)$ (of length $N/8$), …, and the last two scalar signals are $f'_{N/2} = f'_{N/2,0} = f_0 - f_1 + f_2 - f_3 + \cdots - f_{N-1}$ and $f'_0 = f'_{0,0} = f_0 + f_1 + f_2 + \cdots + f_{N-1}$. The 8-point DPT of the signal in matrix form is calculated by

$$[\chi'_8]f = \begin{bmatrix} f'_{1,0} \\ f'_{1,1} \\ f'_{1,2} \\ f'_{1,3} \\ f'_{2,0} \\ f'_{2,2} \\ f'_{4,0} \\ f'_{0,0} \end{bmatrix} = \begin{bmatrix} 1 & 0 & 0 & 0 & -1 & 0 & 0 & 0 \\ 0 & 1 & 0 & 0 & 0 & -1 & 0 & 0 \\ 0 & 0 & 1 & 0 & 0 & 0 & -1 & 0 \\ 0 & 0 & 0 & 1 & 0 & 0 & 0 & -1 \\ 1 & 0 & -1 & 0 & 1 & 0 & -1 & 0 \\ 0 & 1 & 0 & -1 & 0 & 1 & 0 & -1 \\ 1 & -1 & 1 & -1 & 1 & -1 & 1 & -1 \\ 1 & 1 & 1 & 1 & 1 & 1 & 1 & 1 \end{bmatrix} \begin{bmatrix} f_0 \\ f_1 \\ f_2 \\ f_3 \\ f_4 \\ f_5 \\ f_6 \\ f_7 \end{bmatrix}. \tag{5.7}$$

Figure 5.2 shows the signal of length 512 in part (a) and all splitting-signals together in part (b).

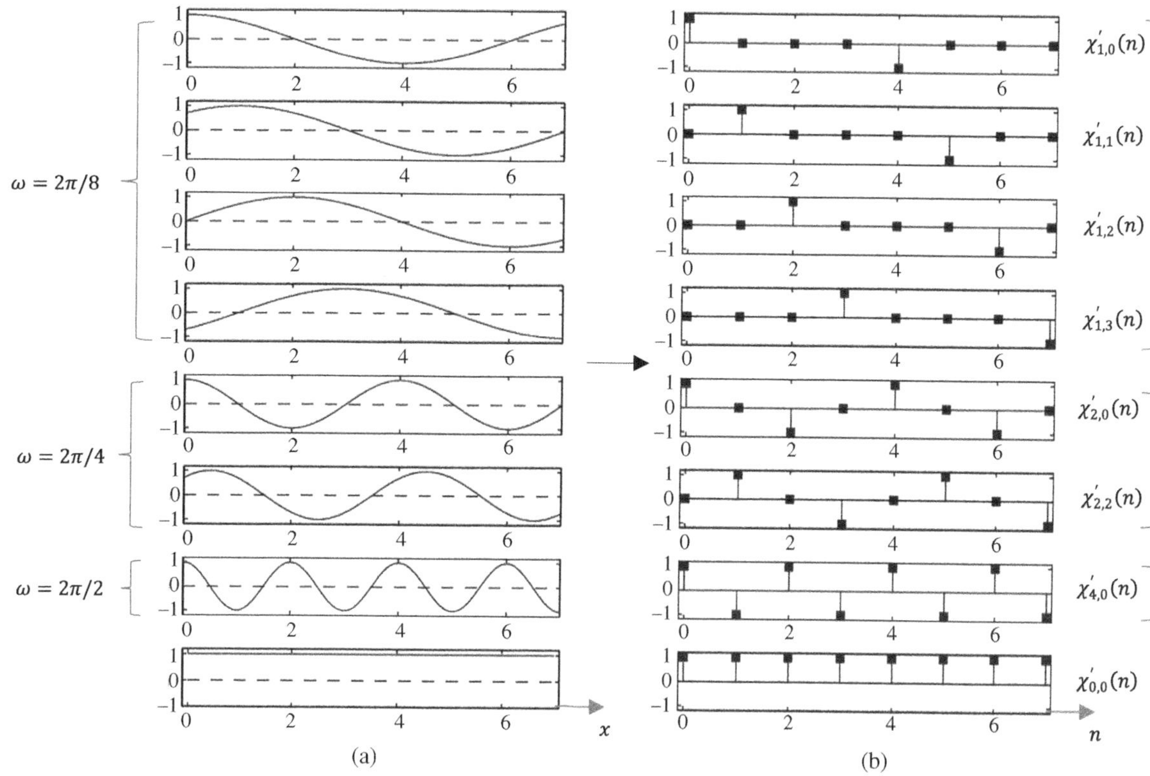

Fig. 5.1 (a) The cosine waves in the interval [0,7] and (b) the corresponding basis paired functions.

For $N = 2$, 4 and 8, the signal-flow graphs of the fast DPT are shown in Fig. 5.3 [10].

One can see that the 4- and 8-point DPTs can be calculated by 3 and 7 butterflies χ'_2. The normalized basis functions of the paired transform are calculated as

$$\chi'_{2^k,2^k t}(n) \to \frac{1}{\sqrt{2^{k+1}}}\chi'_{2^k,2^k t}(n), \quad k = 0, 1, ..., (r-1), \tag{5.8}$$

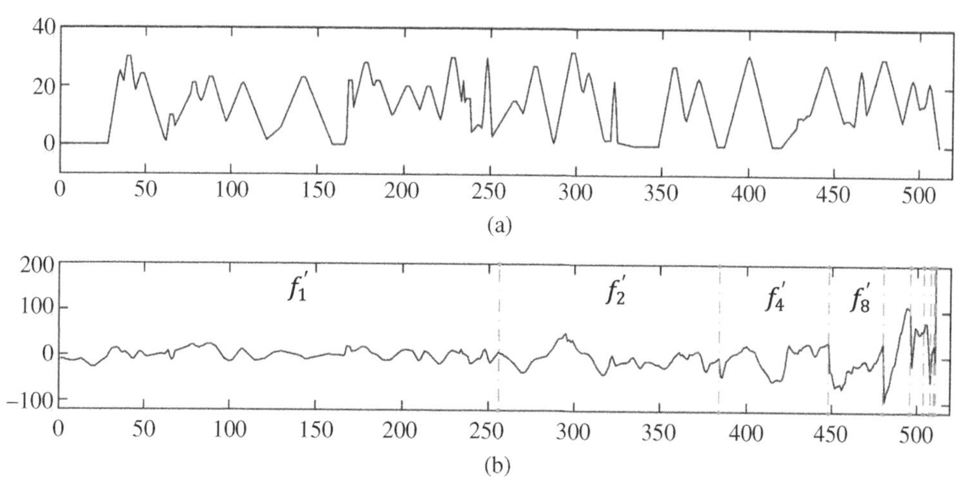

Fig. 5.2 (a) The signal of length 512 and (b) its paired transform (10 splitting-signals).

$$A_2 = \begin{bmatrix} 1 & -1 \\ 1 & 1 \end{bmatrix}.$$

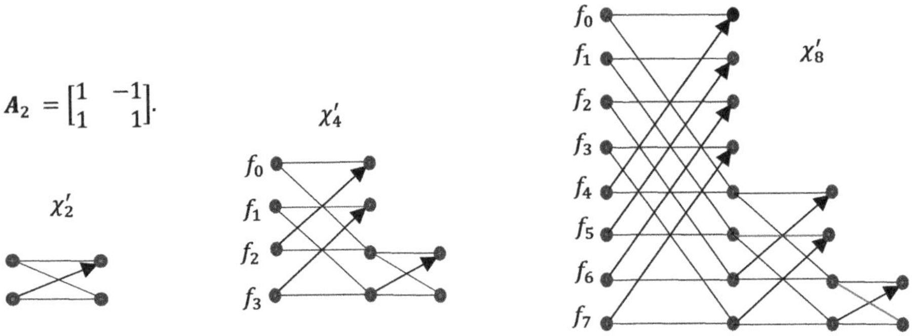

Fig. 5.3 Signal-flow graphs of the 2-, 4-, and 8-point DPTs.

and $\chi'_{0,0}(n) = 1/\sqrt{N}$. For instance, when $N = 8$, the normalization of the matrix is performed by multiplication of the matrix $[\chi'_8]$ in Eq. 5.1 by the diagonal matrix $D = \text{diag}(\underbrace{1/\sqrt{2}, 1/\sqrt{2}, 1/\sqrt{2}, 1/\sqrt{2}}, \underbrace{1/2, 1/2}, \underbrace{1/\sqrt{8}}, \underbrace{1/\sqrt{8}})$. The matrix of the 8-point DPT is

$$[\chi'_8] = D[\chi'_8] = \begin{bmatrix} \frac{1}{\sqrt{2}} & 0 & 0 & 0 & -\frac{1}{\sqrt{2}} & 0 & 0 & 0 \\ 0 & \frac{1}{\sqrt{2}} & 0 & 0 & 0 & -\frac{1}{\sqrt{2}} & 0 & 0 \\ 0 & 0 & \frac{1}{\sqrt{2}} & 0 & 0 & 0 & -\frac{1}{\sqrt{2}} & 0 \\ 0 & 0 & 0 & \frac{1}{\sqrt{2}} & 0 & 0 & 0 & -\frac{1}{\sqrt{2}} \\ \frac{1}{2} & 0 & -\frac{1}{2} & 0 & \frac{1}{2} & 0 & -\frac{1}{2} & 0 \\ 0 & \frac{1}{2} & 0 & -\frac{1}{2} & 0 & \frac{1}{2} & 0 & -\frac{1}{2} \\ \frac{1}{\sqrt{8}} & -\frac{1}{\sqrt{8}} & \frac{1}{\sqrt{8}} & -\frac{1}{\sqrt{8}} & \frac{1}{\sqrt{8}} & -\frac{1}{\sqrt{8}} & \frac{1}{\sqrt{8}} & -\frac{1}{\sqrt{8}} \\ \frac{1}{\sqrt{8}} & \frac{1}{\sqrt{8}} & \frac{1}{\sqrt{8}} & \frac{1}{\sqrt{8}} & \frac{1}{\sqrt{8}} & \frac{1}{\sqrt{8}} & \frac{1}{\sqrt{8}} & \frac{1}{\sqrt{8}} \end{bmatrix}. \tag{5.9}$$

This matrix has determinant 1 and is unitary, that is, its inverse matrix equals the transpose matrix. In Eq. 5.7, this matrix is given in integer form, that is, without normalization. Therefore, it is orthogonal only by rows.

In the general case, the DPT can be calculated by $(N-1)$ butterflies, or the matrices $A_2 = [\chi'_2]$. The matrices of the high-order DPTs can be defined in recursive form as [9]

$$[\chi'_{16}] = \begin{bmatrix} I_8 & -I_8 \\ [\chi'_8] & [\chi'_8] \end{bmatrix}, \qquad [\chi'_{32}] = \begin{bmatrix} I_{16} & -I_{16} \\ [\chi'_{16}] & [\chi'_{16}] \end{bmatrix}, \ldots, \tag{5.10}$$

where I_M denotes the identity matrix $M \times M$. The above matrices are considered without the normalized coefficients. The normalized matrix $1/\sqrt{2}A_2$ is unitary, which is important to mention since in quantum computing, the operations on one qubit are described by only unitary transforms. The matrix A_2 can be described by the NOT gate and gate H,

$$A_2 = XH = \begin{bmatrix} 0 & 1 \\ 1 & 0 \end{bmatrix} \frac{1}{\sqrt{2}} \begin{bmatrix} 1 & 1 \\ 1 & -1 \end{bmatrix} = \frac{1}{\sqrt{2}} \begin{bmatrix} 1 & -1 \\ 1 & 1 \end{bmatrix}, \quad \det A_2 = 1. \tag{5.11}$$

The matrix A_2 is the same Hadamard gate, but with a different order of output. Note that $A_2^2 \neq I_2$,

$$A_2^2 = \begin{bmatrix} 0 & -1 \\ 1 & 0 \end{bmatrix} = \begin{bmatrix} 0 & 1 \\ 1 & 0 \end{bmatrix} \begin{bmatrix} 1 & 0 \\ 0 & -1 \end{bmatrix} = XZ,$$

but H is the square root of I_2, that is, $H^2 = I_2$, and $\det H = -1$.

5.2 The Quantum Circuits for the Paired Transform

In this section, we describe the simple quantum schemes for calculation of the paired transform. The quantum paired transform (QPT) is defined as the following transform of states: $\varphi(f)\rangle \rightarrow |\chi'(f)\rangle$. The input is the r-qubit superposition of the signal $\{f_n;\ n = 0:(N-1)\}$,

$$|\varphi(f)\rangle = f_0|0\rangle + f_1|1\rangle + \cdots + f_{N-1}|N-1\rangle. \tag{5.12}$$

It is considered that $|f_0|^2 + |f_1|^2 + \ldots + |f_{N-1}|^2 = 1$. The r-qubit superposition of the QPT is

$$|\chi'(f)\rangle = \sum_{t=0}^{N/2-1} f'_{1,t}|t\rangle + \sum_{t=0}^{N/4-1} f'_{2,2t}|N/2+t\rangle + \sum_{t=0}^{N/8-1} f'_{4,4t}|3N/2+t\rangle + \cdots + f'_{N/2,0}|N-2\rangle + f'_{0,0}|N-1\rangle. \tag{5.13}$$

Example 5.2 3-Qubit QPT
Consider the 3-qubit state of the signal $|\varphi(f)\rangle = f_0|0\rangle + f_1|1\rangle + \ldots + f_7|7\rangle$ and the state of the 3-qubit QPT

$$|\chi'(f)\rangle = [f'_{1,0}|0\rangle + f'_{1,1}|1\rangle + f'_{1,2}|2\rangle + f'_{1,3}|3\rangle] + [f'_{2,0}|4\rangle + f'_{2,2}|5\rangle] + f'_{4,0}|6\rangle + f'_{0,0}|7\rangle.$$

The quantum circuit of the 3-qubit quantum paired transform (QPT) is shown in Fig. 5.4 [10].

The general case when $N = 2^r$, $r > 2$:

The quantum circuit for calculating the r-qubit QPT is shown in Fig. 5.5 [9]. One gate A_2 and $(r-1)$ controlled gates A_2 are used. The input in this circuit is the r-qubit state $|\varphi(f)\rangle$.

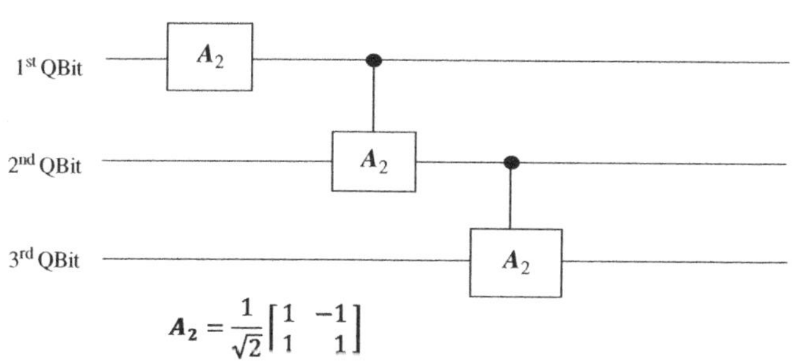

Fig. 5.4 The quantum circuit for the 3-qubit QPT.

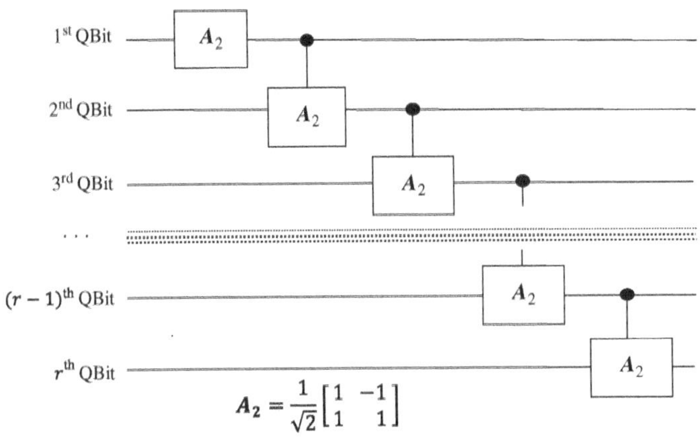

Fig. 5.5 The quantum circuit for the r-qubit QPT.

5.3 The Inverse DPT

The inverse N-point DPT can be written as the sum of all paired signals calculated by [9, 11]

$$f_n = \frac{1}{2}f'_{1,n} + \frac{1}{4}f'_{2,2n} + \frac{1}{8}f'_{4,4n} + \frac{1}{16}f'_{8,8n} + \cdots + \frac{1}{2^r}f'_{2^{r-1},0} + \frac{1}{2^r}f'_{0,0}, \quad n = 0\colon (N-1). \tag{5.14}$$

Here, the second indices are taken modulo N. The inverse paired transform can be presented in matrix form.

Example 5.3 8-Point Inverse DPT

The matrix of the 8-point inverse DPT can be written as

$$[\chi'_8]^{-1} = \frac{1}{\sqrt{2}}\begin{bmatrix} I_4 & I_4 \\ -I_4 & I_4 \end{bmatrix}\left(I_4 \oplus \frac{1}{\sqrt{2}}\begin{bmatrix} I_2 & I_2 \\ -I_2 & I_2 \end{bmatrix}\right)\left(I_6 \oplus \frac{1}{\sqrt{2}}\begin{bmatrix} 1 & 1 \\ -1 & 1 \end{bmatrix}\right), \tag{5.15}$$

or, $[\chi'_8]^{-1} = (B_2 \otimes I_4)(I_4 \oplus (B_2 \otimes I_2))(I_4 \oplus CB_2)$. The signal-flow graph for the 8-point inverse DPT is given in Fig. 5.6.

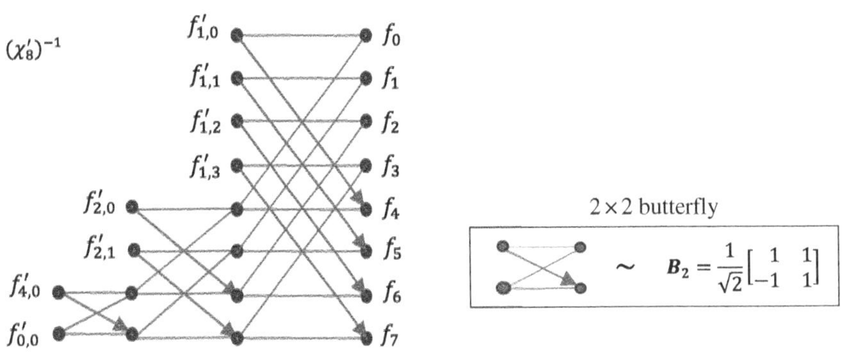

Fig. 5.6 The signal-flow graphs of the 8-point inverse DPT.

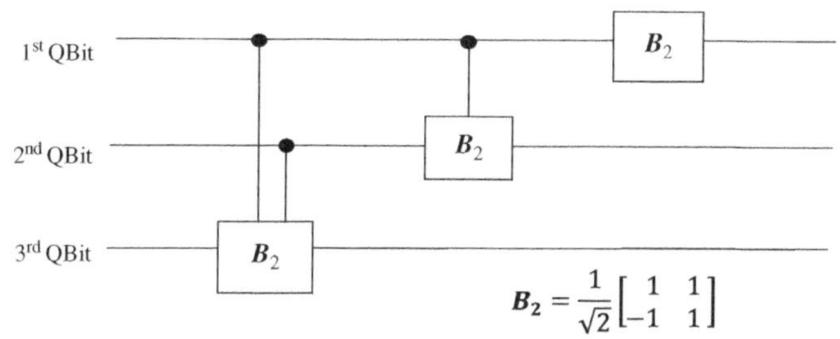

Fig. 5.7 The quantum circuit for the 3-qubit inverse QPT.

The inverse QPT is defined as the following transform of states: $|\chi'(f)\rangle \rightarrow \varphi(f)\rangle$. The quantum circuit for the 3-qubit inverse QPT is given in Fig. 5.7 [9]. Here, the main operation, or the butterfly, is described by the inverse matrix

$$B_2 = A_2^{-1} = A_2^T = \frac{1}{\sqrt{2}}\begin{bmatrix} 1 & 1 \\ -1 & 1 \end{bmatrix} = ZH_2.$$

The first gate B_2 is used with two control qubits, namely, the first two qubits, when both are in the state $|1\rangle$. The second gate B_2 works when the first qubit is in the state $|1\rangle$.

5.3.1 The First Circuit for the Inverse QPT

In the general case $N = 2^r$, $r \geq 2$, the quantum circuit for computing the r-qubit inverse DPT $|\chi'(f)\rangle \rightarrow \varphi(f)\rangle$ is shown in Fig. 5.8. Each gate B_2 (except the last one) is controlled by a few qubits. Namely, gate B_2 which is applied to the kth qubit, where $k \in \{r, r-1, ..., 2\}$, is controlled by $(k-1)$ first qubits when they all are in the state $|1\rangle$.

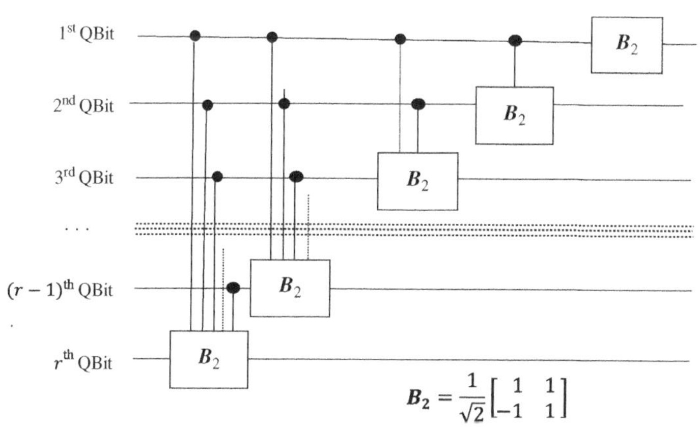

Fig. 5.8 The quantum circuit for the r-qubit inverse QPT.

5.4 Fast Discrete Hadamard Transform

In this section, we describe the analytical and matrix representations of the discrete Hadamard transform (DHdT) [2, 12]. Note that the Walsh–Hadamard transform is considered. As is customary, all matrices are considered equivalent up to permutation of their rows or columns. The core 2×2 matrix is A_2 not H_2. Let $N = 2^r$, $r > 1$, and let A_N be the N-point DHdT whose image over a signal f_n of length N is written as [9, 13]

$$H_p = \sum_{n=0}^{N-1} f_n \alpha_p(n), \qquad p = 0{:}\ (N-1). \tag{5.16}$$

The basis functions of the transform are $\alpha_p(n) = (-1)^{n_0p_0+n_1p_1+\cdots+n_{r-1}p_{r-1}}$, where and n_k and p_k are coefficients of binary representation of numbers expansions of integer numbers n and p.

Example 5.4 Hadamard Matrix 8 × 8

Let A_8 be the 8-point DHdT which is constructed as shown in Table 5.1.

The left least significant bit numbering is used in the table. This Hadamard matrix is different from the Walsh–Hadamard matrix (in Eq. 4.7). The recursive form of Sylvester's construction is used

$$A_2 = \frac{1}{\sqrt{2}}\begin{bmatrix} 1 & -1 \\ 1 & 1 \end{bmatrix}, \quad A_4 = \frac{1}{2}\begin{bmatrix} 1 & -1 & -1 & 1 \\ 1 & 1 & -1 & -1 \\ 1 & -1 & 1 & -1 \\ 1 & 1 & 1 & 1 \end{bmatrix} = \begin{bmatrix} A_2 & -A_2 \\ A_2 & A_2 \end{bmatrix}, \quad A_N = \begin{bmatrix} A_{N/2} & -A_{N/2} \\ A_{N/2} & A_{N/2} \end{bmatrix}.$$

Example 5.5 Decomposition of the Hadamard Matrix 4 × 4

The 4-point DHdT can be decomposed by the 4-point paired transforms as [8]

$$A_4 = (\chi_2' \oplus I)\chi_4' = \begin{bmatrix} 1 & -1 & 0 & 0 \\ 1 & 1 & 0 & 0 \\ 0 & 0 & 1 & 0 \\ 0 & 0 & 0 & 1 \end{bmatrix}\begin{bmatrix} 1 & 0 & -1 & 0 \\ 0 & 1 & 0 & -1 \\ 1 & -1 & 1 & -1 \\ 1 & 1 & 1 & 1 \end{bmatrix} = \begin{bmatrix} 1 & -1 & -1 & 1 \\ 1 & 1 & -1 & -1 \\ 1 & -1 & 1 & -1 \\ 1 & 1 & 1 & 1 \end{bmatrix}.$$

Table 5.1 The 8-point Hadamard transform calculation.

p, n	p_0	p_1	p_2	n_0	n_1	n_2	$n_0p_0 + n_1p_1 + n_2p_2$								$[A_8] = \|(-1)^{n_0p_0+n_1p_1+n_2p_2}\|$							
7	1	1	1	1	1	1	0	1	1	2	1	2	2	3	1	−1	−1	1	−1	1	1	−1
6	0	1	1	0	1	1	0	0	1	1	1	1	2	2	1	1	−1	−1	−1	−1	1	1
5	1	0	1	1	0	1	0	1	0	1	1	2	1	2	1	−1	1	−1	−1	1	−1	1
4	0	0	1	0	0	1	0	0	0	0	1	1	1	1	1	1	1	1	−1	−1	−1	−1
3	1	1	0	1	1	0	0	1	1	2	0	1	1	2	1	−1	−1	1	1	−1	−1	1
2	0	1	0	0	1	0	0	0	1	1	0	0	1	1	1	1	−1	−1	1	1	−1	−1
1	1	0	0	1	0	0	0	1	0	1	0	1	0	1	1	−1	1	−1	1	−1	1	−1
0	0	0	0	0	0	0	0	0	0	0	0	0	0	0	1	1	1	1	1	1	1	1

Example 5.6 Decomposition of the Hadamard matrix 8 × 8
The matrix of the 8-point DHdT is

$$A_8 = \begin{bmatrix} A_4 & -A_4 \\ A_4 & A_4 \end{bmatrix} = \begin{bmatrix} 1 & -1 & -1 & 1 & -1 & 1 & 1 & -1 \\ 1 & 1 & -1 & -1 & -1 & -1 & 1 & 1 \\ 1 & -1 & 1 & -1 & -1 & 1 & -1 & 1 \\ 1 & 1 & 1 & 1 & \underbrace{-1 \quad -1 \quad -1 \quad -1} \\ 1 & -1 & -1 & 1 & 1 & -1 & -1 & 1 \\ 1 & 1 & -1 & -1 & 1 & 1 & -1 & -1 \\ 1 & -1 & 1 & -1 & 1 & -1 & 1 & -1 \\ 1 & 1 & 1 & 1 & 1 & 1 & 1 & 1 \end{bmatrix}. \tag{5.17}$$

When splitting by the paired transform, this matrix is represented as [8]

$$A_8 = P_8 \begin{bmatrix} \begin{bmatrix} [\chi'_2] & & \\ & 1 & \\ & & 1 \end{bmatrix} [\chi'_4] & & & \\ & [\chi'_2] & & \\ & & 1 & \\ & & & 1 \end{bmatrix} [\chi'_8], \qquad ([\chi'_2] = A_2). \tag{5.18}$$

Here, P_8 is the matrix of the permutation $\begin{pmatrix} 01234567 \\ 73516240 \end{pmatrix} = (07)(13)(25)(46)$. The 3-qubit quantum Hadamard transform is defined as the transform of states $|\varphi(f)\rangle \rightarrow |A_8(f)\rangle = H_0|0\rangle + H_1|0\rangle + \ldots + H_p|7\rangle$.

Figure 5.9 shows the circuit for calculating the 8-point DHdT [10].

The equivalent circuit in quantum computation is given in Fig. 5.10.

It will be shown below that the permutation P_8 in this circuit can be accomplished with 3 gates X and the permutation $P_X = P_{(13)(46)}$, as shown in the circuit in Fig. 5.11.

Note that the permutation P_X is the 3-bit SWAP operation $(x, y, z) \rightarrow (z, y, x)$, when $x, y, z \in \{0, 1\}$. It is not the 3-qubit SWAP, $|x\rangle \otimes |y\rangle \otimes |z\rangle \rightarrow |z\rangle \otimes |y\rangle \otimes |x\rangle$, that is, the permutation (142)(356) that can be implemented by 6 and 7 CNOT gates [14]. The same circuit but with Walsh–Hadamard gates H_2 instead of A_2 is shown in Fig. 5.12. Thus, we have two circuits for the same DHdT , one is written for the calculation in the classical computer and another for the calculation in the quantum computer.

Example 5.7 4-Qubit QHdT
For the $N = 8$ case, Figure 5.13-1 shows this circuit for calculating the 4-qubit QHdT. The quantum transformation is defined as

$$\varphi(f)\rangle = f_0|0\rangle + f_1|0\rangle + \cdots + f_p|15\rangle \rightarrow |\, A_8(f)\rangle = H_0|0\rangle + H_1|0\rangle + \cdots + H_p|15\rangle.$$

The coefficients of the 16-point DHdT are calculated in order 15, 7, 11, 3, 13, 5, 9, 1 and 14, 6, 10, 2, 12, 4, 8, 0. Therefore, the following permutation can be used to get the natural order of the coefficients:

$$P_{16} = (0, 15)(1, 7)(2, 11)(4, 13)(6, 9)(8, 14).$$

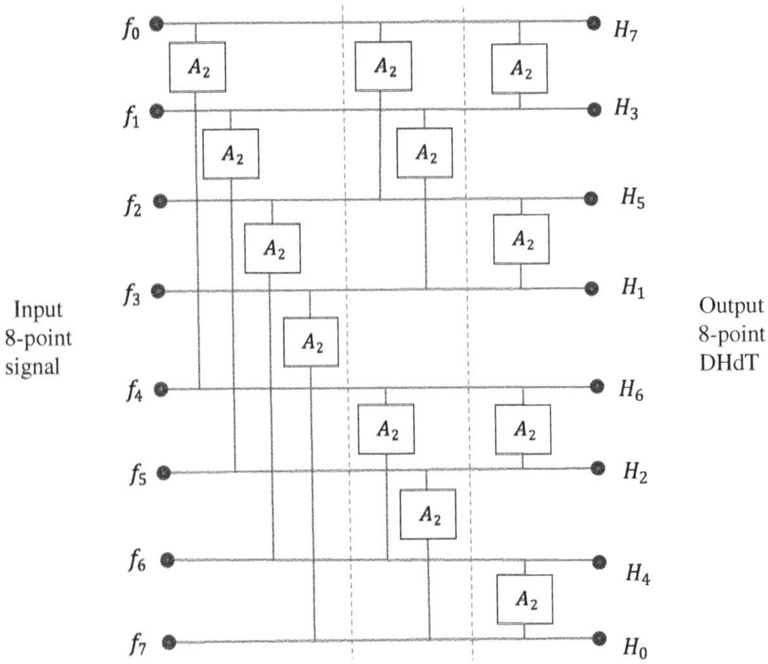

Fig. 5.9 The circuit for calculating the 8-point DHdT with three stages.

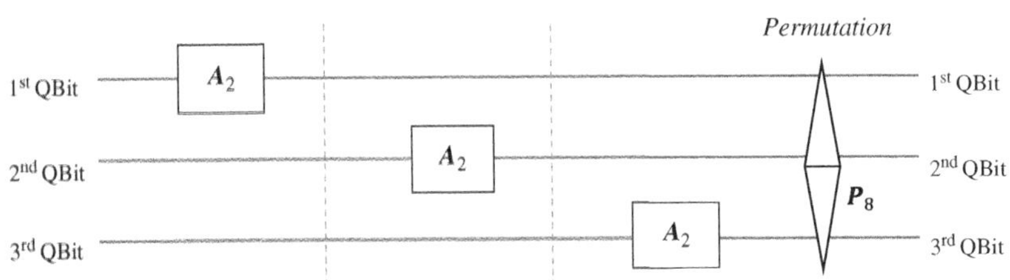

$$A_2 = \frac{1}{\sqrt{2}}\begin{bmatrix} 1 & -1 \\ 1 & 1 \end{bmatrix}, \quad X = \begin{bmatrix} 0 & 1 \\ 1 & 0 \end{bmatrix}, \quad P_8 = (07)(13)(25)(46)$$

Fig. 5.10 The circuit for the 3-qubit direct Hadamard transform by the paired splitting.

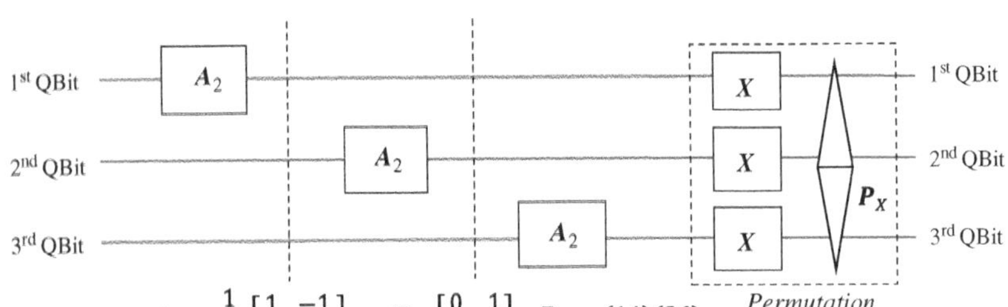

$$A_2 = \frac{1}{\sqrt{2}}\begin{bmatrix} 1 & -1 \\ 1 & 1 \end{bmatrix}, \quad X = \begin{bmatrix} 0 & 1 \\ 1 & 0 \end{bmatrix}, \quad P_X = (14)(36)$$

Fig. 5.11 The circuit for the 3-qubit direct Hadamard transform by the paired splitting.

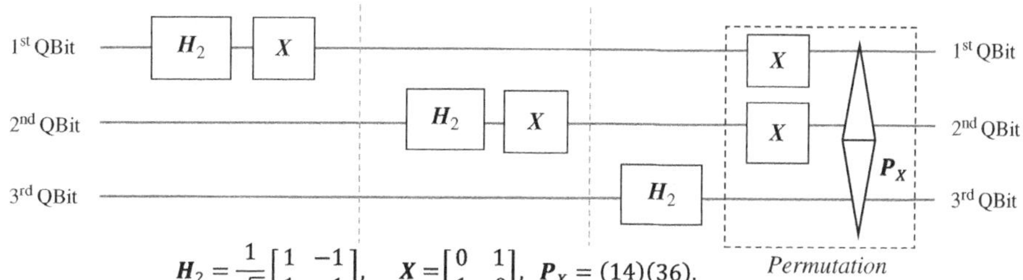

$$H_2 = \frac{1}{\sqrt{2}}\begin{bmatrix} 1 & -1 \\ 1 & 1 \end{bmatrix}, \quad X = \begin{bmatrix} 0 & 1 \\ 1 & 0 \end{bmatrix}, \quad P_X = (14)(36).$$

Fig. 5.12 The circuit for the 3-qubit quantum Hadamard transform (QHdT) by the paired splitting with Walsh–Hadamard gates.

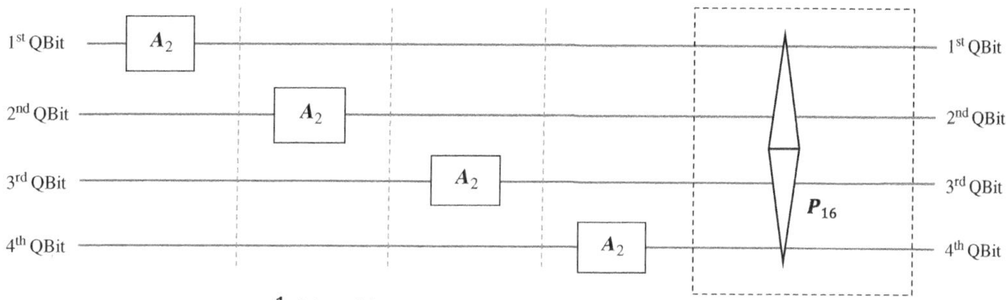

$$A_2 = \frac{1}{\sqrt{2}}\begin{bmatrix} 1 & -1 \\ 1 & 1 \end{bmatrix}, \quad P_X = (0,15)(1,7)(2,11)(4,13)(6,9)(8,14).$$

Fig. 5.13-1 The circuit for the 4-qubit direct QHdT by the paired splitting.

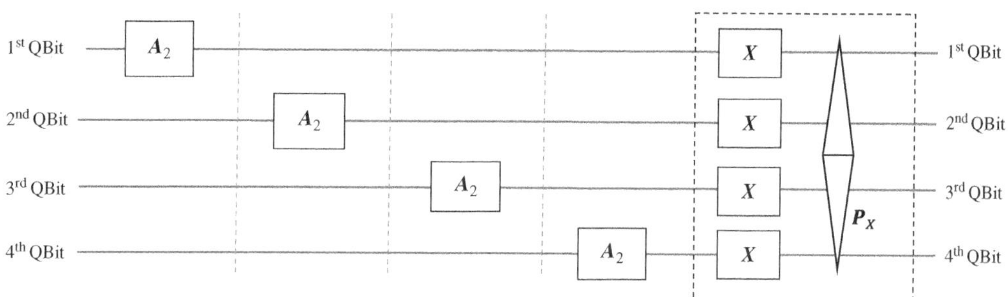

$$A_2 = \frac{1}{\sqrt{2}}\begin{bmatrix} 1 & -1 \\ 1 & 1 \end{bmatrix}, \quad X = \begin{bmatrix} 0 & 1 \\ 1 & 0 \end{bmatrix}, \quad P_X = (1,8)(2,4)(3,12)(5,10)(7,14)(11,13).$$

Fig. 5.13-2 The circuit for the 4-qubit direct QHdT by the paired splitting.

As shown in Fig. 5.13-2, this permutation can be accomplished with 4-bit SWAP operation following $X^{\otimes 4}$ gate:

$$P_X = (1,8)(2,4)(3,12)(5,10)(7,14)(11,13),$$

or

$$(1,8)\colon 00\,01 \to 10\,00, \quad (2,4)\colon 00\,10 \to 01\,00, \quad (3,12)\colon 00\,11 \to 11\,00,$$
$$(5,10)\colon 01\,01 \to 10\,10, \quad (7,14)\colon 01\,11 \to 11\,10, \quad (11,13)\colon 10\,11 \to 11\,01.$$

The equivalent circuit in the classical computer for calculating the 16-point fast Hadamard transform (FHdT) is as shown in Fig. 5.14. For high length $N = 2^r$, $r > 4$, the circuit for computing the N-point Hadamard transform is described in a similar way.

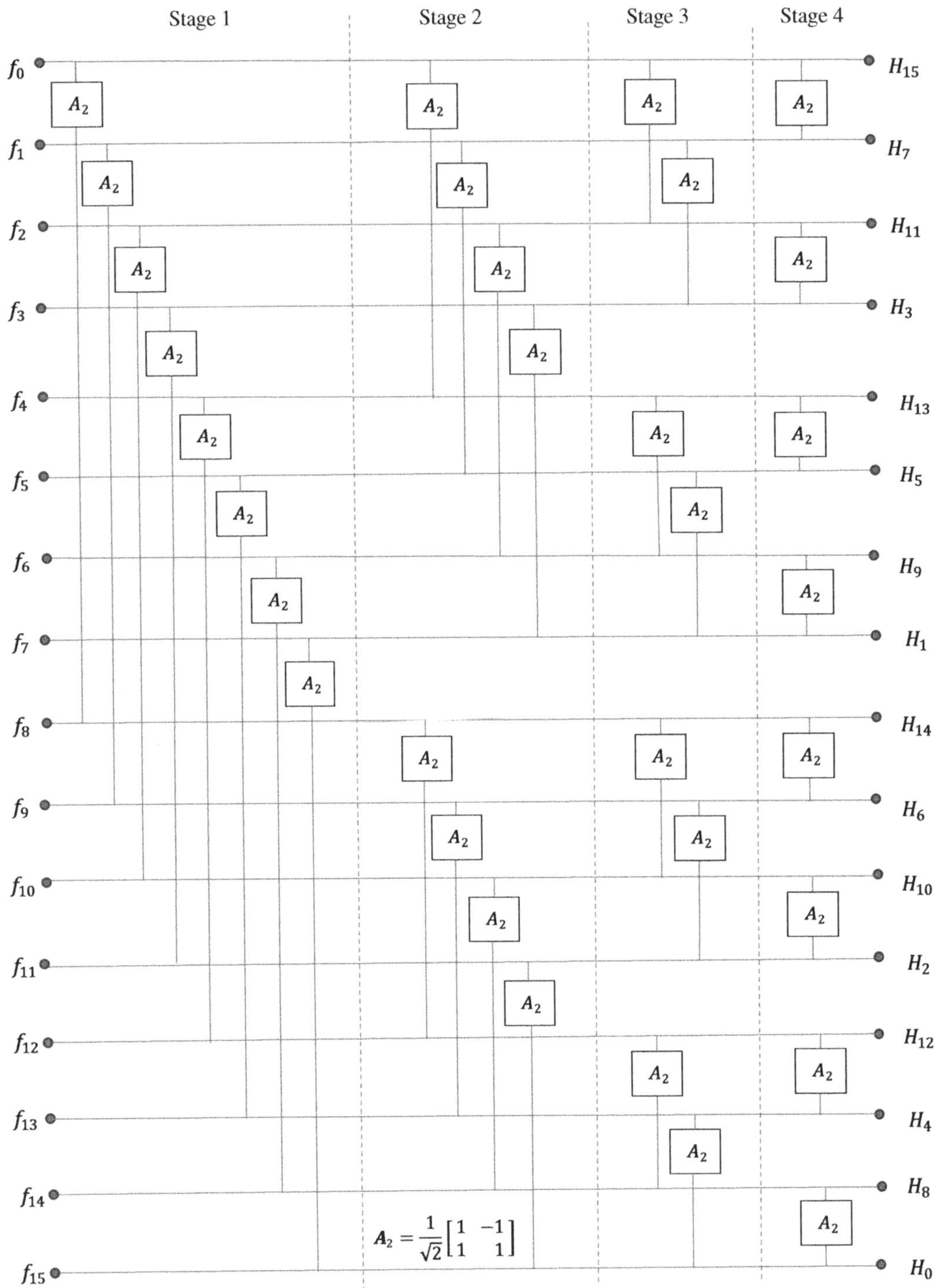

Fig. 5.14 Circuit for computing the quantum 16-point Hadamard transform.

It should be noted that the permutation is not necessary operation in this circuits, as well as in the circuits above for the 8-point DHdT. The inverse Hadamard transform with data given in order 15,7,11,3,13,5,9,1 and 14,6,10,2,12,4,8,0 will result in the natural order of the signal.

5.5 Quantum Fourier Transform

In this section, we describe the circuits for calculating quantum Fourier transform (QFT), which are based on the paired transform-based splitting. In many works in quantum computation [15–18], the N-point discrete Fourier transform (DFT) of the signal f_n is defined as

$$F_p = \frac{1}{\sqrt{N}}\sum_{n=0}^{N-1} f_n e^{i2\pi np/N}, \qquad p = 0, 1, 2, ..., (N-1). \tag{5.19}$$

The case is when $N = 2^r$, $r \geq 1$. The quantum analog of this concept is defined as follows. Consider in the standard basis state the r-qubit superposition

$$|\varphi(f)\rangle = f_0|0\rangle + f_1|1\rangle + f_2|2\rangle + f_3|3\rangle + \cdots + f_{N-1}|N-1\rangle, \tag{5.20}$$

with condition that $|f_0|^2 + |f_1|^2 + |f_2|^2 + ... + |f_{N-1}|^2 = 1$. Thus, the signal is represented as the quantum ket-vector with amplitudes f_n,

$$|\varphi(f)\rangle = \begin{bmatrix} f_0 \\ f_1 \\ f_2 \\ \cdots \\ f_{N-1} \end{bmatrix} = f_0\begin{bmatrix} 1 \\ 0 \\ 0 \\ \cdots \\ 0 \end{bmatrix} + f_1\begin{bmatrix} 0 \\ 1 \\ 0 \\ \cdots \\ 0 \end{bmatrix} + f_2\begin{bmatrix} 0 \\ 0 \\ 1 \\ \cdots \\ 0 \end{bmatrix} + \cdots + f_{N-1}\begin{bmatrix} 0 \\ 0 \\ 0 \\ \cdots \\ 1 \end{bmatrix}. \tag{5.21}$$

The r-qubit QFT is defined as the N-point DFT of the signal in the same computation basis state, that is, it is the r-qubit state, or the quantum state vector

$$|\varphi(F)\rangle = \begin{bmatrix} F_0 \\ F_1 \\ F_2 \\ \cdots \\ F_{N-1} \end{bmatrix} = F_0\begin{bmatrix} 1 \\ 0 \\ 0 \\ \cdots \\ 0 \end{bmatrix} + F_1\begin{bmatrix} 0 \\ 1 \\ 0 \\ \cdots \\ 0 \end{bmatrix} + F_2\begin{bmatrix} 0 \\ 0 \\ 1 \\ \cdots \\ 0 \end{bmatrix} + \cdots + F_{N-1}\begin{bmatrix} 0 \\ 0 \\ 0 \\ \cdots \\ 1 \end{bmatrix}, \tag{5.22}$$

that is,

$$|\varphi(F)\rangle = F_0|0\rangle + F_1|1\rangle + F_2|2\rangle + F_3|3\rangle + \cdots + F_{N-1}|N-1\rangle.$$

The QFT changes the state of the r-qubit; it is the transformation $|\varphi(f)\rangle \to |\varphi(F)\rangle$. If the result of measurement $|\varphi(f)\rangle$ is the state $|n\rangle$ with probability $|f_n|^2$, the QFT may occur or be measured in the same state with probability $|F_n|^2$. The normalized coefficient $1/\sqrt{N}$ is used in Eq. 5.19, in order to have the same condition $|F_0|^2 + |F_1|^2 + ... + |F_{N-1}|^2 = 1$ for the new state $|\varphi(F)\rangle$.

5.5.1 The Paired DFT

The N-point DFT can be split by the paired transform similar to the Hadamard transform. We consider the paired fast Fourier transform (FFT) and the corresponding quantum circuits. The N-point DFT is considered in the traditional form which is used in signal and image processing,

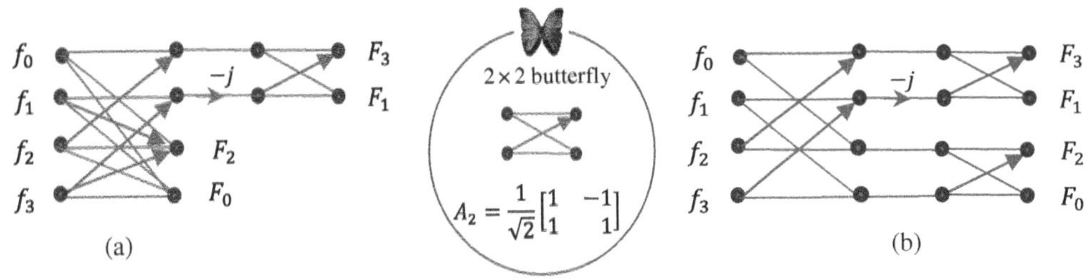

Fig. 5.15 Signal-flow graph of the paired 4-point FFT with (a) two paired transforms and (b) four butterflies A_2.

$$F_p = \frac{1}{\sqrt{N}}\sum_{n=0}^{N-1} f_n W_N^{np} = \frac{1}{\sqrt{N}}\sum_{n=0}^{N-1} f_n e^{-i2\pi np/N}, \quad p = 0, 1, \dots, (N-1). \tag{5.23}$$

Here, $W_N = \exp(-i2\pi/N)$. The r-qubit DFT is defined as in Eq. 5.22, as the transform of the states $|\varphi(f)\rangle \to |\varphi(F)\rangle$.

Example 5.8 4-Point DFT
Consider the $N = 4$ case. The 4-point DFT is calculated by

$$F_p = \frac{1}{2}\sum_{n=0}^{3} f_n W_4^{np} = \frac{1}{2}\sum_{n=0}^{3} f_n e^{-i2\pi np/4}, \quad p = 0, 1, 2, 3, \quad W_4 = -i.$$

The signal-flow graphs of the 4-point DFT with the operations of 2×2 butterfly are shown in Fig. 5.15.

The output of the paired DFT is written with the permuted components. Therefore, one can consider the permutation (0,3) of outputs, which is described by the matrix

$$P = P_{(0,3)} = \begin{bmatrix} 0 & 0 & 0 & 1 \\ 0 & 1 & 0 & 0 \\ 0 & 0 & 1 & 0 \\ 1 & 0 & 0 & 0 \end{bmatrix}.$$

Figure 5.16 shows the circuit of the 4-point DFT with four gates A_2 and one trivial rotation matrix [10].

In the matrix form, the above circuit (with the permutation $P = (0, 3)$) is described as

$$\begin{bmatrix} F_0 \\ F_1 \\ F_2 \\ F_3 \end{bmatrix} = \begin{bmatrix} 0 & 0 & 0 & 1 \\ 0 & 1 & 0 & 0 \\ 0 & 0 & 1 & 0 \\ 1 & 0 & 0 & 0 \end{bmatrix} \times \frac{1}{\sqrt{2}}\begin{bmatrix} 1 & -1 & 0 & 0 \\ 1 & 1 & 0 & 0 \\ 0 & 0 & 1 & -1 \\ 0 & 0 & 1 & 1 \end{bmatrix}\begin{bmatrix} 1 & 0 & 0 & 0 \\ 0 & -i & 0 & 0 \\ 0 & 0 & 1 & 0 \\ 0 & 0 & 0 & 1 \end{bmatrix} \times \frac{1}{\sqrt{2}}\begin{bmatrix} 1 & 0 & -1 & 0 \\ 0 & 1 & 0 & -1 \\ 1 & 0 & 1 & 0 \\ 0 & 1 & 0 & 1 \end{bmatrix}\begin{bmatrix} f_0 \\ f_1 \\ f_2 \\ f_3 \end{bmatrix}.$$

The matrix of the transform is equal to

$$F = \frac{1}{2}\begin{bmatrix} 1 & 1 & 1 & 1 \\ 1 & -i & -1 & i \\ 1 & -1 & 1 & -1 \\ 1 & i & -1 & -i \end{bmatrix} = \begin{bmatrix} 0 & 0 & 0 & 1 \\ 0 & 1 & 0 & 0 \\ 0 & 0 & 1 & 0 \\ 1 & 0 & 0 & 0 \end{bmatrix}\frac{1}{\sqrt{2}}\begin{bmatrix} 1 & -1 & 0 & 0 \\ 1 & 1 & 0 & 0 \\ 0 & 0 & 1 & -1 \\ 0 & 0 & 1 & 1 \end{bmatrix}\begin{bmatrix} 1 & 0 & 0 & 0 \\ 0 & -i & 0 & 0 \\ 0 & 0 & 1 & 0 \\ 0 & 0 & 0 & 1 \end{bmatrix}\frac{1}{\sqrt{2}}\begin{bmatrix} 1 & 0 & -1 & 0 \\ 0 & 1 & 0 & -1 \\ 1 & 0 & 1 & 0 \\ 0 & 1 & 0 & 1 \end{bmatrix}$$

$$= P\,(I_2 \otimes A_2)\left(\overline{S} \oplus I_2\right)(A_2 \otimes I_2).$$

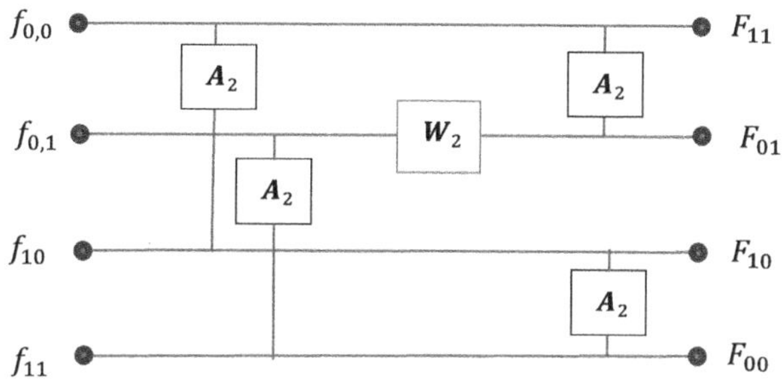

Fig. 5.16 Circuit for the 2-point DFT with three stages.

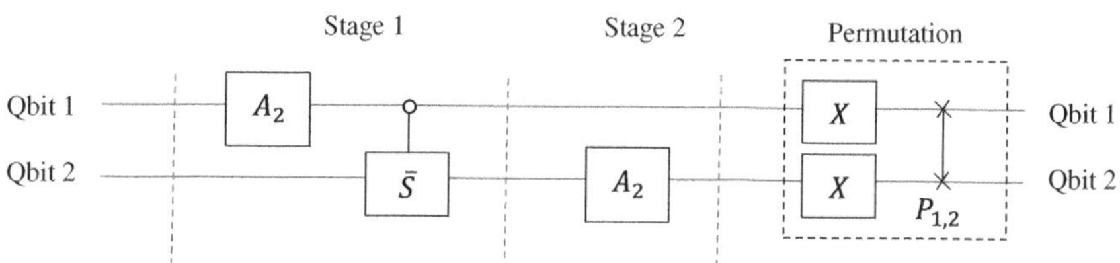

$$A_2 = \frac{1}{\sqrt{2}}\begin{bmatrix}1 & -1\\ 1 & 1\end{bmatrix},\quad \bar{S} = \begin{bmatrix}1 & 0\\ 0 & -i\end{bmatrix},\quad X = \begin{bmatrix}1 & -1\\ 1 & 1\end{bmatrix},\quad P_{1,2} = (1,2).$$

Fig. 5.17 The circuit of the 2-qubit QFT with two butterflies A_2.

The determinant of this unitary matrix equals det $F = i$. The permutation P can be presented as the following product:

$$P_{(0,3)} = \begin{bmatrix}0&0&0&1\\0&1&0&0\\0&0&1&0\\1&0&0&0\end{bmatrix} = \begin{bmatrix}0&0&0&1\\0&0&1&0\\0&1&0&0\\1&0&0&0\end{bmatrix}\begin{bmatrix}1&0&0&0\\0&0&1&0\\0&1&0&0\\0&0&0&1\end{bmatrix} = (X \otimes X)P_{(1,2)}. \tag{5.24}$$

where $P_{(1,2)}$ is the 2-qubit SWAP operation. The quantum circuit of the transform is shown in Fig. 5.17.

The same quantum circuit with the Hadamard matrix H_2 instead of the butterfly A_2 is shown in Fig. 5.18. Note that $A_2 = H_2X$ and $XX = I_4$. Also, det $A_2 = 1$ and det $H_2 = -1$. A_2 is the matrix of rotation by 45°, and the Hadamard matrix $H_2 = A_2X$ is the rotation after the CNOT gate.

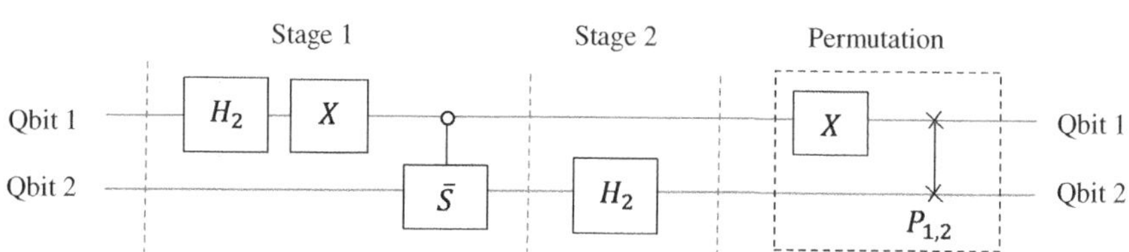

$$H_2 = \frac{1}{\sqrt{2}}\begin{bmatrix}1 & 1\\ 1 & -1\end{bmatrix},\quad \bar{S} = \begin{bmatrix}1 & 0\\ 0 & -i\end{bmatrix},\quad X = \begin{bmatrix}0 & 1\\ 1 & 0\end{bmatrix},\quad P_{1,2} = (1,2)$$

Fig. 5.18 The circuit of the 2-qubit QFT with two Hadamard gates.

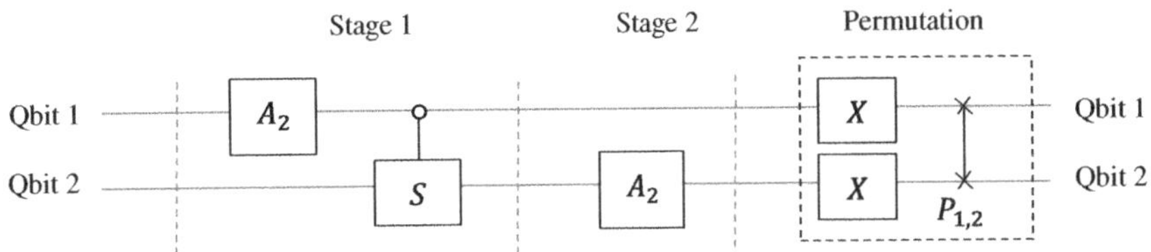

$$A_2=\frac{1}{\sqrt{2}}\begin{bmatrix}1 & -1\\ 1 & 1\end{bmatrix},\quad S=\begin{bmatrix}1 & 0\\ 0 & i\end{bmatrix},\quad X=\begin{bmatrix}1 & -1\\ 1 & 1\end{bmatrix},\quad P_{1,2}=(1,2).$$

Fig. 5.19 The circuit of the 2-qubit inverse QFT with two butterflies A_2.

For the inverse 4-point DFT, the same circuit can be used, only the complex conjugate gate $\overline{S}$ must be changed by S gate, as shown in Fig. 5.19.

Example 5.9 Paired 8-Point FFT

We consider the $N = 8$ case. The signal-flow graph of the 8-point DFT by the paired transforms is shown in Fig. 5.20 [8, 9].

Here, the twiddle factors are $W^1 = W_8^1 = e^{-i2\pi/8} = 0.7071(1-i)$ and $W^3 = W_8^3 = e^{-i2\pi/8} = 0.7071(-1-i)$. The matrix of the 8-point DFT can be written as [9]

$$F_8 = P_8\left[\begin{array}{ccc} \underbrace{\begin{bmatrix}\chi_2' & & \\ & 1 & \\ & & 1\end{bmatrix}\operatorname{diag}\begin{Bmatrix}1\\ -i\\ 1\\ 1\end{Bmatrix}\chi_4'} & & \\ & \chi_2' & \\ & & 1 \\ & & & 1\end{array}\right]\operatorname{diag}\begin{Bmatrix}1\\ W_8^1\\ -i\\ W_8^3\\ 1\\ -i\\ 1\\ 1\end{Bmatrix}\chi_8', \tag{5.25}$$

where the matrix P_8: $(7, 3, 5, 1, 6, 2, 4, 0) \rightarrow (0, 1, 2, 3, 4, 5, 6, 7)$ corresponds to the permutation of outputs.

Using the full graphs of the 8- and 4-point DPTs, we can draw the signal-flow graph of the Fourier transform in Fig. 5.20, as shown in Fig. 5.21 (for more detail see [10]).

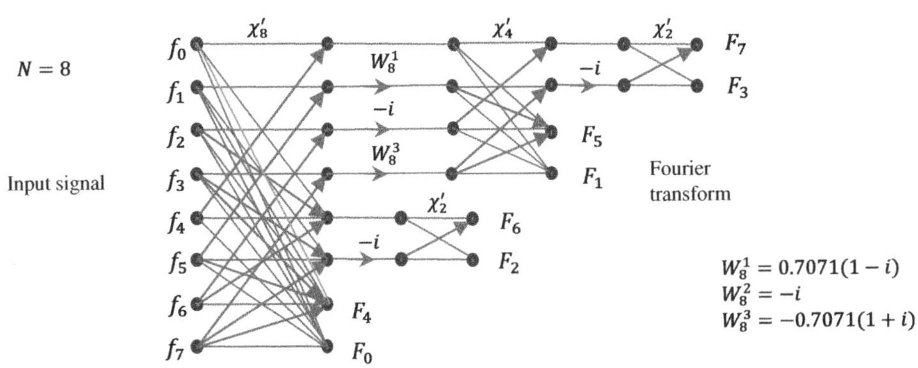

Fig. 5.20 The signal-flow graph of the paired 8-point FFT.

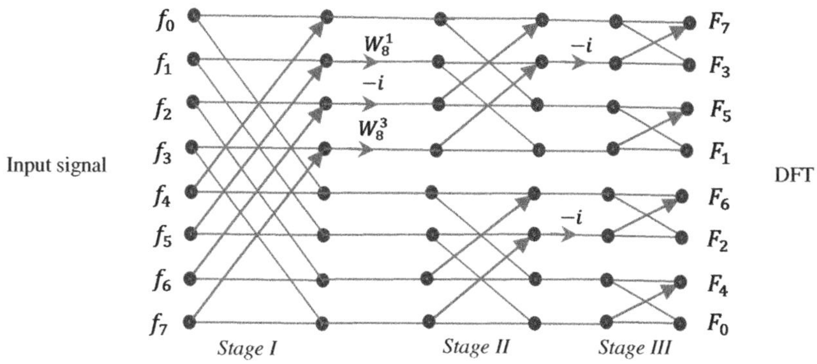

Fig. 5.21 Signal-flow graph of the 8 point DFT with three groups of butterflies.

In the matrix form, this signal-flow graph is described as follows:

$$\Phi_8 = \underbrace{\begin{bmatrix} 1 & -1 & & & & & & \\ 1 & 1 & & & & & & \\ & & 1 & -1 & & & & \\ & & 1 & 1 & & & & \\ & & & & 1 & -1 & & \\ & & & & 1 & 1 & & \\ & & & & & & 1 & -1 \\ & & & & & & 1 & 1 \end{bmatrix}}_{Stage\ 3} \underbrace{\begin{bmatrix} 1 & & & & & & & \\ & -i & & & & & & \\ & & 1 & & & & & \\ & & & 1 & & & & \\ & & & & 1 & & & \\ & & & & & -i & & \\ & & & & & & 1 & \\ & & & & & & & 1 \end{bmatrix} \begin{bmatrix} 1 & & -1 & & & & & \\ & 1 & & -1 & & & & \\ 1 & & 1 & & & & & \\ & 1 & & 1 & & & & \\ & & & & 1 & & -1 & \\ & & & & & 1 & & -1 \\ & & & & 1 & & 1 & \\ & & & & & 1 & & 1 \end{bmatrix}}_{Stage\ 2}$$

$$\times \underbrace{\begin{bmatrix} 1 & & & & & & & \\ & W^1 & & & & & & \\ & & -i & & & & & \\ & & & W^3 & & & & \\ & & & & 1 & & & \\ & & & & & 1 & & \\ & & & & & & 1 & \\ & & & & & & & 1 \end{bmatrix} \begin{bmatrix} 1 & & & & -1 & & & \\ & 1 & & & & -1 & & \\ & & 1 & & & & -1 & \\ & & & 1 & & & & -1 \\ 1 & & & & 1 & & & \\ & 1 & & & & 1 & & \\ & & 1 & & & & 1 & \\ & & & 1 & & & & 1 \end{bmatrix}}_{Stage\ 1}.$$

Figure 5.22 shows the block scheme of the paired 8-point DFT with 12 butterflies A_2 and 5 twiddle coefficients (3 of them are trivial, $W^2 = -i$). The circuit is given in the traditional form with the input and output components, f_n and F_p, respectively.

This circuit makes it easy to draw the quantum circuit for the 3-qubit Fourier transform, which we describe in detail here. In the first two stages, a few butterflies A_2 are performed following the multiplication by twiddle coefficients. In the last stage, only four butterflies A_2 work.

1) The matrix of the calculation in the first stage is described as

$$W_I A_I = W_I(A_2 \otimes I_4) = \left(\begin{bmatrix} 1 & 0 & 0 & 0 \\ 0 & W^1 & 0 & 0 \\ 0 & 0 & W^2 & 0 \\ 0 & 0 & 0 & W^3 \end{bmatrix} \oplus I_4 \right) (A_2 \otimes I_4). \tag{5.26}$$

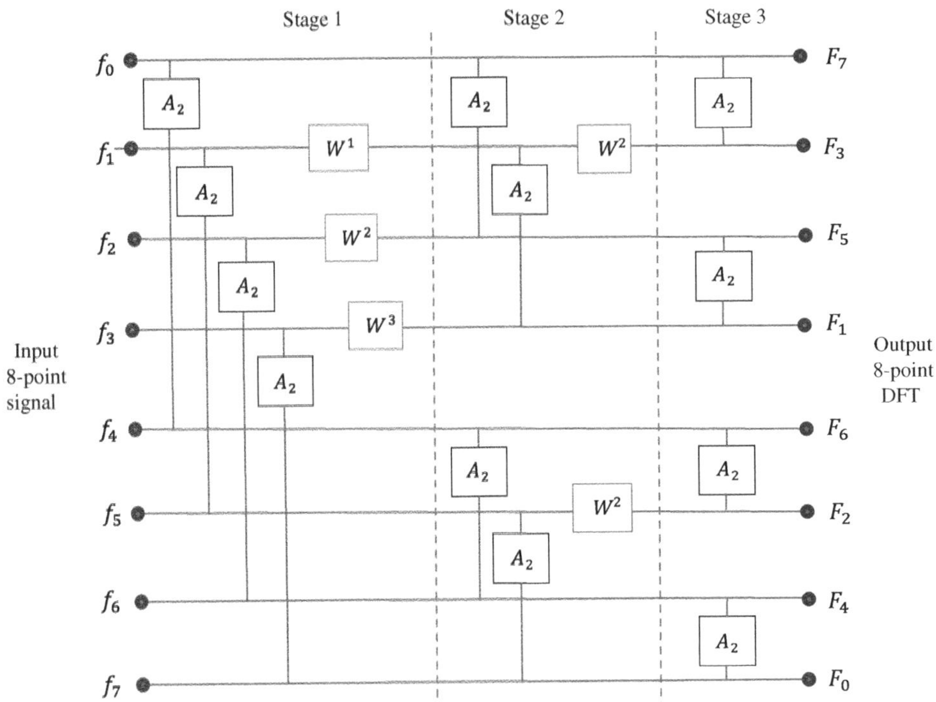

Fig. 5.22 The block diagram for the 8-point complex FFT.

The gate $A_2 \otimes I_4$ represents 4 butterflies A_2 on the bit planes 0 and 4, 1 and 5, 2 and 6, and 3 and 7. The diagonal matrix with the twiddle coefficients can be presented as ($W^2 = -i$)

$$\begin{bmatrix} 1 & 0 & 0 & 0 \\ 0 & W^1 & 0 & 0 \\ 0 & 0 & W^2 & 0 \\ 0 & 0 & 0 & W^3 \end{bmatrix} = \begin{bmatrix} 1 & 0 & 0 & 0 \\ 0 & W^1 & 0 & 0 \\ 0 & 0 & -i & 0 \\ 0 & 0 & 0 & -iW^1 \end{bmatrix} = \begin{bmatrix} 1 & 0 \\ 0 & -i \end{bmatrix} \otimes \begin{bmatrix} 1 & 0 \\ 0 & W^1 \end{bmatrix} = \overline{S} \otimes \overline{T}.$$

Here, the gate $T = T(\pi/4)$ and $\overline{T} = T(-\pi/4)$. We denote by $\overline{S}$ the gate that adds a relative shift of $-\pi/4$, that is,

$$\overline{S} = \begin{bmatrix} 1 & 0 \\ 0 & -i \end{bmatrix} = \begin{bmatrix} 1 & 0 \\ 0 & -1 \end{bmatrix} \begin{bmatrix} 1 & 0 \\ 0 & i \end{bmatrix} = ZS = SZ.$$

Thus, the matrix in Eq. 5.26 can be written as

$$W_I A_I = \left(\left[\overline{S} \otimes \overline{T}\right] \oplus I_4\right)(A_2 \otimes I_4)$$

with the circuit shown in two forms in Fig. 5.23-1 in part (a). These two forms of the circuit are considered equivalent. The gate W_I is the 0-controlled gate $C_0\left(\overline{S} \otimes \overline{T}\right)$.

Note that the above matrix with the exponential coefficients can also be written as

$$\begin{bmatrix} 1 & 0 & 0 & 0 \\ 0 & W^1 & 0 & 0 \\ 0 & 0 & W^2 & 0 \\ 0 & 0 & 0 & W^3 \end{bmatrix} = \begin{bmatrix} 1 & 0 & 0 & 0 \\ 0 & W^1 & 0 & 0 \\ 0 & 0 & 1 & 0 \\ 0 & 0 & 0 & W^1 \end{bmatrix} \begin{bmatrix} 1 & 0 & 0 & 0 \\ 0 & 1 & 0 & 0 \\ 0 & 0 & -i & 0 \\ 0 & 0 & 0 & -i \end{bmatrix}$$

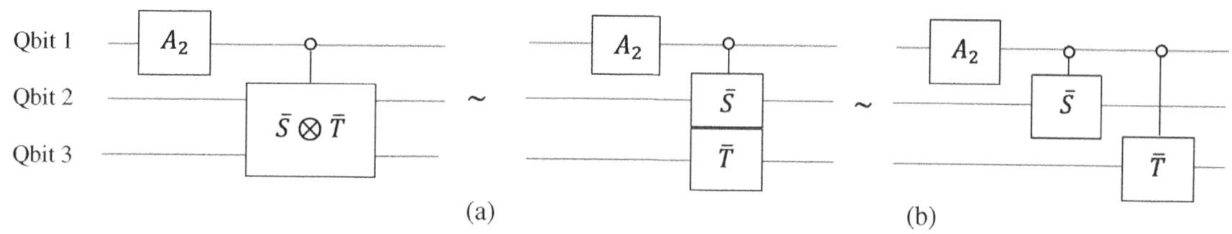

Fig. 5.23-1 The equivalent drawing of the circuit in the first stage of calculations: (a) with 0-controlled gate $\bar{S} \otimes \bar{T}$ and (b) with two 0-controlled gates $\bar{S}$ and $\bar{T}$.

and

$$W_I = \left(\begin{bmatrix} 1 & 0 & 0 & 0 \\ 0 & W^1 & 0 & 0 \\ 0 & 0 & 1 & 0 \\ 0 & 0 & 0 & W^1 \end{bmatrix} \oplus I_4\right)\left(\begin{bmatrix} 1 & 0 & 0 & 0 \\ 0 & 1 & 0 & 0 \\ 0 & 0 & -i & 0 \\ 0 & 0 & 0 & -i \end{bmatrix} \oplus I_4\right)$$
$$= \left(\left(I_2 \otimes \begin{bmatrix} 1 & 0 \\ 0 & W^1 \end{bmatrix}\right) \oplus I_4\right)\left(\left(\begin{bmatrix} 1 & 0 \\ 0 & -i \end{bmatrix} \otimes I_2\right) \oplus I_4\right).$$

Thus, the matrix $W_I = \left((I_2 \otimes \overline{T}) \oplus I_4\right)\left((\overline{S} \otimes I_2) \oplus I_4\right)$. The circuits for the gate $W_I A_I$ with this decomposition of W_I are shown in Fig. 5.23-1 in part (b). The matrices $\overline{T}$ and $\overline{S}$ are complex conjugate to the matrices T and S, respectively.

Note that the order of the 0-contrlolled gates $\overline{T}$ and $\overline{S}$ in the circuit in part (b) is not important. These two gates are described by the diagonal matrices. If we change the order of these two gates, we get the same circuit, as shown in Fig. 5.23-2.

2) In the second stage, the calculations are described by the following matrix:

$$W_{II}A_{II} = W_{II}(I_2 \otimes A_2 \otimes I_2) = \left(I_2 \otimes \begin{bmatrix} 1 & 0 & 0 & 0 \\ 0 & -i & 0 & 0 \\ 0 & 0 & 1 & 0 \\ 0 & 0 & 0 & 1 \end{bmatrix}\right)(I_2 \otimes A_2 \otimes I_2)$$

$$= \left(I_2 \otimes \left(\begin{bmatrix} 1 & 0 \\ 0 & -i \end{bmatrix} \oplus I_2\right)\right)(I_2 \otimes A_2 \otimes I_2) = \left(I_2 \otimes (\overline{S} \oplus I_2)\right)(I_2 \otimes A_2 \otimes I_2) \tag{5.27}$$

The corresponding circuit is shown in Fig. 5.24.

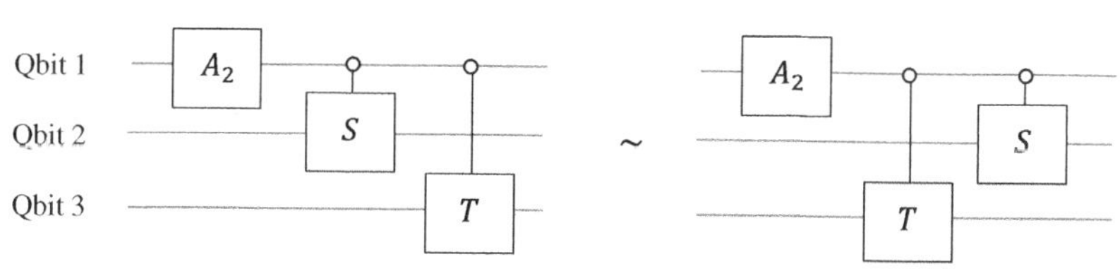

Fig. 5.23-2 The circuit with two 0-controlled gates on 3 qubits.

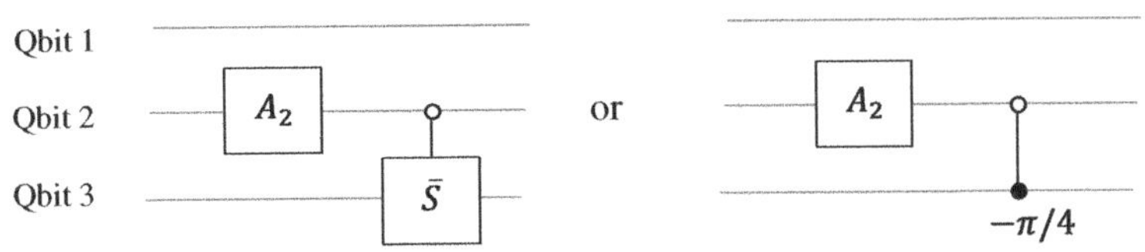

Fig. 5.24 The circuit of the second stage of calculations.

3) In the last stage, the matrix with 4 butterflies is

$$A_{III} = I_4 \otimes A_2 = \begin{bmatrix} A_2 & & & \\ & A_2 & & \\ & & A_2 & \\ & & & A_2 \end{bmatrix}.$$

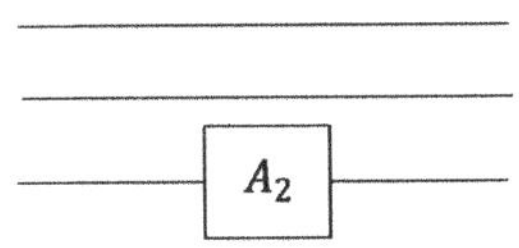

In the 8-point paired DFT, the coefficients of the transform are obtained in the order 7, 3, 5, 1, 6, 2, 4, and 0. Therefore, if necessary, we can perform the following permutation:

$$P_8 = \begin{pmatrix} 01234567 \\ 73516240 \end{pmatrix} = (07)(13)(25)(46).$$

This permutation can be accomplished by using two permutations, $X \otimes X \otimes X$ and the 3-bit reversal (shuffling),

$$X \otimes X \otimes X = \begin{pmatrix} 01234567 \\ 76543210 \end{pmatrix} \text{ and } P_X = \begin{pmatrix} 01234567 \\ 04261537 \end{pmatrix} = (1,4)(3,6).$$

Indeed,

$$P_8\colon \begin{bmatrix} 0 \\ 1 \\ 2 \\ 3 \\ 4 \\ 5 \\ 6 \\ 7 \end{bmatrix} = \begin{bmatrix} 0 & 0 & 0 \\ 0 & 0 & 1 \\ 0 & 1 & 0 \\ 0 & 1 & 1 \\ 1 & 0 & 0 \\ 1 & 0 & 1 \\ 1 & 1 & 0 \\ 1 & 1 & 1 \end{bmatrix} \xrightarrow{X \otimes X \otimes X} \begin{bmatrix} 1 & 1 & 1 \\ 1 & 1 & 0 \\ 1 & 0 & 1 \\ 1 & 0 & 0 \\ 0 & 1 & 1 \\ 0 & 1 & 0 \\ 0 & 0 & 1 \\ 0 & 0 & 0 \end{bmatrix} \xrightarrow{P_X} \begin{bmatrix} 1 & 1 & 1 \\ 0 & 1 & 1 \\ 1 & 0 & 1 \\ 0 & 0 & 1 \\ 1 & 1 & 0 \\ 0 & 1 & 0 \\ 1 & 0 & 0 \\ 0 & 0 & 0 \end{bmatrix} = \begin{bmatrix} 7 \\ 3 \\ 5 \\ 1 \\ 6 \\ 2 \\ 4 \\ 0 \end{bmatrix}. \tag{5.28}$$

Thus, the matrix of the 8-point DFT can be written as

$$\Phi_8 = \underbrace{P_X X^{\otimes 3}}_{permutation} \cdot \underbrace{A_{III}}_{stage\ 3} \cdot \underbrace{W_{II} A_{II}}_{stage\ 2} \cdot \underbrace{W_I A_I}_{stage\ 1}. \tag{5.29}$$

The corresponding quantum circuit for this transform is shown in Fig. 5.25.

If the use the Hadamard gates H_2, the circuit above must be changed as shown in Fig. 5.26. The A_2 gate is the composition of two gates, $A_2 = XH_2$. Also, one Pauli NOT was removed in the permutation, because $XA_2 = XXH_2 = H_2$.

The circuit for the inverse 8-point, or 3-qubit, QFT is described similarly. The twiddle coefficients only must be changed by their complex conjugates. It means that in the above circuits, we need only change the gates $\bar{S}$ and $\bar{T}$ by

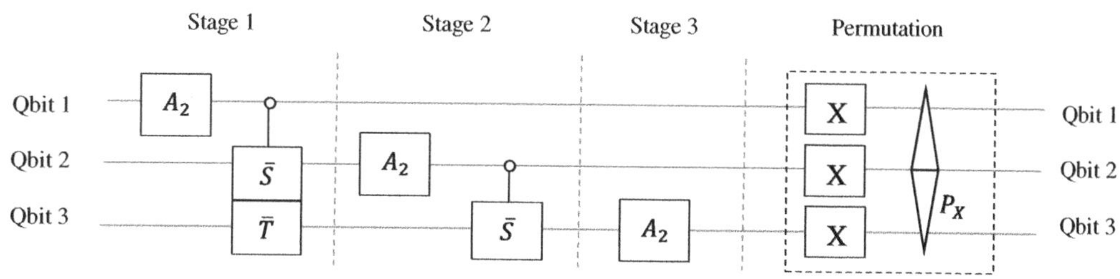

$$A_2 = \frac{1}{\sqrt{2}}\begin{bmatrix}1 & -1\\ 1 & 1\end{bmatrix},\quad \bar{S} = \begin{bmatrix}1 & 0\\ 0 & -i\end{bmatrix},\quad \bar{T} = \begin{bmatrix}1 & 0\\ 0 & e^{-i\pi/4}\end{bmatrix}\quad X = \begin{bmatrix}0 & 1\\ 1 & 0\end{bmatrix},\quad P_X = (1,4)(3,6)$$

Fig. 5.25 The circuit with two 0-controlled gates for the 3-qubit QFT by the paired splitting.

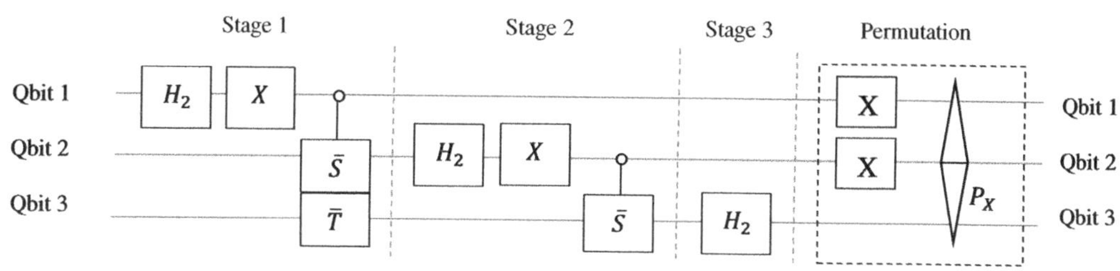

$$H_2 = \frac{1}{\sqrt{2}}\begin{bmatrix}1 & 1\\ 1 & -1\end{bmatrix},\quad \bar{S} = \begin{bmatrix}1 & 0\\ 0 & -i\end{bmatrix},\quad \bar{T} = \begin{bmatrix}1 & 0\\ 0 & e^{-i\pi/4}\end{bmatrix}\quad X = \begin{bmatrix}0 & 1\\ 1 & 0\end{bmatrix},\quad P_X = (1,4)(3,6)$$

Fig. 5.26 The circuit for the direct 3-qubit QFT by the paired splitting.

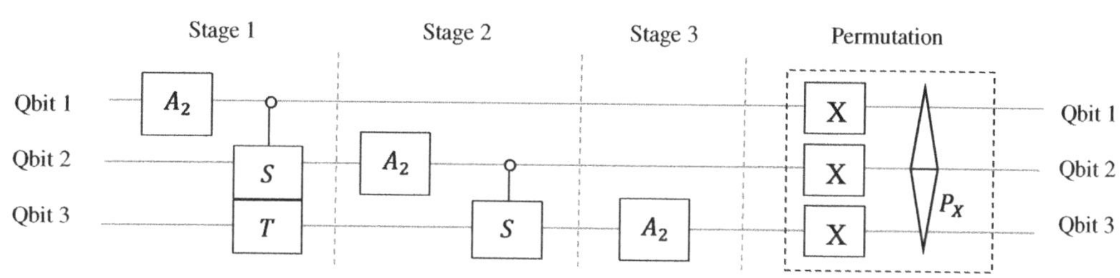

$$A_2 = \frac{1}{\sqrt{2}}\begin{bmatrix}1 & -1\\ 1 & 1\end{bmatrix},\quad S = \begin{bmatrix}1 & 0\\ 0 & i\end{bmatrix},\quad T = \begin{bmatrix}1 & 0\\ 0 & e^{i\pi/4}\end{bmatrix}\quad X = \begin{bmatrix}0 & 1\\ 1 & 0\end{bmatrix},\quad P_X = (1,4)(3,6)$$

Fig. 5.27 The circuit for the inverse 3-qubit QFT by the paired splitting.

the gates S and T, respectively. The corresponding quantum circuit for this transform with butterfly A_2 is shown in Fig. 5.27.

Figure 5.28 shows the traditional block diagram of 16-point DFT, which is prepared for building the quantum circuit for calculating the 4-qubit QFT [10]. It will be shown that the performance of 32 butterflies A_2 is equivalent to 4 gates A_2 in the quantum circuit. As in the case of 3-qubit QFT, the permutation of outputs is possible, if necessary, work with outputs in a natural order. If we keep the order as it is and perform the inverse Fourier transform by using a similar circuit, then the signal will be received in a natural order. Now, we build the quantum for this transform on 4-qubit input.

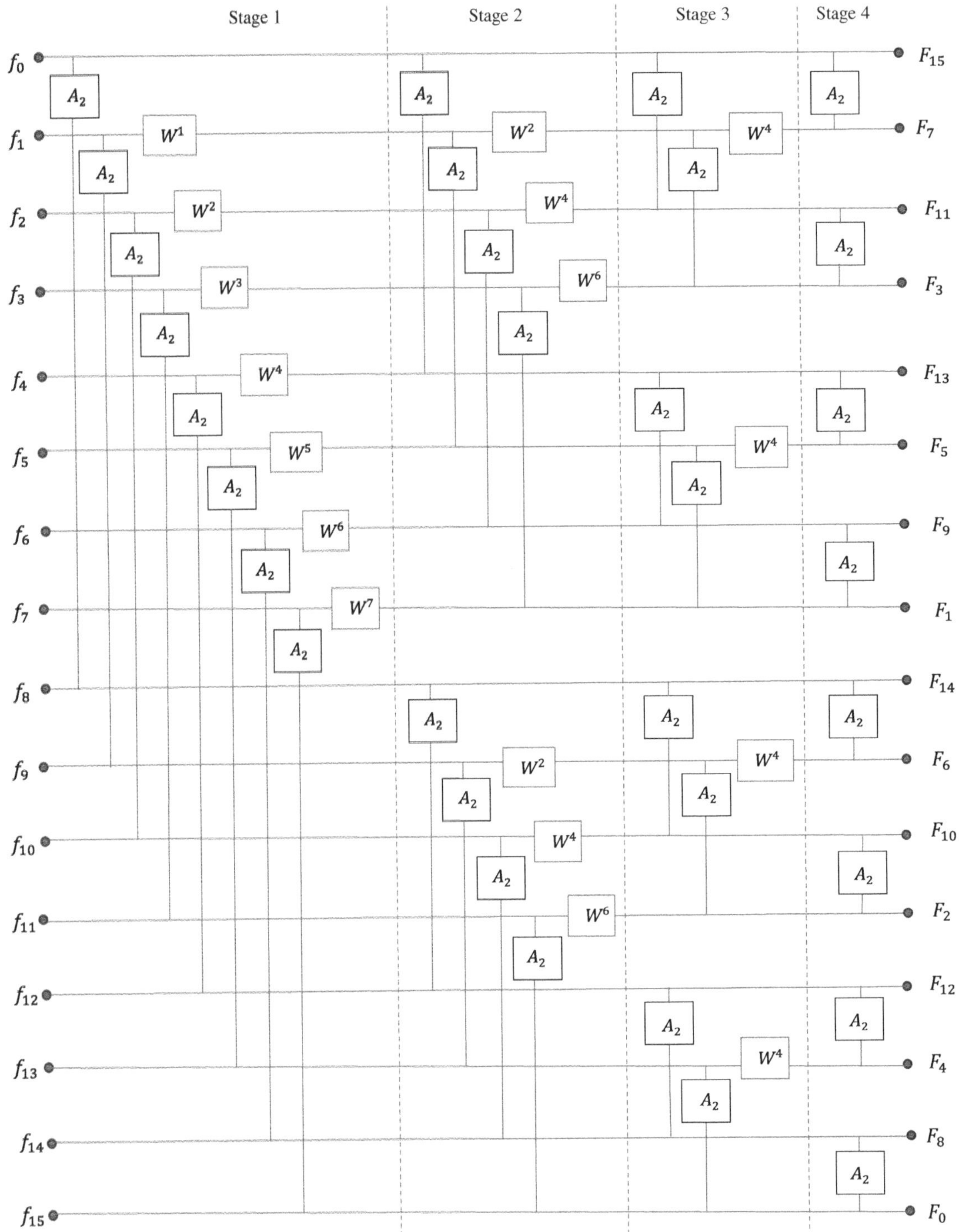

Fig. 5.28 The block scheme of the circuit for computing the 16-point DFT or 4-qubit QFT.

5.5.2 Algorithm of the 4-Qubit QFT

Stage 1: The first 8-butterfly operation is described by the matrix $A_I = (A_2 \otimes I_2 \otimes I_2 \otimes I_2)$. The next diagonal matrix of twiddle coefficients equals to

$$W_I = \left(\begin{bmatrix} 1 & 0 & 0 & 0 & 0 \\ 0 & W^1 & 0 & 0 & 0 \\ 0 & 0 & W^2 & 0 & 0 \\ & & & \cdots & \\ 0 & 0 & 0 & & W^7 \end{bmatrix} \oplus I_8 \right) = \left(\begin{bmatrix} 1 & 0 \\ 0 & W^4 \end{bmatrix} \otimes \begin{bmatrix} 1 & 0 \\ 0 & W^2 \end{bmatrix} \otimes \begin{bmatrix} 1 & 0 \\ 0 & W^1 \end{bmatrix} \right) \oplus I_8. \tag{5.30}$$

Here, $W^1 = \exp(-i2\pi/16) = \exp(-i\pi/8)$. Therefore, using the phase shift gates, we can write

$$W_I = W_{I,8} \oplus I_8 = \left(\overline{P}^4 \otimes \overline{P}^2 \otimes \overline{P} \right) \oplus I_8, \quad \overline{P} = P(-\pi/8) = \begin{bmatrix} 1 & 0 \\ 0 & W^1 \end{bmatrix}.$$

In this stage, the operation $W_I A_I$ can be described by the circuit shown in Fig. 5.29-1. This circuit is drawn in three equivalent forms; we will use the second one. Note that in the example for the 3-qubit QFT described above, the gates P^2 and P^4 are the gates T and S, respectively.

Stage 2: The 8-butterfly operation is described by the matrix $A_{II} = I_2 \otimes (A_2 \otimes I_2 \otimes I_2)$. Then, the multiplication by six twiddle coefficients is used, which is described by the diagonal matrix

$$W_{II} = I_2 \otimes (W_{II,4} \oplus I_4) = I_2 \otimes \left(\begin{bmatrix} 1 & 0 & 0 & 0 \\ 0 & W^2 & 0 & 0 \\ 0 & 0 & W^4 & 0 \\ 0 & 0 & 0 & W^6 \end{bmatrix} \oplus I_4 \right) = I_2 \otimes \left(\left(\begin{bmatrix} 1 & 0 \\ 0 & W^4 \end{bmatrix} \otimes \begin{bmatrix} 1 & 0 \\ 0 & W^2 \end{bmatrix} \right) \oplus I_4 \right).$$

Thus, the first two stages of calculation are described by the matrix

$$W_{II} A_{II} = \left[I_2 \otimes \left(\left(\overline{P}^4 \otimes \overline{P}^2 \right) \oplus I_4 \right) \right] \cdot [I_2 \otimes (A_2 \otimes I_2 \otimes I_2)].$$

The corresponding circuit is given in Fig. 5.29-2.

Stage 3: The 8-butterfly operation is described by the following matrix of another local gate: $A_{III} = I_4 \otimes (A_2 \otimes I_2)$. Then, 4 multiplications by the coefficient $W^4 = -i$ are needed. This operation is defined by the diagonal matrix

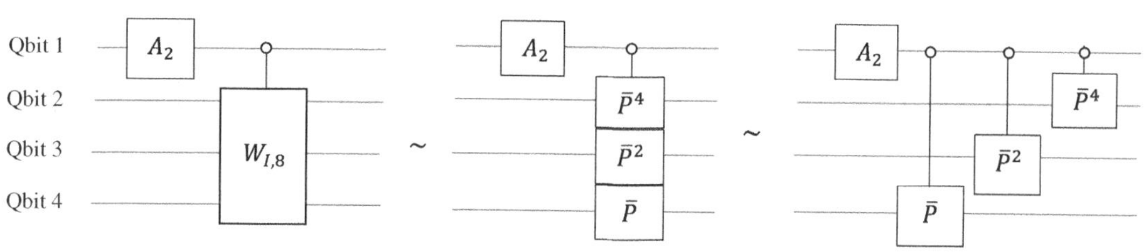

Fig. 5.29-1 The circuit of the first stage of calculations.

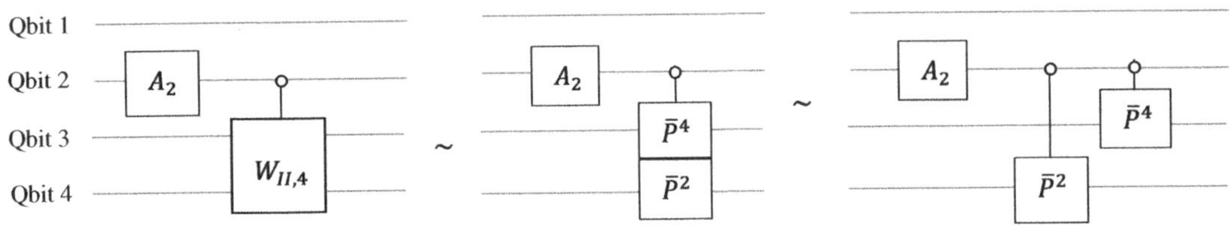

Fig. 5.29-2 The circuit element of the second stage of calculations.

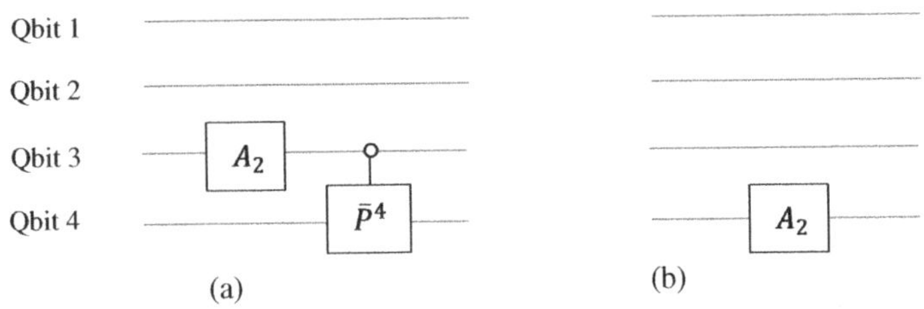

Fig. 5.29-3 The circuit element of (a) the third and (b) fourth stages of calculations.

$$W_{III} = I_4 \otimes \begin{bmatrix} 1 & 0 & 0 & 0 \\ 0 & -i & 0 & 0 \\ 0 & 0 & 1 & 0 \\ 0 & 0 & 0 & 1 \end{bmatrix} = I_4 \otimes \left(\begin{bmatrix} 1 & 0 \\ 0 & -i \end{bmatrix} \oplus I_2 \right) = I_4 \otimes \left(\overline{P}^4 \oplus I_2 \right).$$

Thus, this stage of calculations is described by the matrix

$$W_{III}A_{III} = \left[I_2 \otimes \left(\overline{P}^4 \oplus I_2 \right) \right] \cdot [I_4 \otimes (A_2 \otimes I_2)].$$

Figure 5.29-3 shows the corresponding circuit element in part (a).

Stage 4: The 8-butterfly operation is only used and described by the *local* gate: $A_{IV} = I_8 \otimes A_2$. The gate is shown in Fig. 5.29-3 in part (b).

The coefficients of the 16-point DFT, as the 16-point DHdT, are calculated in order 15, 7, 11, 3, 13, 5, 9, 1 and 14, 6, 10, 2, 12, 4, 8, 0. Therefore, the same P_{16} permutation can be used to get the natural order of the coefficients, if necessary. With the X gates, this permutation will be accomplished by the 4-bit reversal operation. Thus, the 4-qubit QFT is calculated by

$$\Phi_{16} = \left(P_X X^{\otimes 4} \right) A_{IV} \left(W_{III} A_{III} \right) \left(W_{II} A_{II} \right) \left(W_I A_I \right). \tag{5.31}$$

The final quantum circuit for the 4-qubit QFT is given in Fig. 5.30. For the inverse 4-qubit QFT, the similar quantum circuit can be used, only the phase gate must be changed by their conjugate. This circuit is shown in Fig. 5.31. Four A_2 gates, $4 \cdot \frac{3}{2} = 6$ phase shift gates, plus 4 NOT gates, and permutation P_X with 6 exchanges are used.

The circuit in Fig. 5.30 is exactly the quantum analog of the fast 16-point FFT circuit as shown in Fig. 5.28. In other words, we have redrawn the classic paired FFT algorithm for the future quantum computer. On each stage, the gate A_2 describes the work of 8 butterflies A_2, and the set of phase shift gates describes the multiplication by twiddle factors. Thus, the number of operations is exactly the same. We can call this circuit the quantum circuit of the fast Fourier transform. The above circuits for the 2- and 3-qubits also were built as quantum analogs

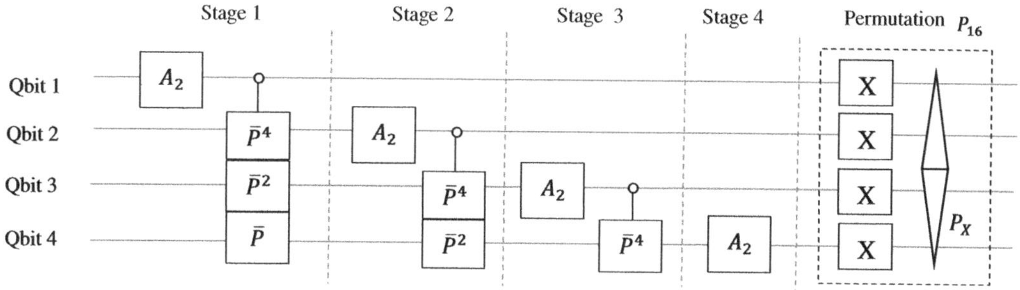

$$A_2 = \frac{1}{\sqrt{2}}\begin{bmatrix}1 & -1\\ 1 & 1\end{bmatrix},\ \bar{P} = \begin{bmatrix}1 & 0\\ 0 & e^{-i\pi/8}\end{bmatrix}\quad X = \begin{bmatrix}0 & 1\\ 1 & 0\end{bmatrix},\quad P_X = (1,8)(2,4)(3,12)(5,10)(7,14)(11,13)$$

Fig. 5.30 The circuit for the 4 qubit QFT by the paired splitting.

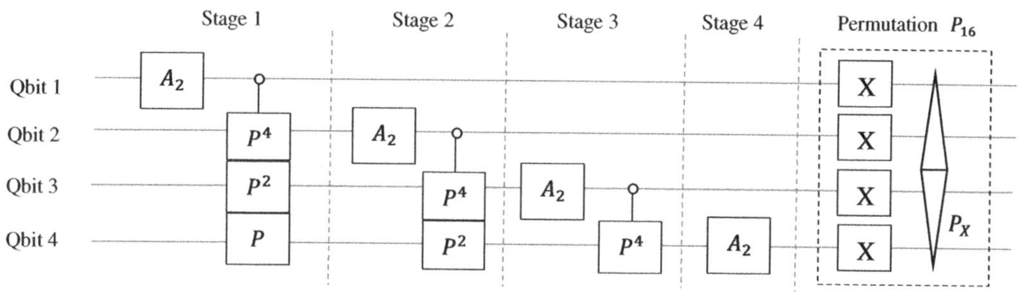

$$A_2 = \frac{1}{\sqrt{2}}\begin{bmatrix}1 & -1\\ 1 & 1\end{bmatrix},\ P = \begin{bmatrix}1 & 0\\ 0 & e^{i\pi/8}\end{bmatrix}\quad X = \begin{bmatrix}0 & 1\\ 1 & 0\end{bmatrix},\quad P_X = (1,8)(2,4)(3,12)(5,10)(7,14)(11,13)$$

Fig. 5.31 The circuit for the inverse 4-qubit QFT by the paired splitting.

of the 4- and 8-point DFT, respectively. In the general case when integer $r > 1$, the quantum circuits for the r-qubit direct and inverse Fourier transforms are described similarly. Will it run faster on a quantum computer than on a classical computer? This is to be expected because we are told that quantum computers are faster. It must be said here that we cannot obtain all coefficients of the Fourier transform, even exactly one of them, in one run of the circuit on a quantum computer. Only one state of the FT quantum superposition can be measured.

As known in the paired algorithm of calculation, the 2^r-point DFT can be used for the 2^r-point DHdT A_N defined by Eq. 5.16 (not the Walsh–Hadamard transform H_N). Only all rotation factors should be removed or considered equal to 1. The above examples of the r-qubit QFT demonstrate the same property. For example, if we remove all phase shift from the circuit in Fig. 5.30, we get the same circuit as in Fig. 5.13-2, for the 4-qubit DHdT.

5.5.3 The Known Algorithm of the QFT

Now, we describe the known algorithm for QFT with another approach, which is based on binary presentation of time and frequency-points. Let us consider the N-point DFT of the signal f_n of length $= 2^r$, $r \geq 1$,

$$F_p = \frac{1}{\sqrt{N}}\sum_{n=0}^{N-1} f_n e^{-i\frac{2\pi}{N}np}, \qquad p = 0, 1, 2, \ldots, (N-1). \tag{5.32}$$

In the same standard computation basis, we define the r-qubit superpositions of the signal and its DFT,

$$|\varphi(f)\rangle = f_0|0\rangle + f_1|1\rangle + f_2|2\rangle + f_3|3\rangle + \cdots + f_{N-1}|N-1\rangle,$$
$$|\varphi(F)\rangle = F_0|0\rangle + F_1|1\rangle + F_2|2\rangle + F_3|3\rangle + \cdots + F_{N-1}|N-1\rangle.$$

under condition that $|f_0|^2 + |f_1|^2 + |f_2|^2 + \cdots + |f_{2^r-1}|^2 = 1$. The r-qubit superposition of the Fourier transform can be written as

$$|\varphi(F)\rangle = \sum_{p=0}^{N-1} F_p|p\rangle = \sum_{p=0}^{N-1}\left(\frac{1}{\sqrt{N}}\sum_{n=0}^{N-1} f_n e^{-i\frac{2\pi}{N}np}\right)|p\rangle = \sum_{n=0}^{N-1} f_n\left(\frac{1}{\sqrt{N}}\sum_{p=0}^{N-1} e^{-i\frac{2\pi}{N}np}|p\rangle\right), \tag{5.33}$$

or

$$|\varphi(F)\rangle = \sum_{n=0}^{N-1} f_n|\varphi_n\rangle. \tag{5.34}$$

Here, $\{|\varphi_n\rangle;\ n = 0{:}\ (N-1)\}$ is the set of r-qubit superpositions

$$|\varphi_n\rangle = \frac{1}{\sqrt{N}}\sum_{p=0}^{N-1} e^{-i\frac{2\pi}{N}np}|p\rangle, \quad n = 0{:}\ (N-1). \tag{5.35}$$

The QFT is defined as the transform on computation basis $|n\rangle \to |\varphi_n\rangle$, $n = 0{:}\ (N-1)$. The calculation of this transform is accomplished by using the binary representation of numbers n and p in Eq. 5.33. We do not describe here detailed calculations for this algorithm of the QFT, since they are very cumbersome (for detail see [15]). We give the quantum scheme for the 4-qubit QFT in Fig. 5.32, wherein the phase shift gates P^k are used instead of their complex conjugates. It is done because the DFT was defined with positive powers of exponential basis functions, that is, with $\exp\left(i\frac{2\pi}{N}np\right)$ in Eq. 5.35.

Here, the notations of matrices were changed, while the matrices of phase shift were originally named as R_k. Namely, in the currently accepted notation

$$R_k = \begin{bmatrix} 1 & 0 \\ 0 & e^{i2\pi/2^k} \end{bmatrix} = P\left(\frac{2\pi}{2^k}\right), \quad k = 1, 2, 3,$$

or

$$R_2 = \begin{bmatrix} 1 & 0 \\ 0 & e^{i\pi/2} \end{bmatrix} = P\left(\frac{\pi}{2}\right) = P^4, \ R_3 = \begin{bmatrix} 1 & 0 \\ 0 & e^{i\pi/4} \end{bmatrix} = P\left(\frac{\pi}{4}\right) = P^2, \ R_4 = \begin{bmatrix} 1 & 0 \\ 0 & e^{i\pi/8} \end{bmatrix} = P \triangleq P\left(\frac{\pi}{8}\right).$$

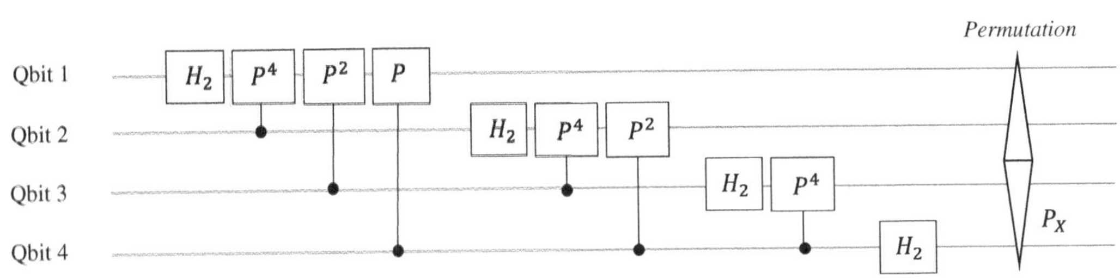

$$A_2 = \frac{1}{\sqrt{2}}\begin{bmatrix} 1 & -1 \\ 1 & 1 \end{bmatrix}, \ P = \begin{bmatrix} 1 & 0 \\ 0 & e^{i\pi/8} \end{bmatrix}, \ P_X = (1,8)(2,4)(3,12)(5,10)(7,14)(11,13)$$

Fig. 5.32 The scheme for the 4-qubit quantum Fourier transform (in definition of Eq. 5.38).

The scheme in Fig. 5.32 is similar to the scheme in Fig. 5.31 (or Fig. 5.30). The same set of phase shift gates is used, only with different control bits. In the schemes with the paired algorithm, the operation of butterfly A_2 and the 0-controlled phase shift gates is used. In the scheme in Fig. 5.32, these butterfly gets are the Hadamard gates H_2, and the 1-controlled gates are used. Now let us take a closer look at a similar scheme for the 3-qubit QFT.

Example 5.10 3-Qubit QFT Circuit

The quantum scheme for 3-qubit QFT is shown in Fig. 5.33.

Here, the coefficients of the transform are calculated in the order $\{F_0, F_4, F_6, F_2, F_7, F_3, F_5, F_1\}$. Therefore, the following permutation is required: $P = (1, 4, 7)(2, 6, 5, 3)$. The first stage of calculation can be written as

$$X_I = \underbrace{(I_4 \oplus [I_2 \otimes T])}\underbrace{(I_4 \oplus [S \otimes I_2])}\underbrace{(H_2 \otimes I_4)}. \tag{5.36}$$

The second stage is described by the matrix

$$X_{II} = \underbrace{(I_2 \otimes [I_2 \oplus S])}\underbrace{(I_2 \otimes H_2 \otimes I_2)}. \tag{5.37}$$

The last stage with only one locale gate is described by the matrix $X_{III} = (I_4 \otimes H_2)$. Also, the permutation is required with the following matrix representation:

$$P = P_X = (1,4,7)(2,6,5,3) = \begin{bmatrix} 1 & 0 & 0 & 0 & 0 & 0 & 0 & 0 \\ 0 & 0 & 0 & 0 & 0 & 0 & 0 & 1 \\ 0 & 0 & 0 & 1 & 0 & 0 & 0 & 0 \\ 0 & 0 & 0 & 0 & 0 & 1 & 0 & 0 \\ 0 & 1 & 0 & 0 & 0 & 0 & 0 & 0 \\ 0 & 0 & 0 & 0 & 0 & 0 & 1 & 0 \\ 0 & 0 & 1 & 0 & 0 & 0 & 0 & 0 \\ 0 & 0 & 0 & 0 & 1 & 0 & 0 & 0 \end{bmatrix}.$$

Note that this is not the 3-qubit SWAP operation, or permutation (1, 4, 2)(3, 5, 6). Thus, the scheme above is described as $F = P_X X_{III} X_{II} X_I$. In fact, it is the scheme to calculate the inverse 3-qubit QFT, since as mentioned above, the exponential basis functions are used in Eq. 5.32 not as $e^{-i\frac{2\pi}{N}np}$, but $e^{i\frac{2\pi}{N}np}$. In other words, the QFT for the scheme in Fig. 5.33 is considered as

$$|\varphi_n\rangle = \frac{1}{\sqrt{N}} \sum_{p=0}^{N-1} e^{i\frac{2\pi}{N}np}|p\rangle, \quad n = 0{:}\ (N-1). \tag{5.38}$$

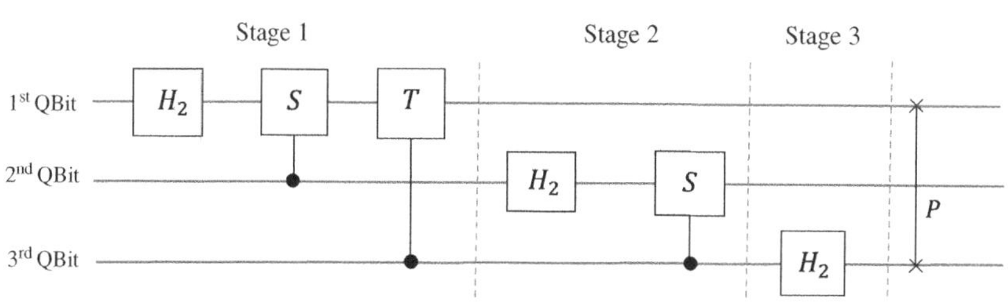

$T = \begin{bmatrix} 1 & 0 \\ 0 & e^{i\pi/4} \end{bmatrix}$, $H_2 = \frac{1}{\sqrt{2}}\begin{bmatrix} 1 & 1 \\ 1 & -1 \end{bmatrix}$, $S = \begin{bmatrix} 1 & 0 \\ 0 & i \end{bmatrix}$, $P = (1,4,7)(2,6,5,3)$

Fig. 5.33 The known circuit for the 3-qubit QFT (in the definition of Eq. 5.38).

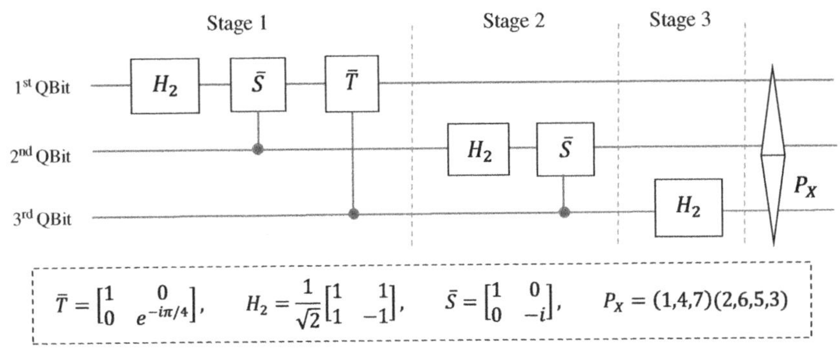

Fig. 5.34 The second circuit for the 3-qubit QFT.

For the direct 3-qubit QFT defined by Eq. 5.35, the gate T and S must be considered as $\overline{T}$ and $\overline{S}$, respectively. As a result, we obtain the quantum scheme for 3-qubit QFT, which is given in Fig. 5.34.

For the permutation P_X, the following sets of rotations (SR) are used in QD-decomposition by the heap transforms:

$$\left\{\begin{array}{l} SR_1 = \{R_{-90^\circ;5,6}, R_{-90^\circ;4,5}, R_{-90^\circ;3,4}\} \\ SR_2 = \{R_{-90^\circ;4,5}, R_{-90^\circ;3,4}, R_{-90^\circ;2,3}, R_{-90^\circ;1,2}\} \\ SR_3 = \{R_{90^\circ;2,3}\} \\ SR_4 = \{R_{-90^\circ;2,3}, R_{-90^\circ;1,2}, R_{-90^\circ;0,1}\} \\ SR_5 = \{R_{90^\circ;0,1}, R_{90^\circ;1,2}\} \end{array}\right\} \tag{5.39}$$

with the diagonal matrix $B = \text{diag}\{1, -1, 1, 1, -1, 1, 1, -1\}$, which can be written as $B = Z \oplus I_2 \oplus [-Z] \oplus Z$, where the gate $-Z = R_{-90^\circ}X$. The block scheme of the permutation P_X is given in Fig. 5.35. Here, two gates of rotation by $\pm 90^\circ$ are

$$R_{90^\circ} = \begin{bmatrix} 0 & -1 \\ 1 & 0 \end{bmatrix} \quad \text{and} \quad R_{-90^\circ} = \begin{bmatrix} 0 & 1 \\ -1 & 0 \end{bmatrix} = -R_{90^\circ}.$$

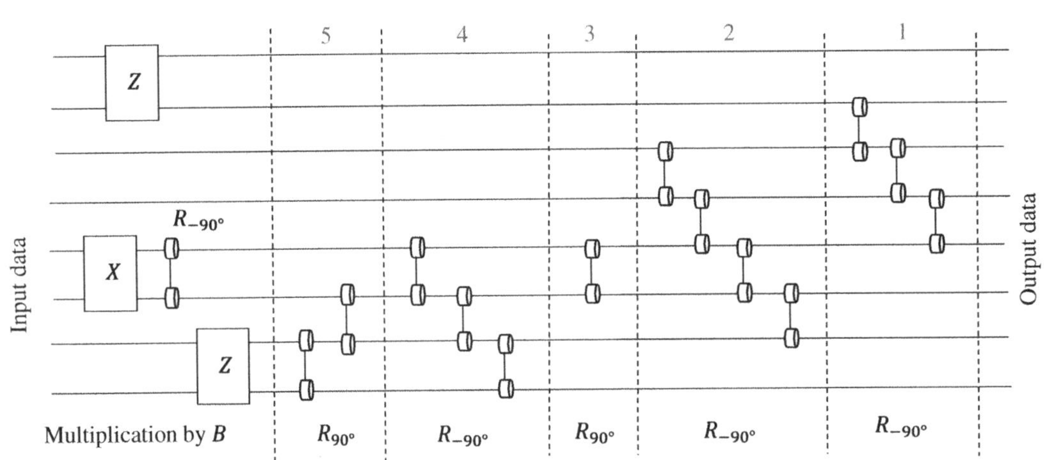

Fig. 5.35 The traditional block scheme of the permutation P_X.

Thus,

$$P_X = B\cdot(R_{90°;0,1}R_{90°;1,2})\cdot(R_{-90°;2,3}R_{-90°;1,2}R_{-90°;0,1})\cdot R_{90°;2,3}$$
$$\cdot(R_{-90°;4,5}R_{-90°;3,4}R_{-90°;2,3}R_{-90°;1,2})\cdot(R_{-90°;5,6}R_{-90°;4,5}R_{-90°;3,4}).$$
$$B = Z\oplus I_2\oplus(R_{-90°}X)\oplus Z.$$

This example illustrates well how complex the scheme for implementing permutation in quantum circuits is.

5.6 Method of 1D Quantum Convolution for Phase Filters

In this section, the method of implementation of the classical diagram $\{f_n\}\to\{F_p\}\to\{Y_p = H_pF_p\}\to\{y_n\}$ for the circular convolution in the frequency domain is presented [19]. The general case is considered; the signal f_n is complex or real. Here, F_p and H_p are the DFTs of the input signal f_n and the impulse characteristic h_n of a linear time-invariant system or filter, respectively. The length $N = 2^r$, $r > 1$. The convolution (circular)

$$y_n = f_n\circledast h_n = (f\circledast h)_n = \sum_{k=0}^{N-1} f_k h_{(k-n)\bmod N},\qquad n = 0{:}\ (N-1). \tag{5.40}$$

The indices are considered by modulo N.

First, we assume that both signal f_n and the characteristic H_p have unit energy, that is,

$$\sum_{n=0}^{N-1}|f_n|^2 = \sum_{p=0}^{N-1}|H_p|^2 = 1.$$

The pair of direct and inverse N-point DFTs, $\mathcal{F}_N$ and $\mathcal{F}_N^{-1}$, are considered in a traditional in the DSP form

$$F_p = \frac{1}{\sqrt{N}}\sum_{n=0}^{N-1} f_n W^{np},\quad f_n = \frac{1}{\sqrt{N}}\sum_{p=0}^{N-1} F_p W^{-np},\quad n,p = 0{:}\ (N-1), \tag{5.41}$$

where $W = \exp(-i2\pi/N)$. We also assume that the characteristic of the filter $H_p \neq 0$, for all frequency-points $p = 0{:}$ $(N-1)$. Otherwise, if this frequency characteristic has zeros, for instance, at frequency-points 1 and 2, then one additional qubit can be used to compose the superposition with vector $\left\{\underbrace{F_0, F_1, F_2, F_3, \ldots, F_{N-1}}, \underbrace{0, 0, 0, 0, \ldots, 0}\right\}$. Then, the permutation can be used to move these components to the second part as $\left\{\underbrace{F_0, 0, 0, F_3, \ldots, F_{N-1}}, \underbrace{0, F_1, F_2, 0, \ldots, 0}\right\}$. Subsequent processing will only be on the first part of this vector.

The calculation of the circular convolution $y_n = f_n\circledast h_n$ can be described by the following steps.

Step 1: Consider the following $(r+1)$ quantum mixed-type superposition of states:

$$|\varphi\rangle = \frac{1}{\sqrt{2}}\left(\sum_{n=0}^{N-1} f_n|n\rangle + \sum_{p=0}^{N-1} H_p|N+p\rangle\right) = \frac{1}{\sqrt{2}}\sum_{n=0}^{N-1} f_n|0\rangle|n\rangle + \frac{1}{\sqrt{2}}\sum_{p=0}^{N-1} H_p|1\rangle|p\rangle. \tag{5.42}$$

In the second part of this equation, the basis states $|n\rangle$ and $|p\rangle$ are from the basis $B_r = \{|0\rangle, |1\rangle, \ldots, |N-1\rangle\}$. Thus, $|\varphi\rangle = 1/\sqrt{2}[|0\rangle|\varphi(f)\rangle + |1\rangle|\Phi(H)\rangle]$, where the r-qubit superpositions

$$|\varphi(f)\rangle = \sum_{n=0}^{N-1} f_n|n\rangle \quad\text{and}\quad \Phi(H)\rangle = \sum_{p=0}^{N-1} H_p|p\rangle. \tag{5.43}$$

For ease of calculation, we will omit the normalization coefficient $1/\sqrt{2}$ from the superposition $|\varphi\rangle$, as well as such coefficients from the following superpositions. Only at the last stage of calculations, when observing/measuring the final result, will the corresponding normalizing coefficient be used.

Step 2: Perform the r-qubit QFT of only the superposition of the signal $|\varphi(f)\rangle$. For that, we can use the first qubit as a 0-control qubit. The result is the $(r+1)$-qubit superposition $|\Psi\rangle = (\mathcal{F}_N \oplus I_N)|\varphi\rangle$, or

$$|\Psi\rangle = |0\rangle|\Phi(F)\rangle + |1\rangle|\Phi(H)\rangle = |0\rangle \sum_{p=0}^{N-1} F_p|p\rangle + |1\rangle \sum_{p=0}^{N-1} H_p|p\rangle = \sum_{p=0}^{N-1} F_p|p\rangle + \sum_{p=0}^{N-1} H_p|N+p\rangle. \tag{5.44}$$

Here, $\mathcal{F}_N$ stands for the matrix of the N-point DFT. The realization of the r-qubit QFT can be accomplished by one of the quantum circuits described in Section 5, for instance, the paired transform-based algorithm.

On this stage, we obtain the state of the $2N$-D vector (the unit vector if considering the coefficient $1/\sqrt{2}$)

$$\boldsymbol{A} = (F_0, F_1, F_2, \ldots, F_{N-2}, F_{N-1}, H_0, H_1, H_2, \ldots, H_{N-2}, H_{N-1})'. \tag{5.45}$$

Step 3: Process each state $|\phi_p\rangle = |0\rangle F_p + |1\rangle H_p$ by the corresponding diagonal matrix

$$\boldsymbol{V}_p = \begin{bmatrix} H_p & 0 \\ 0 & H_p^{-1} \end{bmatrix}, \qquad \det(V_P) = 1. \tag{5.46}$$

Applying this matrix on $|\phi_p\rangle$, we obtain the superposition

$$\boldsymbol{V}_p\left|\phi_p\right\rangle = \begin{bmatrix} H_p & 0 \\ 0 & H_p^{-1} \end{bmatrix}\begin{bmatrix} F_p \\ H_p \end{bmatrix} = \begin{bmatrix} H_pF_p \\ 1 \end{bmatrix} = H_pF_p\begin{bmatrix} 1 \\ 0 \end{bmatrix} + \begin{bmatrix} 0 \\ 1 \end{bmatrix} = H_pF_p|0\rangle + |1\rangle.$$

Note that in order to use $\boldsymbol{V}_p$ matrices in quantum computation, they should be unitary. Therefore, it is considered that the frequency characteristic of the filter has values only on the unit circle. In other words, $H_p = e^{i\theta_p}$, where angles $\theta_p \in [0, 2\pi)$. For instance, we can consider the linear phase filters with unit amplitude multiplier, $H_p = e^{ip\tau/2}$, where τ is an integer number and $\tau/2$ is the group delay [20].

Since, the components F_p and H_p are located far from each other at "a distance" of N in $\boldsymbol{A}$, we will consider the permutation

$$\boldsymbol{B} = P\boldsymbol{A} = \left(\underbrace{F_0, H_0,}\underbrace{F_1, H_1,}\underbrace{F_2, H_2,}\ldots,\underbrace{F_{N-2}H_{N-2},}\underbrace{F_{N-1}, H_{N-1}}\right)' \tag{5.47}$$

instead of this vector $\boldsymbol{A}$. The new superposition is $|\Psi_B\rangle = P|\Psi\rangle$,

$$|\Psi_B\rangle = \sum_{p=0}^{N-1} (F_p|2p\rangle + H_p|2p+1\rangle). \tag{5.48}$$

For the $N = 8$ case, this permutation is the following:

$$P = \begin{pmatrix} 0 & 1 & 2 & 3 & 4 & 5 & 6 & 7 & 8 & 9 & 10 & 11 & 12 & 13 & 14 & 15 \\ 0 & 2 & 4 & 6 & 8 & 10 & 12 & 14 & 1 & 3 & 5 & 7 & 9 & 11 & 13 & 15 \end{pmatrix},$$

or $P = (1, 2, 4, 8)(3, 6, 12, 9)(5, 10)(7, 14, 13, 11)$. In the general case, the permutation can be described as

$$P: x \to 2x \bmod(2N-1), \quad \text{for}\, x = 0: (2N-2), \quad \text{and} \quad P(2N-1) = 2N-1. \tag{5.49}$$

Then, the following Kronecker sum of the diagonal matrices is considered:

$$\boldsymbol{V} = \boldsymbol{V}_0 \oplus \boldsymbol{V}_1 \oplus \boldsymbol{V}_2 + \cdots + \oplus \boldsymbol{V}_{N-2} \oplus \boldsymbol{V}_{N-1}. \tag{5.50}$$

It is the $(2N) \times (2N)$ diagonal matrix

$$\boldsymbol{V} = \begin{bmatrix} H_0 & & & & & & & & \\ & H_0^{-1} & & & & & & & \\ & & H_1 & & & & & & \\ & & & H_1^{-1} & & & & & \\ & & & & \ddots & & & & \\ & & & & & H_{N-2} & & & \\ & & & & & & H_{N-2}^{-1} & & \\ & & & & & & & H_{N-1} & \\ & & & & & & & & H_{N-1}^{-1} \end{bmatrix}, \quad \det(V) = 1.$$

The result of multiplication of this matrix on the vector of $|\Psi_B\rangle$ defines the following $(r+1)$-qubit superposition:

$$\left|\breve{\Psi}\right\rangle = \boldsymbol{V}|\Psi_B\rangle = \sum_{p=0}^{N-1}\left(F_p H_P|2p\rangle + |2p+1\rangle\right). \tag{5.51}$$

For the $r = 6$ case, the 7-qubit quantum circuit with 1 and 0-control bits for the operation $\boldsymbol{V}$ is shown in Fig. 5.36. After applying the inverse permutation P^{-1}, we obtain the superposition

$$P^{-1}\left|\breve{\Psi}\right\rangle = |0\rangle \sum_{p=0}^{N-1} H_p F_p|p\rangle + |1\rangle \sum_{p=0}^{N-1}|p\rangle. \tag{5.52}$$

For the above $N = 8$ case, this permutation is $P^{-1} = (1, 8, 4, 2)(3, 6, 9, 12)(5, 10)(7, 11, 13, 14)$.

We need to carry out all necessary calculations at the first control qubit in state of $|0\rangle$. Therefore, all we have to do is to calculate the r-qubit inverse QFT (IQFT) and then make a measurement, as shown in Fig. 5.37.

Step 4: When the first qubit is in the state of $|0\rangle$, apply the r-qubit inverse QFT of the obtained quantum superposition $P^{-1}|\breve{\Psi}\rangle$. This operation can be written as

$$|\psi\rangle = \left(\mathcal{F}_N^{-1} \oplus I_N\right)\left(P^{-1}|\breve{\Psi}\rangle\right) = \left(\mathcal{F}_N^{-1} \oplus I_N\right)\left(|0\rangle \sum_{p=0}^{N-1} H_p F_p|p\rangle + |1\rangle \sum_{p=0}^{N-1}|p\rangle\right) = |0\rangle \sum_{n=0}^{N-1} y_n|n\rangle + |1\rangle \sum_{p=0}^{N-1}|p\rangle.$$

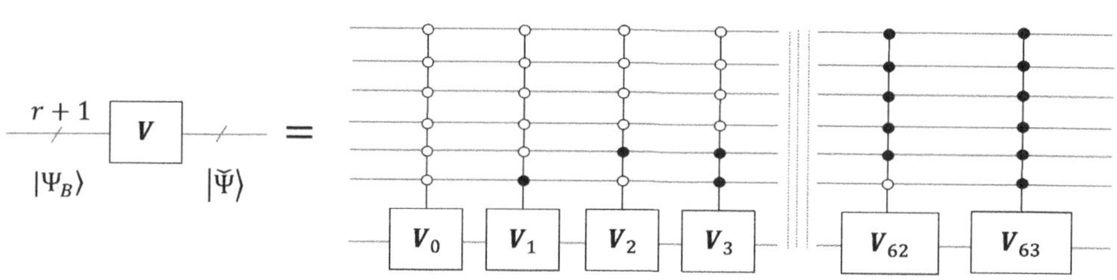

Fig. 5.36 The circuit element for the operator $\boldsymbol{V}$ with the equivalent circuit when $N = 64 = 2^6$.

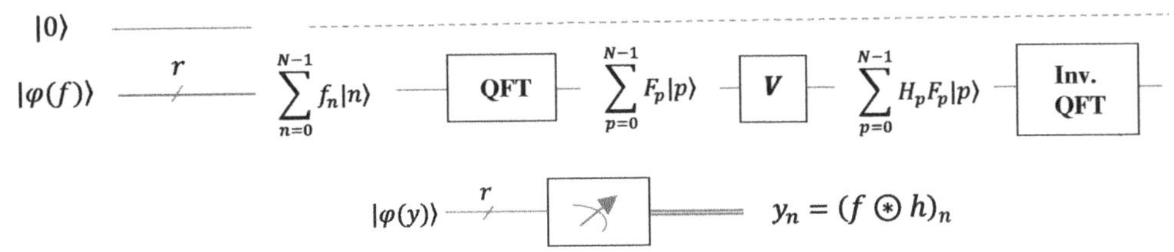

Fig. 5.37 The simplified quantum circuit with $(r+1)$ qubits for the circular convolution y_n, when the first qubit of $|\varphi\rangle$ is $|0\rangle$.

Note that on this stage, it is also possible to apply the r-qubit inverse QFT to both parts of the superposition $P^{-1}|\check{\Psi}\rangle$, to obtain the following:

$$|\psi_1\rangle = (\mathcal{F}_N^{-1} \oplus \mathcal{F}_N^{-1})\left(P^{-1}|\check{\Psi}\rangle\right) = (\mathcal{F}_N^{-1} \oplus \mathcal{F}_N^{-1})\left(|0\rangle \sum_{p=0}^{N-1} H_pF_p|p\rangle + |1\rangle \sum_{p=0}^{N-1}|p\rangle\right) = |0\rangle \sum_{n=0}^{N-1} y_n|n\rangle + |1\rangle\sqrt{N}|0\rangle^{\oplus r}.$$

Indeed, the superposition $(|0\rangle + |1\rangle + |2\rangle + \cdots + |N-1\rangle)/\sqrt{N}$ corresponds to the vector $(1, 1, 1, \ldots, 1)'$ which is the DFT (or Hadamard transform) of the unit impulse $\delta_n = (1, 0, 0, 0, \ldots, 0)'$, or the Kronecker product $|0\rangle^{\oplus r} = |000\ldots0\rangle$.

Step 5: Measure (M) the first qubit of the superposition $|\psi\rangle$. It will be $M_0 = |0\rangle$ or $M_1 = |1\rangle$. If it is in the state $|0\rangle$, then the superposition of the circular convolution is observed. Thus, we can write

$$M|\psi\rangle = (1 - M_0)|\varphi(y)\rangle + M_0\left(\frac{1}{\sqrt{N}}\sum_{p=0}^{N-1}|p\rangle\right), \tag{5.53}$$

where the superposition of the convolution (with the normalization factor K) is

$$|\varphi(y)\rangle = \frac{1}{K}\sum_{n=0}^{N-1} y_n|n\rangle, \quad K = \sqrt{\sum_{n=0}^{N-1} y_n^2}. \tag{5.54}$$

Figure 5.38 shows the quantum scheme of the circular convolution by the above algorithm.

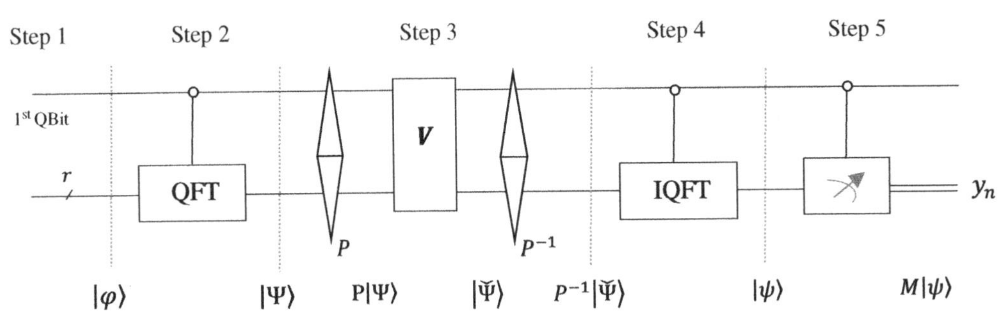

Fig. 5.38 The quantum circuit for the circular convolution y_n of the signal.

References

1 Cooley, J.W. and Tukey, J.W. (1965). An algorithm for the machine calculation of complex Fourier Series. *Mathematics of Computation* 19 (90): 297–301. https://doi.org/10.2307/2003354.

2 Ahmed, N. and Rao, K.R. (1975). *Orthogonal Transforms for Digital Signal Processing*. Berlin, Heidelberg: Springer.

3 Nussbaumer, H.J. (1982). *Fast Fourier Transform and Convolution Algorithms*. Berlin, Heidelberg: Springer.

4 Grigoryan, A.M. (1986). New algorithms for calculating discrete Fourier transforms. *USSR Computational Mathematics and Mathematical Physics* 26 (5): 84–88.

5 Grigoryan, A.M. (1988). An algorithm of computation of the one-dimensional discrete Fourier transform. *Izvestiya VUZov SSSR, Radioelectronica* 31 (5): 47–52.

6 Grigoryan, A.M. (1991). An algorithm of computation of the one-dimensional discrete Hadamard transform. *Izvestiya VUZov SSSR Radioelectronica, USSR* 34 (8): 100–103.

7 Grigoryan, A.M. (2001). 2-D and 1-D multi-paired transforms: frequency-time type wavelets. *IEEE Transactions on Signal Processing* 49 (2): 344–353.

8 Grigoryan, A.M. and Agaian, S.S. (2000). Split manageable efficient algorithm for Fourier and Hadamard transforms. *IEEE Transactions on Signal Processing* 48 (1): 172–183. https://doi.org/10.1109/78.815487.

9 Grigoryan, A.M. and Grigoryan, M.M. (2009). *Brief Notes in Advanced DSP: Fourier Analysis with MATLAB*. CRC Press, Taylor and Francis Group.

10 Grigoryan, A.M. and Agaian, S.S. (2019). Paired quantum Fourier transform with $\log_2 N$ Hadamard gates. *Quantum Information Processing* 18 (217): 26. https://doi.org/10.1007/s11128-019-2322-6.

11 Grigoryan, A.M. (2020). Resolution map in quantum computing: signal representation by periodic patterns. *Quantum Information Processing* 19 (177): 21. https://doi.org/10.1007/s11128-020-02685-7.

12 Agaian, S.S. (1985). *Hadamard Matrices and Their Applications*, Lecture Notes in Mathematics, vol. 1168. New York: Springer – Verlag.

13 Grigoryan, A.M. and Agaian, S.S. (2003). *Multidimensional Discrete Unitary Transforms: Representation, Partitioning, and Algorithms*. New York: Marcel Dekker.

14 Barenco, A., Bennett, C.H., Cleve, R. et al. (1995). Elementary gates for quantum computation. *Physical Review A* 52 (2): 3457–3467.

15 Nielsen, M. and Chuang, I. (2001). *Quantum Computation and Quantum Information*, 2e. Cambridge University Press.

16 Cleve, R. and Watrous J. (2000). Fast parallel circuits for the quantum Fourier transform. *Proceedings of IEEE 41st Annual Symposium on Foundations of Computer Science*, 526–536, Redondo Beach, CA, USA.

17 Browne, D.E. (2007). Efficient classical simulation of the semi-classical quantum Fourier transform. *New Journal of Physics* 9: 146.

18 Li, H.S., Fan, P., Xia, H. et al. (2018). The quantum Fourier transform based on quantum vision representation. *Quantum Information Processing* 17 (333): 25.

19 Grigoryan, A.M. and Agaian, S.S. (2020). 1-D Convolution circuits in quantum computation. *International Journal of Scientific & Engineering Research* 11 (8): 912–916.

20 Orfanidis, S.J. (1996). *Introduction to Signal Processing*. Upper Saddle River, NJ: Prentice-Hall, (chapter 10).

6

Quantum Signal-Induced Heap Transform

In this section, we briefly describe the quantum signal-induced heap transform (QsiHT) related to the so-called discrete signal-induced heap transform (DsiHT), which was successfully used for image enhancement, classification, compression, and filtration [1–4]. This transform is generated by any signal of arbitrary length. This transform is also used for different QR-decompositions of square matrices, real and complex [5–7]. In the case when the matrix is unitary, such decompositions are considered as matrix factorization. The DsiHT has simple algorithms, it is fast and allows for building different quantum circuits for 2-qubit multiplication with 4, 5, and 6 gates of rotation. In the following equations, letter H will be used for the DsiHT. For instance, H_8 is used for the 8-point DsiHT, not for the 8-point discrete Hadamard transforms.

The section is structured to provide a comprehensive understanding of the DsiHT and its applications. It begins with a formal definition and description of the DsiHT, followed by algorithms for initializing quantum states using this transform. Section 6.2 discusses applying DsiHT-based factorization techniques for real matrices, detailing how these methods can be implemented within quantum circuits to perform essential computations efficiently. This includes the construction of quantum circuits for discrete cosine transforms of types II and IV, and the discrete Hartley transform, showcasing the broad applicability of DsiHT in quantum signal processing. Finally, Section 6.3 explores the extension of DsiHT to complex matrices, expanding its utility and effectiveness in handling a more comprehensive range of quantum computing tasks. This chapter provides a foundation for understanding how DsiHT and QR-decompositions can be leveraged to improve quantum algorithms' performance and accuracy, paving the way for more sophisticated applications in quantum image processing and other computational domains.

6.1 Definition

The concept of the discrete heap transformation, or discrete signal-induced transform, was developed by Grigoryan in 2006 to process signals and images by transformations that are generated by one or more pre-selected signals, which are called the generators [8–10]. Here, we consider the case with one generator-signal $\boldsymbol{x} = (x_0, x_1, x_2, \ldots, x_{N-1})$, $N > 1$. The choice of generator depends on the task. The DsiHT is a two-level transform. At each stage of calculation, a basic unitary transformation, T_k, is generated and then applied to the input signal $\boldsymbol{z} = (z_0, z_1, z_2, \ldots, z_{N-1})$. For example, the transformation T_k can be applied to only two components of the processed signal, and one of them will be updated in the next calculation step. The DsiHT uses a path or order in which signal-generator and input signal components are processed. These paths can be different and therefore the DsiHT depends on the path (for detail, see [5]).

As example, Fig. 6.1 shows three different signal processing paths. In part (a), starting from the last components of the 8-point input (generator), two components are processed and the first output will be used as one of the inputs for the next transformation. The DsiHT with this path is called the strong DsiHT, or, *the DsiHT with strong 2-wheel*

Quantum Image Processing in Practice: A Mathematical Toolbox, First Edition. Artyom M. Grigoryan and Sos S. Agaian.

Companion website: www.wiley.com/go/grigoryan/quantumimageprocessing

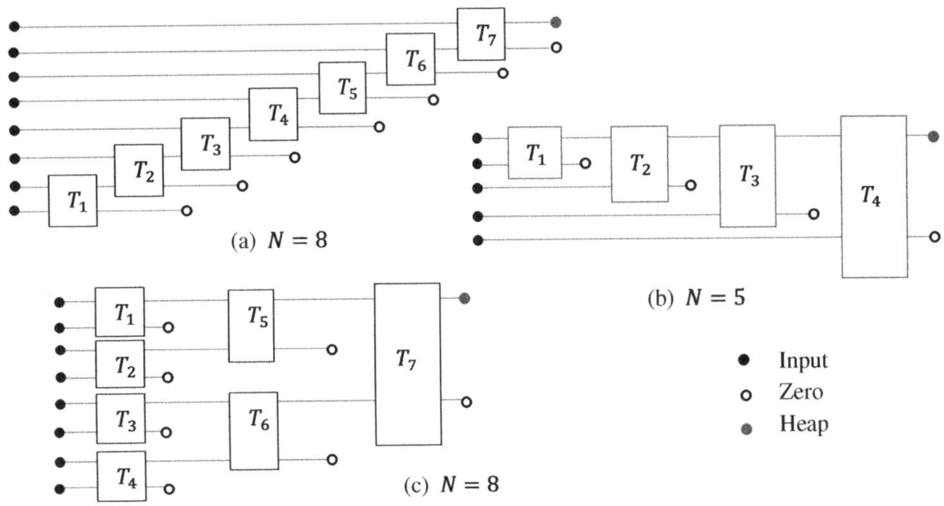

Fig. 6.1 Three different paths of the N-point DsiHT, when (a) $N = 8$, (b) $N = 5$, and (c) $N = 8$.

carriage (for more detail, see [1, 7]). In part (b), the path of the weak DsiHT, or *the DsiHT with weak 2-wheel carriage*, is illustrated for a 5-point signal. The first output of the transform is renewed at each stage of calculation. In part (c), the path is illustrated as it is used for the discrete 8-point Haar transform. The path is an important characteristic of the DsiHT.

First, we start with the real signals. The case is considered, when the whole energy of the signal generator is transferred to the first component of the transform. So, we collect in one heap the whole energy of the signal. The simple form of the basic transformation can be used, for example, the Givens rotation. Let (x_0, x_1) be the given vector and T the transformation defined by the following conditions:

$$T\begin{bmatrix} x_0 \\ x_1 \end{bmatrix} = \begin{bmatrix} x_0' \\ 0 \end{bmatrix} \text{ and } |x_0'|^2 = x_0^2 + x_1^2. \tag{6.1}$$

With the Givens rotation, the matrix of this transform is calculated by

$$T = T_\varphi = W(\varphi) = \begin{bmatrix} \cos\varphi & -\sin\varphi \\ \sin\varphi & \cos\varphi \end{bmatrix}, \varphi = -\arctan\left(\frac{x_1}{x_0}\right). \tag{6.2}$$

If $x_0 = 0$, the angle of rotation is considered $\varphi = -\pi/2$, or $\pi/2$. Thus, the angle of rotation is defined from the condition $x_0 \sin\varphi + x_1 \cos\varphi = 0$ and then applied to calculate the first output x_0'.

The N-point DsiHT uses $(N-1)$ basic transformations $T_{\varphi_k}, k = 1: N-1$, regardless of the path of the transform. The transform of the signal-generator equals

$$H: \boldsymbol{x} = (x_0, x_1, x_2, \ldots, x_{N-1}) \to (y_0, 0, 0, \ldots, 0), \tag{6.3}$$

where y_0 is equal to plus/minus energy $E[\boldsymbol{x}] = \|\boldsymbol{x}\|$ of the signal. Since $y_0 = \pm\sqrt{x_0^2 + x_1^2 + \ldots + x_{N-1}^2}$, we always can consider the sign plus for y_0. Indeed, the DsiHT is composed by basis transforms of type T given in Eq. 6.1. We may consider that $|x_0'| = +\sqrt{x_0^2 + x_1^2}$, since $x_0' = x_0 \cos\varphi - x_1 \sin\varphi$ and if $x_0' < 0$, then the angle of rotation φ can be changed by $(\varphi + \pi)$. Therefore, on each stage of DsiHT calculation, the angle φ_k can be changed by $(\varphi_k + \pi)$, if necessary.

Thus, the transformation as well as the signal-generator are uniquely determined by the data of $E[\boldsymbol{x}]$ and angles of rotation. In other words, the transformation is decoded into the vector $(\|\boldsymbol{x}\|, \varphi_1, \varphi_2, \ldots, \varphi_{N-1})$. For the signal-generator with norm 1, the angular representation holds, $\boldsymbol{x} \to A_{\boldsymbol{x}} = \{\varphi_1, \varphi_2, \ldots, \varphi_{N-1}\}$.

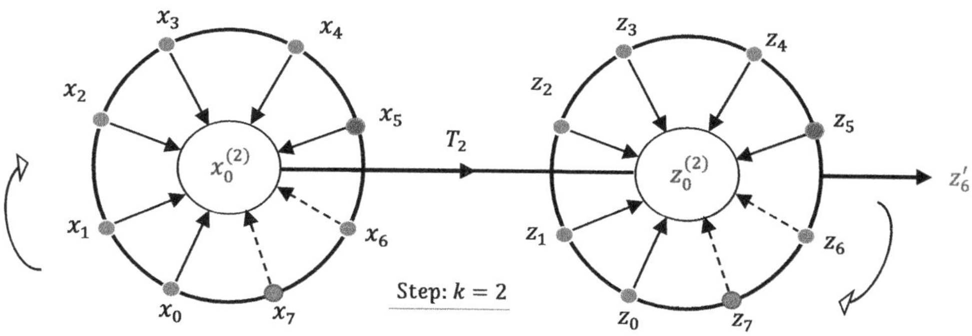

Fig. 6.2 The two-wheel carriage of the 8-point strong DsiHT.

We now describe the DsiHT with a strong two-wheel carriage, which we consider the simplest in constructing a quantum circuit for this transformation. Such a carriage is illustrated in Fig. 6.2, for the $N = 8$ case. The first wheel rotates the signal generator data, and the input signal data rotates synchronously on the second wheel. In the beginning, in the center of each wheel, the last components of the signals are placed. Then, while the wheel makes one revolution, the components of the signal-generator are processed sequentially in order x_{N-1} with x_{N-2}, then x_{N-3}, ..., x_1, x_0. The components of the input signal z are processed in the same way.

6.1.1 The Algorithm of the Strong DsiHT

Step $k = 1$: $\left(x_0^{(0)} = x_{N-1}, z_0^{(0)} = z_{N-1}\right)$.

- The pair of components of the signal-generator is x_{N-2} and x_{N-1}. Calculate the first angle $\varphi_1 = -\arctan(x_{N-1}/x_{N-2})$. Calculate the first output of the transform

$$T_{\varphi_1}: (x_{N-2}, x_{N-1}) \to \left(x_0^{(1)}, 0\right).$$

- Process the components of the input signal,

$$T_{\varphi_1}: (z_{N-2}, z_{N-1}) \to \left(z_0^{(1)}, z'_{N-1}\right).$$

Step $k = 2$:

- The pair of components is x_{N-3} and $x_0^{(1)}$. Calculate the angle $\varphi_2 = -\arctan\left(x_0^{(1)}/x_{N-3}\right)$. Calculate the first output of the transform

$$T_{\varphi_2}: \left(x_{N-3}, x_0^{(1)}\right) \to \left(x_0^{(2)}, 0\right).$$

- Process the components of the input signal by the transform

$$T_{\varphi_2}: \left(z_{N-3}, z_0^{(1)}\right) \to \left(z_0^{(2)}, z'_{N-2}\right).$$

Steps $k = 3, 4, ..., (N-1)$:

- The pair of components is $x_{N-(k+1)}$ and $x_0^{(k-1)}$. Calculate the angle $\varphi_k = -\arctan\left(x_0^{(k-1)}/x_{N-(k+1)}\right)$. Calculate the first output of the transform

$$T_{\varphi_k}: \left(x_{N-(k+1)}, x_0^{(k-1)}\right) \to \left(x_0^{(k)}, 0\right).$$

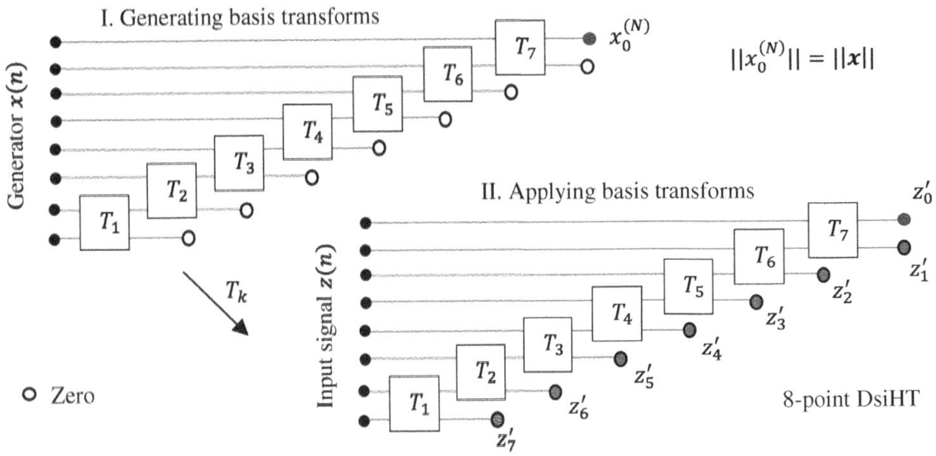

Fig. 6.3 The 8-point strong DsiHT generated and then applied to the signal z_n, $n = 0:7$.

- Process the components of the input signal,

$$T_{\varphi_1}: \left(z_{N-(k+1)}, z_0^{(k-1)}\right) \rightarrow \left(z_0^{(k)}, z'_{N-k}\right).$$

The output is the transform

$$T(\mathbf{z}) = \left(z'_0, z'_1, z'_2, \ldots, z'_{N-2}, z'_{N-1}\right), z'_0 = z_0^{(N-1)}. \tag{6.4}$$

This algorithm is illustrated in Fig. 6.3, for the $N = 8$ case.

Note that the above algorithm can be split by two parts. In the first part, all angles of rotation can be calculated from the signal generator. This is about the angular representation of the generator, $A(\mathbf{x}) = \{\varphi_1, \varphi_2, \ldots, \varphi_{N-1}\}$. In the second part, the DsiHT of the input signal is calculated, $T(\mathbf{z})$.

The DsiHT is a composite transform, H_N. It can be written in the matrix form as

$$H_N = R_{\varphi_{N-1};(0,1)} R_{\varphi_{N-2};(1,2)} \cdots R_{\varphi_2;(N-3,N-2)} R_{\varphi_1;(N-2,N-1)}. \tag{6.5}$$

Here, $R_{\varphi_k;(i,j)}$, when $j = i + 1$, denotes the unit diagonal matrix with the matrix of rotation R_{φ_k} in the cell $(i, j) \times (i, j)$, that is,

$$R_{\varphi_k;(i,j)} = I_i \oplus R_{\varphi_k} \oplus I_{N-i-2} \text{ and } R_{\varphi_k;(0,1)} = R_{\varphi_k} \oplus I_{N-2}. \tag{6.6}$$

We can say $R_{\varphi_k;(i,j)}$ represents the rotation on planes i and j. For instance, in the $N = 4$ case,

$$R_{\varphi_1;(1,2)} = \begin{bmatrix} 1 & 0 & 0 & 0 \\ 0 & \cos\varphi_1 & -\sin\varphi_1 & 0 \\ 0 & \sin\varphi_1 & \cos\varphi_1 & 0 \\ 0 & 0 & 0 & 1 \end{bmatrix} = 1 \oplus R_{\varphi_1} \oplus 1.$$

The 4-point strong DsiHT can be written in the matrix form as

$$H_4 = R_{\varphi_3;(0,1)} R_{\varphi_2;(1,2)} R_{\varphi_1;(2,3)} = \left(R_{\varphi_3} \oplus I_2\right)\left(1 \oplus R_{\varphi_2} \oplus 1\right)\left(I_2 \oplus R_{\varphi_1}\right). \tag{6.7}$$

It should be noted that, as shown in the example above, when calculating the strong DsiHT, only matrices of the type $R_{\varphi_k;(i,i+1)}$ are used, that is, when $j = i + 1$. This is about rotations on adjacent planes. If we use other paths in the

DsiHT, the coefficients of the rotation matrices R_{φ_k} in $R_{\varphi_k;(i,j)}$ will be separated by $(j-i-1)$ columns and rows, when $j > i+1$. It is rotation on planes i and j, which are separated from each other. Such matrices cannot be written by the Kronecker sum of matrices, as in Eq. 6.6. For instance, when using the 4-point DsiHT with the weak carriage, on the second and third computation steps, the following matrices of rotation are used:

$$R_{\varphi_2;(0,2)} = \begin{bmatrix} \cos\varphi_2 & 0 & -\sin\varphi_2 & 0 \\ 0 & 1 & 0 & 0 \\ \sin\varphi_2 & 0 & \cos\varphi_2 & 0 \\ 0 & 0 & 0 & 1 \end{bmatrix} \text{ and } R_{\varphi_3;(0,3)} = \begin{bmatrix} \cos\varphi_3 & 0 & 0 & -\sin\varphi_3 \\ 0 & 1 & 0 & 0 \\ 0 & 0 & 1 & 0 \\ \sin\varphi_3 & 0 & 0 & \cos\varphi_3 \end{bmatrix}.$$

Example 6.1 Vector-generator (1, – 1, 1, 1)

Consider the vector $q = (1, -1, 1, 1)$. The DsiHT with this vector-generator has the matrix

$$H_4 = H_{4;s} = \frac{1}{2}\begin{bmatrix} 1 & -1 & 1 & 1 \\ 1.7321 & 0.5774 & -0.5774 & -0.5774 \\ 0 & 1.6330 & 0.8165 & 0.8165 \\ 0 & 0 & -1.4142 & 1.4142 \end{bmatrix} = \frac{1}{2}D\begin{bmatrix} 1 & -1 & 1 & 1 \\ 3 & 1 & -1 & -1 \\ 0 & 2 & 1 & 1 \\ 0 & 0 & -1 & 1 \end{bmatrix}, \tag{6.8}$$

if we use the strong wheel-carriage. Here, the diagonal matrix $D = \{1, 0.5774, 0.8165, 1.4142\}$. The angles for this transform are $\{\varphi_k;\ k = 1:3\} = \{-45°, 54.7356°, 60°\}$. As follows from Eq. 6.7, the transform is described by the following composition of three rotations:

$$H_4 = \underbrace{\begin{bmatrix} \cos\varphi_3 & -\sin\varphi_3 & 0 & 0 \\ \sin\varphi_3 & \cos\varphi_3 & 0 & 0 \\ 0 & 0 & 1 & 0 \\ 0 & 0 & 0 & 1 \end{bmatrix}\begin{bmatrix} 1 & 0 & 0 & 0 \\ 0 & \cos\varphi_2 & -\sin\varphi_2 & 0 \\ 0 & \sin\varphi_2 & \cos\varphi_2 & 0 \\ 0 & 0 & 0 & 1 \end{bmatrix}\begin{bmatrix} 1 & 0 & 0 & 0 \\ 0 & 1 & 0 & 0 \\ 0 & 0 & \cos\varphi_1 & -\sin\varphi_1 \\ 0 & 0 & \sin\varphi_1 & \cos\varphi_1 \end{bmatrix}}_{\varphi_1 = -45°,\ \varphi_2 = 54.74°,\ \varphi_3 = 60°}.$$

With the permutations $P_2 = (0, 3, 2, 1)$ and $P_2' = (0, 1, 2, 3)$, the second matrix can be written as $R_{\varphi_2;(1,2)} = P_2 (I_2 \oplus R_{\varphi_2})P_2'$, that is,

$$R_{\varphi_2;(1,2)} = \begin{bmatrix} 1 & 0 & 0 & 0 \\ 0 & \cos\varphi_2 & -\sin\varphi_2 & 0 \\ 0 & \sin\varphi_2 & \cos\varphi_2 & 0 \\ 0 & 0 & 0 & 1 \end{bmatrix} = \begin{bmatrix} 0 & 1 & 0 & 0 \\ 0 & 0 & 1 & 0 \\ 0 & 0 & 0 & 1 \\ 1 & 0 & 0 & 0 \end{bmatrix}\begin{bmatrix} 1 & 0 & 0 & 0 \\ 0 & 1 & 0 & 0 \\ 0 & 0 & \cos\varphi_2 & -\sin\varphi_2 \\ 0 & 0 & \sin\varphi_2 & \cos\varphi_2 \end{bmatrix}\begin{bmatrix} 0 & 0 & 0 & 1 \\ 1 & 0 & 0 & 0 \\ 0 & 1 & 0 & 0 \\ 0 & 0 & 1 & 0 \end{bmatrix}.$$

Therefore, the matrix of the 4-point DsiHT can be described by controlled gates as

$$H_4 = \left(R_{\varphi_3} \oplus I_2\right)\underbrace{P_2\left(I_2 \oplus R_{\varphi_2}\right)P_2'}\left(I_2 \oplus R_{\varphi_1}\right). \tag{6.9}$$

The quantum scheme of the 4-point DsiHT, or the 2-qubit quantum signal-induced heap transform, is given in Fig. 6.4.

Thus, we have the quantum scheme for the 2-qubit gate H_4, and $H_4q' = H_4(1, -1, 1, 1)' = (2, 0, 0, 0)' = 2(1, 0, 0, 0)'$. If we normalize the generator, $q = q/2$, then $H_4q' = (1, 0, 0, 0)'$. For the inverse transform, $H_4'(1, 0, 0, 0)' = q'$. The inverse matrix H_4^{-1} is the transpose to H_4, that is, $H_4^{-1} = H_4'$. Consider the 2-qubit superposition

$$H_s = (R_{\varphi_3} \oplus I_2)\, \underbrace{P_2(I_2 \oplus R_{\varphi_2})P_2'}\, (I_2 \oplus R_{\varphi_1})$$

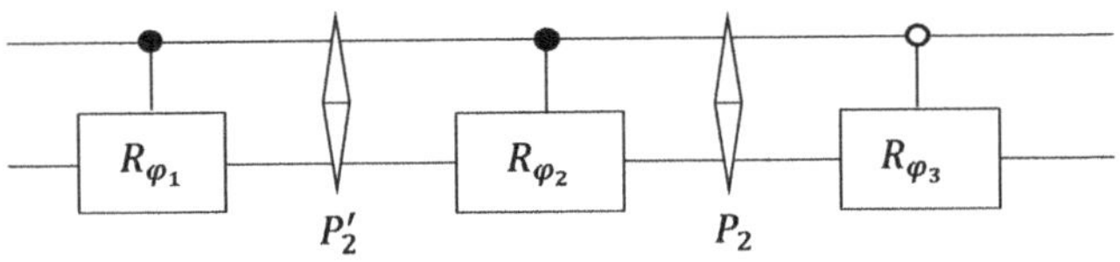

Fig. 6.4 The quantum scheme of the strong wheel-carriage 2-qubit QsiHT.

$$|\varphi(q)\rangle = (I_2 \oplus R_{-\varphi_1})\, \underbrace{P_2(I_2 \oplus R_{-\varphi_2})P_2'}\, (R_{-\varphi_3} \oplus I_2)|0\rangle^{\otimes 2}$$

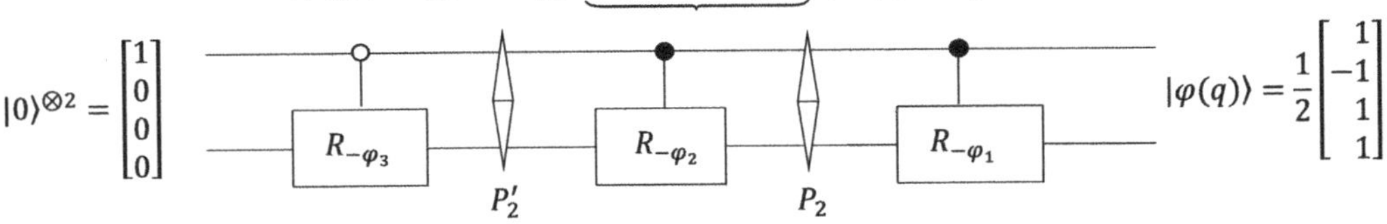

Fig. 6.5 The quantum scheme for preparation the superposition $|\varphi(q)\rangle$ by the 2-qubit inverse QsiHT.

$|\varphi(q)\rangle = (|00\rangle - |01\rangle + |10\rangle + |11\rangle)/2$. Then, to prepare this superposition from the first basis state as $H_4'|00\rangle = |\varphi(q)\rangle$, the quantum circuit in Fig. 6.5 can be used. The transpose matrix is calculated by rotations as follows:

$$H_4' = \left[(R_{\varphi_3} \oplus I_2)\underbrace{P_2(I_2 \oplus R_{\varphi_2})P_2'}(I_2 \oplus R_{\varphi_1})\right]' = (I_2 \oplus R_{\varphi_1})'\left(\underbrace{P_2(I_2 \oplus R_{\varphi_2})P_2'}\right)'(R_{\varphi_3} \oplus I_2)'$$
$$= (I_2 \oplus R_{-\varphi_1})\left(\underbrace{P_2(I_2 \oplus R_{-\varphi_2})P_2'}\right)(R_{-\varphi_3} \oplus I_2).$$

When using the weak wheel-carriage DsiHT by the same generator, the transform matrix equals to

$$H_{4;w} = \frac{1}{2}\begin{bmatrix} 1 & -1 & 1 & 1 \\ 1.4142 & 1.4142 & 0 & 0 \\ -0.8165 & 0.8165 & 1.6330 & 0 \\ -0.5774 & 0.5774 & -0.5774 & 1.7321 \end{bmatrix} = \frac{1}{2}D_1\begin{bmatrix} 1 & -1 & 1 & 1 \\ 1 & 1 & 0 & 0 \\ -1 & 1 & 2 & 0 \\ 1 & -1 & 1 & 3 \end{bmatrix}, \tag{6.10}$$

Here, the diagonal matrix $D_1 = \{1, 1.4142, 0.8165, -0.5774\}$ and the angles of rotations are $\{\varphi_k;\, k = 1:3\} = \{45°, -35.2644°, -30°\}$. Note that the angles change as $\varphi_1 \to \varphi_1 + 90°$ and $\varphi_k \to \varphi_k - 90°$, $k = 2, 3$ The matrix D_1 is the permutation $P_{(1,3)}$ of the matrix D of the DsiHT with the strong wheel-carriage in Eq. 6.8. The composition of the weak wheel-carriage DsiHT is described by the following three matrices of rotations:

$$H_{4;w} = R_{\varphi_3;(0,3)}R_{\varphi_2;(0,2)}R_{\varphi_1;(0,1)}$$
$$= \underbrace{\begin{bmatrix} \cos\varphi_3 & 0 & 0 & -\sin\varphi_3 \\ 0 & 1 & 0 & 0 \\ 0 & 0 & 1 & 0 \\ \sin\varphi_3 & 0 & 0 & \cos\varphi_3 \end{bmatrix}\begin{bmatrix} \cos\varphi_2 & 0 & -\sin\varphi_2 & 0 \\ 0 & 1 & 0 & 0 \\ \sin\varphi_2 & 0 & \cos\varphi_2 & 0 \\ 0 & 0 & 0 & 1 \end{bmatrix}\begin{bmatrix} \cos\varphi_1 & -\sin\varphi_1 & 0 & 0 \\ \sin\varphi_1 & \cos\varphi_1 & 0 & 0 \\ 0 & 0 & 1 & 0 \\ 0 & 0 & 0 & 1 \end{bmatrix}}_{\varphi_1 = 45°,\, \varphi_2 = -35.26°,\, \varphi_3 = -30°}.$$

The first two matrices of rotations can be presented by the 0-controlled gates as

$$R_{\varphi_2;(0,2)} = \begin{bmatrix} 1 & 0 & 0 & 0 \\ 0 & 0 & 1 & 0 \\ 0 & 1 & 0 & 0 \\ 0 & 0 & 0 & 1 \end{bmatrix} \begin{bmatrix} \cos\varphi_2 & -\sin\varphi_2 & 0 & 0 \\ \sin\varphi_2 & \cos\varphi_2 & 0 & 0 \\ 0 & 0 & 1 & 0 \\ 0 & 0 & 0 & 1 \end{bmatrix} \begin{bmatrix} 1 & 0 & 0 & 0 \\ 0 & 0 & 1 & 0 \\ 0 & 1 & 0 & 0 \\ 0 & 0 & 0 & 1 \end{bmatrix},$$

$$R_{\varphi_3;(0,3)} = \begin{bmatrix} 1 & 0 & 0 & 0 \\ 0 & 0 & 1 & 0 \\ 0 & 0 & 0 & 1 \\ 0 & 1 & 0 & 0 \end{bmatrix} \begin{bmatrix} \cos\varphi_3 & -\sin\varphi_3 & 0 & 0 \\ \sin\varphi_3 & \cos\varphi_3 & 0 & 0 \\ 0 & 0 & 1 & 0 \\ 0 & 0 & 0 & 1 \end{bmatrix} \begin{bmatrix} 1 & 0 & 0 & 0 \\ 0 & 0 & 0 & 1 \\ 0 & 0 & 1 & 0 \\ 0 & 1 & 0 & 0 \end{bmatrix}.$$

Here, the matrices of permutations $P_{(1,2)} = (1, 2)$, $P_{(1,3,2)} = (1, 3, 2)$, and $P_{(1,3)} = (1, 3)$ are used. Thus,

$$R_{\varphi_2;(0,2)} = P_{(1,2)} R_{\varphi_2;(0,1)} P_{(1,2)} \text{ and } R_{\varphi_3;(0,3)} = P_{(1,3,2)} R_{\varphi_3;(0,1)} P_{(1,3)}.$$

The matrix of the 4-point DsiHT can be written as

$$H_{4;w} = \underbrace{P_{(1,3,2)} R_{\varphi_3;(0,1)} P_{(1,3)}}\, \underbrace{P_{(1,2)} R_{\varphi_2;(0,1)} P_{(1,2)}}\, \underbrace{R_{\varphi_1;(0,1)}}. \tag{6.11}$$

The product of permutations

$$P_{(1,3)} P_{(1,2)} = \begin{bmatrix} 1 & 0 & 0 & 0 \\ 0 & 0 & 0 & 1 \\ 0 & 0 & 1 & 0 \\ 0 & 1 & 0 & 0 \end{bmatrix} \begin{bmatrix} 1 & 0 & 0 & 0 \\ 0 & 0 & 1 & 0 \\ 0 & 1 & 0 & 0 \\ 0 & 0 & 0 & 1 \end{bmatrix} = \begin{bmatrix} 1 & 0 & 0 & 0 \\ 0 & 0 & 0 & 1 \\ 0 & 1 & 0 & 0 \\ 0 & 0 & 1 & 0 \end{bmatrix} = P_{(1,2,3)}.$$

Therefore, we can write the matrix of the DsiHT by three rotations and three permutations,

$$H_{4;w} = P_{(1,3,2)} R_{\varphi_3;(0,1)} P_{(1,2,3)} R_{\varphi_2;(0,1)} P_{(1,2)}\, R_{\varphi_1;(0,1)} = P_{(1,3,2)} \left(R_{\varphi_3} \oplus I_2\right) P_{(1,2,3)} \left(R_{\varphi_2} \oplus I_2\right) P_{(1,2)} \left(R_{\varphi_1} \oplus I_2\right).$$

The quantum scheme of the 2-qubit QsiHT with the weak wheel-carriage is given in Fig. 6.6 in part (b). The block scheme of this transform with three 0-controlled gates is shown in part (a).

For both transforms, $H_{4;\,s}q' = H_{4;\,w}q' = (2, 0, 0, 0)'$ and $\|q\| = 2$. One can note that the vector-generator $q = (1, -1, 1, 1)$ lies in the first row of these matrices $H_{4,s}$ and $H_{4,w}$, as the first basis function. This is also true in the general case of the N-point DsiHT. If normalize the vector, $q = q/\|q\|$, then $H_4 q' = (1, 0, 0, 0)'$. The matrix

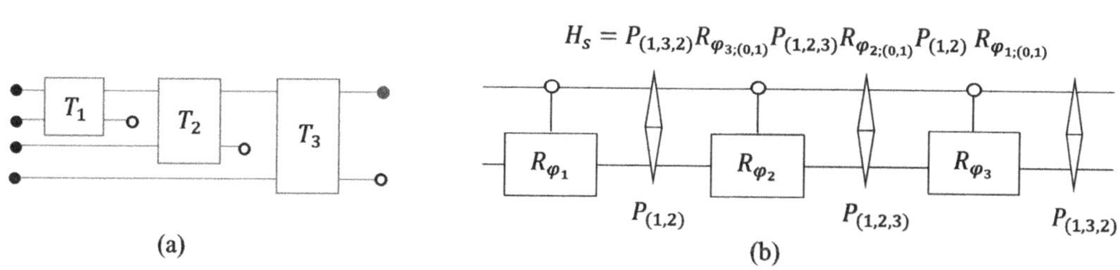

Fig. 6.6 (a) The block-scheme of the 4-point weak DsiHT and (b) the quantum scheme of the 2-qubit weak QsiHT.

of the DsiHT is unitary, $H_4^{-1} = H_4'$, and therefore, $q' = H_4'(1,0,0,0)'$. Considering that the unit vector q represents the 2-qubit superposition $|\varphi\rangle = a\,|\,0\rangle + b\,|\,1\rangle + c\,|\,2\rangle + d\,|\,3\rangle$, the DsiHT maps this superposition to the first computational basis state,

$$H_4\,|\,\varphi\rangle = |\,0\rangle, \text{or } H_4\begin{bmatrix} a\\ b\\ c\\ d\end{bmatrix} = \begin{bmatrix} 1\\ 0\\ 0\\ 0\end{bmatrix}, \text{and } H_4'\begin{bmatrix} 1\\ 0\\ 0\\ 0\end{bmatrix} = \begin{bmatrix} a\\ b\\ c\\ d\end{bmatrix}. \tag{6.12}$$

For the week DsiHT, the matrix of the inverse transform can be written as

$$\begin{aligned} H'_{4;w} &= \left(R_{\varphi_3;(0,1)}R_{\varphi_2;(1,2)}R_{\varphi_1;(2,3)}\right)' = R'_{\varphi_1;(2,3)}R'_{\varphi_2;(1,2)}R'_{\varphi_3;(0,1)} = R_{-\varphi_1;(2,3)}R_{-\varphi_2;(1,2)}R_{-\varphi_3;(0,1)} \\ &= \left(I_2 \oplus R_{-\varphi_1}\right)\left(1 \oplus R_{-\varphi_2} \oplus 1\right)\left(R_{-\varphi_3} \oplus I_2\right), \end{aligned}$$

or it is

$$H'_{4;w} = \begin{bmatrix} 1 & 0 & 0 & 0\\ 0 & 1 & 0 & 0\\ 0 & 0 & \cos\varphi_1 & \sin\varphi_1\\ 0 & 0 & -\sin\varphi_1 & \cos\varphi_1 \end{bmatrix}\begin{bmatrix} 1 & 0 & 0 & 0\\ 0 & \cos\varphi_2 & \sin\varphi_2 & 0\\ 0 & -\sin\varphi_2 & \cos\varphi_2 & 0\\ 0 & 0 & 0 & 1 \end{bmatrix}\begin{bmatrix} \cos\varphi_3 & \sin\varphi_3 & 0 & 0\\ -\sin\varphi_3 & \cos\varphi_3 & 0 & 0\\ 0 & 0 & 1 & 0\\ 0 & 0 & 0 & 1 \end{bmatrix}.$$

Thus, $|\,\varphi\rangle = H_4'\,|\,0\rangle$, which means that any 2-qubit superposition can be obtained from the zero state $|00\rangle$ by the unitary heap transforms. This is true also for any r-qubit superposition $|\varphi\rangle$, where $r > 1$.

6.1.2 Initialization of the Quantum State by the DsiHT

Many quantum circuits start to work with the input superposition being the zero state, $|000...0\rangle = |0\rangle^{\otimes r}$, when $r \geq 1$. Consider a quantum superposition

$$|\,\varphi\rangle = \sum_{n=0}^{2^r-1} f_n|\,n\rangle, |f_0|^2 + |f_1|^2 + ... + |f_{2^r-1}|^2 = 1,$$

with real or complex amplitudes f_n. For example, let the amplitudes be real and f be the 2^r-D vector $(f_0, f_1, ..., f_{2^r-1})'$. If H is the strong DsiHT generated by this vector, then

$$Hf = H(f_0, f_1, ..., f_{2^r-1})' = (1, 0, ..., 0)' = |\,0\rangle^{\otimes r}. \tag{6.13}$$

The DsiHT is the unitary transform. In the considered case, it is the real transform. Therefore, the inverse DsiHT is described by the matrix transpose to the matrix of the DsiHT. For the inverse DsiHT,

$$H^{-1}(|\,0\rangle^{\otimes r}) = H'(1, 0, ..., 0)' = (f_0, f_1, ..., f_{2^r-1})' = f.$$

Example 6.2 Vector-generator (1, − 1, 1, 1, − 1, 1, 1, 1)
Consider the 8D unit vector $q = (1, -1, 1, 1, -1, 1, 1, 1)'/\sqrt{8}$.

A) For the weak DsiHT with this generator, Eq. 6.13 is written as

$$H_8 q = \left(R_{\psi_7;(0,7)}R_{\psi_6;(0,6)}R_{\psi_5;(0,5)}R_{\psi_4;(0,4)}R_{\psi_3;(0,3)}R_{\psi_2;(0,2)}R_{\psi_1;(0,1)}\right)q$$

which is equal to

$$H_8 q = D\underbrace{\begin{bmatrix} 1 & -1 & 1 & 1 & -1 & 1 & 1 & 1 \\ 1 & 1 & 0 & 0 & 0 & 0 & 0 & 0 \\ 1 & -1 & -2 & 0 & 0 & 0 & 0 & 0 \\ 1 & -1 & 1 & -3 & 0 & 0 & 0 & 0 \\ 1 & -1 & 1 & 1 & 4 & 0 & 0 & 0 \\ 1 & -1 & 1 & 1 & -1 & -5 & 0 & 0 \\ 1 & -1 & 1 & 1 & -1 & 1 & -6 & 0 \\ 1 & -1 & 1 & 1 & -1 & 1 & 1 & -7 \end{bmatrix}}_{8\times 8 \text{ matrix } H_8} \times \frac{1}{\sqrt{8}}\begin{bmatrix} 1 \\ -1 \\ 1 \\ 1 \\ -1 \\ 1 \\ 1 \\ 1 \end{bmatrix} = \begin{bmatrix} 1 \\ 0 \\ 0 \\ 0 \\ 0 \\ 0 \\ 0 \\ 0 \end{bmatrix}. \tag{6.14}$$

Here, the diagonal matrix

$$D = \text{diag}\{0.3536, 0.7071, -0.4082, -0.2887, 0.2236, -0.1826, -0.1543, -0.1336\}.$$

All numbers in this and other examples in the book are printed with precision of four digits after decimal point. With this diagonal matrix, the DsiHT matrix is written in the integer form. The first row of this matrix is the non-normalized generator $q = (1, -1, 1, 1, -1, 1, 1, 1)'$ with energy $E[q] = 8$, and the last row differs only in the last component, -7, where there a splash of energy occurs ($-7 = 1 - E[g]$). This DsiHT is performed by seven rotations by the following angles (in degrees):

$$\{\psi_k; k = 1{:}\ 7\} = \{45°, -35.2644°, -30°, 26.5651°, -24.0948°, -22.2077°, -20.7048°\}.$$

For the inverse transform

$$H_8'(|0\rangle^{\otimes 3})0 = \left(R_{-\psi_{1;(0,1)}} R_{-\psi_{2;(0,2)}} R_{-\psi_{3;(03)}} R_{-\psi_{4;(0,4)}} R_{-\psi_{5;(0,5)}} R_{-\psi_{6;(0,6)}} R_{-\psi_{7;(0,7)}}\right)(|0\rangle^{\otimes 3}),$$

we have

$$H_8'\begin{bmatrix} 1 \\ 0 \\ 0 \\ 0 \\ 0 \\ 0 \\ 0 \\ 0 \end{bmatrix} = \underbrace{\begin{bmatrix} 1 & 1 & 1 & 1 & 1 & 1 & 1 & 1 \\ -1 & 1 & -1 & -1 & -1 & -1 & -1 & -1 \\ 1 & 0 & -2 & 1 & 1 & 1 & 1 & 1 \\ 1 & 0 & 0 & -3 & 1 & 1 & 1 & 1 \\ -1 & 0 & 0 & 0 & 4 & -1 & -1 & -1 \\ 1 & 0 & 0 & 0 & 0 & -5 & 1 & 1 \\ 1 & 0 & 0 & 0 & 0 & 0 & -6 & 1 \\ 1 & 0 & 0 & 0 & 0 & 0 & 0 & -7 \end{bmatrix}}_{8\times 8 \text{ matrix of the inverse } H_8'} D \times \begin{bmatrix} 1 \\ 0 \\ 0 \\ 0 \\ 0 \\ 0 \\ 0 \\ 0 \end{bmatrix} = \frac{1}{\sqrt{8}}\begin{bmatrix} 1 \\ -1 \\ 1 \\ 1 \\ -1 \\ 1 \\ 1 \\ 1 \end{bmatrix}. \tag{6.15}$$

Thus, to prepare the 3-qubit superposition

$$|\varphi\rangle = \frac{1}{\sqrt{8}}(|0\rangle - |1\rangle + |2\rangle + |3\rangle - |4\rangle + |5\rangle + |6\rangle + |7\rangle),$$

the unitary transform with H_8' matrix can be used with seven rotations, $H_8'(|0\rangle^{\otimes 3}) = |\varphi\rangle$.

B) For the strong DsiHT with the same generator, the transform

$$H_8 q = \left(R_{\psi_{7;(0,1)}} R_{\psi_{6;(1,2)}} R_{\psi_{5;(2,3)}} R_{\psi_{4;(3,4)}} R_{\psi_{3;(4,5)}} R_{\psi_{2;(5,6)}} R_{\psi_{1;(6,7)}}\right) q$$

can be written as

$$H_8 q = D \underbrace{\begin{bmatrix} 1 & -1 & 1 & 1 & -1 & 1 & 1 & 1 \\ -7 & -1 & 1 & 1 & -1 & 1 & 1 & 1 \\ 0 & 6 & 1 & 1 & -1 & 1 & 1 & 1 \\ 0 & 0 & -5 & 1 & -1 & 1 & 1 & 1 \\ 0 & 0 & 0 & -4 & -1 & 1 & 1 & 1 \\ 0 & 0 & 0 & 0 & 3 & 1 & 1 & 1 \\ 0 & 0 & 0 & 0 & 0 & -2 & 1 & 1 \\ 0 & 0 & 0 & 0 & 0 & 0 & -1 & 1 \end{bmatrix}}_{8 \times 8 \text{ matrix } H_8} \times \frac{1}{\sqrt{8}} \begin{bmatrix} 1 \\ -1 \\ 1 \\ 1 \\ -1 \\ 1 \\ 1 \\ 1 \end{bmatrix} = \begin{bmatrix} 1 \\ 0 \\ 0 \\ 0 \\ 0 \\ 0 \\ 0 \\ 0 \end{bmatrix}. \tag{6.16}$$

The diagonal matrix

$$D = \text{diag}\{0.3536, -0.1336, 0.1543, 0.1826, -0.2236, 0.2887, 0.4082, 0.7071\}.$$

Seven rotations by the following angles (in degrees) are used for this DsiHT:

$$\{\psi_k; k = 1:7\} = \{69.2952^\circ, 67.7923^\circ, -65.9052^\circ, 63.4349^\circ, 60^\circ, -54.7356^\circ, -45^\circ\}.$$

The inverse matrix is $H_8' = \left(R_{-\psi_{1;(6,7)}} R_{-\psi_{2;(5,6)}} R_{-\psi_{3;(4,5)}} R_{-\psi_{4;(3,4)}} R_{-\psi_{5;(2,3)}} R_{-\psi_{6;(1,2)}} R_{-\psi_{7;(0,1)}}\right)$ and we obtain

$$H_8' \begin{bmatrix} 1 \\ 0 \\ 0 \\ 0 \\ 0 \\ 0 \\ 0 \\ 0 \end{bmatrix} = \underbrace{\begin{bmatrix} 1 & -7 & 0 & 0 & 0 & 0 & 0 & 0 \\ -1 & -1 & 6 & 0 & 0 & 0 & 0 & 0 \\ 1 & 1 & 1 & -5 & 0 & 0 & 0 & 0 \\ 1 & 1 & 1 & 1 & -4 & 0 & 0 & 0 \\ -1 & -1 & -1 & -1 & -1 & 3 & 0 & 0 \\ 1 & 1 & 1 & 1 & 1 & 1 & -2 & 0 \\ 1 & 1 & 1 & 1 & 1 & 1 & 1 & -1 \\ 1 & 1 & 1 & 1 & 1 & 1 & 1 & 1 \end{bmatrix}}_{8 \times 8 \text{ matrix of the inverse } H_8'} D \times \begin{bmatrix} 1 \\ 0 \\ 0 \\ 0 \\ 0 \\ 0 \\ 0 \\ 0 \end{bmatrix} = \frac{1}{\sqrt{8}} \begin{bmatrix} 1 \\ -1 \\ 1 \\ 1 \\ -1 \\ 1 \\ 1 \\ 1 \end{bmatrix} = |\varphi\rangle. \tag{6.17}$$

Any r-qubit superposition can be obtained from the zero state $|0\rangle^{\otimes r}$ by the unitary heap transform, as shown in Fig. 6.7. The DsiHT requires only $(2^r - 1)$ rotations and it is generated (induced) by the vector of amplitudes of the superposition. For superpositions with complex amplitudes, the complex DsiHT can be used in a similar way. These transforms are described in Section 6.3.

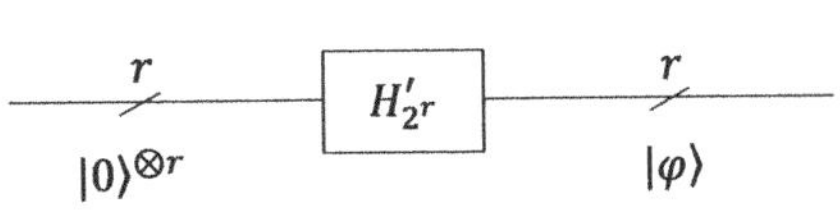

Fig. 6.7 The use of inverse DsiHT to initiate the n-qubit superposition.

Comments

It should be noted that any basis state can be used to initiate the required superposition. The DsiHT can be defined with a heap in any position or a simple rearrangement of its two rows. For instance, in the example above with $N = 4$, the 2-qubit superposition can be obtained from the computational state $|1\rangle$, as

$$|\varphi\rangle = A'\,|1\rangle = \frac{1}{2}\begin{bmatrix} 1.7321 & 0.5774 & -0.5774 & -0.5774 \\ 1 & -1 & 1 & 1 \\ 0 & 1.6330 & 0.8165 & 0.8165 \\ 0 & 0 & -1.4142 & 1.4142 \end{bmatrix}|1\rangle = \frac{1}{2}\begin{bmatrix} 1.7321 & 1 & 0 & 0 \\ 0.5774 & -1 & 1.6330 & 0 \\ -0.5774 & 1 & 0.8165 & -1.4142 \\ -0.5774 & 1 & 0.8165 & 1.4142 \end{bmatrix}|1\rangle.$$

Here, matrix A is composed of matrix $H_{4;s}$ by permuting the first and second rows; the determinant of this matrix is -1. The amplitudes of the 2-qubit superposition $|\varphi\rangle$ are on the second column of the matrix A'.

6.2 DsiHT-Based Factorization of Real Matrices

In this section, we describe QR-decomposition of a square matrix, which is based on the DsiHTs. Different methods of QR decompositions exist [11–13], and the DsiHT-based QR decomposition was shown is the one of the most effective methods in matrix factorization [5–7]. This method can be used for the DsiHT with any path in calculation. The basis 2-point transforms are Givens rotations R_φ. As an example, we consider 4×4 matrix QR-decomposition. Let A be the 4×4 square real matrix

$$A = \begin{bmatrix} a_0 & b_0 & c_0 & d_0 \\ a_1 & b_1 & c_1 & d_1 \\ a_2 & b_2 & c_2 & d_2 \\ a_3 & b_3 & c_3 & d_3 \end{bmatrix} \text{ and } \det(A) \neq 0.$$

In QR decomposition, the matrix is represented as $A = QR$, where Q is a unitary matrix and R is a triangular matrix. If the matrix A is unitary, the matrix R is diagonal (D), with coefficients ±1 of the diagonal, with det $D = \pm1$. This is why, we can call such QR-decomposition the QD-decomposition. The case with complex matrices is similar to this case, only the basis transform R_φ must be changed by complex unitary transforms, which will be described in the next section.

The QD-decomposition of the 4×4 unitary matrix A is implemented in the following three stages.

Stage 1. The vector-generator is the first column of the matrix, $x_1 = (a_0, a_1, a_2, a_3)'$. The 4-point DsiHT, H_4, applying on this vector results in the vector $H_4x_1 = (\|x_1\|, 0, 0, 0)'$. The matrix A is unitary, therefore, $\|x_1\| = 1$. The multiplication of the matrix of the transform H_4 on the matrix A is the matrix with the first column H_4x_1, other three columns will be changed. We denote the obtained new matrix as

$$T_1 = H_4A = \begin{bmatrix} 1 & 0 & 0 & 0 \\ 0 & \hat{b}_1 & \hat{c}_1 & \hat{d}_1 \\ 0 & \hat{b}_2 & \hat{c}_2 & \hat{d}_2 \\ 0 & \hat{b}_3 & \hat{c}_3 & \hat{d}_3 \end{bmatrix}.$$

Stage 2. The 3×3 submatrix of T_1 is renewed. The vector-generator is $x_2 = \left(\hat{b}_1, \hat{b}_2, \hat{b}_3\right)'$ with norm $\|x_2\| = 1$. The corresponding 3-point DsiHT, H_3, applying on this vector results in the vector $H_3x_2 = (\pm1, 0, 0)'$. Calculate the matrix multiplication

$$T_2 = \begin{bmatrix} 1 & 0 \\ 0 & H_3 \end{bmatrix} T_1 = (1 \oplus H_3)T_1 = \begin{bmatrix} 1 & 0 & 0 & 0 \\ 0 & \pm 1 & 0 & 0 \\ 0 & 0 & \tilde{c}_2 & \tilde{d}_2 \\ 0 & 0 & \tilde{c}_3 & \tilde{d}_3 \end{bmatrix}.$$

Stage 3. The 2×2 submatrix of T_2 is renewed. The vector-generator is $x_3 = (\tilde{c}_2, \tilde{c}_3)'$ with norm $\|x_3\| = 1$. The corresponding 2-point DsiHT, H_2, applying on this vector results in the vector $H_2 x_3 = (\pm 1, 0)'$. Calculate the matrix multiplication, to obtain the diagonal matrix D,

$$T_3 = \begin{bmatrix} I_2 & 0 \\ 0 & H_2 \end{bmatrix} T_2 = (I_2 \oplus H_2)T_2 = D = \begin{bmatrix} 1 & 0 & 0 & 0 \\ 0 & \pm 1 & 0 & 0 \\ 0 & 0 & \pm 1 & 0 \\ 0 & 0 & 0 & \pm 1 \end{bmatrix}. \tag{6.18}$$

The QD-decomposition, or diagonalization $T_3 : A \to D$, can be written as

$$T_3 = (I_2 \oplus H_2)(1 \oplus H_3)H_4. \tag{6.19}$$

Thus, we obtain the following decomposition of the matrix:

$$A = T_3' D = H_4'(1 \oplus H_3')(I_2 \oplus H_2')D. \tag{6.20}$$

We mention that the DsiHT can be used for factorization of any unitary matrix. The following assertion is true and is written here for matrix factorization by the strong DsiHTs.

Assertion 6.1
For a real unitary $N \times N$ matrix A, the following diagonalization holds:

$$D = (I_{N-2} \oplus H_2)(I_{N-3} \oplus H_3) \cdots (I_2 \oplus H_{N-2})(1 \oplus H_{N-1})H_N A, \tag{6.21}$$

and the matrix A can be represented by $(N-1)$ matrices of the strong DsiHTs, namely

$$A = H_N'(1 \oplus H_{N-1}')(I_2 \oplus H_{N-2}') \cdots (I_{N-3} \oplus H_3')(I_{N-2} \oplus H_2')D. \tag{6.22}$$

Here, each DsiHT, H_k, $k = 1 : (N-1)$, is induced by the k-point signal-generator determined during factorization. D is the diagonal matrix with coefficients ± 1 and $\det D = \pm 1$.

6.2.1 Quantum Circuits for DCT-II

The $N \times N$ matrix of the N-point DCT of type II can be written as [14]

$$C_N = \sqrt{\frac{2}{N}} \operatorname{diag}\left\{\frac{1}{\sqrt{2}}, 1, 1, \ldots, 1\right\} \left[\cos\left(\frac{\pi}{N}(n + 0.5)k\right)\right]_{n,k = 0:(N-1)}. \tag{6.23}$$

This matrix is unitary. The order of the transform is considered $N = 2^r$, $r > 1$.Next, we consider the examples of the 2^r-point DCT and its analogs, the r-qubit quantum cosine transform (QCT).

Example 6.3 4-Point DCT, or 2-Qubit QCT, of Type II
The matrix of the 4-point inverse DCT is (all numbers are rounded to four decimal places)

$$C_4 = \begin{bmatrix} 0.5 & 0.5 & 0.5 & 0.5 \\ 0.6533 & 0.2706 & -0.2706 & -0.6533 \\ 0.5 & -0.5 & -0.5 & 0.5 \\ 0.2706 & -0.6533 & 0.6533 & -0.2706 \end{bmatrix}, \det C_4 = 1, C_4 C_4' = I_4. \tag{6.24}$$

Stage 1: Angles of rotations at this stage: $\varphi_1 = -0.4961$, $\varphi_2 = -0.7161$, and $\varphi_3 = -1.0472$. The 4-point strong DsiHT has the matrix

$$H_4 = R_{\varphi_3;(0,1)} R_{\varphi_2;(1,2)} R_{\varphi_1;(2,3)} = \begin{bmatrix} 0.5 & 0.6533 & 0.5 & 0.2706 \\ -0.8660 & 0.3772 & 0.2887 & 0.1562 \\ 0 & -0.6565 & 0.6634 & 0.3590 \\ 0 & 0 & -0.4760 & 0.8795 \end{bmatrix}.$$

Stage 2: Angles of rotations at this stage: $\varphi_1 = -0.4249$ and $\varphi_2 = -0.9553$. The 3-point DsiHT has the matrix which is shown below in the 4×4 matrix

$$1 \oplus H_3 = R_{\varphi_2;(1,2)} R_{\varphi_1;(2,3)} = \begin{bmatrix} 1 & 0 & 0 & 0 \\ 0 & 0.5774 & 0.7439 & 0.3366 \\ 0 & -0.8165 & 0.5260 & 0.2380 \\ 0 & 0 & -0.4122 & 0.9111 \end{bmatrix}.$$

Stage 3: Angles of rotations at this stage: $\varphi_1 = -0.7854$ and the 2-point DsiHT has the matrix shown below in the block 2×2,

$$I_2 \oplus H_2 = R_{\varphi_1;(2,3)} = \begin{bmatrix} 1 & 0 & 0 & 0 \\ 0 & 1 & 0 & 0 \\ 0 & 0 & 0.7071 & 0.7071 \\ 0 & 0 & -0.7071 & 0.7071 \end{bmatrix}.$$

The 4-point DCT matrix QD-decomposition is written as $(I_2 \oplus H_2)(1 \oplus H_3)H_4C_4 = D$, where the diagonal matrix $D = \mathrm{diag}\{1, -1, 1, -1\} = I_2 \otimes Z$ and $D^2 = I_4$. Therefore, the matrix of the 4-point DCT can be factorized by

$$C_4 = H_4'(1 \oplus H_3')(I_2 \oplus H_2')(I_2 \otimes Z). \tag{6.25}$$

The set of six angles used in this decomposition in radians and degrees is shown in Table 6.1. Note that the angles are shown for the inverse transforms H_k', $k = 2, 3, 4$, not for H_k, that is, the angles are $-\varphi_k$.

Thus, the following matrix QD-decomposition by six rotations holds:

$$C_4 = \underbrace{\begin{bmatrix} 1 & 0 & 0 & 0 \\ 0 & 1 & 0 & 0 \\ 0 & 0 & \cos\varphi_1 & -\sin\varphi_1 \\ 0 & 0 & \sin\varphi_1 & \cos\varphi_1 \end{bmatrix} \begin{bmatrix} 1 & 0 & 0 & 0 \\ 0 & \cos\varphi_2 & -\sin\varphi_2 & 0 \\ 0 & \sin\varphi_2 & \cos\varphi_2 & 0 \\ 0 & 0 & 0 & 1 \end{bmatrix} \begin{bmatrix} \cos\varphi_3 & -\sin\varphi_3 & 0 & 0 \\ \sin\varphi_3 & \cos\varphi_3 & 0 & 0 \\ 0 & 0 & 1 & 0 \\ 0 & 0 & 0 & 1 \end{bmatrix}}_{\varphi_1 = 28.42^\circ,\ \varphi_2 = 41.03^\circ,\ \varphi_3 = 60^\circ}$$

Table 6.1 Table of angles in QD-decomposition of the 4-point DCT of type II.

H_4'	0.4961, 0.7161, 1.0472	28.4221°, 41.0319°, 60°
H_3'	0.4249, 0.9553	24.3429°, 54.7356°
H_2'	0.7854	45°

$$\underbrace{\begin{bmatrix} 1 & 0 & 0 & 0 \\ 0 & 1 & 0 & 0 \\ 0 & 0 & \cos\varphi_1 & -\sin\varphi_1 \\ 0 & 0 & \sin\varphi_1 & \cos\varphi_1 \end{bmatrix} \begin{bmatrix} 1 & 0 & 0 & 0 \\ 0 & \cos\varphi_2 & -\sin\varphi_2 & 0 \\ 0 & \sin\varphi_2 & \cos\varphi_2 & 0 \\ 0 & 0 & 0 & 1 \end{bmatrix}}_{\varphi_1 = 24.34^\circ,\ \varphi_2 = 54.74^\circ}$$

$$\underbrace{\begin{bmatrix} 1 & 0 & 0 & 0 \\ 0 & 1 & 0 & 0 \\ 0 & 0 & \cos\varphi_1 & -\sin\varphi_1 \\ 0 & 0 & \sin\varphi_1 & \cos\varphi_1 \end{bmatrix}}_{\varphi_1 = 45^\circ} \begin{bmatrix} 1 & 0 & 0 & 0 \\ 0 & -1 & 0 & 0 \\ 0 & 0 & 1 & 0 \\ 0 & 0 & 0 & -1 \end{bmatrix}.$$

As we know,

$$R_{\varphi_{2;(1,2)}} = \begin{bmatrix} 1 & 0 & 0 & 0 \\ 0 & \cos\varphi_2 & -\sin\varphi_2 & 0 \\ 0 & \sin\varphi_2 & \cos\varphi_2 & 0 \\ 0 & 0 & 0 & 1 \end{bmatrix} = \begin{bmatrix} 0 & 1 & 0 & 0 \\ 0 & 0 & 1 & 0 \\ 0 & 0 & 0 & 1 \\ 1 & 0 & 0 & 0 \end{bmatrix} \begin{bmatrix} 1 & 0 & 0 & 0 \\ 0 & 1 & 0 & 0 \\ 0 & 0 & \cos\varphi_2 & -\sin\varphi_2 \\ 0 & 0 & \sin\varphi_2 & \cos\varphi_2 \end{bmatrix} \begin{bmatrix} 0 & 0 & 0 & 1 \\ 1 & 0 & 0 & 0 \\ 0 & 1 & 0 & 0 \\ 0 & 0 & 1 & 0 \end{bmatrix},$$

or $R_{\varphi_{2;(1,2)}} = P_2\left(I_2 \oplus R_{\varphi_2}\right)P_2'$. Therefore, the matrix of the 4-point DCT can be factorized by controlled gates as

$$C_4 = \underbrace{(I_2 \oplus R_{28.42^\circ})P_2(I_2 \oplus R_{41.03^\circ})P_2'(R_{60^\circ} \oplus I_2)} \cdot \underbrace{(I_2 \oplus R_{24.34^\circ})P_2(I_2 \oplus R_{54.74^\circ})P_2'} \cdot \underbrace{I_2 \oplus R_{45^\circ}}(I_2 \otimes Z).$$

The permutations above have the matrices

$$P_2 = \begin{bmatrix} 0 & 1 & 0 & 0 \\ 0 & 0 & 1 & 0 \\ 0 & 0 & 0 & 1 \\ 1 & 0 & 0 & 0 \end{bmatrix} \text{ and } P_2' = \begin{bmatrix} 0 & 0 & 0 & 1 \\ 1 & 0 & 0 & 0 \\ 0 & 1 & 0 & 0 \\ 0 & 0 & 1 & 0 \end{bmatrix}.$$

On bit planes {00,01,10,11}, these permutations can be written as $P_2(x, y) = (x \oplus y \oplus 1,\ y \oplus 1) \bmod 2$ and $P_2'(x, y) = (x \oplus y, y \oplus 1) \bmod 2$, where $x, y \in \{0, 1\}$.

The quantum scheme of the 2-qubit QCT is given in Fig. 6.8. Six controlled gates of rotations, Z gate, and four permutations are used.

$$C_4 = \underbrace{(I_2 \oplus R_{28.42^\circ})\, \underbrace{P_2(I_2 \oplus R_{41.03^\circ})P_2'}\, (R_{60^\circ} \oplus I_2)}\, (I_2 \oplus R_{24.34^\circ})\, \underbrace{P_2(I_2 \oplus R_{54.74^\circ})P_2'}\, \underbrace{I_2 \oplus R_{45^\circ}}\, (I_2 \otimes Z).$$

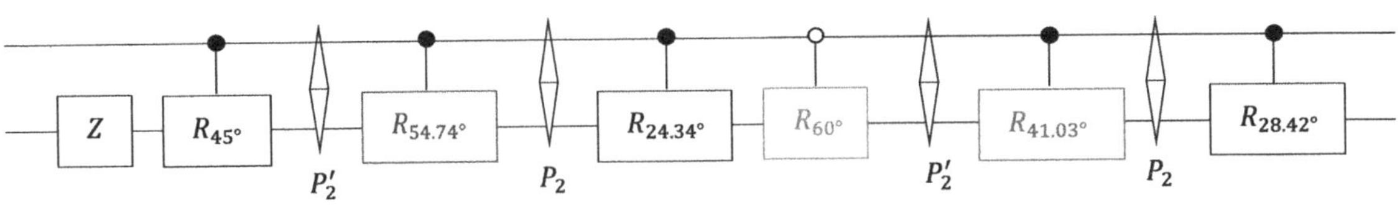

Fig. 6.8 The quantum scheme of the 2-qubit QCT of type II.

Example 6.4 8-Point DCT or 3-Qubit QCT of Type II
In the $N = 8$ case, the QD-decomposition of the 8×8 matrix C_8 by the strong DsiHTs can be written as

$$(I_6 \oplus H_2)(I_5 \oplus H_3) \cdots (I_2 \oplus H_6)(1 \oplus H_7) H_8 C_8 = D. \tag{6.26}$$

Here, the diagonal matrix is $D = (I_4 \otimes Z)$. The set of 28 angles that are used in this decomposition is shown in Table 6.2. The matrix of the 8-point DCT can be factorized by 28 rotations,

$$C_8 = H_8'\left(1 \oplus H_7'\right)\left(I_2 \oplus H_6'\right) \cdots \left(I_5 \oplus H_3'\right)\left(I_6 \oplus H_2'\right)(I_4 \otimes Z). \tag{6.27}$$

Note that any other real unitary transformation differs from the 8-point DCT only by such table of angles. Each $k \times k$ matrix of the DsiHT in Eq. 6.26 can be written as

$$H_k = R_{\varphi_{k-1};(0,1)} \cdots R_{\varphi_3;(k-4,k-3)} R_{\varphi_2;(k-3,k-2)} R_{\varphi_1;(k-2,k-1)}, \; k = 8, 7, \ldots, 2. \tag{6.28}$$

The following sets of rotations (SRs) are used in the DsiHTs:

$$\left\{\begin{array}{l} SR\left(H_8'\right) = \left\{\underline{R_{\varphi_1;6,7}}, R_{\varphi_2;5,6}, \underline{R_{\varphi_3;4,5}}, R_{\varphi_4;3,4}, \underline{R_{\varphi_5;2,3}}, R_{\varphi_6;1,2}, \underline{R_{\varphi_7;0,1}}\right\} \\ SR\left(H_7'\right) = \left\{\underline{R_{\varphi_1;6,7}}, R_{\varphi_2;5,6}, \underline{R_{\varphi_3;4,5}}, R_{\varphi_4;3,4}, \underline{R_{\varphi_5;2,3}}, R_{\varphi_6;1,2}\right\} \\ SR\left(H_6'\right) = \left\{\underline{R_{\varphi_1;6,7}}, R_{\varphi_2;5,6}, \underline{R_{\varphi_3;4,5}}, R_{\varphi_4;3,4}, \underline{R_{\varphi_5;2,3}}\right\} \\ SR\left(H_5'\right) = \left\{\underline{R_{\varphi_1;6,7}}, R_{\varphi_2;5,6}, \underline{R_{\varphi_3;4,5}}, R_{\varphi_4;3,4}\right\} \\ SR\left(H_4'\right) = \left\{\underline{R_{\varphi_1;6,7}}, R_{\varphi_2;5,6}, \underline{R_{\varphi_3;4,5}}\right\} \\ SR\left(H_3'\right) = \left\{\underline{R_{\varphi_1;6,7}}, R_{\varphi_2;5,6}\right\} \\ SR\left(H_2'\right) = \left\{\underline{R_{\varphi_1;6,7}}\right\} \end{array}\right\} \tag{6.29}$$

Here, the angles φ_k in each row are taken from Table 6.2. Sixteen rotations are underlined and 12 not. First, we consider four types of rotation matrices among the first 16. Considering the basis matrix of Givens rotation

$$R_{\varphi_k} = \begin{bmatrix} c_k & -s_k \\ s_k & c_k \end{bmatrix} \triangleq \begin{bmatrix} \cos\varphi_k & -\sin\varphi_k \\ \sin\varphi_k & \cos\varphi_k \end{bmatrix}, k = 1, 2, \ldots, 7,$$

Table 6.2 Table of 28 angles in QR decomposition of the 8-point DCT.

H_8'	27.0123°, 37.7096°, 44.8028°, 50.1609°, 54.5554°, 58.3821°, 69.2952°
H_7'	16.7179°, 26.7817°, 35.3736°, 43.5195°, 51.9625°, 67.7923°
H_6'	13.5916°, 23.6909°, 33.9119°, 45.5436°, 65.9052°
H_5'	13.0841°, 24.6490°, 38.6671°, 63.4349°
H_4'	14.7121°, 30.6626°, 60°
H_3'	20.2477°, 54.7356°
H_2'	45°

we have the following matrices (the bit planes are also shown):

$$R_{\varphi_1;6,7}=\begin{bmatrix}1&&&&&&&\\&1&&&&&&\\&&1&&&&&\\&&&1&&&&\\&&&&1&&&\\&&&&&1&&\\&&&&&&c_1&-s_1\\&&&&&&s_1&c_1\end{bmatrix}=I_4\oplus CR_{\varphi_1}=C(CR_{\varphi_1}),BP=\begin{bmatrix}0&0&0\\0&0&1\\0&1&0\\0&1&1\\1&0&0\\1&0&1\\1&1&0\\1&1&1\end{bmatrix},$$

$$R_{\varphi_3;4,5}=\begin{bmatrix}1&&&&&&&\\&1&&&&&&\\&&1&&&&&\\&&&1&&&&\\&&&&c_3&-s_3&&\\&&&&s_3&c_3&&\\&&&&&&1&\\&&&&&&&1\end{bmatrix}=I_4\oplus C_0R_{\varphi_3}=C(C_0R_{\varphi_3}),BP=\begin{bmatrix}0&0&0\\0&0&1\\0&1&0\\0&1&1\\1&0&0\\1&0&1\\1&1&0\\1&1&1\end{bmatrix},$$

$$R_{\varphi_5;2,3}=\begin{bmatrix}1&&&&&&&\\&1&&&&&&\\&&c_5&-s_5&&&&\\&&s_5&c_5&&&&\\&&&&1&&&\\&&&&&1&&\\&&&&&&1&\\&&&&&&&1\end{bmatrix}=CR_{\varphi_5}\oplus I_4=C_0(CR_{\varphi_5}),BP=\begin{bmatrix}0&0&0\\0&0&1\\0&1&0\\0&1&1\\1&0&0\\1&0&1\\1&1&0\\1&1&1\end{bmatrix},$$

$$R_{\varphi_7;0,1}=\begin{bmatrix}c_7&-s_7&&&&&&\\s_7&c_7&&&&&&\\&&1&&&&&\\&&&1&&&&\\&&&&1&&&\\&&&&&1&&\\&&&&&&1&\\&&&&&&&1\end{bmatrix}=C_0R_{\varphi_7}\oplus I_4=C_0(C_0R_{\varphi_7}),BP=\begin{bmatrix}0&0&0\\0&0&1\\0&1&0\\0&1&1\\1&0&0\\1&0&1\\1&1&0\\1&1&1\end{bmatrix}.$$

These matrices are 3-qubit gate matrices with two control bits, as shown in Fig. 6.9.

The other three types of rotation $R_{\varphi_2;5,6}, R_{\varphi_4;3,4}$, and $R_{\varphi_6;1,2}$ require additional permutations to be described by 3-qubit gates with two control bits. For instance, the matrix of rotation $R_{\varphi_2;5,6}$ can be decomposed as follows:

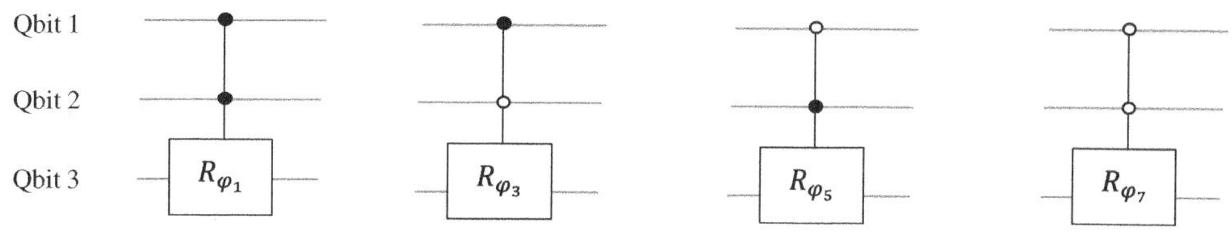

(a) $R_{\varphi_{1;6,7}}$ (b) $R_{\varphi_{3;4,5}}$ (c) $R_{\varphi_{5;2,3}}$ (d) $R_{\varphi_{7;0,1}}$

Fig. 6.9 Four 3-qubit gates of rotation with 2 control qubits. The rotation gates applied to the third qubit with the first two qubits in (a) 1 and 1 states, (b) 1 and 0 states, (c) in 0 and 1 states, and (d) in 0 and 0 states.

$$R_{\varphi_2;\,5,6} = \begin{bmatrix} 1 &&&&&&& \\ & 1 &&&&&& \\ && 1 &&&&& \\ &&& 1 &&&& \\ &&&& 1 &&& \\ &&&&& c_2 & -s_2 & \\ &&&&& s_2 & c_2 & \\ &&&&&&& 1 \end{bmatrix} = I_4 \oplus \begin{bmatrix} 1 &&& \\ & c_2 & -s_2 & \\ & s_2 & c_2 & \\ &&& 1 \end{bmatrix}.$$

With permutations P_2 and P_2', we can move the rotation from the second and third planes to the third and fourth planes. Indeed, the following matrix decomposition holds:

$$\begin{bmatrix} 1 &&& \\ & c_2 & -s_2 & \\ & s_2 & c_2 & \\ &&& 1 \end{bmatrix} = \begin{bmatrix} 0 & 1 & 0 & 0 \\ 0 & 0 & 1 & 0 \\ 0 & 0 & 0 & 1 \\ 1 & 0 & 0 & 0 \end{bmatrix} \begin{bmatrix} 1 &&& \\ & 1 && \\ && c_2 & -s_2 \\ && s_2 & c_2 \end{bmatrix} \begin{bmatrix} 0 & 0 & 0 & 1 \\ 1 & 0 & 0 & 0 \\ 0 & 1 & 0 & 0 \\ 0 & 0 & 1 & 0 \end{bmatrix}.$$

Therefore, the above matrix $R_{\varphi_2;\,5,6}$ can be written as

$$R_{\varphi_2;\,5,6} = I_4 \oplus \left(P_2\left(C_0 R_{\varphi_2}\right)P_2'\right) = \left(I_4 \oplus P_2\right)\left(I_4 \oplus C_0 R_{\varphi_2}\right)\left(I_4 \oplus P_2'\right) = C(P_2) \cdot C\left(C_0 R_{\varphi_2}\right) \cdot C\left(P_2'\right). \tag{6.30}$$

With two permutations, the rotation gate $R_{\varphi_2;\,5,6}$ can be described by the circuit which is shown in Fig. 6.10. On the bit planes, the permutations $C(P_2)$ and $C\left(P_2'\right)$ can be written as

$$C(P_2)\colon (x,y,z) \to (x, y\oplus z\oplus 1, z\oplus 1),\; C\left(P_2'\right)\colon (x,y,z) \to (x, y\oplus z, z\oplus 1), x,y,z \in \{0,1\}.$$

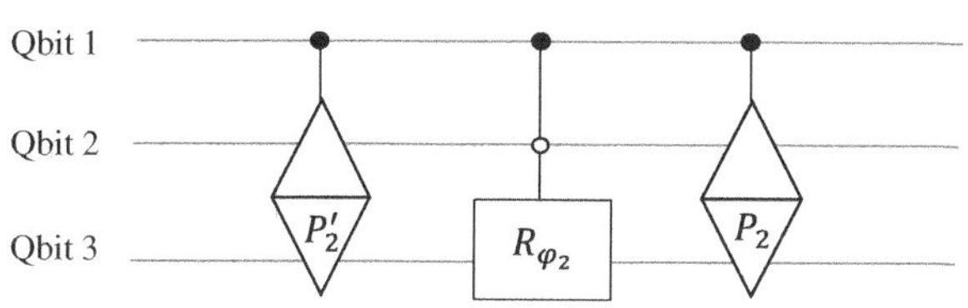

Fig. 6.10 Circuit for the 3-qubit gate $R_{\varphi_2;\,5,6}$.

Now, we consider the matrix of rotation $R_{\varphi_4;3,4}$ and its decomposition as shown

$$R_{\varphi_4;3,4} = \begin{bmatrix} 1 & & & & & & & \\ & 1 & & & & & & \\ & & 1 & & & & & \\ & & & c_4 & -s_4 & & & \\ & & & s_4 & c_4 & & & \\ & & & & & 1 & & \\ & & & & & & 1 & \\ & & & & & & & 1 \end{bmatrix} = I_2 \oplus \begin{bmatrix} 1 & & & \\ & c_4 & -s_4 & \\ & s_4 & c_4 & \\ & & & 1 \end{bmatrix} \oplus I_2 = I_2 \oplus \underbrace{\left(P_2\left(CR_{\varphi_4}\right)P_2'\right)} \oplus I_2.$$

Therefore,

$$R_{\varphi_4;3,4} = (I_2 \oplus P_2 \oplus I_2)\left(I_2 \oplus CR_{\varphi_4} \oplus I_2\right)\left(I_2 \oplus P_2' \oplus I_2\right) \tag{6.31}$$

where the matrix in the middle is

$$\left(I_2 \oplus CR_{\varphi_4} \oplus I_2\right) = \begin{bmatrix} 1 & & & & & & & \\ & 1 & & & & & & \\ & & 1 & & & & & \\ & & & 1 & & & & \\ & & & & c_4 & -s_4 & & \\ & & & & s_4 & c_4 & & \\ & & & & & & 1 & \\ & & & & & & & 1 \end{bmatrix} = I_4 \oplus C_0R_{\varphi_4} = C\left(C_0R_{\varphi_4}\right).$$

The matrices of permutations (2543, 2345) are used in Eq. 6.31,

$$P_{2543} = (I_2 \oplus P_2 \oplus I_2) = \begin{bmatrix} 1 & & & & & & & \\ & 1 & & & & & & \\ & & 0 & 1 & 0 & 0 & & \\ & & 0 & 0 & 1 & 0 & & \\ & & 0 & 0 & 0 & 1 & & \\ & & 1 & 0 & 0 & 0 & & \\ & & & & & & 1 & \\ & & & & & & & 1 \end{bmatrix} = (2543)$$

$$\left(I_2 \oplus P_2' \oplus I_2\right) = \begin{bmatrix} 1 & & & & & & & \\ & 1 & & & & & & \\ & & 0 & 0 & 0 & 1 & & \\ & & 1 & 0 & 0 & 0 & & \\ & & 0 & 1 & 0 & 0 & & \\ & & 0 & 0 & 1 & 0 & & \\ & & & & & & 1 & \\ & & & & & & & 1 \end{bmatrix} = (2345)$$

With these two permutations, the rotation gate $R_{\varphi_4;3,4}$ can be described by the circuit which is shown in Fig. 6.11.

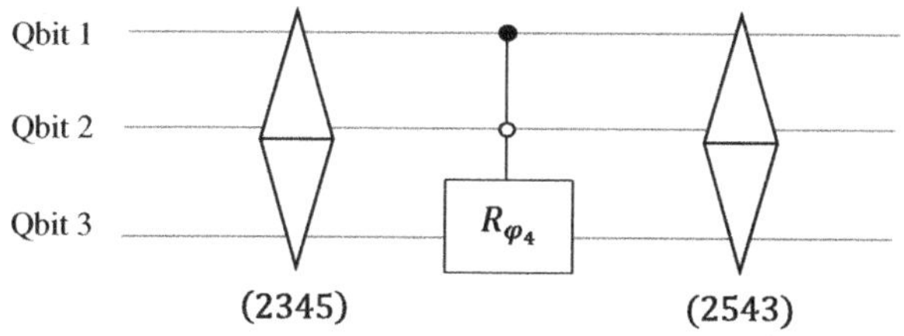

Fig. 6.11 Circuit for the 3-qubit gate $R_{\varphi_4;3,4}$.

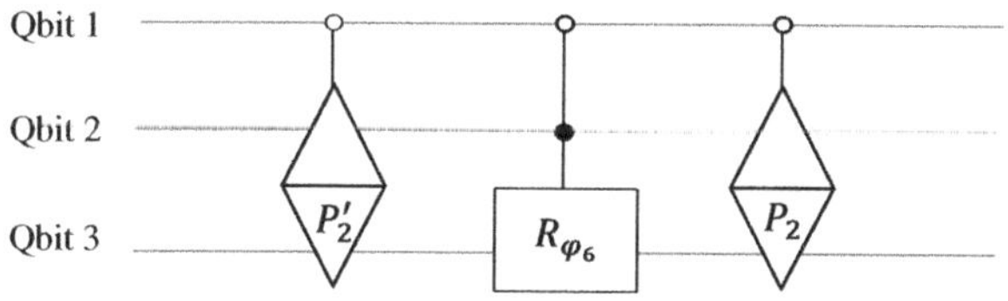

Fig. 6.12 Circuit for the 3-qubit gate $R_{\varphi_6;1,2}$.

Now, we consider the matrix of rotation $R_{\varphi_6;1,2}$ and its decomposition as shown

$$R_{\varphi_6;1,2} = \begin{bmatrix} 1 & & & & & & & \\ & c_6 & -s_6 & & & & & \\ & s_6 & c_6 & & & & & \\ & & & 1 & & & & \\ & & & & 1 & & & \\ & & & & & 1 & & \\ & & & & & & 1 & \\ & & & & & & & 1 \end{bmatrix} = \begin{bmatrix} 1 & & & \\ & c_6 & -s_6 & \\ & s_6 & c_6 & \\ & & & 1 \end{bmatrix} \oplus I_4 = \underbrace{\left(P_2\left(CR_{\varphi_6}\right)P_2'\right)} \oplus I_4.$$

The matrix can be written as

$$R_{\varphi_6;1,2} = (P_2 \oplus I_4)\left(CR_{\varphi_6} \oplus I_4\right)\left(P_2' \oplus I_4\right) = C_0(P_2) \cdot C_0\left(CR_{\varphi_6}\right) \cdot C_0\left(P_2'\right) \tag{6.32}$$

The circuit for this rotation $R_{\varphi_6;1,2}$ is shown in Fig. 6.12.

All described above 8×8 matrices and their circuits can be used to build the quantum circuit for the 8-point DCT of type II. As follows from the set of rotations in Eq. 6.24, this circuit will include 16 rotation gates with two control bits (shown in Fig. 6.9) and 12 gates of rotations, each of which is calculated by 2 permutations and 1 gate of rotation with 2 control bits (shown in Figs. 6.10–6.12). Thus, a total of 28 rotations with 2 control bits plus $2(12) = 24$ gates of permutation are required for the 8-point DCT. As mentioned above, any 8-point unitary transform can be described by the same circuit; only 28 angles will be changed. Note also that one can find more complicated paths for DsiHTs that reduce the number of such permutations in the QR factorization to zero.

6.2.2 Quantum Circuits for the DCT-IV

The $N \times N$ matrix of the N-point DCT of type IV is calculated by [1]

$$\boldsymbol{C}_N = \sqrt{\frac{2}{N}}\left[\cos\left(\frac{\pi}{N}(n + 0.5)(k + 0.5)\right)\right]_{n,k\,=\,0:(N-1)}. \tag{6.33}$$

This matrix is orthogonal with determinant 1, thus $C_N(C_N)' = I_N$. The DCT-IV of the signal $\{f_n;\ n = 0:N-1\}$ is calculated by

$$C_k = \sqrt{\frac{2}{N}}\sum_{n=0}^{N-1} f_n \cos\left(\frac{\pi(n+1/2)(k+1/2)}{N}\right),\ k = 0{:}(N-1).$$

Example 6.5 The 4-Point DCT and 2-Qubit QCT of Type IV
The matrix of the 4-point DCT-IV is equal to (all numbers are rounded to four decimal places)

$$\boldsymbol{C}_4 = \frac{1}{\sqrt{2}}\begin{bmatrix} 0.9808 & 0.8315 & 0.5556 & 0.1951 \\ 0.8315 & -0.1951 & -0.9808 & -0.5556 \\ 0.5556 & -0.9808 & 0.1951 & 0.8315 \\ 0.1951 & -0.5556 & 0.8315 & -0.9808 \end{bmatrix},\ \det \boldsymbol{C}_4 = 1,\ \boldsymbol{C}_4\boldsymbol{C}_4' = I_4. \tag{6.34}$$

Similar to the 4-point DCT-II, the above matrix can be factorized as

$$C_4 = H_4'(1 \oplus H_3')(I_2 \oplus H_2')(I_2 \otimes Z). \tag{6.35}$$

The only difference in the set of angles of rotations in the matrices H_4', H_3', and H_2'. The set of six angles used in this decomposition is given in Table 6.3.

The matrix of the 4-point DCT-IV is factorized by controlled gates as

$$\boldsymbol{C}_4 = \underbrace{(I_2 \oplus R_{19.35^\circ})P_2(I_2 \oplus R_{35.31^\circ})P_2'(R_{46.09^\circ} \oplus I_2)}\cdot\left(\underbrace{I_2 \oplus R_{14.11^\circ})P_2(I_2 \oplus R_{35.31^\circ})P_2'}\cdot\underbrace{I_2 \oplus R_{19.35^\circ}}(I_2 \otimes Z).\right.$$

The quantum scheme of the 2-qubit QCT is given in Fig. 6.13. Six controlled gates of rotations, Z gate, and four permutations are used.

Table 6.3 Table of angles in QD-decomposition of the 4-point DCT of type IV.

H_4'	0.3377, 0.6162, 0.8044	19.3489°, 35.3053°, 46.0906°
H_3'	0.2463, 0.6162	14.1091°, 35.3053°
H_2'	0.3377	19.3489°

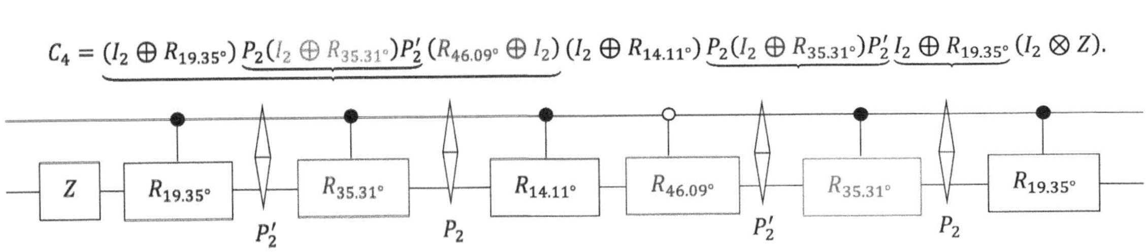

Fig. 6.13 The quantum scheme of the 2-qubit QCT of type IV.

Example 6.6 The 8-Point DCT and 3-Qubit QCT of Type IV
The 8×8 matrix of the N-point DCT of type IV is calculated by

$$C_8 = \frac{1}{2}\left[\cos\left(\frac{\pi}{N}\left(n+\frac{1}{2}\right)\left(k+\frac{1}{2}\right)\right)\right]_{n,k=0:7}, \tag{6.36}$$

$$C_8 = \frac{1}{2}\begin{bmatrix} 0.9952 & 0.9569 & 0.8819 & 0.7730 & 0.6344 & 0.4714 & 0.2903 & 0.0980 \\ 0.9569 & 0.6344 & 0.0980 & -0.4714 & -0.8819 & -0.9952 & -0.7730 & -0.2903 \\ 0.8819 & 0.0980 & -0.7730 & -0.9569 & -0.2903 & 0.6344 & 0.9952 & 0.4714 \\ 0.7730 & -0.4714 & -0.9569 & 0.0980 & 0.9952 & 0.2903 & -0.8819 & -0.6344 \\ 0.6344 & -0.8819 & -0.2903 & 0.9952 & -0.0980 & -0.9569 & 0.4714 & 0.7730 \\ 0.4714 & -0.9952 & 0.6344 & 0.2903 & -0.9569 & 0.7730 & 0.0980 & -0.8819 \\ 0.2903 & -0.7730 & 0.9952 & -0.8819 & 0.4714 & 0.0980 & -0.6344 & 0.9569 \\ 0.0980 & -0.2903 & 0.4714 & -0.6344 & 0.7730 & -0.8819 & 0.9569 & -0.9952 \end{bmatrix}.$$

The QD-decomposition by the strong DsiHTs results in the following calculation of the matrix by 28 rotations:

$$C_8 = H_8'(1 \oplus H_7')(I_2 \oplus H_6')\cdots(I_5 \oplus H_3')(I_6 \oplus H_2')(I_4 \otimes Z). \tag{6.37}$$

This decomposition is similar to the case of 8-point DCT-II described above, but with different angles. Each $k \times k$ matrix of the DsiHT in this equation is calculated by $(k-1)$ rotations as shown in Eq. 6.5,

$$H_k = R_{\varphi_{k-1};(0,1)}\cdots R_{\varphi_3;(k-4,k-3)}R_{\varphi_2;(k-3,k-2)}R_{\varphi_1;(k-2,k-1)},\ k = 8, 7, \ldots, 2. \tag{6.38}$$

The set of 28 angles that are used in this decomposition for the DCT-IV is shown in Table 6.4.

6.2.3 Quantum Circuits for the Discrete Hartley Transform

Consider the $N \times N$ matrix of the N-point discrete Hartley transform (DHrT) with the matrix [15]

$$T_N = \frac{1}{\sqrt{N}}\left[\cos\left(\frac{2\pi nk}{N}\right) + \sin\left(\frac{2\pi nk}{N}\right)\right]_{n,k=0:(N-1)}. \tag{6.39}$$

$N = 2^r$, $r > 1$. The transform of the signal $\{f_n;\ n = 0 : N-1\}$ is calculated by

$$F_k = \frac{1}{\sqrt{N}}\sum_{n=0}^{N-1} f_n\left[\cos\left(\frac{2\pi nk}{N}\right) + \sin\left(\frac{2\pi nk}{N}\right)\right],\ k = 0{:}(N-1).$$

Table 6.4 28 angles in the QD-decomposition of the 8-point DCT-IV.

H_8'	18.6577°, 33.0220°, 41.5482°, 47.6375°, 52.4486°, 56.5227°, 60.1592°
H_7'	10.5019°, 22.0248°, 31.1855°, 39.4537°, 47.6594°, 56.5227°
H_6'	8.1388°, 18.7011°, 28.7126°, 39.4537°, 52.4486°
H_5'	7.5436°, 18.7011°, 31.1855°, 47.6375°
H_4'	8.1388°, 22.0248°, 41.5482°
H_3'	10.5019°, 33.0220°
H_2'	18.6577°

Next, we describe in examples the calculation of the N-point DHrT and its analog r-qubit quantum Hartley transform (QHrT).

Example 6.7 The 8-Point DHrT or 3-Qubit QHrT
We consider the $N = 8$ case,

$$T_8 = \frac{1}{\sqrt{8}}\begin{bmatrix} 1 & 1 & 1 & 1 & 1 & 1 & 1 & 1 \\ 1 & \sqrt{2} & 1 & 0 & -1 & -\sqrt{2} & -1 & 0 \\ 1 & 1 & -1 & -1 & 1 & 1 & -1 & -1 \\ 1 & 0 & -1 & \sqrt{2} & -1 & 0 & 1 & -\sqrt{2} \\ 1 & -1 & 1 & -1 & 1 & -1 & 1 & -1 \\ 1 & -\sqrt{2} & 1 & 0 & -1 & \sqrt{2} & -1 & 0 \\ 1 & -1 & -1 & 1 & 1 & -1 & -1 & 1 \\ 1 & 0 & -1 & -\sqrt{2} & -1 & 0 & 1 & \sqrt{2} \end{bmatrix}.$$

This matrix T_N is unitary, $\det T_N = -1$, and $T_N^2 = I_N$, that is, $T_N = T_N'$. QD-decomposition of this matrix by the strong DsiHTs can be written as in Eq. 6.26,

$$(I_6 \oplus H_2)(I_5 \oplus H_3)\cdots(I_2 \oplus H_6)(1 \oplus H_7)H_8T_8 = D. \tag{6.40}$$

Here, the diagonal matrix is $D = \text{diag}\{1, -1, 1, -1, 1, -1, 1, 1\}$ and it can be written with the phase shift gate $P(\pi)$ as

$$D = (I_4 \otimes Z)(I_4 \oplus I_2 \oplus P(\pi)) = (I_4 \otimes Z) \cdot C(CP(\pi)).$$

The matrix of the 8-point DHrT can be factorized by 28 rotations,

$$T_8 = H_8'(1 \oplus H_7')(I_2 \oplus H_6')(I_3 \oplus H_5')(I_4 \oplus H_4')(I_5 \oplus H_3')(I_6 \oplus H_2')D. \tag{6.41}$$

The set of 28 angles (in degrees) that are used in this diagonalization is shown in Table 6.5. One angle is 0, that is, 27 rotations are made during diagonalization. Note that the last angle in each row is the same as for the 8-point DCT-II.

Note the 4-point discrete Hartley transform has the Hadamard-type matrix

$$T_4 = \frac{1}{2}\begin{bmatrix} 1 & 1 & 1 & 1 \\ 1 & 1 & -1 & -1 \\ 1 & -1 & 1 & -1 \\ 1 & -1 & -1 & 1 \end{bmatrix}, \det T_4 = -1, T_4T_4' = I_4. \tag{6.42}$$

Table 6.5 Table of 28 angles in QD decomposition of the 8-point DHrT.

H_8'	45°, 54.7356°, 60°, 63.4349°, 65.9052°, 67.7923°, 69.2952°
H_7'	43.4495°, 80.6605°, – 53.7712°, 40.0594°, 50.0395°, 67.7923°
H_6'	49.8606°, – 86.8589°, – 62.2966°, 40.0594°, 65.9052°
H_5'	0°, – 86.8589°, – 53.7712°, 63.4349°
H_4'	–49.8606°, 80.6605°, 60°
H_3'	–43.4495°, 54.7356°
H_2'	–45°

Table 6.6 The set of angles in QD-decomposition of the 4-point DHrT.

H'_4	0.7854, 0.9553, 1.0472	45°, 54.7356°, 60°
H'_3	0, 0.9553	0°, 54.7356°
H'_2	−0.7854	−45°

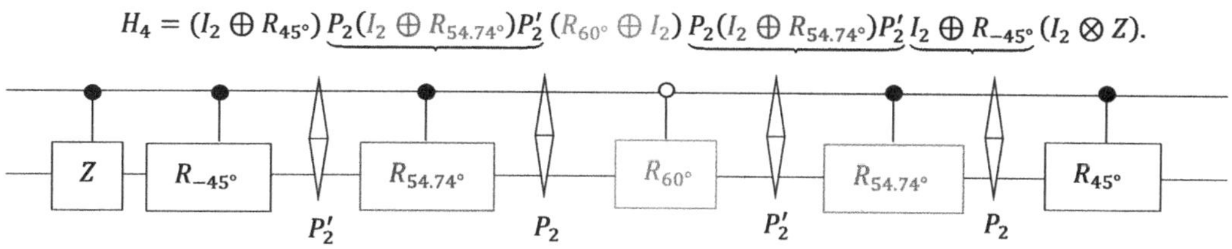

Fig. 6.14 The quantum scheme of the 2-qubit QHrT.

The QD-decomposition can be described similarly to the 4-point DCT of type II above. The difference is the set of rotation angles, which is shown in Table 6.6 (one angle is zero).

The resultant diagonal matrix is $B = \text{diag}\{1, -1, 1, 1\} = Z \oplus I_2$. Therefore, the following matrix QD-decomposition by five rotations holds:

$$H_4 = \underbrace{\begin{bmatrix} 1 & 0 & 0 & 0 \\ 0 & 1 & 0 & 0 \\ 0 & 0 & \cos\varphi_1 & -\sin\varphi_1 \\ 0 & 0 & \sin\varphi_1 & \cos\varphi_1 \end{bmatrix} \begin{bmatrix} 1 & 0 & 0 & 0 \\ 0 & \cos\varphi_2 & -\sin\varphi_2 & 0 \\ 0 & \sin\varphi_2 & \cos\varphi_2 & 0 \\ 0 & 0 & 0 & 1 \end{bmatrix} \begin{bmatrix} \cos\varphi_3 & -\sin\varphi_3 & 0 & 0 \\ \sin\varphi_3 & \cos\varphi_3 & 0 & 0 \\ 0 & 0 & 1 & 0 \\ 0 & 0 & 0 & 1 \end{bmatrix}}_{\varphi_1 = 45°,\, \varphi_2 = 54.74°,\, \varphi_3 = 60°}$$

$$\underbrace{\begin{bmatrix} 1 & 0 & 0 & 0 \\ 0 & \cos\varphi_2 & -\sin\varphi_2 & 0 \\ 0 & \sin\varphi_2 & \cos\varphi_2 & 0 \\ 0 & 0 & 0 & 1 \end{bmatrix}}_{\varphi_2 = 54.74°} \underbrace{\begin{bmatrix} 1 & 0 & 0 & 0 \\ 0 & 1 & 0 & 0 \\ 0 & 0 & \cos\varphi_1 & -\sin\varphi_1 \\ 0 & 0 & \sin\varphi_1 & \cos\varphi_1 \end{bmatrix}}_{\varphi_1 = -45°} \begin{bmatrix} 1 & 0 & 0 & 0 \\ 0 & -1 & 0 & 0 \\ 0 & 0 & 1 & 0 \\ 0 & 0 & 0 & 1 \end{bmatrix}.$$

The matrix of the 4-point DHrT can be factorized by controlled gates as

$$H_4 = \underbrace{(I_2 \oplus R_{45°})P_2(I_2 \oplus R_{54.74°})P'_2(R_{60°} \oplus I_2)} \cdot \underbrace{(P_2(I_2 \oplus R_{54.74°})P'_2} \cdot \underbrace{I_2 \oplus R_{-45°}}(Z \oplus I_2).$$

The quantum scheme of the 2-qubit QHrT is given in Fig. 6.14. Five controlled gates of rotations, the controlled Z gate, and four permutations are used.

6.3 Complex DsiHT

In this section, briefly the concept of the complex DsiHT which is used in analyzing complex matrices and complex transform. In the QR decomposition of a complex matrix by the DsiHT, the basic 2D transformations T are defined in the following way. Consider a complex vector $\boldsymbol{x} = (x_0, x_1)'$ as the generator of the transformation T. The transform of a complex input $(z_0, z_1)'$ is calculated in the matrix form as follows [7]:

$$\boldsymbol{T}: \begin{bmatrix} z_1 \\ z_0 \end{bmatrix} \to \frac{\text{sign}(\text{Real}(x_0))}{\sqrt{|x_0|^2 + |x_1|^2}} \begin{bmatrix} \bar{x}_0 & \bar{x}_1 \\ -x_1 & x_0 \end{bmatrix} \begin{bmatrix} z_1 \\ z_0 \end{bmatrix}. \tag{6.43}$$

In the case of generator, that is, when $(z_0, z_1)' = (x_0, x_1)'$, the transform is

$$\boldsymbol{T} \cdot (x_0, x_1)' = \left(\text{sign}(\text{Real}(x_0)) \sqrt{|x_0|^2 + |x_1|^2}, 0 \right)'. \tag{6.44}$$

The energy of the vector-generator is collected in the first output of the transform.

We consider that the complex vector $\boldsymbol{x}$ represents a unit quaternion $q = (a_1, b_1, c_1, d_1)$, i.e., $x_0 = (a_1 + ib_1)$ and $x_1 = (c_1 + id_1)$. In the polar form, these two complex numbers are $x_0 = |x_0| e^{i\vartheta_0}$ and $x_1 = |x_1| e^{i\vartheta_1}$, where the angles $\vartheta_0, \vartheta_1 \in [0, 2\pi]$. Then, $|x_0|^2 + |x_1|^2 = 1$ and the matrix of the transformation above can be written as

$$T = \text{sign}(a_1) \begin{bmatrix} |x_0| e^{-i\vartheta_0} & |x_1| e^{-i\vartheta_1} \\ -|x_1| e^{i\vartheta_1} & |x_0| e^{i\vartheta_0} \end{bmatrix}. \tag{6.45}$$

With three rotations, we obtain the following representation of this matrix:

$$T = \text{sgn}(a_1) \begin{bmatrix} e^{-i\phi_1} & 0 \\ 0 & e^{i\phi_1} \end{bmatrix} \begin{bmatrix} |x_0| & |x_1| \\ -|x_1| & |x_0| \end{bmatrix} \begin{bmatrix} e^{-i\phi_2} & 0 \\ 0 & e^{i\phi_2} \end{bmatrix}, \tag{6.46}$$

where the angles $\phi_1 = (\vartheta_0 + \vartheta_1)/2$ and $\phi_2 = (\vartheta_0 - \vartheta_1)/2$. The number sign$(a_1)$ can be written as $e^{i\phi_0}$. This global phase $\phi_0 = 0$, if $\vartheta_0 \in [-\pi/2, \pi/2]$ and $\phi_0 = \pi$, otherwise. Calculating the new angle ϑ by

$$|x_0| = \cos(\vartheta), \; |x_1| = \sin(\vartheta), \vartheta \in [0, \pi/2], \tag{6.47}$$

the above matrix can be written as

$$T = e^{i\phi_0} \begin{bmatrix} e^{-i\phi_1} & 0 \\ 0 & e^{i\phi_1} \end{bmatrix} \begin{bmatrix} \cos(\vartheta) & \sin(\vartheta) \\ -\sin(\vartheta) & \cos(\vartheta) \end{bmatrix} \begin{bmatrix} e^{-i\phi_2} & 0 \\ 0 & e^{i\phi_2} \end{bmatrix}. \tag{6.48}$$

Here, the Givens rotation by angle $-\vartheta$ and two matrices of Z-rotation are used, that is,

$$T = e^{i\phi_0} R_z(2\phi_1) W(-\vartheta) R_z(2\phi_2). \tag{6.49}$$

The global phase as the operator can be written with the Pauli shift and Z-rotation gates as follows:

$$e^{i\phi_0} \begin{bmatrix} 1 & 0 \\ 0 & 1 \end{bmatrix} = \begin{bmatrix} 1 & 0 \\ 0 & e^{i2\phi_0} \end{bmatrix} \begin{bmatrix} e^{i\phi_0} & 0 \\ 0 & e^{-i\phi_0} \end{bmatrix} = P(2\phi_0) R_z(-2\phi_0), \tag{6.50}$$

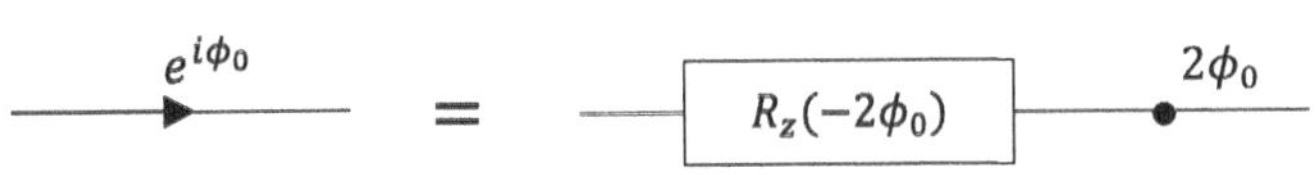

for which we will use the following symbol in circuits (see Table 2.1):

Therefore, the gate T can be calculated by five gates as

$$T = P(2\phi_0) R_z(-2\phi_0) R_z(2\phi_1) W(-\vartheta) R_z(2\phi_2) = P(2\phi_0) R_z(2\phi_1 - 2\phi_0) W(-\vartheta) R_z(2\phi_2), \tag{6.51}$$

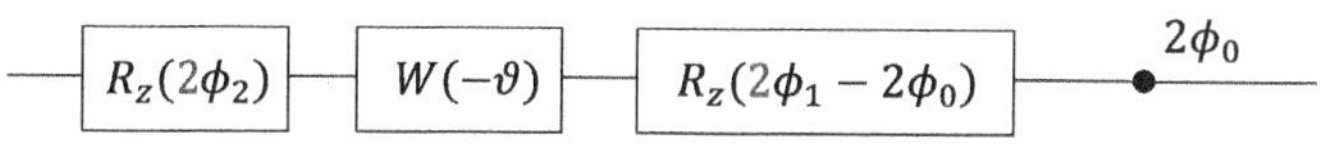

Fig. 6.15 The circuit for the gate *T*.

or

$$T = \begin{bmatrix} 1 & 0 \\ 0 & e^{i2\phi_0} \end{bmatrix} \begin{bmatrix} e^{-i(\phi_1-\phi_0)} & 0 \\ 0 & e^{i(\phi_1-\phi_0)} \end{bmatrix} \begin{bmatrix} \cos(\vartheta) & \sin(\vartheta) \\ -\sin(\vartheta) & \cos(\vartheta) \end{bmatrix} \begin{bmatrix} e^{-i\phi_2} & 0 \\ 0 & e^{i\phi_2} \end{bmatrix}.$$

The quantum circuit for this gate is shown in Fig. 6.15.

The number of such gates T in QD-decomposition of the $N \times N$ complex matrix is estimated as $N(N-1)/2$. For the case when $N = 2^r$, $r > 1$, the number of basic gates in QD-decomposition can be calculated as follows:

$$\mu(N) \leq 2^{r-1}(2^r - 1) \times \begin{cases} 1 & \text{(Givens rotation)} \\ 2 & \text{(Z rotation)} \\ 1 & \text{(Pauli shift).} \end{cases} \tag{6.52}$$

For instance, the number rotation for a 4×4 unitary matrix is no greater than $\mu(4) = 6$, and for 8×8 and 16×16 unitary matrices, these numbers are $\mu(8) \leq 28$ and $\mu(16) \leq 120$. In addition, we may count $2^{r-1}(2^{r-1} - 1)$ permutations, to move some Givens rotations to the next planes, as illustrated for the 8-point DCT of type II described above. Note also that in the QR decomposition of complex matrices the basis matrices T in Eq. 6.48 can be considered without global phases.

References

1 Grigoryan, A.M. and Grigoryan, M.M. (2009). *Brief Notes in Advanced DSP: Fourier Analysis with MATLAB.* Taylor and Francis Group: CRC Press.

2 Grigoryan, A.M. and Agaian, S.S. (2017). Image processing contrast enhancement. *Wiley Encyclopedia of Electrical and Electronics Engineering* 22: https://doi.org/10.1002/047134608X.W5525.pub2.

3 Jenkinson, J., Grigoryan, A.M., and Agaian, S.S. (2015). Enhancement of galaxy images for improved classification,' [9399-32]. *Proceedings of SPIE vol. 9399, 2015 Electronic Imaging: Image Processing: Algorithms and Systems XIII,* 10–11 February, San Francisco, California. https://doi.org/10.1117/12.2083144.

4 Grigoryan, A.M. and Hajinoroozi, M. (2014). Image and audio signal filtration with discrete Heap transforms. *Applied Mathematics and Sciences: An International Journal* 1 (1): 1–18.

5 Grigoryan, A.M. (2014) 'New method of Givens rotations for triangularization of square matrices,' *Journal of Advances in Linear Algebra & Matrix Theory (ALAMT),* 4 (2), pp. 65–78. https://doi.org/10.4236/alamt.2014.42004.

6 Grigoryan, A.M. (2015). 'Fast Heap transform-based QR-decomposition of real and complex matrices: Algorithms and codes,' [9411-21]. *Proceedings of SPIE vol. 9411, Electronic Imaging: Mobile Devices and Multimedia: Enabling Technologies, Algorithms, and Applications 2015,* San Francisco, California. https://doi.org/10.1117/12.2083404

7 Grigoryan, A.M. (2023) 'Effective methods of QR-decompositions of square complex matrices by fast discrete signal-induced heap transforms,' 12 (4), pp. 87–110, *Advances in Linear Algebra & Matrix Theory (ALAMT).* https://doi.org/10.4236/alamt.2022.124005.

8 Grigoryan, A.M. and Grigoryan, M.M. (2006). Nonlinear approach of construction of fast unitary transforms. *2006 40th Annual Conference on Information Sciences and Systems,* Princeton, NJ, USA, pp. 1073–1078. https://doi.org/10.1109/CISS.2006.286625.

9 Grigoryan, A.M. and Grigoryan, M.M. (2007). Discrete unitary transforms generated by moving waves. *Proceedings of the International Conference: Wavelets XII, SPIE: Optics + Photonics 2007*, vol. 6701, article id. 670125. https://doi.org/10.1117/12.728383.

10 Grigoryan, A.M. and Grigoryan, M.M. (2007). New discrete unitary Haar-type heap transforms. *Proc. SPIE* 6701, *Wavelets XII*, 670126. https://doi.org/10.1117/12.728386

11 Householder, A.S. (1958). Unitary triangulation of a nonsymmetric matrix. *Journal ACM* 5 (4): 339–342.

12 Bindel, D., Demmel, J., Kahan, W., and Margues, O. (2001). 'On Computing Givens rotations reliably and efficiently,' *LAPACK Working Note* 148, University of Tennessee, UT-CS-00-449.

13 Alonso, P., Peña, J.M., and Serrano, M.L. (2017). 'QR decomposition of almost strictly sign regular matrices. *Journal of Computational and Applied Mathematics* 318: 646–657.

14 Ahmed, N. and Rao, K.R. (1975). *Orthogonal Transforms for Digital Signal Processing*. Berlin, Heidelberg: Springer.

15 Bracewell, R.N. (1986). *The Hartley Transform*. New York: Oxford University Press.

Part II

Applications in Image Processing

7

Quantum Image Representation with Examples

Efficient quantum image processing requires representing images on a quantum computer, processing them with quantum algorithms, and converting the processed images back into classical form. This process involves storing and manipulating images within a quantum system, a challenge that has led to various quantum image representation methods. While quantum computers offer unprecedented capabilities for complex problems, they are not intended to replace classical computers, which are incredibly portable devices. Instead, in the future, they will likely operate in specialized centers, solving complex 2D and 3D image processing problems that are beyond the capabilities of today's technology. The future of quantum computing in image processing holds great promise for revolutionizing the field. Development will focus on understanding and unifying existing techniques while exploring new possibilities enabled by quantum mechanics. Representing images through quantum superpositions is a key area, with several proposed approaches, each with strengths and weaknesses. The choice of quantum image representation directly impacts the algorithms, efficiency, and design of quantum circuits used in processing. As research progresses, selecting the appropriate representation will be crucial for optimizing quantum image processing tasks and harnessing the full potential of quantum computing.

An abstract diagram of quantum computing under the control of a classical computer is shown in Fig. 7.1. Preparation of the image, its multiple copies, calculation of angles of rotations for many gates, image reconstruction after multiple measurements, and many other operations during the work of a quantum computer will be done, as we understand, by the classical computer. In each task of image processing, the place of the quantum computation should be defined. Of course, future quantum computers will never be able to replace classical computers, especially portable computers that we use in the workplace and in everyday life. Such quantum computers will work in special centers and perform tasks of a future unknown to us now. Undoubtedly, many of the tasks of such computers will be in the field of 2D and 3D image processing. We think about very complex and time-consuming tasks for the classical computer. To date, we are not aware of such examples in image processing; we are waiting for them. The prospects for developing new methods using future quantum computers are great; the main thing here is to thoroughly study the methods developed to date and present a unified theory of image processing. A special place here is occupied by the representation of images through quantum superpositions. Quite a few of them have already been developed. Each of them has its own disadvantages and advantages and examples of successful application. The choice of such a representation will depend on the task and is very important, since the form of representation of a quantum image changes the algorithms, efficiency, and circuits of quantum computing.

In this chapter, we describe the basics of quantum computing in grayscale and color imaging. In classical computers, they are processed in discrete form as matrices. For instance, the grayscale $2^r \times 2^s$-pixel image is the $2^r \times 2^s$ matrix with 2^{r+s} coefficients. The representation of this image is unique. In quantum computation, the same image can be presented in many different ways, which are called *quantum image representations*. They may be composed of $(r+s)$, $(r+s+1)$, $(r+s+2)$, or even more qubits. The preparation of such representations and processing requires good understanding of elementary operations on qubits and multi-qubits. The many examples

Quantum Image Processing in Practice: A Mathematical Toolbox, First Edition. Artyom M. Grigoryan and Sos S. Agaian.

Companion website: www.wiley.com/go/grigoryan/quantumimageprocessing

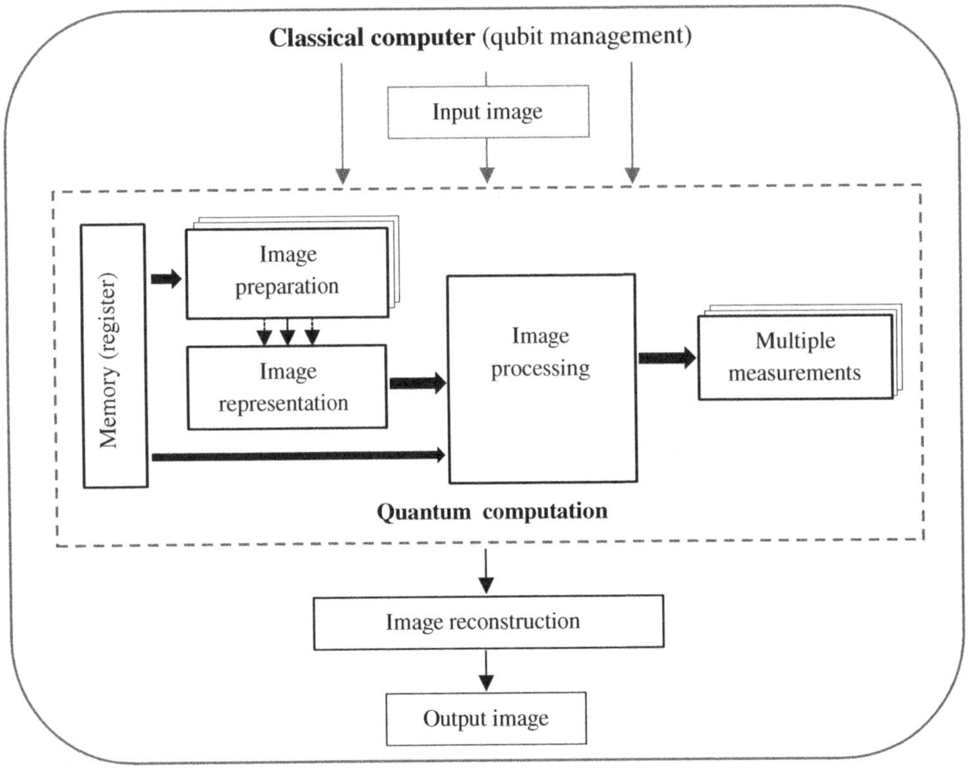

Fig. 7.1 The abstract scheme of the quantum-in-classical computer.

with illustrations and visual mathematics presented in this book will allow readers, we believe, to master the technique of constructing several quantum image processing circuits.

We describe briefly a few models of discrete image representation in the form of multi-qubit superpositions. The image will be considered as a 2D matrix, as well as 1D vector composed by rows or columns of the image. A long list of different approaches exists for quantum signal and image representation [1–4]. Let us immediately note that there is no best quantum image representation that would work most efficiently for all or many complex problems. They all have positive and negative sides. There are many of them, and this is the difference between quantum computing and classical computing in image processing.

7.1 Models of Representation of Grayscale Images

7.1.1 Quantum Pixel Model (QPM)

Consider a discrete grayscale image $f = \{f_{n,m}\}$ with intensities in the interval of integers $[0, L]$. We consider that the range $L + 1$ is a power of 2. For instance, $L = 255$. The image is of size $N \times M$ pixels, where $N = 2^r$, $M = 2^s$, $r, s > 1$. The quantum pixel (n, m) is described by a single qubit with superposition defined as [5]

$$f_{n,m} \rightarrow |f_{n,m}\rangle = \left[\sqrt{1-\frac{f_{n,m}}{L}}|0\rangle + \sqrt{\frac{f_{n,m}}{L}}|1\rangle\right] = \frac{1}{\sqrt{L}}\left[\sqrt{L-f_{n,m}}|0\rangle + \sqrt{f_{n,m}}|1\rangle\right]. \tag{7.1}$$

The amplitudes of this qubit are defined by the gray intensity of the image at this pixel. We may assume that the basis state $|0\rangle$ represents the black color and the state $|1\rangle$ the white color, or vice versa. The color black, 0, will be observed with the probability $(1-f_{n,m}/L)$ and the white color, L, with the probability $f_{n,m}/L$, when measuring this qubit.

We can define the following angle in the interval $[0, \pi/2]$:

$$\vartheta_{n,m} = \cos^{-1}\left(\sqrt{1-\frac{f_{n,m}}{L}}\right). \tag{7.2}$$

The quantum pixel is a single qubit with the superposition $|\varphi(f_{n,m})\rangle = \cos\vartheta_{n,m}|0\rangle + \sin\vartheta_{n,m}|1\rangle$. The image intensities are mapped, or encoded $(f_{n,m} \to \vartheta_{n,m})$, into the interval of angles $[0, \pi/2]$. For the $L = 255$ case, this mapping is illustrated in Fig. 7.2 together with the linear transform

$$f_{n,m} \to \phi_{n,m} = \frac{\pi}{2}\frac{f_{n,m}}{255}, f_{n,m} \in \{0, 1, 2, \ldots, 255\}.$$

This transform is considered in many models of quantum image representation. In this case, the calculation of the cosine function at new angles $\phi_{n,m}$ is required.

If the interval of the image intensity is [0,1], that is, $L = 1$, the quantum pixel is the qubit

$$|\varphi(f_{n,m})\rangle = \sqrt{1-f_{n,m}}|0\rangle + \sqrt{f_{n,m}}|1\rangle = \cos\vartheta_{n,m}|0\rangle + \sin\vartheta_{n,m}|1\rangle. \tag{7.3}$$

If $f_{n,m} = 0$, then the qubit pixel is $|\varphi(f_{n,m})\rangle = |0\rangle$, and when $f_{n,m} = 1$, the qubit pixel is $|1\rangle$.

In general, it is possible to change the range of the image to the interval [0,1], by using the transform $f_{n,m} = f_{n,m}/L$. This is a probabilistic representation of the pixel. The intensity is encoded into the angle $\vartheta_{n,m}$ in the interval $[0, \pi/2]$.

As an example, Fig. 7.3 shows the grayscale image $f_{n,m}$ of size 240×320 pixels and the range $L = 256$ in part (a) and its angular representation $\vartheta_{n,m}$ in part (b). The amplitudes of the basis states $|0\rangle$ and $|1\rangle$ at all pixels are shown in parts (c) and (d), respectively. In part (c), the image composed by the cosine coefficients, or $\sqrt{1-f_{n,m}/L}$, for all pixels is shown, and the image with all sine coefficients, or $\sqrt{f_{n,m}/L}$, is shown in part (d).

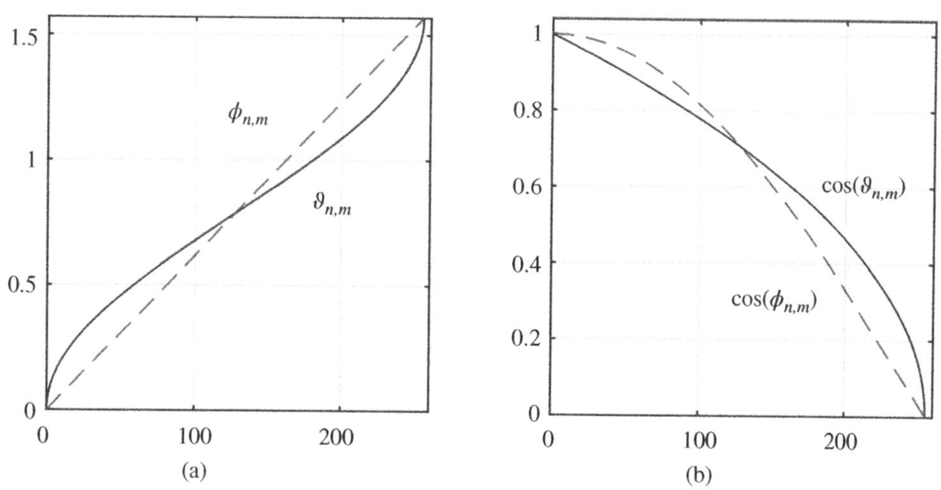

Fig. 7.2 The graphs of (a) two mappings of image intensities into the angles in the interval $[0, \pi/2]$ and (b) their cosine functions.

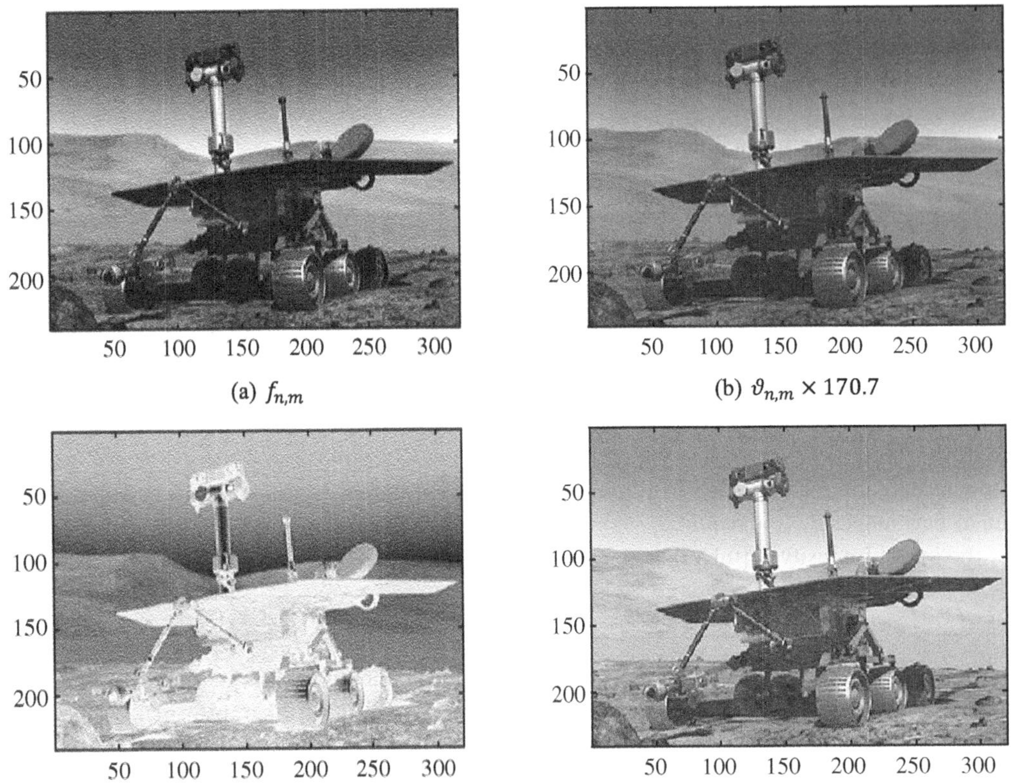

(a) $f_{n,m}$ (b) $\vartheta_{n,m} \times 170.7$

(c) $\cos\vartheta_{n,m} \times 255$ (d) $\sin\vartheta_{n,m} \times 255$

Fig. 7.3 (a) The original image, (b) the image of angles (after multiplying by factor of 170.7), and the image amplitudes at (c) the states $|0\rangle$ and (d) states $|1\rangle$ (after multiplying by factor of 255). *Source:* NASA / Public domain / https://spaceflightnow.com/news/n1007/30spirit/.

The whole image can be represented by the following $(r+s+1)$-qubit superposition:

$$|\varphi(f)\rangle = \frac{1}{\sqrt{NM}}\sum_{n=0}^{N-1}\sum_{m=0}^{M-1}|\varphi(f_{n,m})\rangle|n,m\rangle = \frac{1}{\sqrt{NM}}\sum_{n=0}^{N-1}\sum_{m=0}^{M-1}(\cos\vartheta_{n,m}|0\rangle + \sin\vartheta_{n,m}|1\rangle)|n,m\rangle. \tag{7.4}$$

The states $|n\rangle$ and $|m\rangle$ are the computational basis states and

$$|n,m\rangle = |n\rangle|m\rangle = |n\rangle\otimes|m\rangle = |n_{r-1}n_{r-2}\dots n_1n_0\rangle\otimes|m_{s-1}m_{s-2}\dots m_1m_0\rangle. \tag{7.5}$$

Here, the binary representations $\{n_{r-1}n_{r-2}\dots n_1n_0\}$ and $\{m_{s-1}m_{s-2}\dots m_1m_0\}$ of the numbers n and m are used.

It directly follows from the definition in Eq. 7.3 of the quantum pixel that the same information about the angle $\vartheta_{n,m}$ is stored in amplitudes of the basis states $|0\rangle$ and $|1\rangle$. The probability of a qubit to be in one of the states is defined by the same angle. The probability of measuring the quantum qubit in the state $|0\rangle$ or $|1\rangle$ defines the value of the image.

Note that the order of pixels $|n,m\rangle$ in the sum of Eq. 7.4 corresponds to writing the image line by line, when the pixel (n,m) is numbered as $k = nM + m$, as shown below for the image of 4×4 pixels,

$$\begin{bmatrix} 1 & 5 & 3 & 2 \\ 4 & 6 & . & . \\ . & . & . & . \\ 5 & 7 & 2 & 1 \end{bmatrix} \rightarrow \left[\underbrace{1}_{0000=0},\underbrace{5}_{0001=1},\underbrace{3}_{0010=2},\underbrace{2}_{0011=3},\underbrace{4}_{0100=4},\underbrace{6}_{0101=5},\dots,\underbrace{5}_{1100=12},\underbrace{7}_{1101=13},\underbrace{2}_{1110=14},\underbrace{1}_{1111=15}\right].$$

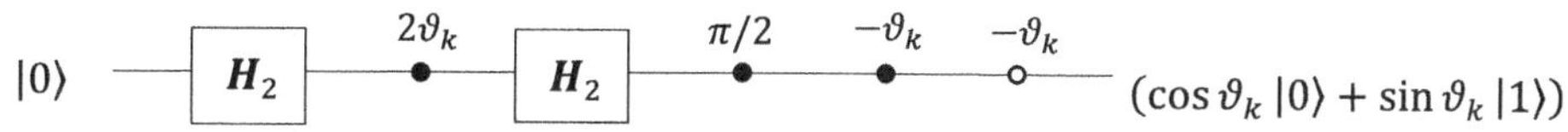

Fig. 7.4 The circuit for preparation of the single qubit.

This is why we can consider the image $f_{n,m}$ as the NM-dimensional vector, or signal, f_k, $k = 0 : (NM - 1)$, and write the superposition in Eq. 7.4 as

$$| \varphi(f)\rangle = \frac{1}{\sqrt{K}} \sum_{k=0}^{K-1} (\cos \vartheta_k | 0\rangle + \sin \vartheta_k | 1\rangle) | k\rangle, \ K = NM. \tag{7.6}$$

Each qubit $|\varphi_k\rangle = \cos \vartheta_k |0\rangle + \sin \vartheta_k |1\rangle$ in this sum can be prepared by the circuit similar to the one given in Fig. 2.5, when $\phi = 0$. Eq. 2.17 can be used for the angles $\vartheta = \vartheta_k$ as (see Fig. 7.4)

$$| \varphi_k\rangle = e^{-i\vartheta_k}[SH_2P(2\vartheta_k)H_2| 0\rangle] = \cos(\vartheta_k)| 0\rangle + \sin(\vartheta_k)| 1\rangle. \tag{7.7}$$

Indeed,

$$e^{-i\vartheta_k}\left(\begin{bmatrix} 1 & 0 \\ 0 & i \end{bmatrix} \times \frac{1}{\sqrt{2}}\begin{bmatrix} 1 & 1 \\ 1 & -1 \end{bmatrix} \times \begin{bmatrix} 1 & 0 \\ 0 & e^{i2\vartheta_k} \end{bmatrix} \times \frac{1}{\sqrt{2}}\begin{bmatrix} 1 & 1 \\ 1 & -1 \end{bmatrix}\begin{bmatrix} 1 \\ 0 \end{bmatrix}\right) = \begin{bmatrix} \cos(\vartheta_k) \\ \sin(\vartheta_k) \end{bmatrix}.$$

The result of this circuit on the input state $|0\rangle$ is the qubit $|\varphi_k\rangle$ without the global phase $-\vartheta_k$, since two phase shift gates P and P_- are used (see Table 2.1). These two phase gates can be substituted by the global phase gate $G(-\vartheta_k)$.

As an example, we consider this representation for the $N = M = 4$, or $K = 16$ case. The image 4×4 is written first into the 16D vector. Considering that $|0\rangle \otimes |k\rangle = |k\rangle$ and $|1\rangle \otimes |k\rangle = |16 + k\rangle$, when $k = 0, 1, 2, \ldots, 15$, the quantum representation of this vector can be written in the 32D row-vector form as

$$\langle\varphi(f) | = \left(\underbrace{\cos \vartheta_0, \cos \vartheta_1, \cos \vartheta_2, \ldots, \cos \vartheta_{15}}, \underbrace{\sin \vartheta_0, \sin \vartheta_1, \sin \vartheta_2, \ldots, \sin \vartheta_{15}}\right)/4, \tag{7.8}$$

or

$$| \varphi(f)\rangle = \frac{1}{4}[\cos \vartheta_0| 0\rangle + \cos \vartheta_1 | 1\rangle + \ldots + \cos \vartheta_{15}| 15\rangle + \sin \vartheta_0 | 16\rangle + \sin \vartheta_1| 17\rangle + \ldots + \sin \vartheta_{15}| 31\rangle].$$

If we denote by C and S two $K \times K$ diagonal matrices with cosine and sine coefficients, and by $\mathbf{C}$ and $\mathbf{S}$ two column vectors composing these coefficients, respectively, then the above superposition can be prepared as follows:

$$| \varphi(f)\rangle = \begin{bmatrix} C & -S \\ S & C \end{bmatrix} \frac{1}{\sqrt{K}} \begin{bmatrix} \mathbf{1} \\ \mathbf{0} \end{bmatrix} = \frac{1}{\sqrt{K}} \begin{bmatrix} \mathbf{C} \\ \mathbf{S} \end{bmatrix}, \text{or} \begin{bmatrix} C & -S \\ S & C \end{bmatrix} \left(| 0\rangle \otimes \frac{1}{\sqrt{K}} \sum_{k=0}^{K-1} | k\rangle\right) = | \varphi(f)\rangle. \tag{7.9}$$

Here, $\mathbf{1}$ and $\mathbf{0}$ denote the K-dimension vectors composed by only 1 and 0, respectively. These two vectors together represent the state $|0\rangle \otimes (|0\rangle + |1\rangle + |2\rangle + \ldots + |K - 1\rangle)$. The matrix $Q = [C, -S; S, C]$ is unitary, has a determinant equal 1, and is composed of K rotations. The $(r + s)$-qubit superposition

$$| P\rangle_{r+s} = H^{\otimes(r+s)}\left(| 0\rangle^{\otimes(r+s)}\right) = \frac{1}{\sqrt{K}} \sum_{k=0}^{K-1} | k\rangle \tag{7.10}$$

describes all pixel positions stored in the register for the $2^r \times 2^s$-pixel image $f_{n,m}$.

Comments 1

In the above model, as well as in a few other models discussed below, the superposition $|\varphi(f)\rangle$ in Eq. 7.6 is defined as the sum of qubits $|q_k\rangle = \cos\vartheta_k|0\rangle + \sin\vartheta_k|1\rangle$ in tensor product with the basis states $|k\rangle$. The tensor multiplication occurs on the right. We will call such superpositions *right-sided*. Together with the superposition $|\varphi(f)\rangle$, we consider *the left-sided superposition* defined as

$$|\,\psi(f)\rangle = \frac{1}{\sqrt{K}}\sum_{k=0}^{K-1}|\,k\rangle|\,q_k\rangle = \frac{1}{\sqrt{K}}\sum_{k=0}^{K-1}|\,k\rangle(\cos\vartheta_k|\,0\rangle + \sin\vartheta_k|\,1\rangle), K = NM. \tag{7.11}$$

In the general case, it will be good to understand the difference between the right- and left-sided superpositions

$$|\,\varphi(f)\rangle = \frac{1}{\sqrt{K}}\sum_{k=0}^{K-1}|\,q_k\rangle|\,k\rangle \text{ and } |\,\psi(f)\rangle = \frac{1}{\sqrt{K}}\sum_{k=0}^{K-1}|\,k\rangle|\,q_k\rangle, \tag{7.12}$$

for a given set of qubits $|q_k\rangle = a_k|0\rangle + b_k|1\rangle$, $k = 0:(K-1)$. The tensor product $|q_k\rangle|k\rangle$ results in the state which can be written as the bra-vector

$$\langle\varphi_k\,| = \left(\underbrace{0,0,\ldots,0,a_k,0,\ldots,0}, \underbrace{0,0,\ldots,0,b_k,0,\ldots,0}\right). \tag{7.13}$$

The vector components are spaced at a distance of K from each other. This was also shown in the above example for the case of $K = 16$ in Eq. 7.8. Now we consider the tensor product $|k\rangle|q_k\rangle$; the result is the bra-vector

$$\langle\psi_k\,| = \left(\underbrace{0,0}_{0},\underbrace{0,0}_{1},\underbrace{0,0}_{2},\ldots\underbrace{0,0}_{k-1},\underbrace{a_k,b_k}_{k},\underbrace{0,0}_{k+1},\ldots,\underbrace{0,0}_{K-1}\right), \tag{7.14}$$

The vector components of qubits $|q_k\rangle$ are placed together. This is why, the superposition $|\psi(f)\rangle$ can be described by the bra-vector

$$\langle\psi(f)\,| = \frac{1}{\sqrt{K}}\left(\underbrace{a_0,b_0}_{0},\underbrace{a_1,b_1}_{1},\underbrace{a_2,b_2}_{2},\ldots\underbrace{a_{k-1},b_{k-1}}_{k-1},\underbrace{a_k,b_k}_{k},\underbrace{a_{k+1},b_{k+1}}_{k+1},\ldots,\underbrace{a_{K-1},b_{K-1}}_{K-1}\right). \tag{7.15}$$

The superposition $|\varphi(f)\rangle$ is described by the bra-vector

$$\langle\varphi(f)\,| = \frac{1}{\sqrt{K}}\left(\underbrace{a_0,a_1,\ldots,a_{k-1},a_k,a_{k+1},\ldots,a_{K-1}},\underbrace{b_0,b_1,\ldots,b_{k-1},b_k,b_{k+1},\ldots,b_{K-1}}\right). \tag{7.16}$$

In the above case with angular representation of qubits, $|q_k\rangle = \cos\vartheta_k|0\rangle + \sin\vartheta_k|1\rangle$, $k = 0:15$, the preparation of the superposition in Eq. 7.15 (for $K = 16$)

$$\langle\psi(f)\,| = \frac{1}{\sqrt{K}}\left(\underbrace{\cos\vartheta_0,\sin\vartheta_0}_{0},\underbrace{\cos\vartheta_1,\sin\vartheta_1}_{1},\underbrace{\cos\vartheta_2,\sin\vartheta_2}_{2},\ldots,\underbrace{\cos\vartheta_{K-1},\sin\vartheta_{K-1}}_{K-1}\right)$$

can be performed by the block-diagonal matrix

$$\boldsymbol{R} = R_{\vartheta_0}\oplus R_{\vartheta_1}\oplus R_{\vartheta_2}\oplus\cdots\oplus R_{\vartheta_{K-2}}\oplus R_{\vartheta_{K-1}},$$

where R_{ϑ_k} are the matrices of rotation by the angles ϑ_k, $k = 0:(K-1)$. Thus, we can obtain the superposition as follows:

$$|\,\psi(f)\rangle = \boldsymbol{R}|\,A\rangle, \text{where } \langle A\,| = \frac{1}{\sqrt{K}}\left(\underbrace{1,0,1,0,\ldots,1,0,1,0}_{2K\text{ times}}\right) = \frac{1}{\sqrt{K}}\left(\underbrace{1,1,\ldots,1,1}_{K\text{ times}}\right)\otimes\langle 0\,|,$$

or

$$| \psi(f)\rangle = \mathbf{R}\left(\left[\frac{1}{\sqrt{K}}\sum_{k=0}^{K-1}| k\rangle\right]\otimes| 0\rangle\right). \tag{7.17}$$

Here, the number A of the state $|A\rangle$ is equal to $(4^{2n} - 1)/3$, $n = \log_2 K$. This preparation is much simpler than preparing the superposition $|\varphi(f)\rangle$ in Eq. 7.9.

Example 7.1 3-Qubit Superposition

Consider the case when $K = 4$, or $n = 2$. The number $A = 85$, the superposition is

$$\langle\psi(f)\,| = \frac{1}{2}\left(\underbrace{\cos\vartheta_0,\ \sin\vartheta_0}_{0},\ \underbrace{\cos\vartheta_1,\ \sin\vartheta_1}_{1},\ \underbrace{\cos\vartheta_2,\ \sin\vartheta_2}_{2},\ \underbrace{\cos\vartheta_3,\ \sin\vartheta_3}_{3}\right),$$

and the block-diagonal matrix is $\mathbf{R} = R_{\vartheta_0} \oplus R_{\vartheta_1} \oplus R_{\vartheta_2} \oplus R_{\vartheta_3}$.

The circuit for superposition $|\psi(f)\rangle$ preparation is given in Fig. 7.5. The 4×4 Hadamard matrix is used, to calculate the superposition $1/2(|0\rangle + |1\rangle + |2\rangle + |3\rangle)$.

The qubit-wise operation on the superposition $|\psi(f)\rangle$ is performed by the block-diagonal matrix $\boldsymbol{U} = U_0 \oplus U_1 \oplus U_2 \oplus \cdots \oplus U_{K-1}$, where U_k, $k = 0:(K-1)$, are 2×2 unitary matrices, or gates. Note that these two superpositions can be obtained from each other by the permutation P. As an example, the circuit for the superposition $|\varphi(f)\rangle$ preparation is given in Fig. 7.6.

For the $K = 4$ case, this permutation is $P = (124)(365)$, and for the $K = 8$ case, this permutation equals to $P = (1, 2, 4, 8)(3, 6, 12, 9)(5, 10)(7, 14, 13, 11)$.

In the general case, the circuit for the $|\psi(f)\rangle$ calculation is given in Fig. 7.7.

As directly follows from Eq. 7.1 (or Eq. 7.3), the simple operation of the negative image, $f_{n,m} \to L - f_{n,m}$, can be performing by the NOT gate as

$$| f_{n,m}\rangle = \sqrt{1-\frac{f_{n,m}}{L}}| 0\rangle + \sqrt{\frac{f_{n,m}}{L}}| 1\rangle \to X\left[\sqrt{1-\frac{f_{n,m}}{L}}| 0\rangle + \sqrt{\frac{f_{n,m}}{L}}| 1\rangle\right] = \sqrt{\frac{f_{n,m}}{L}}| 0\rangle + \sqrt{1-\frac{f_{n,m}}{L}}| 1\rangle.$$

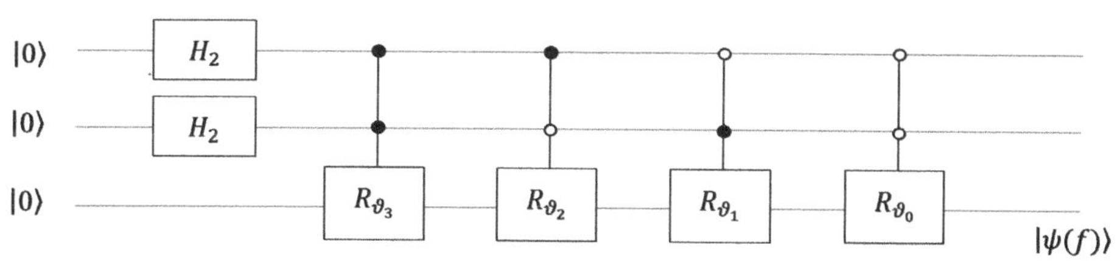

Fig. 7.5 The circuit for calculating the left-sided 3-qubit superposition $|\psi(f)\rangle$.

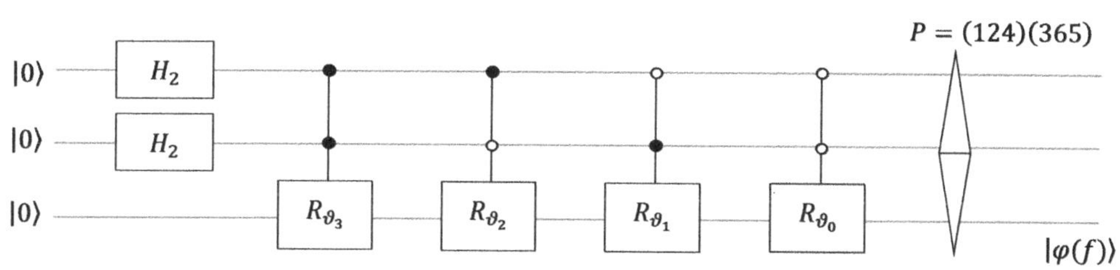

Fig. 7.6 The circuit for calculating the right-sided 3-qubit superposition $|\varphi(f)\rangle$.

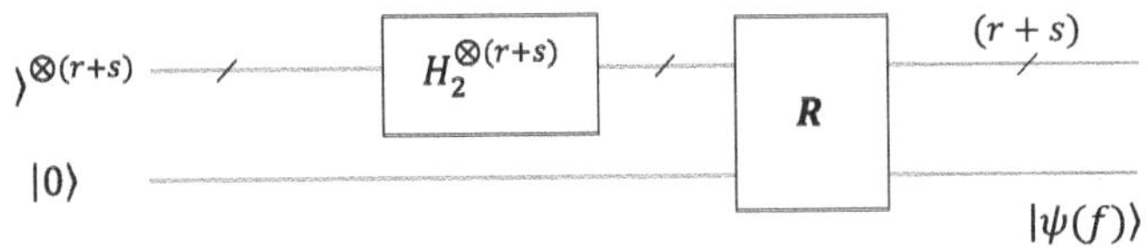

Fig. 7.7 The circuit for calculating the $(r+s+1)$-qubit left-sided superposition $|\psi(f)\rangle$.

Thus, the operation of the superposition is described as

$$|\,\psi(f)\rangle \rightarrow \left(I^{\otimes(r+s)} \otimes X\right)|\,\psi(f)\rangle = X^{\oplus(r+s)}|\,\psi(f)\rangle \tag{7.18}$$

and, when using the right-sided superposition, as follows:

$$|\,\varphi(f)\rangle \rightarrow \left(X \otimes I^{\otimes(r+s)}\right)|\,\varphi(f)\rangle. \tag{7.19}$$

The circuits for this operation, when using the left- and right-sided superpositions for image representation in the pixel model are given in Fig. 7.8 in parts (a) and (b) respectively, with the example of the "cameraman" image of a size of 256×256 pixels (the case when $r = s = 8$).

7.1.2 Qubit Lattice Model (QLM)

In the QLM, the discrete images are represented similarly to the model with quantum pixels. The grayscale or color image $f = f_{n,m}$ of size $N \times M$ pixels, as the $N \times M$ matrix, is presented by the $N \times M$ quantum matrix $\boldsymbol{Q} = \{|q\rangle_{n,m};\ n = 0:(N-1),\ m = 0:(M-1)\}$ with qubits [6]

$$|\,q\rangle_{n,m} = \cos\frac{\vartheta_{n,m}}{2}|\,0\rangle + e^{i\gamma}\sin\frac{\vartheta_{n,m}}{2}|\,1\rangle.$$

Here, the values of grays or colors are encoded in angles $\vartheta_{n,m} \in [0, \pi]$, and γ is a constant in the relative phase $e^{i\gamma}$. The qubit $|q\rangle_{n,m}$ represents the gray $f_{n,m}$ in Hopf coordinates $\cos\frac{\vartheta_{n,m}}{2}$ and $e^{i\gamma}\sin\frac{\vartheta_{n,m}}{2}$ with two degrees of freedom (no global phase). The angles $\vartheta_{n,m}$ are estimated by using a large number of measurements. Each qubit $|q\rangle_{n,m}$ in the matrix $\boldsymbol{Q}$ can be prepared by the circuit similar to the one given in Fig. 7.4, when $\gamma = 0$. In the $\gamma \neq 0$ case, Eq. 2.17 can be used for the angles $\vartheta_k = \vartheta_{n,m}$ as (see Fig. 7.9)

$$|\,q\rangle_k = e^{-i\vartheta_k}[P(\pi/2+\gamma)H_2P(\vartheta_k)H_2|\,0\rangle], \tag{7.20}$$

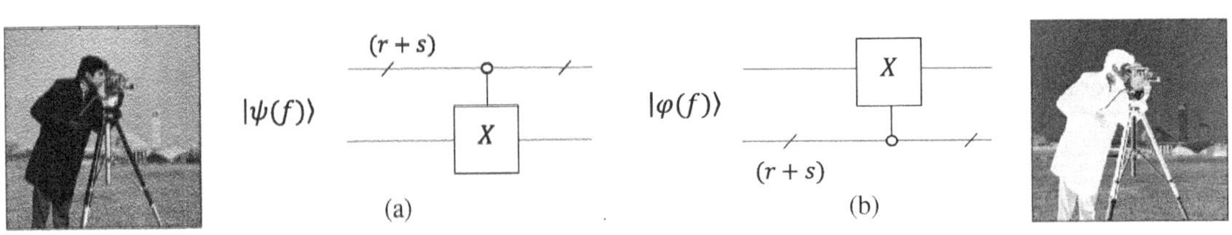

Fig. 7.8 The circuits for calculating the negative image by (a) the left-sided and (b) right-sided superpositions of the image.

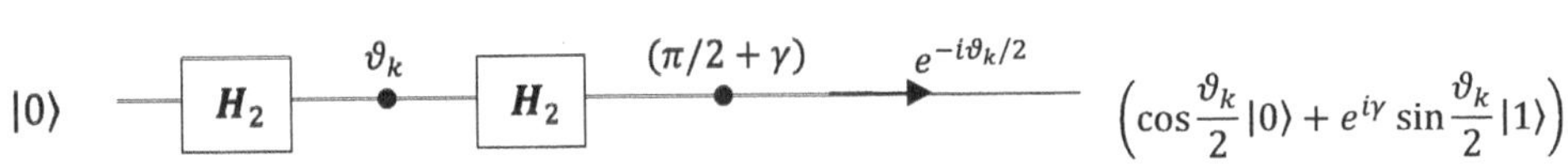

Fig. 7.9 The circuit for preparation of the single qubit with the relative phase.

or,

$$e^{-i\vartheta_k/2}\left(\begin{bmatrix}1 & 0\\ 0 & ie^{i\gamma}\end{bmatrix}\times\frac{1}{\sqrt{2}}\begin{bmatrix}1 & 1\\ 1 & -1\end{bmatrix}\times\begin{bmatrix}1 & 0\\ 0 & e^{i\vartheta_k}\end{bmatrix}\times\frac{1}{\sqrt{2}}\begin{bmatrix}1 & 1\\ 1 & -1\end{bmatrix}\begin{bmatrix}1\\ 0\end{bmatrix}\right)=\begin{bmatrix}\cos\left(\frac{\vartheta_k}{2}\right)\\ e^{i\gamma}\sin\left(\frac{\vartheta_k}{2}\right)\end{bmatrix}.$$

The number of single qubits stored in $\boldsymbol{Q}$ is NM, which is a very large number even for images of small size 256×256. The following $(r+s+1)$-qubit superposition $|\varphi(f)\rangle$ corresponds to the QLM lattice qubit model:

$$|\,\varphi(f)\rangle=\frac{1}{\sqrt{NM}}\sum_{n=0}^{N-1}\sum_{m=0}^{M-1}\left(\cos\frac{\vartheta_{n,m}}{2}|\,0\rangle+e^{i\gamma}\sin\frac{\vartheta_{n,m}}{2}|\,1\rangle\right)|\,n,m\rangle. \tag{7.21}$$

Note that in the model with quantum pixel in Eq. 7.6, the angles are calculated in advance from the values of the image, not estimated by measurements. Here, the angles $\vartheta_{n,m}/2$ have the same range as angles $\vartheta_{n,m}$ in Eq. 7.4. The phase factor $e^{i\gamma}$ is the new parameter in the QLM.

7.1.3 Flexible Representation for Quantum Images

In the flexible representation for quantum images (FRQI) model [7], the discrete integer image of size $2^r\times 2^r$ pixels and range $L=2^l$, $l>1$, is written $(f_{n,m}\to f_k)$ into a 4^r-dimensional (4^r-D) vector and then presented as the following $(2r+1)$-qubit superposition:

$$|\,\varphi(f)\rangle=\frac{1}{2^r}\sum_{k=0}^{4^r-1}|\,q_k\rangle\otimes|\,k\rangle=\frac{1}{2^r}\sum_{k=0}^{4^r-1}(\cos\vartheta_k|\,0\rangle+\sin\vartheta_k|\,1\rangle)\otimes|\,k\rangle. \tag{7.22}$$

The information of image colors is written (or encoded) into angles $\vartheta_k\in[0,\pi/2]$ and the state $|k\rangle$ is the corresponding pixel state $|n,m\rangle$. The interval angle $[0,\pi/2]$ is chosen according to the transform

$$f_k\to\vartheta_k=\frac{\pi}{2}\frac{f_k}{(L-1)},f_k\in\{0,1,2,\ldots,L-1\}.$$

All amplitudes in the image superposition $|\varphi(f)\rangle$ are real and non-negative numbers. This $(2r+1)$-qubit representation is the particular case of the representation in Eq. 7.6 for the case $M=N=2^r$. It is assumed here that the image can be written to the 4^r-D vector not only line by line, $f_{n,m}\to f_k$, $k=nM+m$. This can also be done in 2×2 blocks, for example, or column by column.

The left-sided superposition of the image f in this model will be written as

$$|\,\phi(\psi)\rangle=\frac{1}{2^r}\sum_{k=0}^{4^r-1}|\,k\rangle\otimes(\cos\vartheta_k|\,0\rangle+\sin\vartheta_k|\,1\rangle). \tag{7.23}$$

It should be noted that the reconstruction of the image values $f_{n,m}$ from amplitudes of the qubits in Eq. 7.1 or Eq. 7.3 is the square operation, $\sqrt{1-f_k}\to(1-f_k)$ and $\sqrt{f_k}\to f_k$. In the FRQI model, such reconstruction required the calculation of the arccosine and arcsine functions,

$$\underbrace{\cos\vartheta_k,\ \sin\vartheta_k}\to\vartheta_k\to\underbrace{f_k=\vartheta_k\times\frac{2(L-1)}{\pi}}.$$

In the FRQI model, different geometric transformations can be performed [3]. Such transforms are described as

$$|\,\varphi(f)\rangle\to(I\otimes U)|\,\varphi(f)\rangle=\frac{1}{2^r}\sum_{k=0}^{4^r-1}(\cos\vartheta_k|\,0\rangle+\sin\vartheta_k|\,1\rangle)\otimes U(|\,k\rangle), \tag{7.24}$$

where U is a unitary transform on the pixel-states; $U(|k\rangle) = U(|n, m\rangle)$. As an example, we consider the rotations of the image by 90 and 180 degrees.

The rotation by 90 degree

$$| \varphi(f)\rangle \rightarrow (I \otimes X^{\otimes r} \otimes I^{\otimes r})| \varphi(f)\rangle \rightarrow (I \otimes P_X)(I \otimes X^{\otimes r} \otimes I^{\otimes r})| \varphi(f)\rangle. \tag{7.25}$$

Here, P_X is the coordinate-swapping operation $P_X(|n, m\rangle) = |m, n\rangle$. Figure 7.10 shows the circuit for this operation with an illustration of the 512 × 512-pixel "lake" image.

The rotation by 180 degree can be described as

$$| \varphi(f)\rangle \rightarrow (I \otimes X^{\otimes r} \otimes X^{\otimes r})| \varphi(f)\rangle. \tag{7.26}$$

Figure 7.11 shows the circuit for this operation of rotation with an example on the 512 × 512-pixel 'jet-plane' image.

The operation of thresholding has very simple implementation in the FRQI model, as well. As an example, Figure 7.12 shows the circuit of calculating the binary image after thresholding the image

$$f_{n,m} \rightarrow g_{n,m} = \begin{cases} 1, \text{if } f_{n,m} \geq T; \\ 0, \text{otherwise.} \end{cases}$$

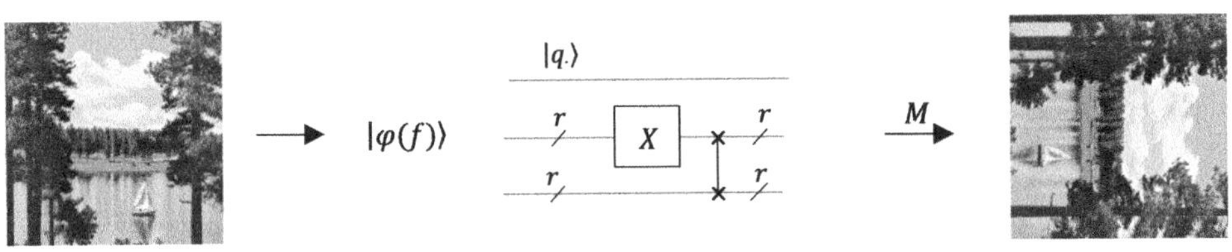

Fig. 7.10 The circuit for rotating the grayscale "lake" image by 90 degree.

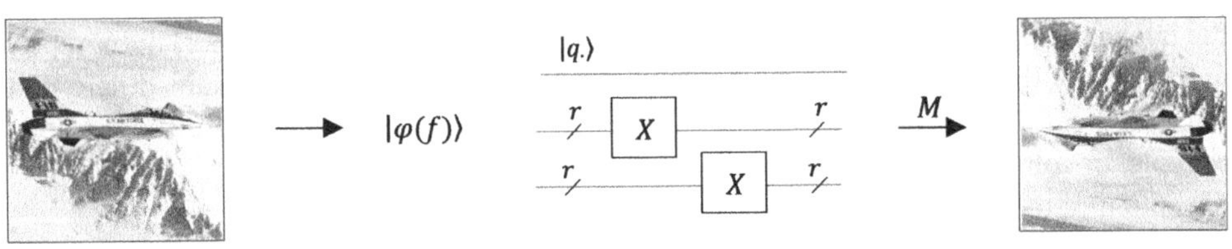

Fig. 7.11 The circuit for rotating the grayscale "jetplane" image by 180 degree.

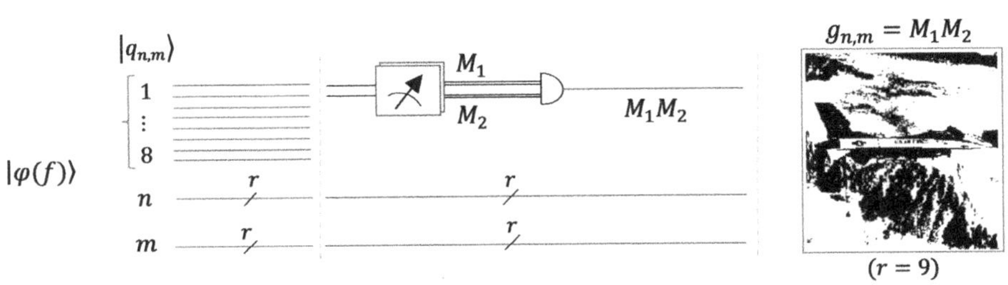

Fig. 7.12 The circuit for thresholding $f_{n,m} \rightarrow g_{n,m}$ the grayscale "jetplane" image by the threshold $T = 128 + 64$.

by the threshold $T = 192$ with an example of the 512×512-pixel 'jetplane' image. Note that $T = 128 + 64$. Therefore, this operation is described by measuring the first and second qubits in the intensity-state $|q_k\rangle = |q_{n,m}\rangle$ at each pixel, as follows:

$$f_{n,m} \to |\, q_{n,m}\rangle = |\, q_{n,m,0}, q_{n,m,1}, q_{n,m,2}, \ldots, q_{n,m,5}, q_{n,m,6}, q_{n,m,7}\rangle \to g_{n,m} = \begin{cases} 1, \text{if } q_{n,m,6} = q_{n,m,7} = 1, \\ 0, \text{if } q_{n,m,6} \text{ or } q_{n,m,7} = 0. \end{cases}$$

The image has the range $L = 2^8$ and its intensity is the 8-qubit state.

The model is designed for images of equal size $2^r \times 2^r$, not for the size $N \times M = 2^r \times 2^s$, when $s \neq r$, what are the dimensions of most images. It should be noted that the output images in the above circuits, as well as in other examples with the FRQI model, can be measured only approximately. The image retrieval is probabilistic; the image cannot be accurately obtained, by using only a small number of measurements. The circuits must be run many times (and this number exceeds hundreds of thousands), and this requires many copies of the input image. This model proposes many processing algorithms, which include changes in both image intensity and coordinates, line detection, histogram equalization, multilevel segmentation, image compression, and image encryption [2–4].

7.1.4 Representation of Amplitudes

Consider a discrete image $f_{n,m}$ of size $2^r \times 2^s$ pixels and the range $L = 2^l$, $l > 1$. The state of pixel (n, m) can be defined as

$$q_{n,m} = \frac{f_{n,m}}{\sqrt{E}}|\, n\rangle|\, m\rangle, E = E(f) = \sum_{n=0}^{N-1}\sum_{m=0}^{M-1} f_{n,m}^2. \tag{7.27}$$

Therefore, the image can be presented by the $(r + s)$-qubit superposition

$$|\, \varphi(f)\rangle = \sum_{n=0}^{N-1}\sum_{m=0}^{M-1} q_{n,m} = \frac{1}{\sqrt{E}}\sum_{n=0}^{N-1}\sum_{m=0}^{M-1} f_{n,m}|\, n, m\rangle. \tag{7.28}$$

This is the simplest image representation model of all known models. It requires $(r + s)$ qubits. Note that in the $L = 8$ case, the discrete image with eight intensity values uses $8NM = 2^{(r+s+3)}$ bits.

If we write the image $f_{n,m}$ into the 1D signal f_k line by line, as shown above in the pixel model, then the presented image representation can be written as

$$|\, \varphi(f)\rangle = \frac{1}{\sqrt{E}}\sum_{k=0}^{NM-1} f_k|\, k\rangle. \tag{7.29}$$

This is the image probabilistic quantum representation, wherein the pixel states $|k\rangle$ have different weights, or probabilities of measurement. The above representation corresponds to our usual representation of a signal (image) as an NM-dimension vector. Indeed, consider the $NM = 8$ case. The superposition can be written as

$$|\, \varphi(f)\rangle = \frac{1}{\sqrt{E}}\left(f_0\begin{bmatrix}1\\0\\0\\\vdots\\0\\0\end{bmatrix} + f_1\begin{bmatrix}0\\1\\0\\\vdots\\0\\0\end{bmatrix} + f_2\begin{bmatrix}0\\0\\1\\\vdots\\0\\0\end{bmatrix} + \ldots + f_6\begin{bmatrix}0\\0\\0\\\vdots\\1\\0\end{bmatrix} + f_7\begin{bmatrix}0\\0\\0\\\vdots\\0\\1\end{bmatrix}\right) = \frac{1}{\sqrt{E}}\begin{bmatrix}f_0\\f_1\\f_2\\\vdots\\f_6\\f_7\end{bmatrix} = \frac{1}{\sqrt{E}}\boldsymbol{f}.$$

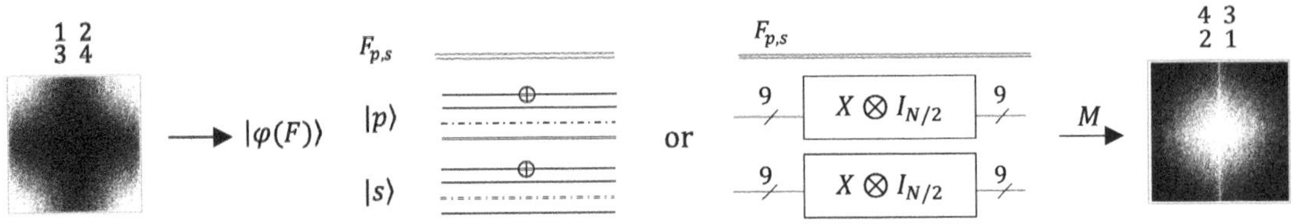

Fig. 7.13 The schemes of cyclic shifting of four parts of the 2-D QFT.

This model with amplitudes requires a minimum number of qubits and is very useful, when using such complex transforms as the quantum Fourier, Hadamard, Hartley, paired transform, cosine, and wavelet transforms. For transforms like the Fourier and Hartley transforms, the operation of cyclic shift in the center is used. For instance, for the 2D $N\times N$-point DFT, this transformation is described as

$$F_{p,s} \rightarrow F_{\left(p-\frac{N}{2}\right)\text{mod } N,\left(s-\frac{N}{2}\right)\text{mod } N}, p, s = 0{:}\ (N-1).$$

In the quantum superposition, this operation is described as

$$|\varphi(F)\rangle = \frac{1}{\sqrt{E}}\sum_{p=0}^{N-1}\sum_{s=0}^{N-1} F_{p,s}|p,s\rangle \rightarrow \frac{1}{\sqrt{E}}\sum_{p=0}^{N-1}\sum_{s=0}^{N-1} F_{p,s}\left|\left(p-\frac{N}{2}\right)\text{mod } N, \left(s-\frac{N}{2}\right)\text{mod } N\right\rangle.$$

Figure 7.13 shows the circuit of this operation on the 2D QFT of the 'jetplane' image of size $N\times N = 2^9\times 2^9$. Let us analyze the amplitude representation in the example with an integer signal of length $N = 8$.

Example 7.2 Superposition Without State $|1\rangle$
We consider the real signal or vector $f = (f_0, f_1, f_2, \ldots, f_7) = (1, 0, 1, 1, 2, 4, 3, 2)$. The energy of the signal $E(f) = 1 + 0 + 1 + 1 + 4 + 16 + 9 + 4 = 36$, and $\sqrt{E(f)} = 6$. Therefore,

$$|\varphi(f)\rangle = \frac{1}{6}\sum_{k=0,k\neq 1}^{7} f_k|k\rangle = \frac{1}{6}|0\rangle + \frac{1}{6}|2\rangle + \frac{1}{6}|3\rangle + \frac{1}{3}|4\rangle + \frac{2}{3}|5\rangle + \frac{1}{2}|6\rangle + \frac{1}{3}|7\rangle.$$

The state $|k\rangle$ can be observed (measured) in this superposition with probability $p_k = f_k^2/E$, $k \in \{0, 2, 3, \ldots, 7\}$. There is no state $|1\rangle$ in this representation. The highest value of the image, 4 at the state $|5\rangle$, has the highest probability of 4/9. Is this superposition represent the signal well? As mentioned in [8], this is a misrepresentation of the real signal. All values of the signal are important and must be treated in the same way. In general, the signal may have many zeros.

Now, let us consider the new signal $g = f - 1 = (0, -1, 0, 0, 1, 3, 2, 1)$, the energy equals $E(g) = 16$. Therefore, the representation of the new signal in quantum state space is

$$|\varphi(g)\rangle = -\frac{1}{4}|1\rangle + \frac{1}{4}|4\rangle + \frac{3}{4}|5\rangle + \frac{1}{2}|6\rangle + \frac{1}{4}|7\rangle.$$

The highest probability of 9/16 is at the basis state $|5\rangle$. This simple operation significantly changes the superposition amplitudes, that is, the probabilities of observing them. Three states, $|0\rangle$, $|2\rangle$, and $|3\rangle$, are missing and the new state appears, $|1\rangle$, in the signal g. That is, in such an image model, the state can disappear and then appear from nowhere.

Note that for real images, many of the basis states will be missing from this superposition, namely those pixels where the image is zero, $f_k = 0$. In other words, in this superposition such states do not exist, it is incomplete. And there can be many such points in the image, especially for the binary images or images after thresholding. If some states are missing in the quantum superposition, this means, oddly enough, that there are no several pixels in the image. In other words, the quantum superposition cannot be measured at many states.

When comparing with the pixel model described by Eq. 7.6, we note that each zero value ($f_k = 0$) of the image is written in the one-state-qubit $\cos\vartheta_k|0\rangle + \sin\vartheta_k|1\rangle = 1|0\rangle + 0|1\rangle = |0\rangle$, and in the superposition $|\varphi(f)\rangle$, it will be written in the state $|k\rangle$, where $k \in \{1, 2, \ldots, K\}$, of the 2^{r+s+1}-D computational basis B_{r+s+1}. The highest values of the image ($f_k = L-1$) are encoded in the angle $\vartheta_k = \pi/2$ with qubit $\cos\vartheta_k|0\rangle + \sin\vartheta_k|1\rangle = 0|0\rangle + 1|1\rangle = |1\rangle$. In the superposition $|\varphi(f)\rangle$, it will be written in the state $|k+K\rangle$ of the basis B_{r+s+1}, where $k \in \{1, 2, \ldots, K\}$. Therefore, Eq. 7.6 for the pixel model can be written as

$$|\varphi(f)\rangle = \frac{1}{\sqrt{K}}\left[\sum_{\{k;f_k \neq 0, L-1\}} (\cos\vartheta_k|0\rangle + \sin\vartheta_k|1\rangle)|k\rangle + \sum_{\{k;f_k=0\}} |k\rangle + \sum_{\{k;f_k=L-1\}} |k+K\rangle\right]. \tag{7.30}$$

These states with $f_k = 0$ (darkest pixels) and $f_k = 0$ (brightest pixels), if such pixels exist, will have the same highest amplitude $1/\sqrt{K}$, or the highest probability of $1/K$, when observed (measured). This may be contrary to our expectations when we are working with grayscale images. For instance, when the image has a large number of zero pixels and only several brightest pixels. Note that with such division of pixels, the model in Eq. 7.29, for the same image, can be written in the following form:

$$|\varphi_A(f)\rangle = \frac{1}{\sqrt{E}}\sum_{\{k;f_k \neq 0, L-1\}} f_k|k\rangle + \frac{L-1}{\sqrt{E}}\sum_{\{k;f_k=L-1\}} |k\rangle, \tag{7.31}$$

or, for the integer image,

$$|\varphi_A(f)\rangle = \frac{1}{\sqrt{E}}\sum_{j=0}^{L-1} j\left[\sum_{\{k;f_k=j\}} |k\rangle\right].$$

Example 7.3 FRQI of a Binary Image

Consider a binary image f_k, that is, when $L = 2$. Then, in the FRQI model, the image superposition is

$$|\varphi(f)\rangle = \frac{1}{\sqrt{K}}\left[\sum_{\{k;f_k=0\}} |k\rangle + \sum_{\{k;f_k=1\}} |k+K\rangle\right] = \frac{1}{\sqrt{K}}\sum_{k=0}^{K-1} |k+f_kK\rangle.$$

The quantum image representation by amplitudes equals to

$$|\varphi_A(f)\rangle = \frac{1}{\sqrt{E}}\sum_{\{k;f_k=1\}} |k\rangle = \frac{1}{\sqrt{E}}\sum_{k=0}^{K-1} f_k|k\rangle.$$

In both models, the information of the binary image is written into the computational basis states, but FRQI superposition is in the basis twice larger. Moreover, the normalized coefficient $\sqrt{E}$ corresponds to the energy of the image, whereas the coefficient $\sqrt{K} > \sqrt{E}$ is defined by the size of the image. In the model with amplitudes, the basis states will be observed only at pixels with value 1, with the high probability $1/E$. In the FRQI model, the basis states at pixels with 0 and 1 values will be observed with the smaller probability $1/K < 1/E$.

In Section 5, the quantum representation by amplitudes was used to calculate the quantum Fourier, Hartley, paired, and cosine transforms. Many other operations can also be considered in this model, including the gradient operators are calculated as the 2D liner windowed convolution of the image with a given mask.

7.1.5 Gradient and Sum Operators

The simple gradient operations in the horizontal and vertical directions are the following differencing operators, respectively: $G_x : f_{n,m} \rightarrow G_x(f)_{n,m} = f_{n+1,\,m} - f_{n,m}$ and $G_y : f_{n,m} \rightarrow G_y(f)_{n,m} = f_{n,m+1} - f_{n,m}$. The images $G_x(f)$ and $G_y(f)$ are called *the horizontal* (or *row*) *gradient* and *the vertical* (or *column*) *gradient* of the image, respectively. It is common to represent the gradient operators by matrices (or masks). Let us consider the following four points on the image that are the neighbors to (n, m):

$$W_{2\times 2} = W_{2\times 2}(n,m) = \begin{bmatrix} \underline{f_{n,m}} & f_{n,m+1} \\ f_{n+1,m} & f_{n+1,m+1} \end{bmatrix}.$$

The center of this window is at the point (n, m); therefore, the element $f_{n,m}$ is underlined. Now, we consider two matrices that represent the above gradient operators

$$[G_x] = \begin{bmatrix} \underline{1} & 0 \\ -1 & 0 \end{bmatrix} \text{ and } [G_y] = [G_x]' = \begin{bmatrix} \underline{1} & -1 \\ 0 & 0 \end{bmatrix}. \tag{7.32}$$

Here, we also mention the following 2×2 and 3×3 masks, which usually written in the form of matrices:

$$R_{2,2} = \frac{1}{2}\begin{bmatrix} 1 & 0 \\ 0 & -1 \end{bmatrix}, R_{3,3} = \frac{1}{3}\begin{bmatrix} 1 & 0 & -1 \\ 1 & 0 & -1 \\ 1 & 0 & -1 \end{bmatrix}, S_{3,3} = \frac{1}{4}\begin{bmatrix} 1 & 0 & -1 \\ 2 & 0 & -2 \\ 1 & 0 & -1 \end{bmatrix}, P_{3,3} = \frac{1}{5}\begin{bmatrix} 1 & 1 & -1 \\ 1 & -2 & -1 \\ 1 & 1 & -1 \end{bmatrix}.$$

These are the masks of the Robert, Robinson, Sobel, and Prewitt operators for edge extraction along the horizontal direction [9]. The corresponding gradient operators along the vertical directions are described by the transpose masks. The coefficients at matrices are defined to preserve the range of the image in the absolute scale. Many other gradient operators are also used in image processing for edge extraction in different directions [10]. In the Robert operators, the differences in the image are calculated along the diagonals. The main operation in other gradients is described as $f_k - f_{k+1}$ or $f_k - f_{k+2}$, if using the 1D row-wise (or column-wise) representation of the image, $f_{n,m} \rightarrow f_k$. In the Robinson gradient, such operations are repeated along three rows and columns, and in the Sobel gradient, along two rows and columns. Therefore, a few copies of the input image are required to perform these operations at each pixel, if we want to make a quantum circuit, to calculate such gradient images.

The above masks, as matrices, are not orthogonal and have determinant 0. Probably, for quantum edge detection, it would be good to define masks, or matrices of order 2×2 and 4×4, not and 3×3. There are real unitary matrices that involve calculating such data differences, as well as performing other operations such as sums. Such examples are described in Section 12, when unitary transforms perform both gradient and average operations. Here, we stand on the gradients in Eq. 7.32 and consider the following 2×2 masks:

$$M_2 = \frac{1}{2}\begin{bmatrix} 1 & -1 \\ 1 & 1 \end{bmatrix} \text{ and } M_2' = \frac{1}{2}\begin{bmatrix} 1 & 1 \\ -1 & 1 \end{bmatrix}, \text{ or } M_2' = \frac{1}{2}\begin{bmatrix} 1 & 1 \\ 1 & -1 \end{bmatrix}.$$

The result of applying the first mask on the 1D data f_k is described as

$$\frac{1}{2}\begin{bmatrix} 1 & -1 \\ 1 & 1 \end{bmatrix}\begin{bmatrix} f_k \\ f_{k+1} \end{bmatrix} = \frac{1}{2}\begin{bmatrix} f_k - f_{k+1} \\ f_k + f_{k+1} \end{bmatrix},$$

and, on the image pixel, it is

$$\frac{1}{2}\begin{bmatrix} 1 & -1 \\ 1 & 1 \end{bmatrix} * \begin{bmatrix} \underline{f_{n,m}} & f_{n,m+1} \\ f_{n+1,m} & f_{n+1,m+1} \end{bmatrix} = \frac{1}{2}\begin{bmatrix} f_{n,m} - f_{n,m+1} \\ f_{n+1,m} + f_{n+1,m+1} \end{bmatrix}.$$

The mask M_2 is the orthogonal matrix A_2, and the second matrix M_2' is the Hadamard matrix H_2. Here the equality is accurate up to a normalized coefficient (1/2 instead of $1/\sqrt{2}$). Therefore, on the image in quantum representation $|\varphi_A(f)\rangle$ and its cyclic shifting $|\varphi_A(f_X)\rangle$ along the horizontal direction, the following operator can be applied:

$$V = I_{K/2} \otimes A_2 = \frac{1}{\sqrt{2}} \begin{bmatrix} 1 & -1 & 0 & 0 & 0 & 0 & 0 \\ 1 & 1 & 0 & 0 & 0 & 0 & 0 \\ 0 & 0 & 1 & -1 & 0 & 0 & 0 \\ 0 & 0 & 1 & 1 & 0 & 0 & 0 \\ 0 & 0 & 0 & 0 & \ddots & 0 & 0 \\ 0 & 0 & 0 & 0 & 0 & 1 & -1 \\ 0 & 0 & 0 & 0 & 0 & 1 & 1 \end{bmatrix}. \tag{7.33}$$

This operation on the original superposition will result in the following superposition:

$$V(\varphi_A(f)\rangle) = |G\rangle_1 + |S\rangle_1 = \frac{1}{\sqrt{2E}} \left[\sum_{k=0:2:K-2} (f_k - f_{k+1})|k\rangle + \sum_{k=0:2:K-2} (f_k + f_{k+1})|k+1\rangle \right]$$

and on the second copy of the image, we obtain ($f_K = f_0$)

$$G(\varphi_A(f)\rangle_X) = |G\rangle_2 + |S\rangle_2 = \frac{1}{\sqrt{2E}} \left[\sum_{k=0:2:K-2} (f_{k+1} - f_{k+2})|k\rangle + \sum_{k=0:2:K-2} (f_{k+1} + f_{k+2})|k+1\rangle \right].$$

If we measure the first qubits in both superpositions in state 0, we get the amplitude of all differences $(f_k - f_{k+1})$, $k = 0:(K-1)$. The measurement of the first qubits at state 1 results in the sums $(f_k + f_{k+1})$, $k = 0:(K-1)$. The abstract circuit for calculating the superpositions $|G\rangle_1$ and $|S\rangle_1$ from the original image and the similar pair $|G\rangle_2$ and $|S\rangle_2$ from the shifted image f_X is given in Fig. 7.14. The superpositions $|G\rangle_1$ and $|G\rangle_2$ together define a gradient image, and $|S\rangle_1$ and $|S\rangle_2$ define a smooth image. These two images are the 2D linear windowed convolutions of the image. The circuit is illustrated with the example for the 512 × 512-pixel "pepper" image, assuming the ideal case of measuring each part of the circuit. The circuit for cyclic shifting is described in the next section and not shown in Fig. 7.14. Sections 12.6 and 12.7 describe gradient operations in more detail with many examples.

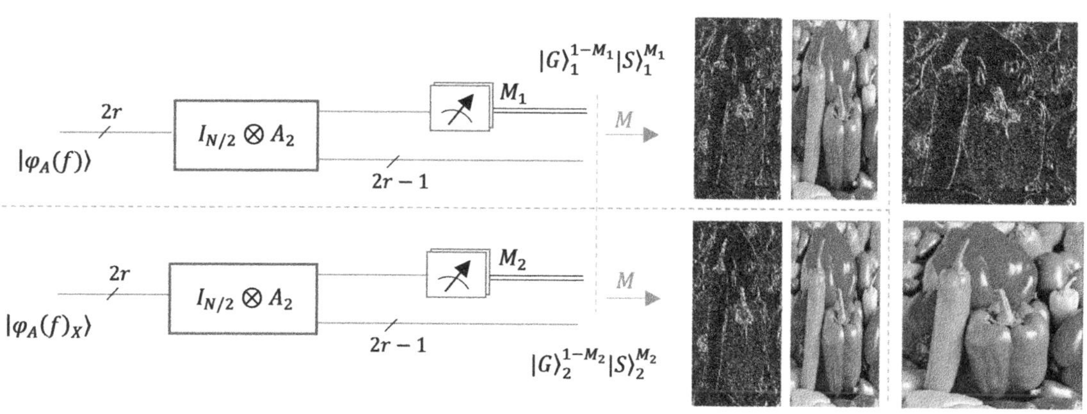

Fig. 7.14 The circuit for calculating two parts of the gradient and smooth images, when applying the Kronecker sum of A_2 matrices.

7.1.6 Real Ket Model

In the real ket model (RKM), the discrete image of square size $N \times N = 2^r \times 2^r$ is divided consequently down into four equal parts and the image is presented as the quantum superposition [11]

$$|\varphi(f)\rangle\rangle = \frac{1}{\sqrt{E(f)}} \sum_{n_1, n_2, \ldots, n_r = 0,1,2,3} c_{n_1,n_2,\ldots,n_r} |(n_1 n_2 \ldots n_r)\rangle\rangle. \tag{7.34}$$

The coefficients $c_{n_1,n_2,\ldots,n_r}$ are values of the image $f_{n,m}$ in different numbering. The numbering of pixels $(n, m) \to (n_1, n_2, \ldots, n_r)$ requires only r numbers from 0, 1, 2, and 3. The notation $|(n_1 n_2 \ldots n_r)\rangle\rangle$ is used for the basis state which is defined by the 2-bit binary representation of numbers $n_1, n_2, \ldots, n_r$. This is an important fact; $|(n_1 n_2 \ldots n_r)\rangle\rangle$ is not the tensor product $|n_1\rangle \otimes |n_2\rangle \otimes \ldots \otimes |n_r\rangle$. The quantum representation $|\varphi(f)\rangle\rangle$ of the image requires $2r$ qubits.

Example 7.4 Model of 8 × 8 Pixel Image
The numbering of the coefficients for the 8 × 8 image is illustrated in Fig. 7.15. For example, the coefficient in the pixel (2, 1) is numbered as (0, 2, 1), and the coefficient in the pixel (5, 4) is numbered as (3, 0, 2). Therefore, the values $c_{0,2,1} = f_{2,1}$ and $c_{3,0,2} = f_{5,4}$ are amplitudes at the computational basis states $|(021)\rangle\rangle$ and $|302\rangle$, respectively. The image is represented by the superposition

$$|\varphi(f)\rangle\rangle = \frac{1}{\sqrt{E(f)}}\left[f_{0,0}|(000)\rangle\rangle + f_{0,1}|(001)\rangle\rangle + f_{0,2}|(010)\rangle\rangle + \ldots + f_{7,6}|(332)\rangle\rangle + f_{7,7}|(333)\rangle\rangle\right]. \tag{7.35}$$

The energy of this image is $E(f) = 224$. Let us consider the state at pixel (2,1), that is, $|(021)\rangle\rangle$. With two bits, we have $0 \to |00\rangle$, $2 \to |10\rangle$, and $1 \to |01\rangle$. Therefore, (021) is the string of bits 001001 which is number 9 and $|(021)\rangle\rangle = |9\rangle$. For the last pixel state number (333), it will be the string 111111, or the number 63. So, the state $|(333)\rangle\rangle$ is $|63\rangle$. Six qubits are required for presentation of this image in Eq. 7.35. Consider that $|021\rangle = |1000\rangle \otimes |0010\rangle \otimes |0100\rangle$ is the state with $4^3 = 2^6$ bits, or the 6-qubit basis state.

Note that the above numbering relates to the well-known quaternary numeral system. The superposition in Eq. 7.35 is in fact a superposition written in a different way with the basis states $B_{64} = \{|0\rangle, |1\rangle, \ldots, |63\rangle\}$,

$$|\varphi(f)\rangle\rangle = \frac{1}{\sqrt{E(f)}}\left[f_{0,0}|0\rangle + f_{0,1}|1\rangle + f_{0,2}|2\rangle + \ldots + f_{7,6}|62\rangle + f_{7,7}|63\rangle\right]. \tag{7.36}$$

It is the image representation by amplitude in Eq. 7.29.

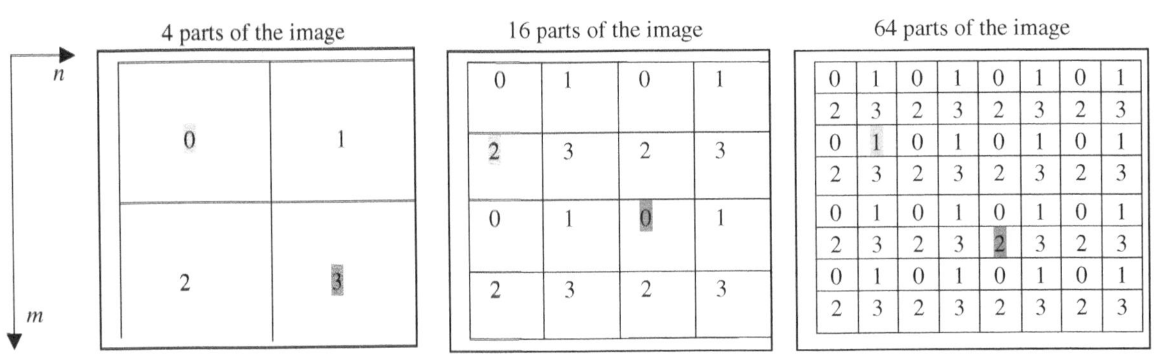

Fig. 7.15 Scheme of splitting an image into blocks and numbering for an 8 × 8 image.

7.1.7 General and Novel Enhanced Quantum Representations (GQIR and NEQR)

In *the novel enhanced quantum representation* model (NEQR) [12], a discrete integer-valued image $f_{n,m}$ of size $N \times M = 2^r \times 2^r$ pixels is considered with the range, or the number of grays, $L = 2^l$, where $l > 1$. This method was extended for other size $N \times M = 2^r \times 2^s$ of images by *the generalized quantum image representation* model (GQIR) [13, 14]. The qubit states of pixels $|n, m\rangle$ are nested into the basis states $|f_{n,m}\rangle$. For instance, the value $f_{1,3} = 37$ of the 4×4 image in the range of 256 can be written as the basis state $|f_{1,3}\rangle|1\rangle|3\rangle = |37\rangle|1\rangle|3\rangle = |0010\ 0101\rangle|01\rangle|11\rangle = |0010\ 0101\rangle|0111\rangle$. Here, $|37\rangle$ is the state in the computational basis $B = \{|0\rangle, |1\rangle, |2\rangle, \ldots, |255\rangle\}$.

The quantum superposition of the image is defined as

$$| \varphi(f)\rangle = \frac{1}{\sqrt{2^{r+s}}}\sum_{n=0}^{N-1}\sum_{m=0}^{M-1} | f_{n,m}\rangle | n, m\rangle. \tag{7.37}$$

Here, $(r+s)$ qubits are used for pixel location $|n, m\rangle$. This quantum superposition requires $(r+s+8)$ qubits for integer-valued images in the range of 256. When the range of the image is 2^l, $l > 1$, then the quantum representation requires $(r+s+l)$ qubits. The $r = s$ case corresponds to NEQR, and the general case of r and s, to the GQIR. More models of quantum image representation can be found in surveys [2, 3].

Example 7.5 Model of 2 × 2 Pixel Image)

Let us consider the small image of 2×2 pixels, $f = \begin{bmatrix} 2 & 4 \\ 6 & 3 \end{bmatrix}$. The range of the image is 2^3. Thus, $r = s = 1$ and $l = 3$. This image is represented as the following quantum 5-qubit superposition:

$$\begin{aligned} | \varphi(f)\rangle &= \frac{1}{2}(| 2\rangle | 0,0\rangle + | 4\rangle | 0,1\rangle + | 6\rangle | 1,0\rangle + | 3\rangle | 1,1\rangle) = \frac{1}{2}(| 010\rangle | 0,0\rangle + | 100\rangle | 0,1\rangle \\ &\quad + | 110\rangle | 1,0\rangle + | 011\rangle | 1,1\rangle) \\ &= \frac{1}{2}(| 01000\rangle + | 10001\rangle + | 11010\rangle + | 01111\rangle) = \frac{1}{2}(| 8\rangle + | 17\rangle + | 26\rangle + | 15\rangle). \end{aligned}$$

It should be noted that the last two bits in each basis state show the position of the pixel. Therefore, when processing the colors of the image in this superposition, these two bits must be preserved for all four states of $|\varphi(g)\rangle$ of a new image g. The first three bits can be only changed. In other words, the superposition of the new image should be in the same form,

$$| \varphi(g)\rangle = \frac{1}{2}[| \cdots 00\rangle + | \cdots 01\rangle + | \cdots 10\rangle + | \cdots 11\rangle].$$

The operations on the two last bit planes represent different types of permutations of pixels.

The left-sided representation of the image in this model is defined as

$$| \psi(f)\rangle = \frac{1}{\sqrt{2^{r+s}}}\sum_{n=0}^{N-1}\sum_{m=0}^{M-1} | n, m\rangle | f_{n,m}\rangle = \frac{1}{\sqrt{2^{r+s}}}\sum_{n=0}^{N-1} | n\rangle \sum_{m=0}^{M-1} | m\rangle | f_{n,m}\rangle.$$

For the above example, we obtain

$$\begin{aligned} | \psi(f)\rangle &= \frac{1}{2}(| 0,0\rangle | 010\rangle + | 0,1\rangle | 100\rangle + | 1,0\rangle | 110\rangle + | 1,1\rangle | 011\rangle) = \\ &= \frac{1}{2}(| 00010\rangle + | 01100\rangle + | 10110\rangle + | 11011\rangle) = \frac{1}{2}(| 2\rangle + | 12\rangle + | 22\rangle + | 27\rangle). \end{aligned}$$

If we consider *the left-and right-sided superposition* of the same image, which is defined as

$$|\phi(f)\rangle = \frac{1}{\sqrt{2^{r+s}}}\sum_{m=0}^{N-1}|n\rangle\left[\sum_{m=0}^{M-1}|f_{n,m}\rangle|m\rangle\right],$$

then we obtain the superposition

$$|\phi(f)\rangle = \frac{1}{2}(|0\rangle|010\rangle|0\rangle + |0\rangle|100\rangle|1\rangle + |1\rangle|110\rangle|0\rangle + |1\rangle|011\rangle|1\rangle) =$$

$$= \frac{1}{2}(|00100\rangle + |01001\rangle + |11100\rangle + |10111\rangle) = \frac{1}{2}(|4\rangle + |9\rangle + |28\rangle + |23\rangle).$$

These are three representations of the same image with different basis states.

It should be noted that the right- and left-sided representations will not be changed, when the 2D image is considered as a 1D signal $f = [2, 4, 6, 3]$. In other words,

$$|\varphi(f)\rangle = \frac{1}{2}\sum_{k=0}^{3}|f_k\rangle|k\rangle = \frac{1}{2}(|010\rangle|00\rangle + |100\rangle|01\rangle + |110\rangle|10\rangle + |011\rangle|11\rangle) = \frac{1}{2}(|8\rangle + |17\rangle + |26\rangle + |15\rangle).$$

$$|\psi(f)\rangle = \frac{1}{2}\sum_{k=0}^{3}|k\rangle|f_k\rangle = \frac{1}{2}(|2\rangle + |12\rangle + |22\rangle + |27\rangle).$$

Let us see how these two superpositions change when 1 is added to the image. The new image is $f + 1 = [3, 5, 7, 4]$ and the right-sided superposition equals to

$$|\varphi(f+1)\rangle = \frac{1}{2}(|011\rangle|00\rangle + |101\rangle|01\rangle + |111\rangle|10\rangle + |100\rangle|11\rangle) = \frac{1}{2}(|12\rangle + |21\rangle + |30\rangle + |19\rangle).$$

Here, the basis states changed by four positions (and 4 is the size of the image). For the left-sided superposition, the change is

$$|\phi(f+1)\rangle = \frac{1}{2}(|00\rangle|011\rangle + |01\rangle|101\rangle + |10\rangle|111\rangle + |11\rangle|100\rangle) = \frac{1}{2}(|3\rangle + |13\rangle + |23\rangle + |28\rangle).$$

For the left-sided superposition, one bit is added to the basis states, as expected. The change occurred smoothly, and not with a significant jump as in the case of the right-sided superposition (see Table 7.1). We believe that left-sided superpositions can be used more easily when using other operations as well.

Compared to other known models, the NEQR model probably requires the highest number of qubits to represent an image. The integer-valued image of range 2^l, $l > 1$, at each pixel-state $|n, m\rangle$, is represented by the state of $|f_{n,m}\rangle$ in the basis states of $B = \{|0\rangle, |1\rangle, \ldots, |2^l - 1\rangle\}$. This basis is with l qubits, or l bit-planes. For instance, for the integer images with the range of 256, it uses 8 bit-planes. Note that in the FRQI, the image format is real and only one qubit is defined at each pixel, and the NEQI uses l qubits, which increases the storage space for the representation. Thus,

Table 7.1 The signals and superpositions (without the normalized coefficient ½).

Image	Right-sided superposition	Left-sided superposition
$f - 1 = [1, 3, 5, 2]$	$\|4\rangle + \|13\rangle + \|22\rangle + \|11\rangle$	$\|1\rangle + \|11\rangle + \|21\rangle + \|26\rangle$
$f = [2, 4, 6, 3]$	$\|8\rangle + \|17\rangle + \|26\rangle + \|15\rangle$	$\|2\rangle + \|12\rangle + \|22\rangle + \|27\rangle$
$f + 1 = [3, 5, 7, 4]$	$\|12\rangle + \|21\rangle + \|30\rangle + \|19\rangle$	$\|3\rangle + \|13\rangle + \|23\rangle + \|28\rangle$

there are many possibilities to obtain different operations on the image $f_{n,m}$ by operating on these l bit-planes. As an example, the quantum circuit of calculating a negative image, using the example of the "jetplane" image, is shown in Fig. 7.16. The range of this 512 × 512-pixel image is 255, that is, $l = 8$. The quantum image representation uses $(9 + 9 + 8) = 26$ qubits. Here, the matrix $X^{\otimes l}$ is the antidiagonal unit matrix of size $2^{18} \times 2^{18}$.

The processing of the image $f_{n,m} \to g_{n,m}$ results in only integer-valued image. In other words, only non-negative integer-to-integer conversion operations can be considered in this model. Complex arithmetic on images is difficult to imagine in the model. When measuring, or observing, the image superposition, the state of the image at any pixel-state $|n, m\rangle$ has the same probability equal to $p_{n,m} = 1/(2^{r+s})$. The expansion of the intensity range $2^l \to 2^{l+l_0}, l_0 > 1$, requires additional l_0 qubits; the number $p_{n,m}$ stays the same. The images with high resolutions require many qubits.

The example of the presentation of the 2 × 4-pixel image $\{f_{n,m};\, n = 0, 1, m = 0\colon 3\} = \{2, 5, 4, 1; 3, 4, 7, 3\}$ is shown in Fig. 7.17. This circuit resembles a counting ruler with a binary number system. The black-and-gray-filled circles (buttons) on the top of the scheme show the empty and active operations, respectively. The number $|f_{n,m}\rangle$ of the basis state is dialed by changing the colors of circles (for C-NOT gates) on the control line. Changing (replacing) the colors of these buttons will allow us to get any 2 × 4-pixel image with the range of 8. Changing the color means removing or adding a CNOT gate on the corresponding bit-plane.

The NEQI model is rich in terms of processing techniques. The ability to operate separately or in parallel across different bit planes of intensity and pixel coordinates makes this model very effective in many applications such as

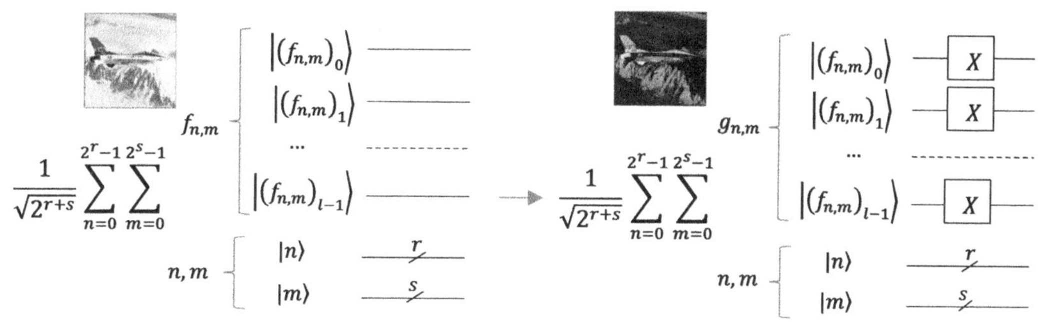

$$|\varphi(f)\rangle \to |\varphi(g)\rangle = (X^{\otimes l} \otimes I^{\otimes r} \otimes I^{\otimes s})|\varphi(f)\rangle$$

Fig. 7.16 NEQI model: transformation of the image $f_{n,m}$ to its negative image $g_{n,m} = 255 - f_{n,m}$.

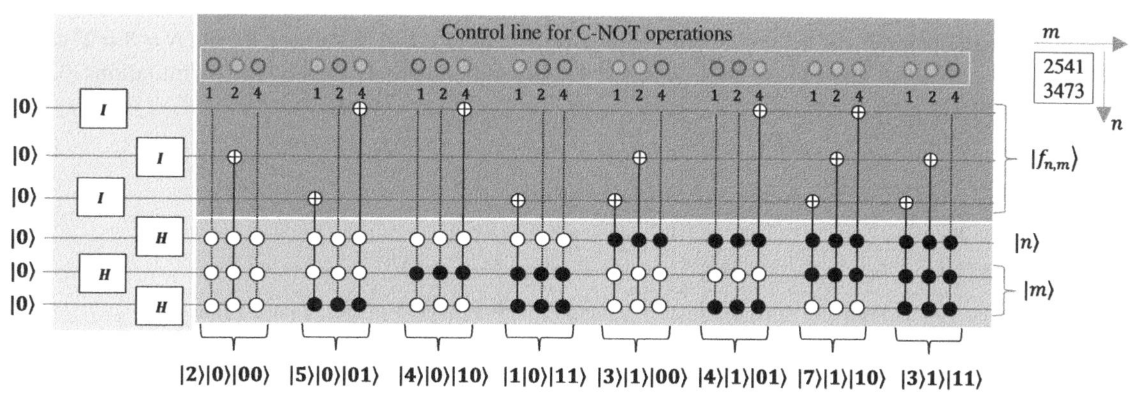

Fig. 7.17 The circuit to preparation of the 2 × 4-pixel image in the NEQI model.

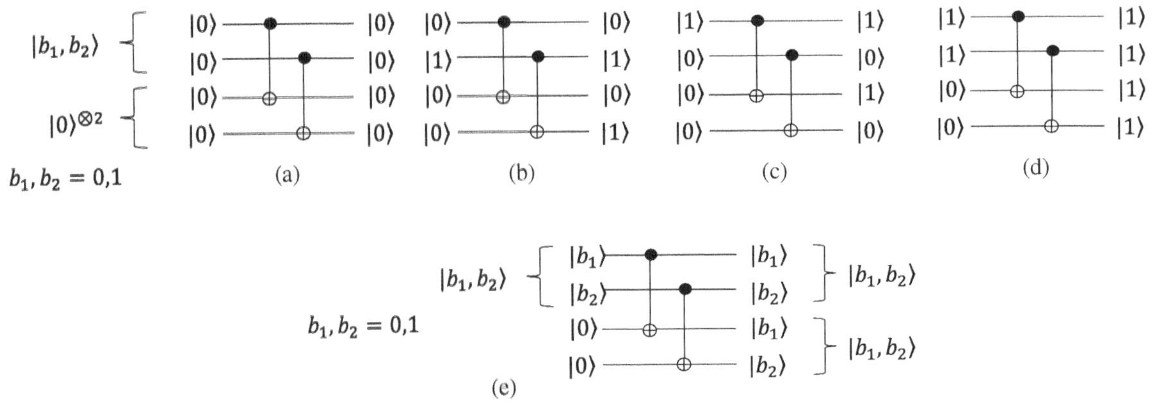

Fig. 7.18 The schemes of copying the 2-qubit sequence with two C-NOT gates. Copying the basis states (a) $|00\rangle$, (b) $|01\rangle$, (c) $|10\rangle$, and (a) $|11\rangle$, and (e) the general basis $|b_1b_2\rangle$.

quantum search algorithm, edge extraction by using the Laplacian, Prewitt, and Sobel operators, image segmentation, watermarking, scrambling, and steganography [4, 15, 16].

The quantum circuit of copying basis states is shown in Fig. 7.18, on the example with 2-qubit states. Here, two additional (or so-called ancillary) qubits $|00\rangle$ are used, to get the copy of the input $|c\rangle = |b_1b_2\rangle$,

$$U_4 : |c\rangle \otimes |00\rangle \rightarrow |c\rangle \otimes |c\rangle, |c\rangle \in B_2 = \{|0\rangle, |1\rangle, |2\rangle, |3\rangle\}.$$

The similar circuit with control l-qubit CNOT gates can be used to copy the information of the image $|f_k\rangle$ in the image representation $|\varphi(f)\rangle$.

The cyclic shift operators along the coordinates (Y-direction) and ordinates (X-direction) by one position to the right (or left) are described as

$$|\varphi(f_X)\rangle_{\pm} = \frac{1}{2^r}\sum_{n=0}^{N-1}\sum_{m=0}^{N-1} |f_{n,m}\rangle |(n \pm 1) \bmod N\rangle |m\rangle \tag{7.38}$$

and

$$|\varphi(f_Y)\rangle_{\pm} = \frac{1}{2^r}\sum_{n=0}^{N-1}\sum_{m=0}^{N-1} |f_{n,m}\rangle |n\rangle |(m \pm 1) \bmod N\rangle. \tag{7.39}$$

The circuit for such operations along the n-coordinates is shown in Fig. 7.19 in part (a), for the $N = 8 = 2^3$ case, when the state $|n\rangle = |n_2n_1n_0\rangle$ maps to the state $|n_1n_0n_2\rangle$. This operation is performed by the permutations (3, 7),

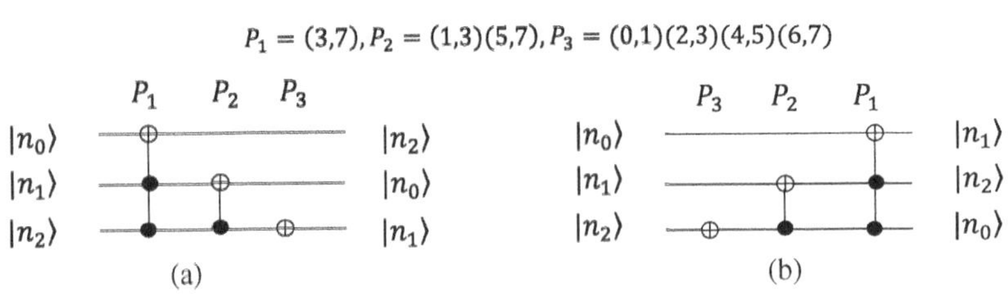

Fig. 7.19 The schemes of cyclic shifting the coordinates to the (a) right and (b) to the left (for $N = 8$).

$P_1 = (7,15)$, $P_2 = (3,7)(11,15)$, $P_3 = (1,3)(5,7)(9,11)(13,15)$, $P_4 = (0,1)(2,3)(4,5) \dots (14,15)$

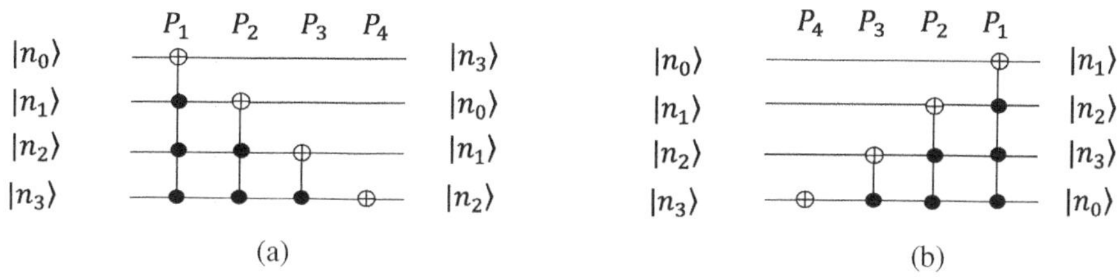

Fig. 7.20 The schemes of cyclic shifting the coordinates to the right (a) and (b) to the left (for $N = 16$).

(1, 3)(5, 7), and (0, 1)(2, 3)(4, 5)(6, 7). The circuit for the inverse operation $(n - 1) \bmod 4$, when the state $|n_2n_1n_0\rangle$ maps to the state $|n_0n_2n_1\rangle$, is shown in part (b).

Figure 7.20 shows the similar shifting operations for the $N = 16$ case, when the 4-qubit state $|n\rangle = |n_3n_2n_1n_0\rangle$ maps to the states $|n_0n_3n_2n_1\rangle$ and $|n_2n_1n_0n_3\rangle$ in parts (a) and (b), respectively.

7.2 Color Image Quantum Representations

Different models are used for color images [9] and in this section, we briefly describe a few such models. Each color model is defined by its primary colors. All other image colors are determined by the amount of primaries in the color. The color image for instance with three primary colors presents itself three grayscale images, or three color channels. Thus, the quantum representations of grayscale images described above can be applied to each color channel separately, or all three color channels can be combined (united) into one large-sized grayscale image.

7.2.1 Quantum Color Pixel in the RGB Model

We consider color images in the RGB color space with the primary colors red (R), green (G), and blue (B) at each pixel. The illustration of the RGB cube with primary colors is given in Fig. 7.21.

The discrete color image $f_{n,m}$ of size $N \times M$ pixels is considered as a 3D data and defined as

$$f_{n,m} = \left\{ (f_R)_{n,m}, (f_G)_{n,m}, (f_B)_{n,m} \right\}. \tag{7.40}$$

Consider $N = 2^r$ and $M = 2^s$, $r, s > 1$. The color image $f = f_{n,m}$ is the triplet of red, green, and blue color components. As an example, Fig. 7.22 shows the color image and the reg, green, and blue color components.

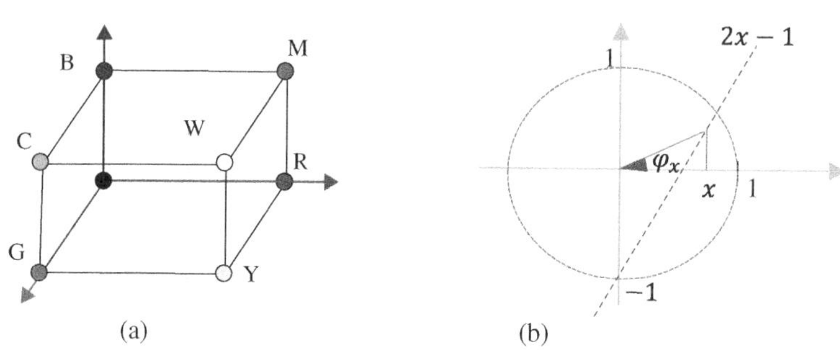

Fig. 7.21 (a) Color cube with primary colors in the RGB model. (b) Angle calculation in Eq. 7.41.

(a) (b) Red (c) Green (d) Blue

Fig. 7.22 (a) The original color image 'house" and its color channels; (b) red, (c) green, and (d) blue (refer colour representation in online version).

Let $f_{n,m}$ be written into a 2^{r+s}-dimensional "color" vector $f = \{((f_R)_k, (f_G)_k, (f_B)_k); k = 0: 2^{r+s} - 1\}$. Also, assume that at each pixel number k, the red, green, and blue components have been normalized and are in the interval [0, 1]. For the image with the range of 256, the normalization coefficient is 255,

$$(f_R)_k \to \frac{(f_R)_k}{255}, (f_G)_k \to \frac{\left(f_g\right)_k}{255}, (f_B)_k \to \frac{(f_B)_k}{255}.$$

7.2.1.1 3-Color Quantum Qubit Model

We can select two components, for instance, the red and green components, and move them into the interval $[-1, 1]$, as $R \to 2R - 1$ and $G \to 2G - 1$. To encode these two colors, the corresponding angles in the interval $[-\pi/2, \pi/2]$ are calculated at each pixel

$$\varphi_R = \sin^{-1}(2R - 1) \text{ and } \varphi_G = \sin^{-1}(2G - 1). \tag{7.41}$$

The geometry of these angles is illustrated in Fig. 7.21 in part (b). The qubit color pixel is defined as [8, 17]

$$|\, q_k\rangle = z_{k,0}|\, 0\rangle + z_{k,1}|\, 1\rangle,\ n = 0: (2^{r+s} - 1). \tag{7.42}$$

where the amplitudes of this qubit are $z_{k,0} = \sqrt{1 - B^2}e^{i\varphi_R}$ and $z_{k,1} = Be^{i\varphi_G}$. The probability of measuring the basis states $|0\rangle$ or $|1\rangle$ in this qubit is defined by the blue color; $|z_{n,0}|^2 = 1 - B^2$ or $|z_{n,1}|^2 = B^2$, respectively. The image-vector f can be represented by the following $(r + s + 1)$-qubit quantum state:

$$|\, \varphi(f)\rangle = \frac{1}{2^{r+s}} \sum_{k=0}^{2^{r+s}-1} (z_{k,0}|\, 0\rangle + z_{k,1}|\, 1\rangle) \otimes |\, k\rangle. \tag{7.43}$$

The amplitudes $z_{k,1}$ and $z_{k,2}$ are complex numbers. It is important to note that the transformation of each color into the vector of amplitudes of the single qubit $(r, g, b) \to (z_0, z_1)$ is invertible. Indeed, the colors can be calculated as follows:

$$\begin{cases} b = |\, z_{k,1}\, | \neq 0, 1, \\ g = \dfrac{1}{2}(\sin\varphi_G + 1), \text{where } \sin\varphi_G = \text{Imag}\dfrac{z_{k,1}}{B}, \\ r = \dfrac{1}{2}(\sin\varphi_R + 1), \text{where } \sin\varphi_R = \text{Imag}\dfrac{z_{k,0}}{\sqrt{1 - B^2}}. \end{cases} \tag{7.44}$$

The cases when $B = 0$ and 1 are considered separately. Thus, the RGB color image can be represented by $(r + s + 1)$ qubits. At each pixel, the triplet $\{(f_R)_k, (f_G)_k, (f_B)_k\}$ of primary colors is encoded into a single qubit $|q_n\rangle$.

7.2.2 NASS Representation

In the normal arbitrary superposition state (NASS), a color image $f_{n,m} = (r_{n,m}, g_{n,m}, b_{n,m})$ of size $2^r \times 2^r$ pixels in the RGB model is considered as the number in the 256th representation [18],

$$c(n, m) = 256^2 r_{n,m} + 256 g_{n,m} + b_{n,m}. \tag{7.45}$$

The base of this presentation is 256 for images in the standard range of 256 of grays. This representation corresponds to the system with a 24-bit, or 3-byte memory word. The image is presented as the grayscale image with the large range of intensity $L = 256^3$. The coefficients $c(n, m)$ can be large. Therefore, they can be scaled.

The grayscale image $c(n, m)$ of size $2^r \times 2^r$ pixels, as the 4^r-D vector with components $c(k) = c(n, m)$, where $k = n2^r + m$, is represented by the following $2r$-qubit superposition:

$$|\varphi(f)\rangle = \frac{1}{A}\sum_{k=0}^{4^r-1} c(k)\,|k\rangle, c(k) \in \{0, 1, 2, ..., 2^{24} - 1\}. \tag{7.46}$$

The normalization coefficient is $A = \sqrt{c^2(0) + c^2(1) + ... + c(4^r - 1)}$. The amplitudes of states determine a probabilistic representation of the quantum image. It is not difficult to see from the 256th representation of colors (which is given in Eq. 7.45) that the colors with red components, even of low intensity, have a very high probability of measurement compared to green and blue components of the image (see more in [8]).

7.2.3 NASSTC Model

In the NASS model with three components (NASSTC) [18, 19], the standard basis states of pixels $|i\rangle$ are united with an incomplete 4-state-based superpositions

$$|\check{f}_C\rangle = \frac{1}{A}\sum_{k=0}^{4^r-1} (r(k)|\,10\rangle + g(k)|\,01\rangle + b(k)|\,11\rangle)|\,k\rangle\,. \tag{7.47}$$

It is considered that the coefficients of the red, green, and blue components, $r_{n,m}$, $g_{n,m}$, and $b_{n,m}$, are written into the 4^r-D vectors with components $r(k)$, $g(k)$, and $b(k)$, respectively. The normalized coefficient equals to

$$A = \sqrt{\sum_{k=0}^{4^r-1} \left[r^2(k) + g^2(k) + b^2(k)\right]}. \tag{7.48}$$

This representation of the RGB color image requires $(2r + 2)$ qubits. The primary colors are presented by 2-qubit superposition at each pixel. In the next sections, we discuss in detail more general models based on 2-qubit representations in the quaternion algebra.

7.2.4 Novel Quantum Representation of Color Images (NCQI)

RCQI model is a generalization of the NEQR model applied for color images in the RGB color model [20]. The color image with the primary red, green, and blue colors $f = (r, g, b)$ is represented by the quantum superposition

$$|\varphi(f)\rangle = \frac{1}{2^r}\sum_{k=0}^{4^r-1} |\,r(k)\rangle|\,g(k)\rangle|\,b(k)\rangle|\,k\rangle. \tag{7.49}$$

The image is integer. The number of qubits in this model increases with the range of the image. If the range of color components is $256 = 2^8$, then at pixel number k, the primary colors are presented by 8-qubit state each. For instance, the red color $r = 108$ is represented by the 8-bit state $|r\rangle = |108\rangle = |0110\ 1100\rangle$. The state

$|r(k)\rangle|g(k)\rangle|b(k)\rangle = |r(k)\rangle \otimes |g(k)\rangle \otimes |b(k)\rangle$ is the $3\times 8 = 24$-qubit state in the above superposition. This quantum $(2r+24)$-qubit superposition is composed of 4^r states of $4^r \times 2^{2r}$ basis states.

In the general case, for images of higher range $L = 2^l$, $l > 8$,the image is presented by a $(2r+3l)$-qubit superposition with only 2^{2r} basis states. Compared to other known models, this model probably requires the highest number of qubits to represent an RGB color image.

This model was modified by *the quantum representation of color images* (QRCI) which uses bit-plane information of colors [21]. For the image of range 2^l, $l > 1$,the binary representation of each color is considered,

$$r(k) = (r_{k,l-1}, r_{k,l-2}, \ldots, r_{k,2}, r_{k,1}, r_{k,0}), g(k) = (g_{l-1}, g_{l-2}, \ldots, g_2, g_1, g_0), b(k) = (b_{l-1}, b_{l-2}, \ldots, b_2, b_1, b_0).$$

Then, the bit-wise color information is encoded into the sequence of l states, each from the eight basis states of the basis $B_3 = \{|000\rangle, |001\rangle, |010\rangle, \ldots, |111\rangle\}$. Namely, the following states are calculated:

$$C_i(k) = |\, r_{k,i} g_{k,i} b_{k,i}\rangle = |\, r_{k,i}\rangle \otimes |\, g_{k,i}\rangle \otimes |\, b_{k,i}\rangle \in B_3, i = 0, 1, \ldots, l-1. \tag{7.50}$$

The color image is represented by the $(3+l+2r)$-qubit superposition with only $4^r l$ basis states,

$$|\, \phi(f)\rangle = \frac{1}{2^r\sqrt{2^l}} \sum_{i=0}^{l-1} \sum_{k=0}^{4^r-1} |\, C_i(k)\rangle |\, i\rangle |\, k\rangle. \tag{7.51}$$

In the case when the image range varies in the interval [0,255], that is, $l = 3$, $|\phi(f)\rangle$ is the $(2r+6)$-qubit superposition. This superposition is illustrated in Fig. 7.23 in part (a) with two examples of color processing on bit-planes. In part (b), the circuit for the color complement of each primary color is shown, and the 3-gate swap operation is shown in part (c). The color complement operation on the above superposition is accomplished by the transform

$$U_{crgb}\colon |\, \phi(f)\rangle \rightarrow \left(X^{\otimes l} \otimes I^{\otimes(2r+l)}\right) |\, \phi(f)\rangle. \tag{7.52}$$

The red and blue color swapping operation is performed by the transform

$$U_{rb}\colon |\, \phi(f)\rangle \rightarrow \left(A \otimes I^{\otimes(2r+l)}\right) |\, \phi(f)\rangle, \tag{7.53}$$

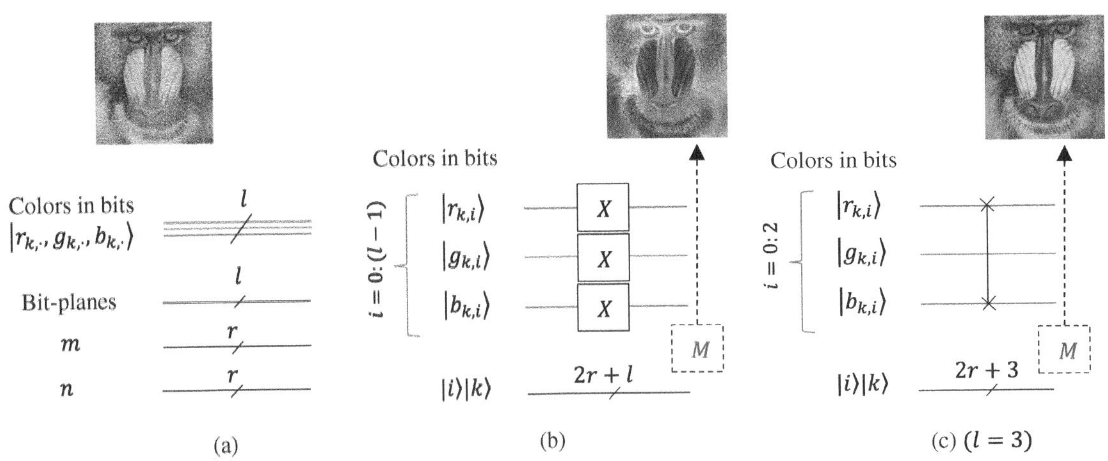

Fig. 7.23 The circuit element for (a) the image superposition RCQI model, (b) complement of colors, and (c) swap of colors (red and blue) (refer colour representation in online version).

where A is the matrix of the 3-qubit swap operation, or permutation $P = (1, 4)(3, 6)$, described in Example 4.1. The ideal results of these two operations are shown in the example with the color "mandrill' image. The results are theoretical, since to achieve such images by these quantum circuits, it is required to run and the measurements (M) performed multiple times.

7.2.5 Multi-channel Representation of Images (MCRI)

The color image can be represented by using the concept of the quantum pixel described in Eq. 7.3, but for the color pixel. This color pixel in the RGB model can be written as the tensor product of "color" qubits represented by angles from the interval $[0, \pi/2]$, as

$$|c(n,m)\rangle = \cos\vartheta_{c;n,m}|0\rangle + \sin\vartheta_{c;n,m}|1\rangle,$$

for primary colors $c = r$, g, and b. Therefore, the color is represented by the following $(r+s+8)$-qubit superposition [22, 23]:

$$|\varphi(f)\rangle = \frac{1}{\sqrt{NM}}\sum_{n=0}^{N-1}\sum_{m=0}^{M-1}|\cos\vartheta_{r;n,m}|0\rangle + \sin\vartheta_{r;n,m}|1\rangle\rangle \otimes |\cos\vartheta_{g;n,m}|0\rangle + \sin\vartheta_{g;n,m}|1\rangle\rangle$$
$$\otimes|\cos\vartheta_{b;n,m}|0\rangle + \sin\vartheta_{b;n,m}|1\rangle\rangle|n,m\rangle. \quad (7.54)$$

Other color models can also be used with similar representation. We consider several such models.

1) **XYZ color model**. In this model, the tristimulus values X, Y, and Z are calculated by the linear combinations of the primary colors of the RGB color model as [24]

$$\begin{bmatrix} X \\ Y \\ Z \end{bmatrix} = \boldsymbol{A}\begin{bmatrix} R \\ G \\ B \end{bmatrix} = \begin{bmatrix} 0.49 & 0.31 & 0.2 \\ 0.17697 & 0.8124 & 0.01063 \\ 0 & 0.01 & 0.99 \end{bmatrix}\begin{bmatrix} R \\ G \\ B \end{bmatrix}, \det\boldsymbol{A} = 0.340085.$$

   The sum of the coefficients of matrix $\boldsymbol{A}$ in each row is 1. Z value is very close to blue, B. Figure 7.24 illustrates the transformation of primary colors from RGB color model into XYZ model in parts (a) and (b).

   As an example, Fig. 7.25 shows the original RGB color image in part (a) and the images of the X, Y, and Z components in parts (b), (c), and (d), respectively. The color image in the XYZ format is shown in part (e).

2) **CMY color model**. The primary colors in the CMY model are cyan (C), magenta (M), and yellow (Y) [9]. This model is subtractive, meaning that the primary colors are calculated by $C = 1 - R$, $M = 1 - G$, $Y = 1 - B$. The image with triangles of primary colors in this model is illustrated in Fig. 7.24 in part (c).

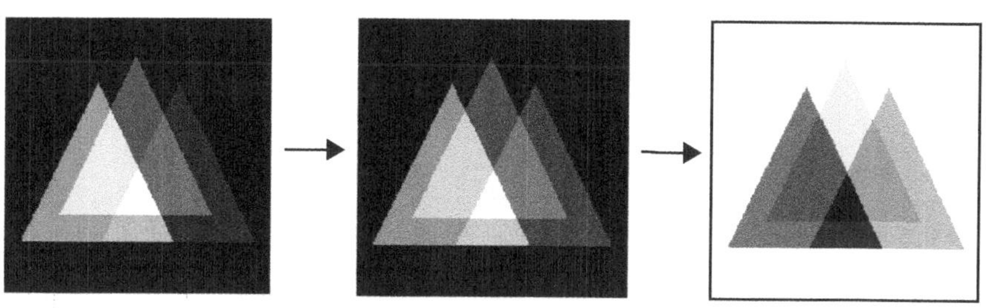

(a) (b) (c)

Fig. 7.24 The image of colors in (a) RGB, (b) XYZ, and (c) CMY color models.

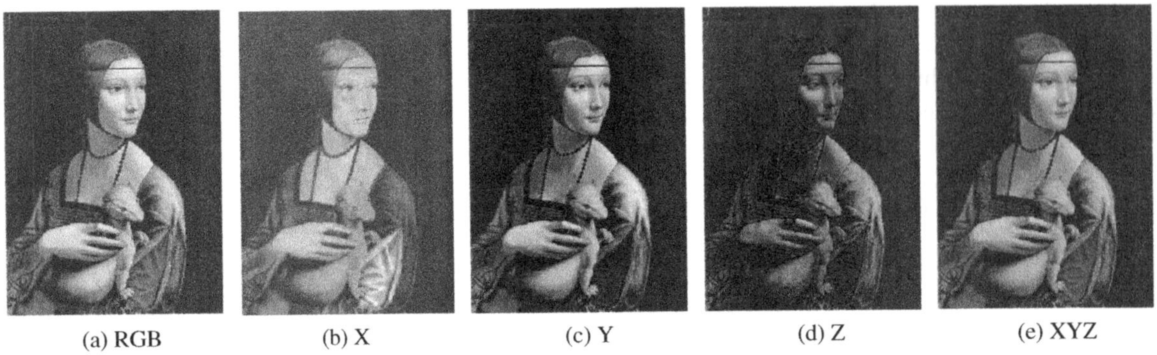

Fig. 7.25 (a) The original color image and its color components (b) X, (c) Y, and (d) Z. (e) The color image in the XYZ model. *Source:* National Museum / Wikimedia Commons / Public domain.

Fig. 7.26 (a) The original RGB color image, and component (b) C, (c) M, and (d) Y. (e) The color image in the CMY model.

As an example, Figure 7.26 shows the RGB color image in part (a) and images of the cyan, magenta, and yellow components in parts (b), (c), and (d), respectively. The color image in the CMY format is shown in part (e).

For these models, in Eqs. 7.40–7.54 for quantum color image representation, we need to change the primary colors (R, G, B) at each pixel by (X, Y, Z) and (C, M, Y) colors, respectively.

3) **HSI color model**. This 3-component model, or the Hue-Saturation-Intensity (HSI) color model is calculated from the RGB model by the following nonlinear transformation at each pixel [9]:

$$H = \begin{cases} \vartheta, & \text{if } b \leq g \\ 360 - \vartheta, & \text{if } b > g \end{cases}, \vartheta = \cos^{-1}\left[\frac{1}{2}\frac{2r-g-b}{\sqrt{(r-g)^2+(r-b)(g-b)}}\right],$$

$$I = \frac{r+g+b}{3},$$

$$S = 1 - \frac{\min(r,g,b)}{I}, (\text{if } I = 0, S = 1).$$

The hue is the angle in the interval [0,360°] and defines the pure color, the saturation determines the amount of white in the color, and the intensity of the color is the gray value of the primary colors. The values of saturation are in the interval [0, 1] and the intensity in the interval [0, 255]. As an example, Fig. 7.27 shows the color "flower" image and its components in the HSI model.

(a) RGB (b) H (c) S (d) I

Fig. 7.27 (a) The original RGB color image and (b)–(d) its components in the HSI model.

4) **HSV color model**. This is a 3-component model, or the Hue-Saturation-Value (HSV) color model. The components of the color image in this model are calculated as follows [24]:
 The value component is $V = V_{n,m} = \max(r_{n,m}, g_{n,m}, b_{n,m})$.
 The saturation determines the amount of white in the color,

$$S = S_{n,m} = 1 - \frac{\min\left(r_{n,m}, g_{n,m}, b_{n,m}\right)}{\max\left(r_{n,m}, g_{n,m}, b_{n,m}\right)} = 1 - \frac{m_{n,m}}{V_{n,m}}, \text{ if } V_{n,m} \neq 0. \tag{7.55}$$

 Here, $m_{n,m} = \min(r_{n,m}, g_{n,m}, b_{n,m})$ and if $V_{n,m} = 0$, $S = 0$.
 The hue component defines the pure color and is calculated in angles in the circle of 360°,

$$H = H_{n,m} = \begin{cases} 60\dfrac{g_{n,m} - b_{n,m}}{\Delta_{n,m}} \bmod 360, \text{ if } r_{n,m} = V_{n,m}, \\ 60\dfrac{b_{n,m} - r_{n,m}}{\Delta_{n,m}} + 120, \text{ if } g_{n,m} = V_{n,m}, \\ 60\dfrac{r_{n,m} - g_{n,m}}{\Delta_{n,m}} + 240, \text{ if } b_{n,m} = V_{n,m}, \end{cases}$$

 where $\Delta_{n,m} = V_{n,m} - m_{n,m}$.
 As an example, Fig. 7.28 shows the color image in part (a) and its components in the HSV model in parts (b)–(d). The image in the HSV color map is shown in part (e).

7.2.6 Quantum Image Representation in HSI Model (QIRHSI)

The NCQIs described above are used for the RGB images. A similar representation was presented for color images in the HSI model [25]. Consider the color image $f_{n,m}$ of size $2^r \times 2^r$-pixels, $r > 1$, which is written line by line into the

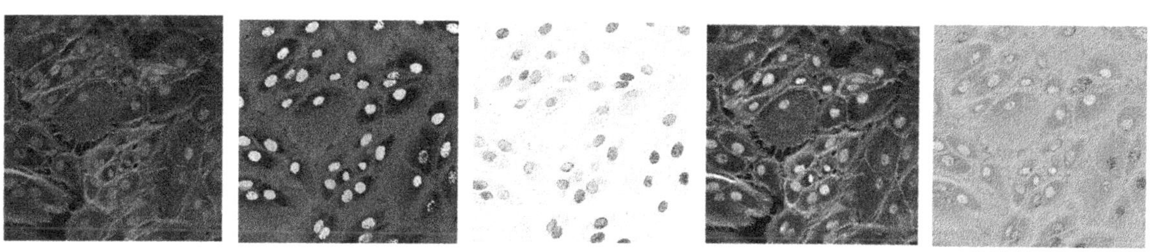

(a) RGB (b) H (c) S (d) V (e) HSV

Fig. 7.28 (a) The original RGB color image, (b–(d) its components in the HSV model, and (e) HSV image. *Source:* National Institutes of Health / Public doamin / https://medialibrary.nei.nih.gov/media/3511 / last accessed on August 08, 2024.

signal f_k, $k = 0{:}(4^r - 1)$. With the range of the image 2^l, $l > 1$, the image is presented as the $(2r + 2 + l)$-qubit superposition

$$| \phi(f)\rangle = \frac{1}{2^r}\sum_{k=0}^{4^r-1} \underbrace{| H_k\rangle | S_k\rangle | I_k\rangle}| k\rangle. \tag{7.56}$$

Here, at each pixel state $|k\rangle$, the hue (H) and saturation (S) components are encoded in 2 angles, $\vartheta_{H,n}$ and $\vartheta_{S,n}$, and one qubit is used for each,

$$| H_k\rangle = \cos(\vartheta_{H,k})| 0\rangle + \sin(\vartheta_{H,k})| 1\rangle, | S_k\rangle = \cos(\vartheta_{S,k})| 0\rangle + \sin(\vartheta_{S,k})| 1\rangle.$$

The intensity (I) is written in the binary form as l-qubit state number I_k, that is, $| I_k\rangle = | I_k^0 I_k^1 \cdots I_k^{l-3} I_k^{l-2} I_k^{l-1}\rangle$.For instance, if $I_k = 108$, then in the range of 256, this number has a binary representation is {00110110}. Thus, $|I_k\rangle = |108\rangle = |00110110\rangle$.

In this model, only the intensity component is processing, $I_n \to I'_n$. It should be noted that all three primary components in the HSI model are dependent. The hue H_n and saturation are the functions of the red, green, and blue colors, that is, $H_n = H_n(R_n, G_n, B_n)$ and $I_n = I_n(R_n, G_n, B_n)$.The same for the saturation is $S_n = 1 - \min\{R_n, G_n, B_n\}/I_n$. Many methods in color image processing (including Retinex-based image enhancement methods) use the HSI or HSV color models and are based on processing only the intensity or value components. This destroys the color model of color images, and the processed images have various artifacts. For the above case, it means that the processed data $\{H_n, S_n, I'_n\}$ do not correspond to a color image in the HSI model.

Also, if only the intensity component is processing, the representation of only the grayscale image can be used

$$| \phi_1(I)\rangle = \frac{1}{2^r}\sum_{k=0}^{4^r-1} \underbrace{| I_k\rangle}| k\rangle,$$

and then, the state components $|H_k\rangle|S_k\rangle$ can be added, to receive the superposition $|\phi(f)\rangle$.

7.2.7 Transformation 2 × 2 Model for Color Images

In this section, we consider the simple model of processing color images. It should be noted that a color image can be processed as a single grayscale image, by using one of the color-to-gray models proposed by Grigoryan [24]. Here, we discuss only one of such models, namely the 2×2 and 1×4 models. In this model, color images are transformed into grayscale images. Then, the enhancement algorithms, which work well for grayscale images, can be used for the newly transformed color-to-gray image. Thus, the color image enhancement mainly consists of the transformation of the color image into the corresponding grayscale image.

A color RGB image of size $N \times M$ pixels with the red (R), green (G), blue (B), plus gray (I) components is presented by a grayscale image of twice size, as shown in Table 7.2. Also, we can consider the mapping shown in Table 7.3.

Here, the gray component is considered to be the average of three primary colors, i.e., $I = (R + G + B)/3$. The luminous of the RGB image, which is calculated by $I = 0.3R + 0.59G + 0.11B$, can also be considered. Thus, the new transformed grayscale image is obtained by arranging side by side 4 values of I, R, G, and B of each pixel. This color-to-gray image transformation (C-2-G IT) is reversible. After processing the color-gray image, a new color image is obtained, by using the inverse gray-to-color image transform (G-2-C IT). As an example, Figure 7.29 shows the original 584×565-pixel image in part (a) and the image of the color-to-gray transform in part (b). This grayscale image was enhanced by the Fourier transform-based 0.84-rooting method. This enhancement is described as the alpha-rooting, which is very effective method in gray and color image enhancement [26, 27],

$$F_{p,s} \to G_{p,s} = F_{p,s}| F_{p,s} |^{1-\alpha}, \alpha = 0.84, p, s = 0{:}\, N-1, (M-1), \tag{7.57}$$

Table 7.2 The 2 × 2-model of color-gray transform image.

I(0,0)	R(0,0)	I(0,1)	R(0,1)	I(0,2)	R(0,2)	...
G(0,0)	B(0,0)	G(0,1)	B(0,1)	G(0,2)	B(0,2)	...
I(1,0)	R(1,0)	I(1,1)	R(1,1)	I(1,2)	R(1,2)	...
G(1,0)	B(1,0)	G(1,1)	B(1,1)	G(1,2)	B(1,2)	...
I(2,0)	R(2,0)	I(2,1)	R(2,1)	I(2,2)	R(2,2)	...
G(2,0)	B(2,0)	G(2,1)	B(2,1)	G(2,2)	B(2,2)	...
...	...	...	...	...	...	...

Table 7.3 The 1×4-model of color-gray transform image.

I(0,0)	R(0,0)	G(0,0)	B(0,0)	I(0,1)	R(0,1)	G(0,1)	B(0,1)	...
I(1,0)	R(1,0)	G(1,0)	B(1,0)	I(1,1)	R(1,1)	G(1,1)	B(1,1)	...
I(2,0)	R(2,0)	G(2,0)	B(2,0)	I(2,1)	R(2,1)	G(2,1)	B(2,1)	...
...	...	...	...	...	...	...	...	...

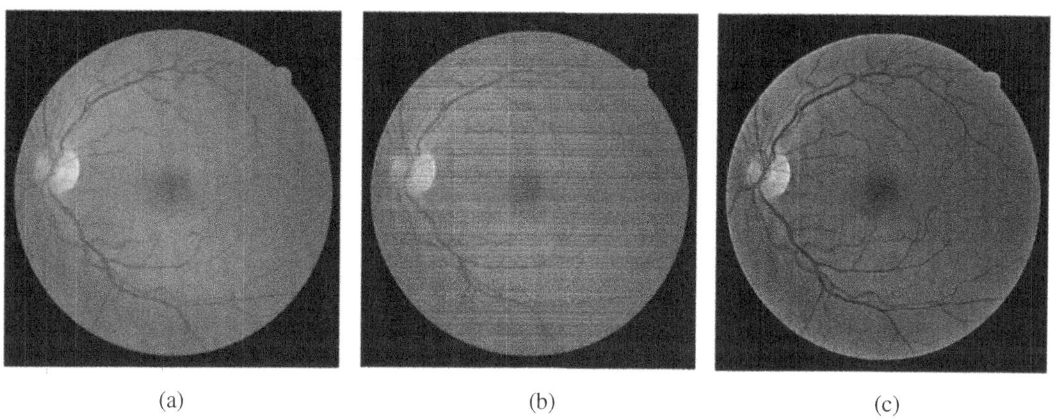

Fig. 7.29 (a) Original RGB color image "eye.tif," (b) the color-to-grayscale image in the 2 × 2 model, and (c) the enhanced image after performing 0.84-rooting on the grayscale image.

followed by the inverse 2D discrete Fourier transform, $\{G_{p,s}\} \to \{g_{n,m};\ n,\ m = 0\!: N-1,\ (M-1)\}$. Then, the new image $g_{n,m}$ was transformed back to the color image which is shown in part (c). One can notice that the image was well enhanced. In this model, the color image at pixel $|n\rangle$ is presented by the 2-qubit superposition

$$|\, f_n\rangle = \frac{1}{\sqrt{E(f_n)}}(i_n|\, 00\rangle + r_n|\, 01\rangle + g_n|\, 10\rangle + b_n|\, 11\rangle).$$

Here, $E(f_k)$ is the energy of the color at the pixel, $E(f_k) = i_k^2 + r_k^2 + g_k^2 + b_k^2$. The image by the following $(r+s+2)$-qubit superposition:

$$|\, \varphi(f)\rangle = \frac{1}{\sqrt{2^{r+s}}} \sum_{n=0}^{2^{r+s}-1} \frac{1}{\sqrt{E(f_n)}}(i_n|\, 00\rangle + r_n|\, 01\rangle + g_n|\, 10\rangle + b_n|\, 11\rangle) \otimes |\, n\rangle. \tag{7.58}$$

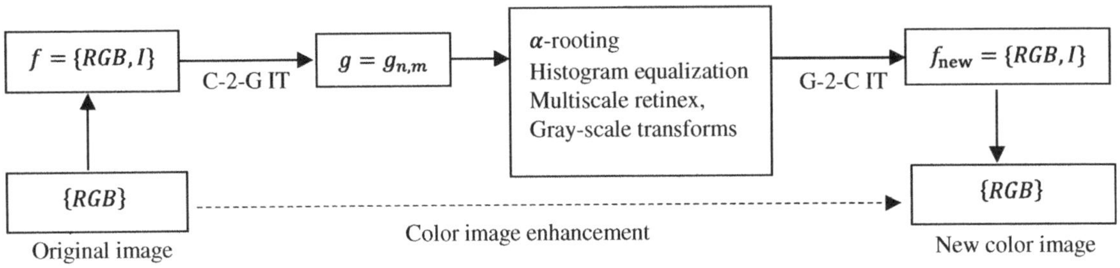

Fig. 7.30 The block-diagram of color image enhancement by enhancing the color-to-gray transformed image in the 2 × 2 model.

The general block-diagram of processing the color image in this model is shown in Fig. 7.30. The enhancement of the color image is reduced to enhancement of the corresponding grayscale image in the 2 × 2 model. This model is simple and effective in color image enhancement.

In this model, the left-sided $(r + s + 2)$-qubit superposition of the image is defined by

$$| \psi(f)\rangle = \frac{1}{\sqrt{2^{r+s}}} \sum_{n=0}^{2^{r+s}-1} | n\rangle \otimes (i_n| 00\rangle + r_n| 01\rangle + g_n| 10\rangle + b_n| 11\rangle) \frac{1}{\sqrt{E(f_n)}}. \tag{7.59}$$

We consider the tensor product of the point and color, $|\psi_n\rangle = |n\rangle|f_n\rangle$, where $| f_n\rangle = (i_n| 00\rangle + r_n| 01\rangle + g_n| 10\rangle + b_n| 11\rangle)\frac{1}{\sqrt{E(f_n)}}$. This product is the bra-vector

$$\langle \psi_n | = \left(\underbrace{0,0,0,0}_{0}, \underbrace{0,0,0,0}_{1}, \ldots, \underbrace{0,0,0,0}_{n-1}, \underbrace{i_n, r_n, g_n, b_n}_{n}, \underbrace{0,0,0,0}_{n+1}, \ldots, \underbrace{0,0,0,0}_{NM-1} \right) \frac{1}{\sqrt{E(f_n)}}. \tag{7.60}$$

The color components at pixel n are placed together. The left-sided superposition $|\psi(f)\rangle$ of the image can be described by the bra-vector

$$\langle \psi(f) | = \left(\underbrace{i_0, r_0, g_0, b_0}_{0}, \underbrace{i_1, r_1, g_1, b_1}_{1}, \ldots, \underbrace{i_n, r_n, g_n, b_n}_{n}, \ldots, \underbrace{i_{NM-1}, r_{NM-1}, g_{NM-1}, b_{NM-1}}_{NM-1} \right) \frac{1}{\sqrt{E(f)}}. \tag{7.61}$$

In comparison, the tensor product $|\varphi_n\rangle = |f_n\rangle|n\rangle$ can be written as the following bra-vector:

$$\langle \varphi_n | = \left(\underbrace{0, \ldots, 0, i_n, 0, \ldots, 0}, \underbrace{0, \ldots, 0, r_n, 0, \ldots, 0}, \underbrace{0, \ldots, 0, g_n, 0, \ldots, 0}, \underbrace{0, \ldots, 0, b_n, 0, \ldots, 0} \right) \frac{1}{\sqrt{E(f_n)}}. \tag{7.62}$$

In this vector, the components of the color are located in series at a distance of NM from each other. The right-sided superposition $|\varphi(f)\rangle$ is described by the bra-vector

$$\langle \varphi(f) | = \left(\underbrace{i_0, i_1, \ldots, i_{NM-1}}, \underbrace{r_0, r_1, \ldots, r_{NM-1}}, \underbrace{g_0, g_1, \ldots, g_{NM-1}}, \underbrace{b_0, b_1, \ldots, b_{NM-1}} \right) \frac{1}{\sqrt{E(f)}}. \tag{7.63}$$

The preparation of the superposition $|\psi(f)\rangle$ in Eq. 7.61 can be performed by the block-diagonal matrix

$$\boldsymbol{U} = U_0 \oplus U_1 \oplus U_2 \oplus \cdots \oplus U_{NM-1}, \tag{7.64}$$

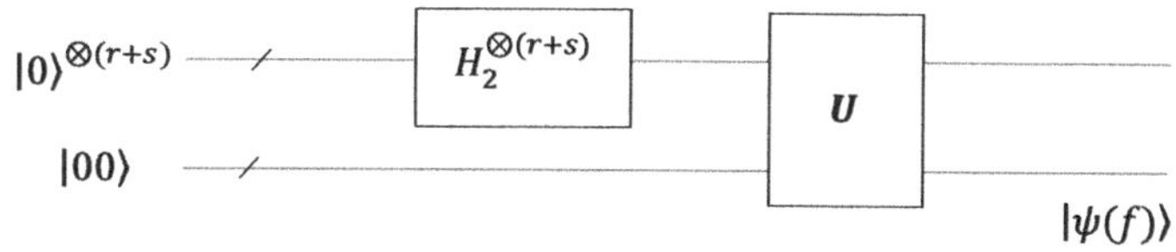

Fig. 7.31 The circuit for calculating the $(r+s+2)$-qubit left-sided superposition $|\psi(f)\rangle$.

where U_k are unitary 4×4 matrices with the first columns equal to $(i_n, r_n, g_n, b_n)'$. Such matrices are easy to construct and will be presented in the next section. Thus, we can obtain the superposition as follows:

$$|\psi(f)\rangle = \boldsymbol{U}|A\rangle, \text{where } \langle A| = \frac{1}{\sqrt{NM}}\left(\underbrace{1,0,0,0,1,0,0,0,\ldots,1,0,0,0,1,0,0,0}_{NM \text{ times}}\right) = \frac{1}{\sqrt{NM}}\left(\underbrace{1,1,\ldots,1,1}_{NM \text{ times}}\right)\otimes\langle 00|,$$

or

$$|\psi(f)\rangle = \boldsymbol{U}\left(\left[\frac{1}{\sqrt{NM}}\sum_{k=0}^{NM-1}|k\rangle\right]\otimes|00\rangle\right). \tag{7.65}$$

The corresponding circuit for the $|\psi(f)\rangle$ calculation is given in Fig. 7.31.

For the right-sided superposition $|\varphi(f)\rangle$ of this color image, this circuit can be used with the additional permutation, $P: |\psi(f)\rangle \to |\varphi(f)\rangle$.

There are many image representations in quantum computation, and we have described only a few of them. In conclusion, we also note that the most effective image enhancement methods are in quaternion algebras, since quaternion images contain color information as a whole and are processed as a whole [24, 28]. Corresponding quaternion quantum models for image representation exist and are discussed in Chapters 11 and 12.

References

1 Mastriani, M. (2017). Quantum image processing? *Quantum Information Processing* 16 (1): 27.

2 Yan, F., Iliyasu, A.M., and Venegas-Andraca, S.E. (2016). A survey of quantum image representations. *Quantum Information Processing* 15 (1): 1–35.

3 Yongquan, C., Xiaowei, L., and Nan, J. (2018). A survey of quantum image representations. *Chinese Journal of Electronics* 27 (4): 9.

4 Yan, F. and Venegas-Andraca, S.E. (2020). *Quantum Image Processing*. Springer Nature Singapore Pte Ltd.

5 Grigoryan, A.M. and Agaian, S.S. (2022) 'Quantum representation of 1-D signals on the unit circle,' *Quantum Information Processing*, 21, 24.

6 Venegas-Andraca, S. and Bose, S. (2003). Storing, processing, and retrieving an image using quantum mechanics. In: *Proceedings of SPIE Conf. Quantum Information and Computation* (4 August 2003), 134–147. SPIE (The International Society for Optics and Photonics). https://doi.org/10.1117/12.485960.

7 Le, P., Dong, F., and Hirota, K. (2011). A flexible representation of quantum images for polynomial preparation, image compression, and processing operations. *Quantum Information Processing* 10 (1): 63–84.

8 Grigoryan, A.M. and Agaian, S.S. (2020) 'New look on quantum representation of images: Fourier transform representation,' *Quantum Information Processing*, 19:148, p. 26. https://doi.org/10.1007/s11128-020-02643-3

9 Gonzalez, R.C. and Woods, R.E. (2002). *Digital Image Processing*. Upper Saddle River, New Jersey: Prentice Hall.

10 Grigoryan, A.M. and Agaian, S.S. (2022) 'Gradients and compass operators: method of rotations,' p. 12. *Proceedings of SPIE, Defense + Commercial Sensing, Multimodal Image Exploitation and Learning 2022*, Orlando, Florida, United States: SPIE. https://doi.org/10.1117/12.2618353

11 Su, J., Guo, X., Liu, C., and Li, L. (2020). A new trend of quantum image representations. *IEEE Access* 8: 18.

12 Zhang, Y., Lu, K., Gao, Y., and Wang, M. (2013). NEQR: A novel enhanced quantum representation of digital images. *Quantum Information Processing* 12 (8): 2833–2860.

13 Jiang, N., Wang, J., and Mu, Y. (2015). Quantum image scaling up based on nearest-neighbor interpolation with integer scaling ratio. *Quantum Information Processing* 14 (11): 4001–4026.

14 Jiang, N., Mu, Y., Wang, J. et al. (2015). Quantum image pseudo color coding based on the density-stratified method. *Quantum Information Processing* 14 (5): 1735–1755.

15 Zhou, R., Sun, Y., and Fan, P. (2014). Quantum image gray-code and bit scrambling. *Quantum Information Processing* 14 (5): 1717–1734.

16 Jian, M., Zhao, N., and Wang, L. (2016). LSB based quantum image steganography algorithm. *International Journal of Theoretical Physics* 55 (1): 107–123.

17 Ulyanov, S. and Petrov, S. (2012). Quantum face recognition and quantum visual cryptography: Models and algorithms. *Electronic Journal, System Analysis in Science and Education* 1: 17.

18 Hai-Sheng, L., Zhu, Q., Zhou, R.-G. et al. (2014). Multidimensional color image storage, retrieval, and compression based on quantum amplitudes and phases. *Information Sciences* 273: 212–232.

19 Hai-Sheng, L., Shuxiang, S., Fan, P. et al. (2019). Quantum vision representations and multi-dimensional quantum transforms. *Information Sciences* 502: 42–58.

20 Sang, J.Z., Wang, S., and Li, Q. (2017). A novel quantum representation of color digital images. *Quantum Information Processing* 16 (42): 42–56.

21 Wang, L., Ran, Q., Ma, J. et al. (2019). QRCI: 'A new quantum representation model of color digital images. *Optic Communications* 438: 147–158.

22 Sun, B., Le, P.Q., Iliyasu, A.M. et al. (2011). A multi-channel representation for images on quantum computers using the RGBα color space. In: *IEEE 7th International Symposium on Intelligent Signal Processing (WISP)*, 1–6. IEEE. 10.1109/WISP.2011.6051718.

23 Sun, B., Iliyasu, A.M., Yan, F. et al. (2013). An RGB multi-channel representation for images on quantum computers. *JACII* 17: 404–417.

24 Grigoryan, A.M. and Agaian, S.S. (2018). *Quaternion and Octonion Color Image Processing with MATLAB. SPIE* PM279: https://doi.org/10.1117/3.2278810.

25 Chen, GL. Song, XH. Venegas-Andraca, S.E. et al. (2022) 'QIRHSI: novel quantum image representation based on HSI color space model,' *Quantum Information Processing*, 21:5, p. 31.

26 Grigoryan, A.M., John, A., and Agaian, S.S. (2018). A Novel color image enhancement method by the transformation of color images to 2-D grayscale images. *International Journal of Signal Processing and Analysis* 2 (1): 18. https://doi.org/10.35840/2631-5114/3502.

27 Grigoryan, A.M., John, A., and Agaian, S.S. (2018). A novel image enhancement method of 3-D medical images by transforming the 3-D images to 2-D grayscale images. In: *Proceedings of SPIE, Defense + Commercial Sensing, Mobile Multimedia/Image Processing, Security, and Applications*, vol. 10668, 10. Orlando, Florida: SPIE https://doi.org/10.1117/12.2304667.

28 Grigoryan, A.M. and Agaian, S.S. (2020). Quaternion quantum image representation: New models. In: *Proceedings of SPIE 11399, Mobile Multimedia/Image Processing, Security, and Applications 2020*, 1139900, p. 6. SPIE https://doi.org/10.1117/12.2557862.

8

Image Representation on the Unit Circle and MQFTR

In this model of quantum image representation, the concept of the Fourier transform qubit representation (FTQR) is used [1, 2]. Information of an image in each pixel is written in the phase of the basis state in the quantum superposition. There is no constraint on the size and range of the signal and image when using the FTQR. It possesses properties, such as the sum, amplification, shifting, that are not available for the known methods of quantum signal and image representation. The calculation of phase-type amplitudes in signal and image representation is simple. These amplitudes can be calculated in advance and used by the lookup table method. In the FTQR, all states in the superposition have an equal probability. FTQR requires $(r + s)$ qubits for images of size $2^r \times 2^s$ pixels.

First, we consider the 1D case. Let f_n be an integer-valued signal of length $N = 2^r$, $r > 1$, and the range L, where $L > 1$ is an integer. Values of the signal f_n can be mapped into the quarter unit circle by the transformation

$$A\colon\ f_n \to A[f_n] = e^{i\alpha f_n}, \quad n = 0\colon (N-1). \tag{8.1}$$

The coefficient $\alpha = 2\pi/(4L)$ is used for this transformation (see Fig. 8.1).

The r-qubit state of the signal is defined as

$$|\psi(f)\rangle = \frac{1}{\sqrt{2^r}}\sum_{n=0}^{2^r-1} e^{i\alpha f_n}|n\rangle = \frac{1}{\sqrt{2^r}}\left(e^{i\alpha f_0}|0\rangle + e^{i\alpha f_1}|1\rangle + \cdots + e^{i\alpha f_{2^r-1}}|2^r-1\rangle\right). \tag{8.2}$$

This representation $|\psi(f)\rangle$ requires r qubits. All values of the signal, including the zeros, are counted. The signal reconstruction from this representation relates to the measurement of the superposition state. When measuring $|\psi(f)\rangle$ and getting the state $|n\rangle$, its amplitude $e^{i\alpha f_n}/\sqrt{2^r}$ allows for calculating the value f_n of the signal. The probability of observing the state $|n\rangle$ is the same, $1/2^r$ for all basis states.

8.1.1 Preparation for FTQR

The algorithm of the FTQR of the signal f_n can be described as follows.

Apply the Hadamard transform $H_2^{\otimes r}$ on r qubits all initialized to $|0\rangle$, to obtain the equal superposition of N basis states,

$$|\psi(0)\rangle = H_2^{\otimes r}|0\rangle^{\otimes r} = \frac{1}{\sqrt{2^r}}\sum_{n=0}^{2^r-1}|n\rangle. \tag{8.3}$$

Apply the phase shift gates ($\phi = \alpha f_n$) to the basis states $|n\rangle$, $n = 0\colon(2^r - 1)$,

$|n\rangle$ 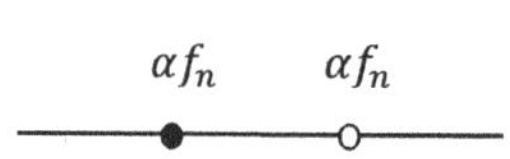 $e^{i\alpha f_n}|n\rangle$ (8.4)

Quantum Image Processing in Practice: A Mathematical Toolbox, First Edition. Artyom M. Grigoryan and Sos S. Agaian.

Companion website: www.wiley.com/go/grigoryan/quantumimageprocessing

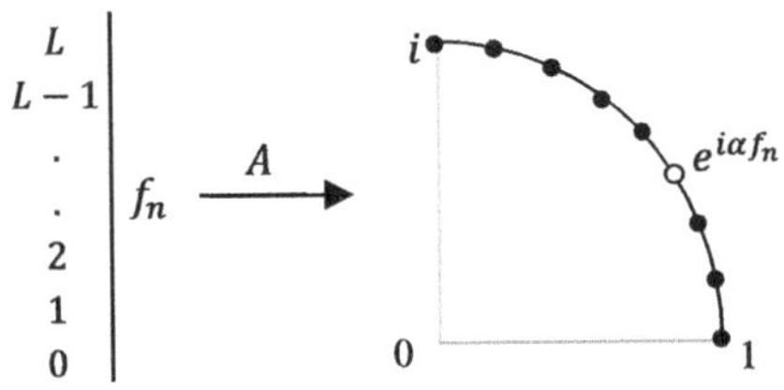

Fig. 8.1 Mapping of the integer interval [0, L] into the L points on the quarter circle.

The global phase gate can also be used, $e^{i\alpha f_n}|n\rangle = G(\alpha f_n)|n\rangle$ (see Table 2.1).

Compose the superposition

$$|\psi(0)\rangle \to |\psi(f)\rangle = \frac{1}{\sqrt{2^r}}\sum_{n=0}^{2^r-1} e^{i\alpha f_n}|n\rangle, \quad \text{where } \alpha = \pi/(2L).$$

The phase gates can be described by the following diagonal $N \times N$ matrices:

$$S_n = \text{diag}\{s_{i,i}\}_{i=0:(2^r-1)}, \quad s_{i,i} = \begin{cases} 1, & \text{if } i \neq n \\ e^{i\alpha f_n}, & \text{if } i = n \end{cases}. \tag{8.5}$$

Therefore, we obtain the required superposition,

$$\left(\prod_{n=0}^{2^r-1} S_n\right)|\psi\rangle = |\psi(f)\rangle. \tag{8.6}$$

Figure 8.2 shows the quantum circuit for preparation of the FTQR [1].

As shown in Section 6.1, any superposition can be obtained by the quantum DsiHT on the input (1, 0, 0, ..., 0). The QsiHT generated by the phase coefficients can be used to prepare the superposition $|\psi(f)\rangle$. The inverse quantum DsiHT results in the input $|\psi(f)\rangle$. The corresponding quantum scheme for preparing this superposition is shown in Fig. 8.3.

8.1.2 Constant Signal and Global Phase

The unit signal $f_n \equiv 1$, $n = 0{:}\,(2^r - 1)$, is represented by the following r-qubit superposition:

$$|\psi(1)\rangle = \frac{1}{\sqrt{2^r}}\sum_{n=0}^{2^r-1} e^{i\alpha 1}|n\rangle = e^{i\alpha}\left(\frac{1}{\sqrt{2^r}}\sum_{n=0}^{2^r-1}|n\rangle\right) = e^{i\alpha}|\psi(0)\rangle.$$

Similarly, for any constant signal $f_n \equiv C$, its superposition $|\psi(C)\rangle = e^{i\alpha C}|\psi(0)\rangle$.

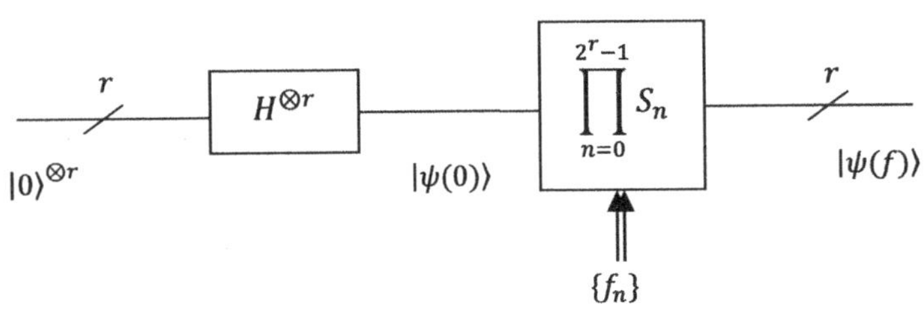

Fig. 8.2 The circuit for the FTQR of the signal.

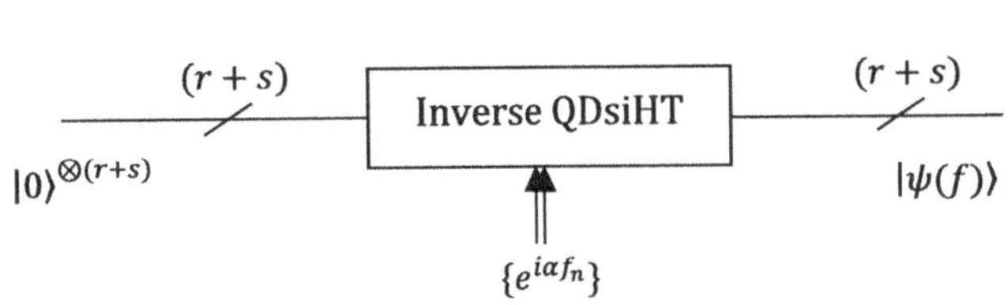

Fig. 8.3 The circuit for the FTQR of the grayscale image by the QsiHT.

8.1.3 Inverse Transform

The value f_k of the signal can be reconstructed from the amplitude of the state of measurement $|n\rangle$ as follows:
Denote by X the amplitude $e^{i\alpha f_n}/\sqrt{2^r}$ and calculate $C_n = \cos(\alpha f_n) = \text{Real}(X\sqrt{2^r})$.

$$\text{Calculate} f_n = \frac{1}{\alpha}\cos^{-1}(C_n).$$

In the general case of $N = 2^r$, values other than $\alpha = 2\pi/(4L)$ can also be used. For long signals, when $2^r \gg 4L$, we can consider $\alpha = \omega_0 = 2\pi/2^r$, as the smallest frequency in the 2^r-point DFT. Then, the signal representation in Eq. 8.2 is

$$\left|\check{f}\right\rangle = |\psi(f)\rangle = \frac{1}{\sqrt{2^r}}\sum_{n=0}^{2^r-1} e^{i\frac{2\pi}{2^r}f_n}|n\rangle = \frac{1}{\sqrt{N}}\sum_{n=0}^{N-1} e^{i\frac{2\pi}{N}f_n}|n\rangle.$$

Note that this representation differs much from the quantum representation with amplitudes

$$|\varphi(f)\rangle = \frac{1}{E}(f_0|0\rangle + f_1|1\rangle + f_2|2\rangle + \ldots + f_{2^r-1}|2^r-1\rangle), \tag{8.7}$$

where the normalized coefficient E is the energy of the signal $E = \sqrt{|f_0|^2 + |f_1|^2 + \ldots + |f_{2^r-1}|^2}$.

Example 8.1 Unit Impluse Superposition
For the unit impulse signal $u_0 = \{1, 0, 0, 0\}$, $L = 2$, and $\alpha = \pi/4$, we obtain the 2-qubit representation

$$|\check{u}_0\rangle = \frac{1}{2}\left[e^{i\pi/4}|00\rangle + |01\rangle + |10\rangle + |11\rangle\right].$$

For the unit impulse signal at point 1, $u_1 = \{0, 1, 0, 0\}$, the superposition is $|\check{u}_0\rangle = \left[|00\rangle + e^{i\pi/4}|01\rangle + |10\rangle + |11\rangle\right]/2$. In the representation by amplitudes in Eq. 8.7, for these two unit signals, we have the following basis states: $|\varphi(u_0)\rangle = |00\rangle$ and $|\varphi(u_1)\rangle = |01\rangle$.

In the space of 2^r-point integer-valued signals with the range $L = 2^l$, $l \geq 1$, the unit impulse $u_k = \{0, 0, \ldots, 0, 1, 0\ldots, 0\}$, $k \in \{0, 2^r - 1\}$, has the representation

$$|\check{u}_0\rangle = \frac{1}{\sqrt{2^r}}\left[|0\rangle + |1\rangle + |2\rangle + \ldots + e^{i\alpha}|k\rangle + \ldots + |2^r-1\rangle\right], \quad (\alpha = \pi/(2L)).$$

Example 8.2 Bell States with Phase Shift
For the signal $f = \{0, 1, 1, 0\}$, we obtain $L = 2$, $\alpha = \pi/4$. Therefore, the 2-qubit representation

$$\left|\check{f}\right\rangle = \frac{1}{2}\left[|00\rangle + e^{i\pi/4}|01\rangle + e^{i\pi/4}|10\rangle + |11\rangle\right] = \frac{|00\rangle + |11\rangle}{2} + e^{i\pi/4}\frac{|01\rangle + |10\rangle}{2}.$$

This 2-qubit contains the Bell states with phase shift $\pi/4$ between them. For comparison, in the representation by amplitudes in Eq. 8.7, we obtain the Bell state $|\varphi(f)\rangle = (|01\rangle + |10\rangle)/\sqrt{2}$.

For the 4-point signal $g = \{1, 2, 2, 1\}$, $L = 4$, and $\alpha = \pi/8$. The 2-qubit representation of the signal is

$$|\check{g}\rangle = \frac{1}{2}\left[e^{i\pi/8}|00\rangle + e^{i\pi/4}|01\rangle + e^{i\pi/4}|10\rangle + e^{i\pi/8}|11\rangle\right] = e^{i\pi/8}\frac{|00\rangle + |11\rangle}{2} + e^{i\pi/4}\frac{|01\rangle + |10\rangle}{2}.$$

8.1.4 Property of Phase

Consider adding the constants ±1 to the signal, $y_n = f_n \pm 1$. The FTQR of the new signal is equal to the original superposition with the global phase,

$$|\check{y}\rangle = \frac{1}{\sqrt{2^r}}\sum_{n=0}^{2^r-1} e^{i\alpha(f_n \pm 1)}|n\rangle = \frac{1}{\sqrt{2^r}}\sum_{n=0}^{2^r-1} e^{\pm i\alpha}e^{i\alpha f_n}|n\rangle = e^{\pm i\alpha}\frac{1}{\sqrt{2^r}}\sum_{n=0}^{2^r-1} e^{i\alpha f_n}|n\rangle = e^{\pm i\alpha}\left|\check{f}\right\rangle.$$

Similarly, $\left|\left(\widecheck{f \pm k}\right)\right\rangle = e^{\pm i\alpha k}\left|\check{f}\right\rangle$, for an integer k. Note that the above operations do not require subsequent normalization. For comparison, with representation in Eq. 8.7, the signal $y_n = f_n + 1$ will be in the superposition

$$|\varphi(f+1)\rangle = \frac{1}{E_1}\sum_{n=0}^{2^r-1}(f_n+1)|0\rangle, \tag{8.8}$$

with the new normalized coefficient calculated by $E_1^2 = |f_0+1|^2 + |f_1+1|^2 + \ldots + |f_{2^r-1}+1|^2$.

8.2 Operations with Kronecker Product

The following properties hold for the tensor product with the FTQR (for more detail see [1]):

(The time-scaling by a factor of 1/2) The product

$$\left|\check{f}\right\rangle \otimes |0\rangle = \left(\frac{1}{\sqrt{N}}\sum_{n=0}^{N-1} e^{i\frac{2\pi}{N}f_n}|n\rangle\right) \otimes |0\rangle = \frac{1}{\sqrt{N}}\sum_{n=0}^{N-1} e^{i\frac{2\pi}{N}f_n}|2n\rangle$$

is the $(r+1)$-qubit representation of the discrete-time signal $y(n) = f(n/2)$, $n = 0, 2, 4, \ldots, (2N-2)$. Here, we use the fact that $|n\rangle \otimes |0\rangle = |2n\rangle$. For instance, $|3\rangle \otimes |0\rangle = [0001]' \otimes [10]' = [00000010]' = |6\rangle$.

(The time-shift and scaling by a factor of 1/2) The product

$$\left|\check{f}\right\rangle \otimes |1\rangle = \left(\frac{1}{\sqrt{N}}\sum_{n=0}^{N-1} e^{i\frac{2\pi}{N}f_n}|n\rangle\right) \otimes |1\rangle = \frac{1}{\sqrt{N}}\sum_{n=0}^{N-1} e^{i\frac{2\pi}{N}f_n}|2n+1\rangle$$

is the $(r+1)$-qubit representation of the discrete-time signal $y(n) = f\left(\frac{n-1}{2}\right)$, $n = 1, 3, 5, \ldots, (2N-1)$. Here, we use the fact that $|n\rangle \otimes |1\rangle = |2n+1\rangle$.

(The product of two representations) The Kronecker product

$$\left|\check{f}\right\rangle \otimes |\check{g}\rangle = \left(\frac{1}{\sqrt{N}}\sum_{n=0}^{N-1} e^{i\alpha f_n}|n\rangle\right) \otimes \left(\frac{1}{\sqrt{N}}\sum_{m=0}^{N-1} e^{i\alpha g_m}|m\rangle\right) = \frac{1}{N}\sum_{n=0}^{N-1}\sum_{m=0}^{N-1} e^{i\alpha(f_n+g_m)}|n\rangle \otimes |m\rangle$$

$$= \underbrace{\frac{1}{N}\sum_{n=0}^{N-1} e^{i\alpha(f_n+g_n)}|n(N+1)\rangle}\; + \frac{1}{N}\sum_{\substack{n,m=0:N-1\\ n\neq m}} e^{i\alpha(f_n+g_m)}|Nn+m\rangle$$

describes the $2r$-qubit quantum superposition. Here, $|n\rangle \otimes |m\rangle = |Nn+m\rangle$, $m = 0:(N-1)$. Therefore, the sum of two signals can be measured in the basis states $|n(N+1)\rangle$, where $n = 0:(N-1)$.

The Fourier transform quantum superposition can be modified (by a permutation) into the r-qubit superposition

$$|\psi_1(f)\rangle = \frac{1}{\sqrt{N}}\sum_{n=0}^{N/2-1}\left[e^{i\alpha f_{2n}}|0\rangle + e^{i\alpha f_{2n+1}}|1\rangle\right]|n\rangle. \tag{8.9}$$

Here, each pair of values f_{2n} and f_{2n+1} is packed into a single qubit. If the signal has equal values in neighbor points, let us say, in $2n_0$ and $(2n_0 + 1)$, then at the basis state $|n_0\rangle$ in the above superposition, the Hadamard state presents with a phase factor,

$$|\psi_1(f)\rangle = \frac{1}{\sqrt{N}}\left(+ \ldots \sqrt{2}\left[e^{i\alpha f_{2n0}} \underbrace{\left(\frac{|0\rangle + |1\rangle}{\sqrt{2}}\right)} \right] |n_0\rangle + \ldots \right).$$

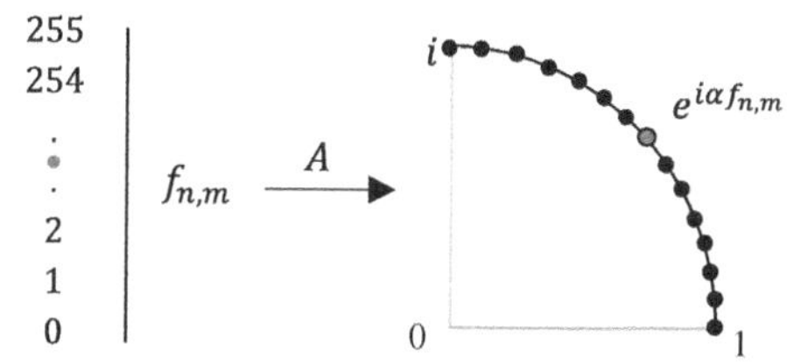

Fig. 8.4 Transforming the integer interval [0,255] into the 256 points on the quarter circle.

8.3 FTQR Model for Grayscale Image

The grayscale integer image $f = f_{n,m}$ of size $N \times M = 2^r \times 2^s$ pixels of the range $L = 256$ is represented by the exponential coefficients in a way that is similar to the 1D case. The image is transforming into a quarter of the unite circle as (see Fig. 8.4)

$$f_{n,m} \to A\left[f_{n,m}\right] = e^{i\alpha f_{n,m}}, \qquad \alpha = \pi/512. \tag{8.10}$$

The $(r + s)$-qubit Fourier transform quantum representation is defined by the following superposition [19]:

$$|\psi(f)\rangle = \frac{1}{\sqrt{NM}} \sum_{m=0}^{M-1} \sum_{n=0}^{N-1} A\left[f_{n,m}\right] |n, m\rangle. \tag{8.11}$$

The image $f_{n,m}$ can be written into the 1D vector f_k, $k = nM + m$, where $n = 0{:}\,(N - 1)$ and $m = 0{:}\,(M - 1)$. Then, the circuit for implementation of the FTQR will be similar to the one shown in Fig. 8.2.

8.4 Color Image FTQR Models

1) Many methods process the color components separately. The above-proposed model MQFTR can be applied for each color [2]. For simplicity of calculations, we consider an RGB color image $f_{n,m} = \{r_{n,m}, g_{n,m}, b_{n,m}\}$ as the 1D vector f_k, $k = 0{:}\,K - 1$, where $K = 2^{r+s}$. The red (r), green (g), and blue (b) components, each in the range L, can be represented by the $(r + s)$-qubit FTQRs as

$$|\psi(c)\rangle = \frac{1}{\sqrt{K}} \sum_{k=0}^{K-1} A[c_k] |k\rangle = \frac{1}{\sqrt{K}} \sum_{k=0}^{K-1} e^{i\alpha c_{n,m}} |k\rangle, \qquad c = r, g, b. \tag{8.12}$$

Here, the mapping on the unit circle $A[x] = e^{i\alpha x}$, when $x = 0{:}\,255$. The constant $\alpha = 2\pi/1024$, considering that $L = 256$.

2) It is possible to unite the color components without additional qubits and develop new models of quantum representation [2]. We describe such a model and introduce the concept of the 4-point discrete Fourier transform of qubits. Consider an RGB color image together with its grayscale component, that is, the image $f_{n,m} = (r_{n,m}, g_{n,m}, b_{n,m}, i_{n,m})$. This image is represented as a multi-qubit state in the following way. Because of four components, the unit circle is divided by 4 arcs, each of 90°, and each color is mapped to one of these arcs, as shown in Fig. 8.5. The values of each color component are considered to be in the integer interval [0,255]. Each pixel is defined with four points on the unit circle, and each point is in its color circular arc. These

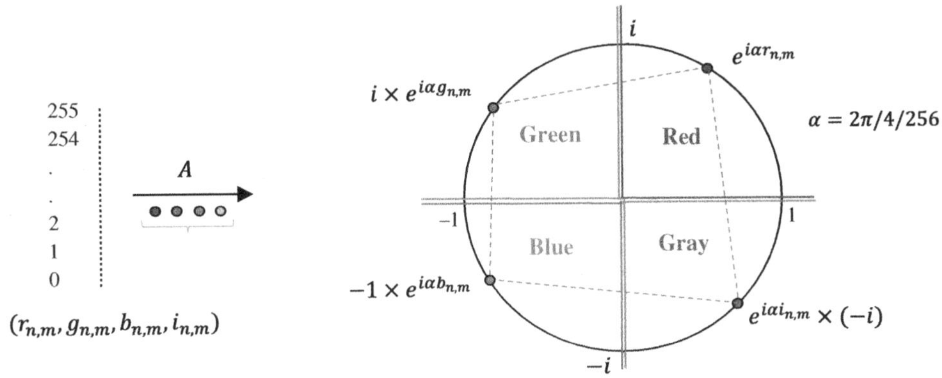

Fig. 8.5 Transformation of colors (for the RGB plus gray colors) (Refer Online version for colour representation).

four points compose a quadrilateral of the RGB plus color. For the grayscale components of the image, these quadrilaterals are squares. This model is convenient for the case with quaternion images, which is described in Section 12.

The case with only 3 RGB colors, when the unit circle is divided by 3 arcs each of 120° is described in detail in [1]. The transformation of colors is shown in Fig. 8.6.

These states can be united into the following $(r+s)$-qubit superposition of states of colors:

$$|\psi(f)\rangle = \frac{1}{B}\sum_{k=0}^{K-1}(A[r_k] + iA[g_k] - A[b_k] - iA[i_k])|k\rangle. \tag{8.13}$$

The normalization coefficient is calculated by

$$B = \sqrt{\sum_{k=0}^{K-1} |A[r_k] + iA[g_k] - A[b_k] - iA[i_k]|^2}.$$

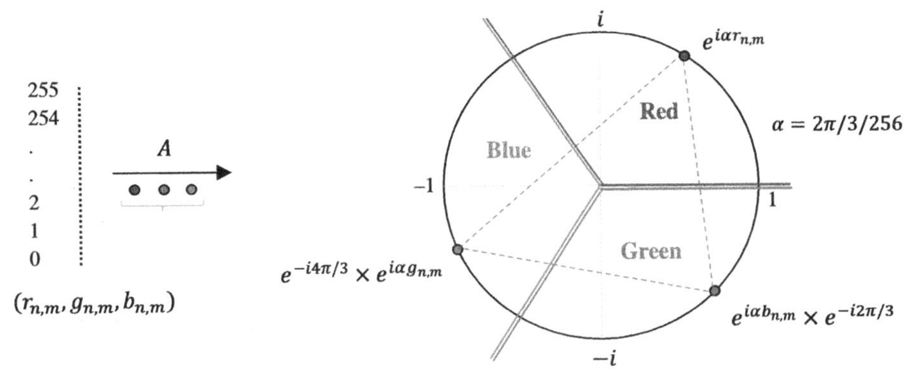

Fig. 8.6 Transformation of colors (for the RGB model) (Refer Online version for colour representation).

It is possible to consider the 2-qubit quantum Fourier transform, or 4-point DFT of the vector-signal $\{A[r_k], A[g_k], A[b_k], A[i_k]\}$,

$$F_{p;k} = A[r_k] + A[g_k]W^p + A[b_k]W^{2p} + A[i_k]W^{3p}, p = 0, 1, 2, 3.$$

Here, $W = \exp(-2\pi i/4)$. Then, the state $A[r_k] + iA[g_k] - A[b_k] - iA[i_k] = F_{3;k}$.

8.5 The 2D Quantum Fourier Transform

The 2D discrete Fourier transform is used for processing images in the frequency domain in many tasks, such as linear filtration, image enhancement, and denoising. Many effective algorithms of the 2D DFT exist [3], but we consider the simple column–row method of calculation for images of size with powers of 2.

Consider the 2D DFT of the image

$$F_{p,s} = (\mathcal{F}_{N,M}f)_{p,s} = \frac{1}{\sqrt{NM}}\sum_{m=0}^{M-1}\sum_{n=0}^{N-1} f_{n,m}W_N^{np}W_M^{ms}, \qquad p,s = 0: (N-1), (M-1). \tag{8.14}$$

Here, $W_N = \exp(-i2\pi/N)$ and $W_M = \exp(-i2\pi/M)$. This transform is separable; it can be calculated by the 1D DFTs by rows and then by columns, and vice versa. We consider $N = M = 2^r$, $r > 1$,and

$$F_{p,s} = \frac{1}{\sqrt{N}}\sum_{m=0}^{N-1}\left[\frac{1}{\sqrt{N}}\sum_{n=0}^{N-1} f_{n,m}W_N^{np}\right]W_N^{ms}, \qquad p,s = 0: (N-1). \tag{8.15}$$

The inverse $N \times M$-point DFT is calculated by

$$f_{n,m} = \frac{1}{\sqrt{N}}\sum_{s=0}^{N-1}\left[\frac{1}{\sqrt{N}}\sum_{p=0}^{N-1} F_{p,s}W_N^{-np}\right]W_N^{-ms}, \qquad n,m = 0: (N-1).$$

To perform this transform in quantum computation, the image is written into the 1D signal by columns, $f_k = f_{Nn+m}$, $k = 0: (N^2 - 1)$. The image-signal is considered to be normalized. The algorithm below first calculates the 1D quantum Fourier transforms of N parts of the signal, followed by permutation. The implementation of 1D QFT is done using the circuits described in Section 5, which are the paired transform-based FFT [4].

8.5.1 Algorithm of the 2D QFT

Prepare sub-superpositions of the signal, each with N amplitudes,

$$|\varphi(f)\rangle = \frac{1}{\sqrt{N}}\left(|\varphi(f)\rangle_0 + |\varphi(f)\rangle_1 + \ldots + |\varphi(f)\rangle_{N-1}\right). \tag{8.16}$$

Namely, the superpositions

$$|\varphi(f)\rangle_0 = \frac{1}{E_0}\left(\underbrace{f_0|0\rangle + f_1|1\rangle + \ldots + f_{N-1}|N-1\rangle}\right),$$

$$|\varphi(f)\rangle_1 = \frac{1}{E_1}\left(\underbrace{f_N|0\rangle + f_{N+1}|1\rangle + \ldots + f_{2N-1}|N-1\rangle}\right), \ldots,$$

$$|\varphi(f)\rangle_{N-1} = \frac{1}{E_{N-1}}\left(\underbrace{f_{N^2-N}|0\rangle + f_{N^2-N+1}|1\rangle + \ldots + f_{N^2-1}|N^2-1\rangle}\right).$$

Here, $E_n, n = 0: (N-1)$, are energies of columns of the image.

Compute the r-qubit QFT on each sub-superposition $|\varphi(f)\rangle_n$, $n = 0{:}\,(N-1)$,

$$|\varphi(F)\rangle_n = F_{n;0}|0\rangle + F_{n;1}|1\rangle + F_{n;2}|2\rangle + \ldots + F_{n;N-1}|N-1\rangle, \tag{8.17}$$

where ($W_N = \exp(-i2\pi/N)$),

$$F_{n;p} = \frac{1}{\sqrt{N}} \sum_{m=0}^{N-1} f_{nN+m} W_N^{mp}, \quad p = 0{:}\,(N-1). \tag{8.18}$$

Compose the N^2-dimension vector $\boldsymbol{V}$ from the calculated r-qubit QFT data $F_{n;p}$,

$$\boldsymbol{V} = \left(\underbrace{F_{0;0}, F_{0;1}, \ldots, F_{0;N-1}}, \underbrace{F_{1;0}, F_{1;1}, \ldots, F_{1;N-1}}, \ldots, \underbrace{F_{N-1;0}, F_{N-1;1}, \ldots, F_{N-1;N-1}}\right)'. \tag{8.19}$$

Calculate the permutation P of the vector $\boldsymbol{V}$ as

$$P : \boldsymbol{V} \rightarrow \boldsymbol{V}_1 = \left(\underbrace{F_{0;0}, F_{N-1;0}, F_{2N-1;0}, \ldots, F_{N^2-1;0}}_{n=0}, \ldots, \underbrace{F_{0;n}, F_{N-1;n}, F_{2N-1;n}, \ldots, F_{N^2-1;n}, \ldots}_{n=1:(N-1)}\right)'. \tag{8.20}$$

1) Perform steps 2 and 3 for the new superposition $|\varphi(\boldsymbol{V}_1)\rangle$.
2) Compute the inverse permutation P^{-1} on the obtained quantum superposition $|\varphi(\boldsymbol{V})\rangle$.

The general block-diagram of the 2D quantum Fourier transform (QFT) of an image of $N \times N$ pixels is shown in Fig. 8.7. We call this quantum transform *the 2D quantum* $r \times r$*-qubit QFT*.

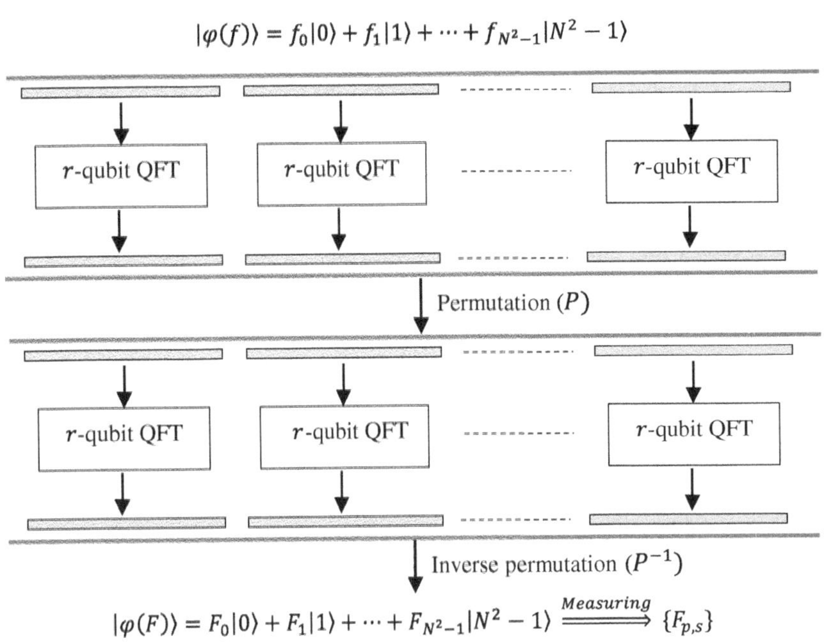

Fig. 8.7 The diagram of calculation of the $r \times r$-qubit QFT of the image superposition.

Example 8.3 2D 1 × 1-Qubit QFT

Consider a 2 × 2 image and the quantum circuit with two qubits, which is given in Fig. 8.8. The 2-point Fourier and Hadamard transforms use the same gate. The 2-D QFT is implemented by two controlled Hadamard gates.

State preparation is done through the use of the quantum 2-qubit signal-induced Heap transform (QsiHT), $\mathcal{H}_4$, using the strong wheel-carriage [6, 7]. Here, the normalized signal $x = (x_0, x_1, x_2, x_3) = (f_0, f_1, f_2, f_3)/\sqrt{E(f)}$ is used as the generator for the DsiHT,

$$(x_2, x_3) \overset{R_{\vartheta_0}}{\rightarrow} \left(y_0^{(1)}, 0\right), \quad \left(y_0^{(1)}, x_1\right) \overset{R_{\vartheta_1}}{\rightarrow} \left(y_0^{(2)}, 0\right), \quad \left(y_0^{(2)}, x_0\right) \overset{R_{\vartheta_2}}{\rightarrow} \left(y_0^{(3)}, 0\right), \quad y_0^{(3)} = 1. \tag{8.21}$$

Thus, $x = \mathcal{H}_4^{-1}(1,0,0,0)'$. The DsiHT uses three rotations $R_{\vartheta_0}, R_{\vartheta_1}$, and R_{ϑ_2} with angles $\{\vartheta_0, \vartheta_1, \vartheta_2\}$ of the angular representation of the generator x. After state preparation, the local gate of the 1-qubit QFT is applied to perform the 2-point DFT over two columns of the image. Then, the permutation is applied to the next stage of operations. This permutation P is defined by

$$P\colon F_0|00\rangle + \underbrace{F_1|01\rangle + F_2|10\rangle} + F_3|11\rangle \rightarrow F_0|00\rangle + \underbrace{F_2|01\rangle + F_1|10\rangle} + F_3|11\rangle. \tag{8.22}$$

The permutation is $P = (1, 2)$, or SWAP gate,

$$P = P_{(1,2)} = \begin{bmatrix} 1 & 0 & 0 & 0 \\ 0 & 0 & 1 & 0 \\ 0 & 1 & 0 & 0 \\ 0 & 0 & 0 & 1 \end{bmatrix}, \quad P = P^{-1}.$$

All calculations for the 2D 1 × 1-qubit QDFT can be written as

$$|\varphi(F)\rangle = \underbrace{P_{(1,2)}(I_2 \otimes \mathcal{F}_2)}\underbrace{P_{(1,2)}(I_2 \otimes \mathcal{F}_2)}\underbrace{\mathcal{H}_4^{-1}\,|\,00\rangle}. \tag{8.23}$$

Here, $\mathcal{H}_4^{-1}\,|00\rangle = |\varphi(f)\rangle$. In the matrix form, these calculations can be written as follows:

$$|\varphi(F)\rangle = \underbrace{\begin{bmatrix} 1 & 0 & 0 & 0 \\ 0 & 0 & 1 & 0 \\ 0 & 1 & 0 & 0 \\ 0 & 0 & 0 & 1 \end{bmatrix} \frac{1}{\sqrt{2}} \begin{bmatrix} 1 & 1 & 0 & 0 \\ 1 & -1 & 0 & 0 \\ 0 & 0 & 1 & 1 \\ 0 & 0 & 1 & -1 \end{bmatrix}} \times \underbrace{\begin{bmatrix} 1 & 0 & 0 & 0 \\ 0 & 0 & 1 & 0 \\ 0 & 1 & 0 & 0 \\ 0 & 0 & 0 & 1 \end{bmatrix} \frac{1}{\sqrt{2}} \begin{bmatrix} 1 & 1 & 0 & 0 \\ 1 & -1 & 0 & 0 \\ 0 & 0 & 1 & 1 \\ 0 & 0 & 1 & -1 \end{bmatrix}} \underbrace{\left[\mathcal{H}_4^{-1}\right] \begin{bmatrix} 1 \\ 0 \\ 0 \\ 0 \end{bmatrix}}.$$

Let the 2 × 2 image be $f_{2\times 2} = \begin{bmatrix} 1 & 2 \\ 3 & 4 \end{bmatrix}$ and $\{f_k;\, k = 0{:}\,3\} = \{1, 3, 2, 4\}$. The DsiHT is calculated by rotations

$$R_{\vartheta_k} = \begin{bmatrix} \cos\vartheta_k & -\sin\vartheta_k \\ \sin\vartheta_k & \cos\vartheta_k \end{bmatrix}, \quad k = 1, 2, 3,$$

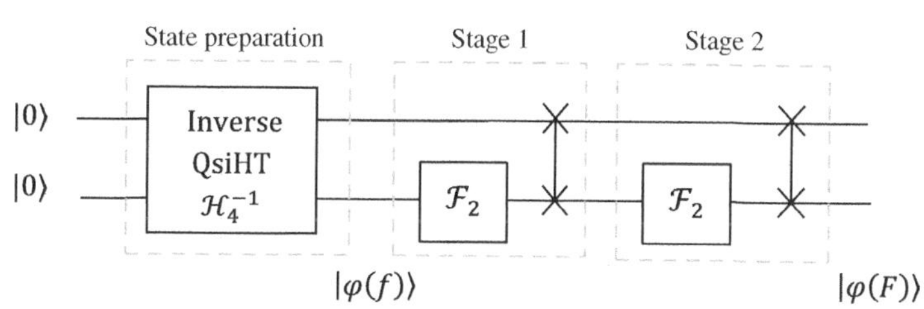

Fig. 8.8 The circuit of the 2D 1 × 1-qubit QFT.

with the angles $\{\vartheta_0, \vartheta_1, \vartheta_2\} = \{-53.1301°, -68.1986°, -79.4803°\}$. The unitary matrix of the heap transform $\mathcal{H}_4$ is calculated as follows:

$$[\mathcal{H}_4] = (R_{\vartheta_2} \oplus I_2)(1 \oplus R_{\vartheta_1} \oplus 1)(I_2 \oplus R_{\vartheta_0}),$$

$$4] = \begin{bmatrix} \cos\vartheta_2 & -\sin\vartheta_2 & 0 & 0 \\ \sin\vartheta_2 & \cos\vartheta_2 & 0 & 0 \\ 0 & 0 & 1 & 0 \\ 0 & 0 & 0 & 1 \end{bmatrix} \begin{bmatrix} 1 & 0 & 0 & 0 \\ 0 & \cos\vartheta_1 & -\sin\vartheta_1 & 0 \\ 0 & \sin\vartheta_1 & \cos\vartheta_1 & 0 \\ 0 & 0 & 0 & 1 \end{bmatrix} \begin{bmatrix} 1 & 0 & 0 & 0 \\ 0 & 1 & 0 & 0 \\ 0 & 0 & \cos\vartheta_0 & -\sin\vartheta_0 \\ 0 & 0 & \sin\vartheta_0 & \cos\vartheta_0 \end{bmatrix}.$$

Thus,

$$[\mathcal{H}_4] = \begin{bmatrix} 0.1826 & 0.3651 & 0.5477 & 0.7303 \\ -0.9832 & 0.0678 & 0.1017 & 0.1356 \\ 0 & -0.9285 & 0.2228 & 0.2971 \\ 0 & 0 & -0.8000 & 0.6000 \end{bmatrix}.$$

The inverse matrix is its transpose matrix equal to

$$[\mathcal{H}_4^{-1}] = (I_2 \oplus R_{-\vartheta_0})(1 \oplus R_{-\vartheta_1} \oplus 1)(R_{-\vartheta_2} \oplus I_2) = \begin{bmatrix} 0.1826 & -0.9832 & 0 & 0 \\ 0.3651 & 0.0678 & -0.9285 & 0 \\ 0.5477 & 0.1017 & 0.2228 & -0.8000 \\ 0.7303 & 0.1356 & 0.2971 & 0.6000 \end{bmatrix}.$$

Example 8.4 2D QFT on 4 × 4 Image

The scheme of calculation of the 2D 2 × 2-qubit QDFT is given in Fig. 8.9. The state preparation is accomplished by the unitary 1-point DsiHT, or 4-qubit QsiHT.

The permutation is $P_x = (1, 12, 3, 4)(2, 8)(5, 13, 15, 7)(6, 9, 14, 11)$. In the matrix form, the quantum scheme for the 2D 2 × 2 qubit QFT can be represented as,

$$|\varphi(F)\rangle = \underbrace{P_x^{-1}(I_4 \otimes \mathcal{F}_4)}\underbrace{P_x(I_4 \otimes \mathcal{F}_4)}\underbrace{\mathcal{H}_4^{-1}\,|\,0^{\otimes 4}\rangle}.$$

Here, the inverse permutation $P_x^{-1} = (1, 4, 3, 12)(2, 8)(5, 7, 15, 13)(6, 11, 14, 9)$.

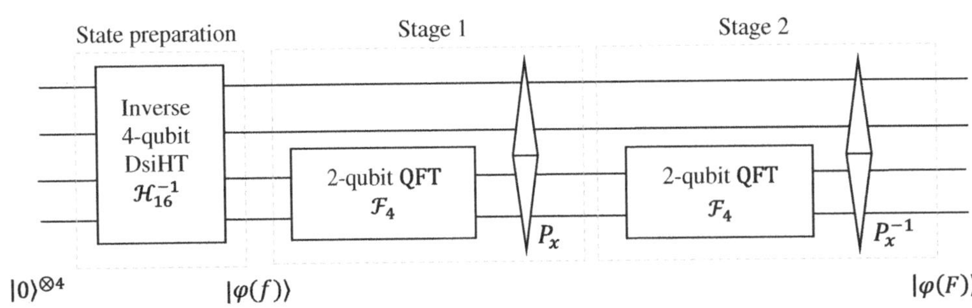

Fig. 8.9 The block-diagram of the 2D 2 × 2-qubit QFT.

8.5.2 Examples in Qiskit

Let the 4×4 image be the following:

$$f_{4\times 4} = \begin{bmatrix} 1 & 2 & 3 & 2 \\ 2 & 3 & 4 & 3 \\ 1 & 5 & 3 & 4 \\ 3 & 1 & 2 & 2 \end{bmatrix}, \quad E = E(f) = \sqrt{\sum_{n=0}^{3}\sum_{m=0}^{3} f_{n,m}^2} = 11.1803.$$

This image, $f_{4\times 4}$, is then converted to a 16D vector f_k column by column, $\{f_k; k = 0{:}\ 15\} = \{1, 2, 1, 3, 2, 3, 5, 1, 3, 4, 3, 2, 2, 3, 4, 2\}$. The image superposition is

$$|\varphi(f)\rangle = \frac{1}{\sqrt{E}}\sum_{k=0}^{15} f_k|k\rangle.$$

To prepare qubits for this superposition, the signal is normalized, $f = f/\sqrt{E} = f/\sqrt{11.1803}$. Then, it is used as the generator for the DsiHT with the strong wheel-carriage. The angles of rotations are

$$\{\vartheta_n; n = 0{:}\ 14\} = \{-53.1301°, -68.1986°, -79.4803°, \ldots, -79.6532°, -84.8685°\}.$$

The inverse $[\mathcal{H}_4^{-1}]$ heap transform with 15 rotations is used for state preparation

$$|\varphi(f)\rangle = [\mathcal{H}_4^{-1}]|0\rangle^{\otimes 4} = 0.0894|0000\rangle + 0.1789|0001\rangle + 0.2682|0010\rangle + \ldots + 0.1789|1111\rangle.$$

Performing the quantum scheme yields the following theoretical state vector:

$$|\varphi(F)\rangle = 0.9168|0000\rangle + (-0.1118 - 0.0894i)|0001\rangle + \ldots + (-0.0224 - 0.0894i)|1111\rangle.$$

The probabilities for various simulations, or shots, for this scheme are demonstrated in Table 8.1, where the ideal number of samples would be around 100,000.

Example 8.5 2D QFT on 3 × 3 Qubit Image

The block-diagram for computing the 2D 3×3-qubit QDFT is shown in Fig. 8.10. The permutation P_x in the set of 64 elements and it inverse can be calculated by Python (for more detail, see [8]). The simulation of this quantum diagram was performed by simulator Qiskit. The Python-based codes written by our undergraduate student Alexis Gomez can be found by the address: https://ceid.utsa.edu/agrigoryan/. Both direct and reversible 3×3-qubit QDFTs are computed in the codes.

Table 8.1 Qiskit measured amplitudes for the 4 × 4 image

Basis states	Probabilities				
	Theoretical	10,000 shots	20,000 shots	50,000 shots	100,000 shots
0000	0.8410	0.8421	0.8361	0.8400	0.8410
0001	0.0205	0.0193	0.0213	0.0204	0.0205
0011	0.0205	0.0194	0.0207	0.0200	0.0205
...	...	...	...	...	...
1110	0.0065	0.0069	0.0077	0.0064	0.0065
1111	0.0085	0.0077	0.0091	0.0091	0.0085

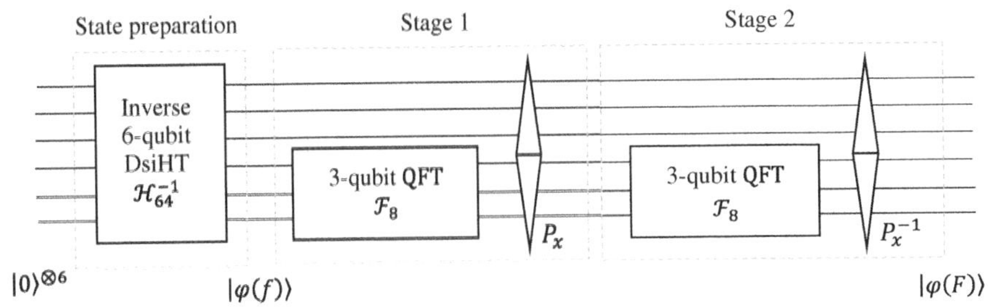

Fig. 8.10 The block-diagram of the 3 × 3-qubit QFT.

$$f = \frac{1}{\sqrt{3443}}\begin{bmatrix} 8 & 7 & 6 & 5 & 4 & 6 & 7 & 8 \\ 5 & 9 & 4 & 5 & 5 & 4 & 9 & 5 \\ 9 & 4 & 6 & 9 & 10 & 6 & 4 & 3 \\ 7 & 5 & 10 & 16 & 14 & 8 & 5 & 6 \\ 6 & 5 & 9 & 13 & 12 & 9 & 5 & 7 \\ 8 & 4 & 6 & 8 & 11 & 6 & 4 & 8 \\ 5 & 9 & 4 & 5 & 7 & 4 & 9 & 5 \\ 8 & 7 & 6 & 5 & 4 & 6 & 7 & 8 \end{bmatrix}.$$

As in the above example with the 2×2-qubit image, the 64-point DsiHT (or 6-qubit QsiHT, H) can be used to initiate the 6-qubit vector representing the image 8×8. The orthogonal heap transform is generated by this image. Thus, $Hf = |0\rangle^{\otimes 6}$, or $f = H^T|0\rangle^{\otimes 6}$. The amplitude representation of the image is considered to be

$$|\varphi(f)\rangle = \sum_{k=0}^{63} f_k|k\rangle = \frac{1}{\sqrt{3443}}(8|0\rangle + 5|1\rangle + 9|2\rangle + 7|3\rangle + \ldots 7|60\rangle + 8|61\rangle + 5|62\rangle + 8|63\rangle).$$

The 3D mesh of the 2D DFT of the image in the absolute scale and cyclically shifted to the center is shown in Fig. 8.11 in part (a). The results $|\varphi(F)\rangle$ of performing the quantum scheme by using Python (for permutations) and Qiskit yields the meshes shown in parts (b) and (c) for 1,000 and 100,000 shots, respectively.

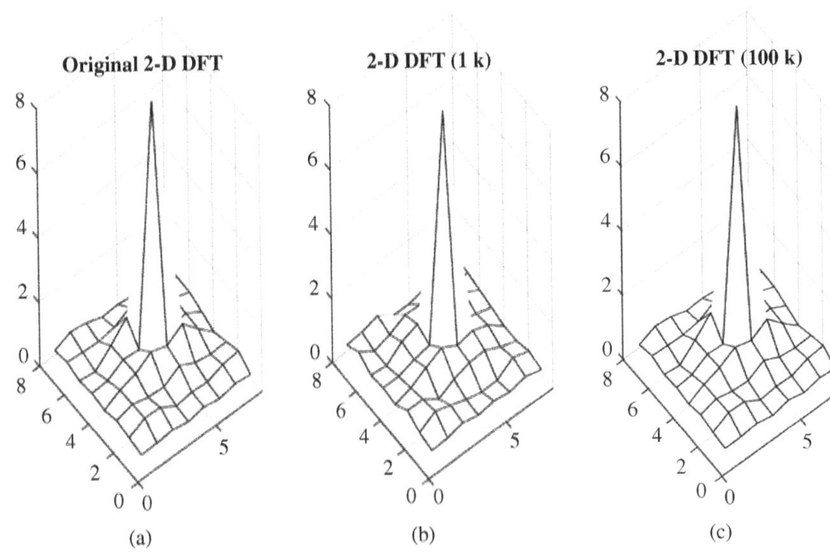

Fig. 8.11 The 2D DFT in the absolute scale for (a) the original image, and after simulation for (b) 1,000 shots and (c) 100,000 shots.

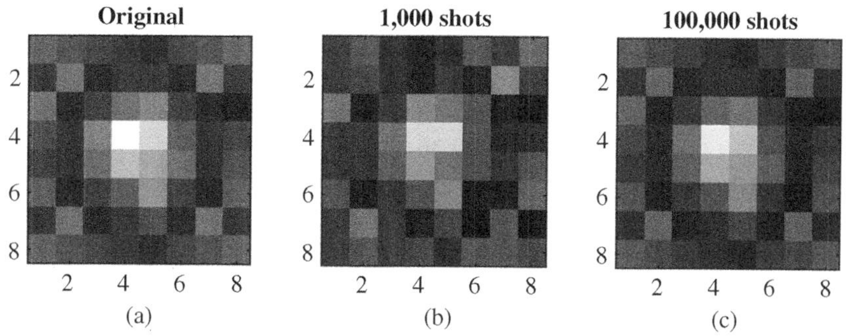

Fig. 8.12 (a) The original grayscale image, and after simulation for (b) 1,000 shots and (c) 100,000 shots.

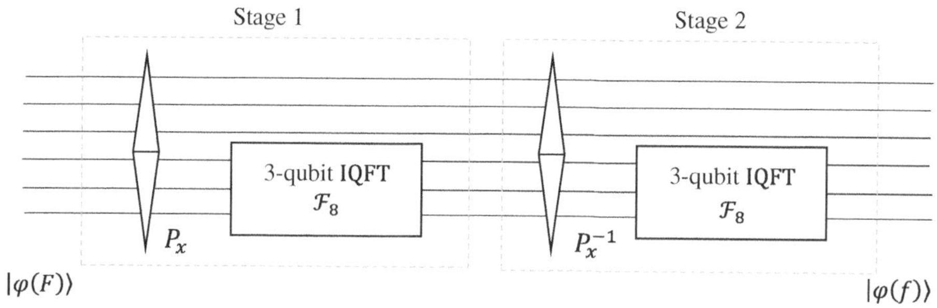

Fig. 8.13 The block-diagram of the 3 × 3-qubit inverse QFT (IQFT).

The mean square errors are equal to 0.1447 and 0,0407 for 1,000 and 100,000 shots, respectively. All circuit elements in the scheme of Fig. 8.10 are gates that are reversible. The results of using the direct 3×3-qubit QFTs followed by performing inverse 3×3-qubit QFT after 1,000 and 100,0000 shots are shown in Fig. 8.12. The image in part (c) after rounding matches the original image exactly.

The scheme of the 3×3-qubit inverse QFT is given in Fig. 8.13.

References

1 Grigoryan, A.M. and Agaian, S.S. (2020). New look on quantum representation of images: Fourier transform representation. *Quantum Information Processing* 19 (148): 26. https://doi.org/10.1007/s11128-020-02643-3.

2 Grigoryan, A.M. and Agaian, S.S. (2022). Quantum representation of 1-D signals on the unit circle. *Quantum Information Processing* 21 (24): 15. https://doi.org/10.1007/s11128-021-03237-3.

3 Grigoryan, A.M. and Agaian, S.S. (2003). *Multidimensional Discrete Unitary Transforms: Representation, Partitioning, and Algorithms*. New York: Marcel Dekker.

4 Grigoryan, A.M. and Grigoryan, M.M. (2009). *Brief Notes in Advanced DSP: Fourier Analysis with MATLAB*. Taylor and Francis Group: CRC Press.

5 Grigoryan, A.M. and Agaian, S.S. (2019) 'Paired quantum Fourier transform with $\log_2 N$ Hadamard gates,' *Quantum Information Processing*, 18: 217, p. 26. https://doi.org/10.1007/s11128-019-2322-6

6 Grigoryan, A.M. (2023) 'Effective methods of QR-decompositions of square complex matrices by fast discrete signal-induced heap transforms,' 12(4), pp. 87–110, *Advances in Linear Algebra & Matrix Theory (ALAMT).* https://doi.org/10.4236/alamt.2022.124005.

7 Grigoryan, A.M. (2014). New method of Givens rotations for triangularization of square matrices. *Journal of Advances in Linear Algebra & Matrix Theory (ALAMT)* 4 (2): 65–78. https://doi.org/10.4236/alamt.2014.42004.

8 Grigoryan, A.M., Alexis, A.G., and Agaian, S.S. (2024). Methods of calculation of the 2-D quantum Fourier transform of images. In: *Proceedings of SPIE 13033 Conference, Defense + Commercial Sensing 2024*, 13033-22, p. 12. Maryland, USA.

9

New Operations of Qubits

In a quantum computer, mathematicians do not have much freedom in their work. Therefore, working here is much more interesting, a lot needs to be done within a very narrow framework and you need to think differently here. Many tasks seem impossible. There is no rich arithmetic in quantum computers. All operations are described by permutations and reversible (unitary) transformations. Therefore, all output and input states have the same dimension. Another feature of the qubit is that it is impossible to determine the state of qubits in a computational problem from a single observation. Copying states is not possible, and observing/measuring any calculation state requires a large number of calculation iterations. There is a need to find new efficient operations on qubits, two-qubits, multi-qubits, and quantum superpositions to expand our capabilities in processing signals and images. This is especially true for image processing in the frequency domain, where the main transformation is the quantum Fourier transform (QFT).

In this section, we show that single qubits with real amplitudes can be multiplied and divided similarly to complex numbers [1, 2]. In other words, we describe new operations of qubits and quantum representations, which include (i) multiplication of single qubits and properties; (ii) multiplication, division, and inverse between quantum superpositions; (iii) new representation of the QFT; (iv) ideal low- and high-pass filtration; and (v) qubit quadratic equations. These operations can be used to process quantum computation images and signals.

9.1 Multiplication

For two qubits $|\varphi_1\rangle = a_1|0\rangle + b_1|1\rangle$ and $|\varphi_2\rangle = a_2|0\rangle + b_2|1\rangle$ with real amplitudes, the product of these qubits is defined as

$$|\varphi\rangle \triangleq |\varphi_1\rangle \cdot |\varphi_2\rangle = a|0\rangle + b|1\rangle, \tag{9.1}$$

where the amplitudes $a = a_1a_2 - b_1b_2$ and $b = a_1b_2 + a_2b_1$. This multiplication is written in the matrix form as

$$\begin{bmatrix} a \\ b \end{bmatrix} = \begin{bmatrix} a_1 & -b_1 \\ b_1 & a_1 \end{bmatrix} \begin{bmatrix} a_2 \\ b_2 \end{bmatrix} = (a_1a_2 - b_1b_2)\begin{bmatrix} 1 \\ 0 \end{bmatrix} + (b_1a_2 + a_1b_2)\begin{bmatrix} 0 \\ 1 \end{bmatrix}. \tag{9.2}$$

Here, the 2×2 matrix has determinant equal $a_1^2 + b_1^2 = 1$ and it is the matrix of rotation,

$$R_{|\varphi_1\rangle} = \begin{bmatrix} a_1 & -b_1 \\ b_1 & a_1 \end{bmatrix} = R_\vartheta = \begin{bmatrix} \cos\vartheta & -\sin\vartheta \\ \sin\vartheta & \cos\vartheta \end{bmatrix}, \quad \vartheta = \operatorname{atan}(b_1/a_1).$$

Quantum Image Processing in Practice: A Mathematical Toolbox, First Edition. Artyom M. Grigoryan and Sos S. Agaian.

Companion website: www.wiley.com/go/grigoryan/quantumimageprocessing

For the basis states $|\varphi_1\rangle = |0\rangle$ and $|1\rangle$, the corresponding matrices of rotation are

$$R_{|0\rangle} = R_0 = \begin{bmatrix} 1 & 0 \\ 0 & 1 \end{bmatrix} \quad \text{and} \quad R_{|1\rangle} = R_{\frac{\pi}{2}} = \begin{bmatrix} 0 & -1 \\ 1 & 0 \end{bmatrix} = \begin{bmatrix} 0 & 1 \\ 1 & 0 \end{bmatrix}\begin{bmatrix} 1 & 0 \\ 0 & -1 \end{bmatrix} = XZ.$$

Therefore, $|1\rangle^2 = |1\rangle \cdot |1\rangle = -|0\rangle$. In the general case, the square of the single qubit is calculated by

$$|\varphi_1\rangle^2 \triangleq |\varphi_1\rangle \cdot |\varphi_1\rangle = (a_1^2 - b_1^2)|0\rangle + 2a_1b_1|1\rangle. \tag{9.3}$$

Note that the matrices X and Z are not matrices of multiplication in Eq. 9.2.

The multiplication of qubits is

1) commutative, $|\varphi_1\rangle \cdot |\varphi_2\rangle = |\varphi_2\rangle \cdot |\varphi_1\rangle$;
2) associative, $(|\varphi_1\rangle \cdot |\varphi_2\rangle) \cdot |\varphi_3\rangle = |\varphi_1\rangle \cdot (|\varphi_2\rangle \cdot |\varphi_3\rangle)$;
3) the state $|0\rangle$ is the unit of this multiplication operation, $|0\rangle \cdot |\varphi_1\rangle = |\varphi_1\rangle$;
4) $|1\rangle \cdot |\varphi_1\rangle = R_{\pi/2}|\varphi_1\rangle = -b_1|0\rangle + a_1|1\rangle$.

Example 9.1 Multiplication of Hadamard States

We consider the Hadamard states $|\varphi_1\rangle = (|0\rangle + |1\rangle)/\sqrt{2}$ and $|\varphi_2\rangle = (|0\rangle - |1\rangle)/\sqrt{2}$. The multiplication $|\varphi_1\rangle \cdot |\varphi_2\rangle = |0\rangle$. For the qubit $|\varphi_1\rangle$, we obtain the following square qubit: $|\varphi_1\rangle^2 = |\varphi_1\rangle \cdot |\varphi_1\rangle = |1\rangle$. Also, the square is $|\varphi_2\rangle^2 = |\varphi_2\rangle \cdot |\varphi_2\rangle = -|1\rangle$. The multiplication by the qubit $|\varphi_1\rangle$ is described by the matrix

$$A_2 = R_{|\varphi_1\rangle} = R_{\frac{\pi}{4}} = \frac{1}{\sqrt{2}}\begin{bmatrix} 1 & -1 \\ 1 & 1 \end{bmatrix},$$

and the multiplication by qubit $|\varphi_2\rangle$ is described by the matrix

$$R_{|\varphi_2\rangle} = R_{-\frac{\pi}{4}} = \frac{1}{\sqrt{2}}\begin{bmatrix} 1 & 1 \\ -1 & 1 \end{bmatrix} = \begin{bmatrix} 1 & 0 \\ 0 & -1 \end{bmatrix} \times \frac{1}{\sqrt{2}}\begin{bmatrix} 1 & 1 \\ 1 & -1 \end{bmatrix} = ZH_2.$$

9.1.1 Conjugate Qubit

The qubit conjugate to $|\varphi\rangle = a|0\rangle + b|1\rangle$ is defined as $\overline{|\varphi\rangle} = |\overline{\varphi}\rangle \triangleq a|0\rangle - b|1\rangle$. It is clear that $|\overline{\overline{\varphi}}\rangle = |\varphi\rangle$. The conjugate of basis states $|\overline{0}\rangle = |0\rangle$ and $|\overline{1}\rangle = -|1\rangle$. Also, we obtain

$$|\varphi\rangle \cdot |\overline{\varphi}\rangle = (a^2 + b^2)|0\rangle = |0\rangle. \tag{9.4}$$

The conjugate qubit is obtained from the qubit by using the Pauli-Z gate, that is,

$$|\overline{\varphi}\rangle = Z|\varphi\rangle = \begin{bmatrix} 1 & 0 \\ 0 & -1 \end{bmatrix}\begin{bmatrix} a_1 \\ b_1 \end{bmatrix} = \begin{bmatrix} a_1 \\ -b_1 \end{bmatrix}.$$

The following property is valid $\overline{|\varphi_1\rangle \cdot |\varphi_2\rangle} = \overline{|\varphi_1\rangle} \cdot \overline{|\varphi_2\rangle}$, for any two qubit superpositions $|\varphi_1\rangle$ and $|\varphi_2\rangle$.

9.1.2 Inverse Qubit

The qubit which is inverse to $|\varphi_1\rangle$ is defined from the equation $|\varphi_1\rangle \cdot |\varphi\rangle = |0\rangle$ as follows:

$$|\varphi\rangle = |\varphi_1\rangle^{-1} = \frac{|0\rangle}{|\varphi_1\rangle} \triangleq |0\rangle \cdot |\overline{\varphi}_1\rangle = |\overline{\varphi}_1\rangle. \tag{9.5}$$

The inverse qubit is the complex conjugate qubit. For example, two qubits in the Hadamard superpositions $(|0\rangle + |1\rangle)/\sqrt{2}$ and $(|0\rangle - |1\rangle)/\sqrt{2}$ are the conjugate qubits.

9.1.3 Division of Qubits

The qubits can be divided, by using the operation of multiplication as

$$|\varphi\rangle = \frac{|\varphi_1\rangle}{|\varphi_2\rangle} \triangleq |\varphi_1\rangle \cdot |\varphi_2\rangle^{-1} = |\varphi_1\rangle \cdot |\overline{\varphi}_2\rangle. \tag{9.6}$$

Note that $|1\rangle/|0\rangle = |1\rangle$ and $|0\rangle/|1\rangle = -|1\rangle$. In the matrix form, the operation of division of qubits can be written as

$$\begin{bmatrix} a_1 & -b_1 \\ b_1 & a_1 \end{bmatrix}\begin{bmatrix} a_2 \\ -b_2 \end{bmatrix} = \underbrace{\begin{bmatrix} a_1 & -b_1 \\ b_1 & a_1 \end{bmatrix}\begin{bmatrix} 1 & 0 \\ 0 & -1 \end{bmatrix}}\begin{bmatrix} a_2 \\ b_2 \end{bmatrix} = \begin{bmatrix} a_1 & b_1 \\ b_1 & -a_1 \end{bmatrix}\begin{bmatrix} a_2 \\ b_2 \end{bmatrix}. \tag{9.7}$$

The division of qubits has matrix representation $R_\vartheta Z$ with the determinant $\det R_\vartheta Z = -1$.

Example 9.2 Division of Hadamard States
We consider the Hadamard states $|\varphi_1\rangle = (|0\rangle + |1\rangle)/\sqrt{2}$ and $|\varphi_2\rangle = (|0\rangle - |1\rangle)/\sqrt{2}$. Then, $|\varphi_2\rangle^{-1} = |\overline{\varphi}_2\rangle = |\varphi_1\rangle = (|0\rangle + |1\rangle)/\sqrt{2}$ and $|\varphi_1\rangle/|\varphi_2\rangle = |\varphi_1\rangle \cdot |\overline{\varphi}_2\rangle = |\varphi_1\rangle^2 = |1\rangle$.

For another pair of qubits $|\varphi_1\rangle = (|0\rangle + 2|1\rangle)/\sqrt{5}$ and $|\varphi_2\rangle = (3|0\rangle + |1\rangle)/\sqrt{10}$, we obtain the following, according to Eq. 9.7:

$$\frac{1}{\sqrt{5}}\begin{bmatrix} 1 & -2 \\ 2 & 1 \end{bmatrix} \times \frac{1}{\sqrt{10}}\begin{bmatrix} 3 \\ -1 \end{bmatrix} = \frac{1}{5\sqrt{2}}\begin{bmatrix} 5 \\ 5 \end{bmatrix} = \frac{1}{\sqrt{2}}\begin{bmatrix} 1 \\ 1 \end{bmatrix}.$$

Therefore, $|\varphi_1\rangle/|\varphi_2\rangle = (|0\rangle + |1\rangle)/\sqrt{2}$. Similarly, we obtain

$$\frac{|\varphi_2\rangle}{|\varphi_1\rangle} = |\varphi_2\rangle \cdot |\overline{\varphi}_1\rangle = \frac{1}{\sqrt{10}}\begin{bmatrix} 3 & -1 \\ 1 & 3 \end{bmatrix} \times \frac{1}{\sqrt{5}}\begin{bmatrix} 1 \\ -2 \end{bmatrix} = \frac{1}{5\sqrt{2}}\begin{bmatrix} 5 \\ -5 \end{bmatrix} = \frac{1}{\sqrt{2}}\begin{bmatrix} 1 \\ -1 \end{bmatrix} = \frac{|0\rangle - |1\rangle}{\sqrt{2}}.$$

Thus, we obtain the Hadamard qubit superpositions by dividing two qubits.

9.1.4 Operations on Qubits with Relative Phases

We consider a way to extend the multiplication, inversion, and division operations to the general case of qubits. The arithmetic of single qubits was considered above in the real space, that is, for qubits $|\varphi\rangle = a_1|0\rangle + b_1|1\rangle$ with real amplitudes. Such qubits are used in many models in signal and image processing. In the general case, we can consider qubits with the relative phase, $|\tilde{\varphi}\rangle = a_1|0\rangle + e^{i\phi}b_1|1\rangle$. Here, a_1 and b_1 are real numbers, $a_1^2 + b_1^2 = 1$, and $\phi \in [0, 2\pi]$. These qubits can be presented with phase shift gate as

$$|\tilde{\varphi}\rangle = P(\phi)|\varphi\rangle = \begin{bmatrix} 1 & 0 \\ 0 & e^{i\phi} \end{bmatrix}|\varphi\rangle = \begin{bmatrix} 1 & 0 \\ 0 & e^{i\phi} \end{bmatrix}\begin{bmatrix} a_1 \\ b_1 \end{bmatrix}.$$

The qubit without phase can be calculated from $|\tilde{\varphi}\rangle$ by the phase shift gate as $|\varphi\rangle = P(-\phi)|\tilde{\varphi}\rangle$.The multiplication of two qubits $|\tilde{\varphi}_1\rangle = a_1|0\rangle + e^{i\phi_1}b_1|1\rangle$ and $|\tilde{\varphi}_2\rangle = a_2|0\rangle + e^{i\phi_2}b_2|1\rangle$ is defined as [1]

$$|\tilde{\varphi}_1\rangle \cdot |\tilde{\varphi}_2\rangle = [P(\phi_1)|\varphi_1\rangle] \cdot [P(\phi_2)|\varphi_2\rangle] \triangleq P(\phi_1 + \phi_2)[|\varphi_1\rangle \cdot |\varphi_2\rangle]. \tag{9.8}$$

This operation is

1) Commutative and associative.
2) Inverse qubit is defined as $|\tilde{\varphi}\rangle^{-1} \triangleq P(-\phi)|\overline{\varphi}\rangle = a_1|0\rangle - e^{-i\phi}b_1|1\rangle$.
3) The division of two qubits is defined as

$$\frac{|\tilde{\varphi}_1\rangle}{|\tilde{\varphi}_2\rangle} \triangleq |\tilde{\varphi}_1\rangle \cdot |\tilde{\varphi}_2\rangle^{-1} = [P(\phi_1)|\varphi_1\rangle] \cdot [P(-\phi_2)|\overline{\varphi}_2\rangle] = P(\phi_1 - \phi_2)[|\varphi_1\rangle \cdot |\overline{\varphi}_2\rangle].$$

A detailed discussion of this general case is beyond the scope of this chapter, but this is how we currently consider a way to extend the multiplication, inversion, and division operations to the general case of qubits.

9.1.5 Quadratic Qubit Equations

In this section, we consider a quadratic qubit equation and its solution with an example. By using the above operations of multiplication and division of qubits, we can work with different types of equations with qubits. For instance, it is not difficult to verify the following equation for qubit superposition:

$$(|\varphi_1\rangle + |\varphi_2\rangle)^2 = |\varphi_1\rangle^2 + 2|\varphi_1\rangle|\varphi_2\rangle + |\varphi_2\rangle^2. \tag{9.9}$$

This equation involves the sum and multiplication of qubits. Similarly, $(|\varphi_1\rangle + |\varphi_2\rangle)^3 = |\varphi_1\rangle^3 + 3|\varphi_1\rangle^2|\varphi_2\rangle + 3|\varphi_1\rangle|\varphi_2\rangle^2 + |\varphi_2\rangle^3$, and so on.

Now, we consider an example of the solution of the quadratic qubit equation with the constant coefficients a and b and a given qubit $|\varphi_1\rangle = a_1|0\rangle + b_1|1\rangle$:

$$|\varphi\rangle^2 + a|\varphi\rangle + b|0\rangle = |\varphi_1\rangle. \tag{9.10}$$

To find the solution $|\varphi\rangle = a_0|0\rangle + b_0|1\rangle$, we have to solve the following system of equations:

$$\left.\begin{aligned} 2a_0^2 + aa_0 + b - 1 - a_1 &= 0 \\ (2a_0 + a)b_0 &= b_1. \end{aligned}\right\} \tag{9.11}$$

Therefore, the solutions of Eq. 9.10 are

$$a_0 = \frac{-a \pm \sqrt{a^2 - 8(b - 1 - a_1)}}{4}, \quad b_0 = \frac{b_1}{2a_0 + a}, \tag{9.12}$$

if only $|a_0|^2 + |b_0|^2 = 1$.

Example 9.3 Solutions of Qubit Equation
We consider the case with the qubit in the superposition $|\varphi_1\rangle = a_1|0\rangle + b_1|1\rangle$, where $a_1 = 1/\sqrt{3} = 0.5774$ and $b_1 = -\sqrt{1 - 1/3} = -0.8165$. The equation we want to solve is

$$|\varphi\rangle^2 + a|\varphi\rangle + b|0\rangle = a_1|0\rangle + b_1|1\rangle. \tag{9.13}$$

Such equation can be solved not for all coefficients a and b. Solutions of this equation are qubits $|\varphi\rangle = a_0|0\rangle + b_0|1\rangle$, that is, the condition $|a_0|^2 + |b_0|^2 = 1$ should be considered. As an example, we consider the case when $a = 0.2$ and $b \in [-0.5, 1.6]$. The solutions $(a_{0,1}, b_{0,1})$ and $(a_{0,2}, b_{0,2})$ of Eq. 9.12 are calculated by

$$a_{0,1} = \frac{-1/5 + \sqrt{1/25 - 8(b - 1.5774)}}{4}, \quad b_{0,1} = \frac{-0.8165}{2a_{0,1} + 1/5}, \tag{9.14}$$

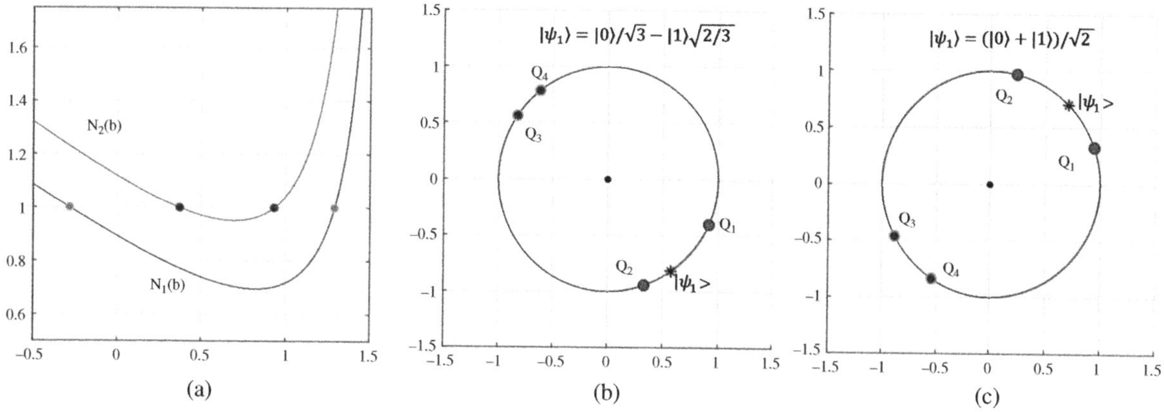

Fig. 9.1 (a) The graphs of functions $N_1(b)$ and $N_2(b)$, for the $a = 1/5$ case and (b) the geometry of the qubit-solutions. (c) The geometry of such qubits for the Hadamard state.

$$a_{0,2} = \frac{-1/5 - \sqrt{1/25 - 8(b - 1.5774)}}{4}, \quad b_{0,2} = \frac{-0.8165}{2a_{0,2} + 1/5}. \tag{9.15}$$

These coefficients are functions of b. We define the functions $N_1(b) = \sqrt{|a_{0,1}|^2 + |b_{0,1}|^2}$ and $N_2(b) = \sqrt{|a_{0,2}|^2 + |b_{0,2}|^2}$. The graphs of these two functions are shown in Fig. 9.1 in part (a). Each graph has two points lying on horizontal 1. Thus, the equation

$$|\varphi\rangle^2 + \frac{1}{5}|\varphi\rangle + b|0\rangle = \frac{1}{\sqrt{3}}|0\rangle - \sqrt{\frac{2}{3}}|1\rangle \tag{9.16}$$

can be solved for coefficients $b = -0.2826$ and 1.2891 wherein $N_1(b) = 1$, and $b = 0.3720$ and 0.9310, wherein $N_2(b) = 1$. These points for $N_1(b)$ and $N_2(b)$ are different; if $N_1(b) = 1$ then $N_2(b) \neq 1$. Thus, for given $a = 0.2$ and the above qubit $|\varphi_1\rangle$, only four quadratic qubit equations $|\varphi\rangle^2 + a|\varphi\rangle + b|0\rangle = |\varphi_1\rangle$ can be solved and each of these equations has only one solution $|\varphi\rangle = |\varphi_b\rangle$. These solutions are

$$Q_1 = |\varphi_{-0.2826}\rangle = 0.9157|0\rangle - 0.4020|1\rangle, \quad Q_2 = |\varphi_{1.2891}\rangle = 0.3329|0\rangle - 0.9429|1\rangle,$$
$$Q_3 = |\varphi_{0.3720}\rangle = -0.8279|0\rangle - 0.5608|1\rangle, \quad Q_4 = |\varphi_{0.9310}\rangle = -0.6207|0\rangle - 0.7841|1\rangle.$$

Here, the amplitudes of the qubits are calculated by Eqs. 9.14 and 9.15. These qubits can be represented by the amplitudes as the points on the unique circle, as shown in Fig. 9.1 in part (b). The qubit $|\varphi_1\rangle = 1/\sqrt{3}|0\rangle - \sqrt{2/3}|1\rangle$ is also shown in this circle. We can say that qubit $|\varphi_1\rangle$ is a state in which four other qubits will enter into the quadratic equation. The quadratic equations with $a = 0.25$ and the Hadamard state $|\varphi_1\rangle = (|0\rangle + |1\rangle)/\sqrt{2}$ are described in detail in [1]. In this case, also only four different equations can be solved. For comparison with the above example, the corresponding four qubits of these solutions are shown together with the Hadamard state in part (c).

9.1.6 Multiplication of *n*-Qubit Superpositions

Consider a right-sided $(r + 1)$-qubit superposition

$$|\varphi_1\rangle = \frac{1}{\sqrt{N}} \sum_{n=0}^{N-1} |\alpha\rangle_n |n\rangle, \qquad |\alpha\rangle_n = a_n|0\rangle + b_n|1\rangle. \tag{9.17}$$

Here, $N = 2^r$, $r > 1$, and $|\alpha\rangle_n$ are single qubits, and $|\alpha\rangle_n|n\rangle$ is the tensor product $|\alpha\rangle_n \otimes |n\rangle$. Such superpositions are considered for instance in such models, as the quantum pixel, flexible representation for quantum image (FRQI), and qubit lattice model described in Section 7.1.3. All qubits $|\alpha\rangle_n$ are considered without relative phases.

The above superposition can also be written as

$$|\varphi_1\rangle = \frac{1}{\sqrt{N}}\left[|0\rangle \sum_{n=0}^{N-1} a_n|n\rangle + |1\rangle \sum_{n=0}^{N-1} b_n|n\rangle\right],$$

or as the following superposition in the computational basis $B_{r+1} = \{|0\rangle, |1\rangle, \ldots, |2N-1\rangle\}$:

$$|\varphi_1\rangle = \frac{1}{\sqrt{N}}\sum_{n=0}^{N-1}(a_n|n\rangle + b_n|N+n\rangle).$$

For instance, when $N = 4$, we have the following 3-qibit superposition:

$$|\varphi_1\rangle = \frac{1}{\sqrt{8}}([a_0|0\rangle + b_0|4\rangle] + [a_1|1\rangle + b_1|5\rangle] + [a_2|2\rangle + b_2|6\rangle] + [a_3|3\rangle + b_3|7\rangle]).$$

After performing the permutation

$$P_1 = (1,2,4)(3,6,5) = \begin{pmatrix} 0\,1\,2\,3\,4\,5\,6\,7 \\ 0\,2\,4\,6\,1\,3\,5\,7 \end{pmatrix},$$

this 3-qubit superposition is written as

$$P|\varphi_1\rangle = \frac{1}{\sqrt{8}}([a_0|0\rangle + b_0|1\rangle] + [a_1|2\rangle + b_1|3\rangle] + [a_2|4\rangle + b_2|5\rangle] + [a_3|6\rangle + b_3|7\rangle]).$$

This quantum superposition represents the 8D vector $(a_0, b_0, a_1, b_1, a_2, b_2, a_3, b_3, a_4, b_4)'$. In fact, it is the left-sided superposition,

$$P|\varphi_1\rangle = \frac{1}{\sqrt{8}}\sum_{n=0}^{7}|n\rangle|\alpha\rangle_n.$$

In the computational basis B_{r+1}, we consider two $(r+1)$-qubit superpositions

$$|\varphi_1\rangle = \frac{1}{\sqrt{N}}\sum_{n=0}^{N-1}|\alpha\rangle_n|n\rangle \quad \text{and} \quad |\varphi_2\rangle = \frac{1}{\sqrt{N}}\sum_{n=0}^{N-1}|\beta\rangle_n|n\rangle. \tag{9.18}$$

Here, $|\alpha\rangle_n$ and $|\beta\rangle_n$ are single qubits for $n = 0 : (N-1)$, and $|\alpha\rangle_n|n\rangle$ is the tensor product $|\alpha\rangle_n \otimes |n\rangle$. The operation of multiplication of these $(r+1)$-qubit superpositions is defined as

$$|\varphi_1\rangle \cdot |\varphi_2\rangle \triangleq \frac{1}{\sqrt{N}}\sum_{n=0}^{N-1}(|\alpha\rangle_n \cdot |\beta\rangle_n)|n\rangle. \tag{9.19}$$

This operation of multiplication of superpositions is commutative and associative. In the case when $|\varphi_2\rangle = |\varphi_1\rangle$, the square of the quantum superposition is

$$|\varphi_1\rangle^2 = |\varphi_1\rangle \cdot |\varphi_1\rangle = \frac{1}{\sqrt{N}}\sum_{n=0}^{N-1}|\alpha\rangle_n^2|n\rangle.$$

Here, the square of the single qubit at basis state $|n\rangle$ is calculated by $|\alpha\rangle_n^2 = (a_n^2 - b_n^2)|0\rangle + 2a_nb_n|1\rangle$, when $|\alpha\rangle_n = a_n|0\rangle + b_n|1\rangle$.

9.1.7 Conjugate Superposition

For $|\varphi_1\rangle$ superposition, the conjugate superposition is defined as

$$|\overline{\varphi}_1\rangle = \frac{1}{\sqrt{N}}\sum_{n=0}^{N-1}|\overline{\alpha}\rangle_n|n\rangle. \tag{9.20}$$

Note that the product of the conjugate superpositions is calculated as follows:

$$|\varphi_1\rangle\cdot|\overline{\varphi}_1\rangle \triangleq \frac{1}{\sqrt{N}}\sum_{n=0}^{N-1}\left(|\alpha\rangle_n\cdot|\overline{\alpha}\rangle_n\right)|n\rangle = \frac{1}{\sqrt{N}}\sum_{n=0}^{N-1}|0\rangle|n\rangle = \frac{1}{\sqrt{N}}\sum_{n=0}^{N-1}|n\rangle. \tag{9.21}$$

In the last sum, $|n\rangle$ are the basis states in the space of $(r+1)$-qubit superpositions. Thus, this product is the same for all superpositions $|\varphi_1\rangle$. One can note that the product contains the Hadamard transform on r qubits initially all instate $|0\rangle$. In other words, $|\varphi_1\rangle\cdot|\overline{\varphi}_1\rangle = |0\rangle\left[H^{\oplus r}|0\rangle^{\oplus r}\right]$.

9.1.8 Division of Multi-qubit Superpositions

The operation of division of $(r+1)$-qubit superpositions

$$|\varphi_1\rangle = \frac{1}{N}\sum_{n=0}^{N-1}|\alpha\rangle_n|n\rangle \quad\text{and}\quad |\varphi_2\rangle = \frac{1}{N}\sum_{n=0}^{N-1}|\beta\rangle_n|n\rangle,$$

is defined as follows [1]:

$$\frac{|\varphi_1\rangle}{|\varphi_2\rangle} \triangleq |\varphi_1\rangle\cdot|\overline{\varphi}_2\rangle = \frac{1}{N}\sum_{n=0}^{N-1}\left(|\alpha\rangle_n\cdot|\overline{\beta}\rangle_n\right)|n\rangle. \tag{9.22}$$

9.1.9 Operations on Left-Sided Superpositions

The above operations were described for the right-sided superposition. The multiplication and division can be also defined for the left-sided superpositions

$$|\varphi_1\rangle = \frac{1}{N}\sum_{n=0}^{N-1}|n\rangle|\alpha\rangle_n \quad\text{and}\quad |\varphi_2\rangle = \frac{1}{N}\sum_{n=0}^{N-1}|n\rangle|\beta\rangle_n,$$

as follows:

$$|\varphi_1\rangle\cdot|\varphi_2\rangle \triangleq \frac{1}{\sqrt{N}}\sum_{n=0}^{N-1}|n\rangle\left(|\alpha\rangle_n\cdot|\beta\rangle_n\right) \quad\text{and}\quad \frac{|\varphi_1\rangle}{|\varphi_2\rangle} \triangleq |\varphi_1\rangle\cdot|\overline{\varphi}_2\rangle = \frac{1}{N}\sum_{n=0}^{N-1}|n\rangle\left(|\alpha\rangle_n\cdot|\overline{\beta}\rangle_n\right). \tag{9.23}$$

This multiplicative arithmetic on superpositions is different from the arithmetic on the right-sided superpositions. The conjugate operation is defined as

$$|\overline{\varphi}_1\rangle = \frac{1}{\sqrt{N}}\sum_{n=0}^{N-1}|n\rangle|\overline{\alpha}\rangle_n \tag{9.24}$$

results in the following product of the conjugate superpositions:

$$|\varphi_1\rangle\cdot|\overline{\varphi}_1\rangle \triangleq \frac{1}{\sqrt{N}}\sum_{n=0}^{N-1}|n\rangle\left(|\alpha\rangle_n\cdot|\overline{\alpha}\rangle_n\right) = \frac{1}{\sqrt{N}}\sum_{n=0}^{N-1}|n\rangle|0\rangle = \frac{1}{\sqrt{N}}\sum_{n=0}^{N-1}|2n\rangle. \tag{9.25}$$

This product is the same for all superpositions but differs from the product in Eq. 9.21.

9.1.10 Quantum Sum of Signals

Consider the FTQR model for grayscale image $\{f_{n,m}\}$ of size $N_1 \times N_2$ pixels and in the range L. This image, written as the 1D signal $f = \{f_n\}$ of length $N = N_1N_2 = 2^{r+s}$, is described by the right-sided $(r+s)$-qubit superposition

$$|\varphi(f)\rangle = \frac{1}{\sqrt{N}}\sum_{n=0}^{N-1} e^{i\alpha f_n}|\,n\rangle, \qquad \alpha = \pi/(2L). \tag{9.26}$$

In this equation, the complex exponential coefficients as 2D vectors present single qubits in the representation

$$e^{i\alpha f_n} = \begin{bmatrix}\cos(\alpha f_n)\\ \sin(\alpha f_n)\end{bmatrix} = \cos(\alpha f_n)|0\rangle + \sin(\alpha f_n)|1\rangle.$$

Therefore, the superposition in Eq. 9.26 can be considered as the $(r+s+1)$-qubit state

$$|\varphi(f)\rangle = \frac{1}{\sqrt{N}}\sum_{n=0}^{N-1}(a_n|0\rangle + b_n|1\rangle)|n\rangle,$$

where $a_n = \cos(\alpha f_n)$ and $b_n = \sin(\alpha f_n)$.

Let us consider another signal $g = \{g_n\}$ with the same range L and the rotations of single qubits in the superposition $|\varphi(f)\rangle$ by the matrix $M = \oplus_{n\in\{0,1,\ldots,N-1\}}R_{\vartheta_n}$, where the 2×2 rotation matrices are $R_{\vartheta_n} = R_{\alpha g_n}$. These rotations are described by the product of $|\varphi(f)\rangle$ and the superposition

$$|\varphi(g)\rangle = \frac{1}{\sqrt{N}}\sum_{n=0}^{N-1}[\cos(\vartheta_n)|0\rangle + \sin(\vartheta_n)|1\rangle]|n\rangle.$$

The result of this multiplication is the $(r+1)$-qubit superposition

$$|\varphi(f)\rangle \cdot |\varphi(g)\rangle = \frac{1}{\sqrt{N}}\sum_{n=0}^{N-1} e^{i\alpha(f_n+g_n)}|n\rangle = |\varphi(f+g)\rangle. \tag{9.27}$$

It is the quantum superposition of the sum of signals, $f+g$. As an example, when $g=f$, we obtain $|\varphi(2f)\rangle = |\varphi(f)\rangle \cdot |\varphi(f)\rangle = |\varphi(f)\rangle^2$ and $|\varphi(3f)\rangle = |\varphi(2f)\rangle \cdot |\varphi(f)\rangle = |\varphi(f)\rangle^3$.

Similarly, the superposition of the difference of signals, $f-g$, is the multiplication of $|\varphi(f)\rangle$ by the superposition which is conjugate to $|\varphi(g)\rangle$,

$$|\varphi(f)\rangle \cdot |\overline{\varphi}(g)\rangle = |\varphi(f-g)\rangle = \frac{1}{\sqrt{N}}\sum_{n=0}^{N-1} e^{i\alpha(f_n-g_n)}|n\rangle. \tag{9.28}$$

Thus, in the FTQR model, the addition and subtraction of images can be described by the operation of multiplication of quantum superpositions. For comparison, we consider the FRQI model of quantum representation for the signal f,

$$\left|\breve{f}\right\rangle = \frac{1}{\sqrt{N}}\sum_{n=0}^{N-1}\left(|a\rangle_n|0\rangle + |b\rangle_n|1\rangle\right)|n\rangle.$$

In this model, it is not possible to express the quantum representation $\left|\left(\widecheck{f+g}\right)\right\rangle$ of the sum of images $(f+g)$ by representations $\left|\breve{f}\right\rangle$ and $|\breve{g}\rangle$.

Example 9.4 Image Processing
There are models of signal and image quantum representation, which use single qubits containing information of the signal or intensities or colors of the image at each point. As an example, we consider the quantum pixel model. Let $f = \{f_{n,m}\}$ be the image of $N \times M$ pixels, where $N = 2^r$, $M = 2^s$, $r, s > 1$, with the range of intensities in the interval

of integers [0,255]. We assume that the basis states $|0\rangle$ and $|1\rangle$ correspond to the black and white colors, respectively. Then, the quantum pixel (n, m) is defined as the single qubit

$$f_{n,m} \to |f_{n,m}\rangle = \sqrt{\frac{f_{n,m}}{255}}|0\rangle + \sqrt{1-\frac{f_{n,m}}{255}}|1\rangle. \tag{9.29}$$

The square roots in this equation can be written as cosine and sine, that is, $|f_{n,m}\rangle = \cos\vartheta_{n,m}|0\rangle + \sin\vartheta_{n,m}|1\rangle$. Then, the image can be presented by the following $(r+s+1)$-qubit superposition:

$$|\check{f}\rangle = \frac{1}{\sqrt{NM}}\sum_{n=0}^{N-1}\sum_{m=0}^{M-1}|f_{n,m}\rangle|n,m\rangle. \tag{9.30}$$

A similar model is used in the flexible representation for quantum images, wherein the information of image colors is written into angles ϑ_n (see Section 7.1.3). The discrete image of size $2^r \times 2^r$ pixel is written into a 4^r-dimensional vector and then presented as the following $(2r+1)$-qubit superposition:

$$|\check{f}\rangle = \frac{1}{2^r}\sum_{n=0}^{4^r-1}(\cos\vartheta_n|0\rangle + \sin\vartheta_n|1\rangle)|n\rangle. \tag{9.31}$$

In these two models, we can define the multiplication and division of quantum image superpositions. A lot of interesting work can be expected here on the application of such operations in signal and image processing. As an example, Section 12 describes such an application in frequency domain image filtering.

9.2 Quantum Fourier Transform Representation

Consider the traditional concept of the QFT in the form of r-qubit superposition

$$|\Psi(F)\rangle = \frac{1}{E[f]}\sum_{p=0}^{N-1}F_p|p\rangle.$$

The normalized coefficient is defined by Parseval's theorem

$$\sum_{p=0}^{N-1}|F_p|^2 = E^2[f] = \sum_{n=0}^{N-1}|f_n|^2,$$

where $E[f]$ denotes the energy of the signal.

Let R_p and I_p be the real and imaginary parts of the DFT, $F_p = R_p + iI_p$. Each coefficient F_p is considered as the 2D vector $(R_p, I_p)'$, which then, after normalization, is presented as a single qubit $|\psi(F_p)\rangle$,

$$F_p = \begin{bmatrix} R_p \\ I_p \end{bmatrix} = R_p\begin{bmatrix} 1 \\ 0 \end{bmatrix} + I_p\begin{bmatrix} 0 \\ 1 \end{bmatrix} = R_p|0\rangle + I_p|1\rangle \approx |\psi(F_p)\rangle = \frac{1}{|F_p|}(R_p|0\rangle + I_p|1\rangle).$$

Therefore, we can consider the following $(r+1)$-qubit right-sided superposition for the QFT:

$$|\Psi(F)\rangle = \frac{1}{\sqrt{N}}\sum_{p=0}^{N-1}|\psi(F_p)\rangle|p\rangle = \frac{1}{\sqrt{N}}\sum_{p=0}^{N-1}\frac{1}{|F_p|}(R_p|0\rangle + I_p|1\rangle)|p\rangle = \frac{1}{\sqrt{N}}\sum_{p=0}^{N-1}\frac{1}{|F_p|}(R_p|p\rangle + I_p|p+N\rangle). \tag{9.32}$$

Another $(r+1)$-qubit superposition for the QFT can be written as

$$|\Psi(F)\rangle = \frac{1}{E[f]}\sum_{p=0}^{N-1}|\psi(F_p)\rangle|p\rangle = \frac{1}{E[f]}\sum_{p=0}^{N-1}(R_p|0\rangle + I_p|1\rangle)|p\rangle. \tag{9.33}$$

For example, when $f_n \equiv 1$, then $F_p = \sqrt{N}\delta_p$, $E[f] = \sqrt{N}$, and

$$|\Psi(F)\rangle = \frac{1}{\sqrt{N}}\sum_{p=0}^{N-1}\sqrt{N}\delta_p|p\rangle = |0\rangle.$$

Here, the delta Kronecker symbol $\delta_p = 1$, if $p = 0$, and $\delta_p = 0$, otherwise.

9.3 Linear Filter (Low-Pass Filtration)

Given frequency characteristic of a linear filter, $Y_p, p = 0, 1, \ldots, N-1$, the filtration is the operation of multiplication $G_p = Y_pF_p$ at each frequency-point p. This is the case of the traditional Fourier transform. In quantum computation, such operation can be described in the following way [1]. The $(r+1)$-qubit right-sided superposition of the N-point DFT of the filtered signal g_n is defined as

$$|\Psi(G)\rangle = \frac{1}{\sqrt{N}}\sum_{p=0}^{N-1}|\psi(G_p)\rangle|p\rangle = \frac{1}{\sqrt{N}}\sum_{p=0}^{N-1}\underbrace{\frac{1}{|G_p|}(\mathrm{Re}[G_p]|0\rangle + \mathrm{Im}[G_p]|1\rangle)}|p\rangle. \tag{9.34}$$

We can obtain this superposition from the superposition $|\Psi(F)\rangle$ in Eq. 9.32, by using the operation of multiplication of single qubits at basis states $|p\rangle$. Indeed, at the frequency-point p, the filter component $Y_p \neq 0$ can be written in the polar form as $Y_p = A_p + jB_p = |Y_p|[\cos\vartheta(p) + j\sin\vartheta(p)]$, where the angle $\vartheta(p) = \mathrm{atan}(B_p/A_p)$. For the complex number Y_p, we can define a single qubit of the filter

$$|Y_p\rangle = \frac{1}{|Y_p|}(A_p|0\rangle + B_p|1\rangle) = \cos\vartheta(p)|0\rangle + \sin\vartheta(p)|1\rangle. \tag{9.35}$$

Then, we need to multiply the qubits $|Y_p\rangle$ and $|\psi(F_p)\rangle$. Such a state-wise multiplication of qubits allows for calculating the states of the filtered signal Y_pF_p in the frequency domain. In other words, we can obtain the qubits $|Y_p\rangle \cdot |\psi(F_p)\rangle$ with the amplitudes calculated by the 2×2 matrix of rotation

$$R_{\vartheta(p)} = \frac{1}{|Y_p|}\begin{bmatrix} A_p & -B_p \\ B_p & A_p \end{bmatrix} = \begin{bmatrix} \cos\vartheta(p) & -\sin\vartheta(p) \\ \sin\vartheta(p) & \cos\vartheta(p) \end{bmatrix}. \tag{9.36}$$

Indeed,

$$\underbrace{\frac{1}{|Y_p|}(A_p|0\rangle + B_p|1\rangle)}\cdot\underbrace{\frac{1}{|F_p|}(R_p|0\rangle + I_p|1\rangle)} = \underbrace{\frac{1}{|Y_p|}\begin{bmatrix} A_p & -B_p \\ B_p & A_p \end{bmatrix}}\times\underbrace{\frac{1}{|F_p|}\begin{bmatrix} R_p \\ I_p \end{bmatrix}} = \frac{1}{|Y_pF_p|}\begin{bmatrix} A_pR_p - B_pI_p \\ B_pR_p + A_pI_p \end{bmatrix}$$

Thus, at the basis state $|p\rangle$,we obtain the qubit in the superposition

$$R_{\vartheta(p)} \times \frac{1}{|F_p|}\begin{bmatrix} R_p \\ I_p \end{bmatrix} = \frac{1}{|Y_pF_p|}[\mathrm{Re}(Y_pF_p)|0\rangle + \mathrm{Im}(Y_pF_p)|1\rangle]. \tag{9.37}$$

This single qubit corresponds to the normalized filtered component $G_p = Y_pF_p$ in the superposition of Eq. 9.34. Normalization of the superposition amplitudes is necessary in quantum computing, and this is the biggest difference from the traditional computing. The normalization of the filter characteristic in Eq. 9.35 shows that we can consider examples with the ideal low-pass, high-pass, and band-pass filters Y_p with phase $\vartheta(p)$ or without it at the frequency-point where $Y_p \neq 0$. In such basis states $|p\rangle$, $|Y_p| = 1$. At frequency-points where $\vartheta(p)=0$, the above matrix of rotation $R_{\vartheta(p)}$ is the identity matrix.

Example 9.5 4-Point Low-Pass Filter
Let us consider an ideal filter with a phase, which is defined as

$$\{Y_p; p = 0, 1, 2, 3\} = \{e^{i\vartheta_0}, e^{i\vartheta_1}, 0, e^{i\vartheta_3}\}, \quad \vartheta_0, \vartheta_1, \vartheta_3 \neq 0.$$

This example is for $N = 4$ and a signal with the energy $E[f] = 2$.If this filter has a real impulse response, then we assume $\vartheta_3 = -\vartheta_1$. The $N = 4$ case is considered only to shorten the formulas here. The 4-point DFT represented by the 3-qubit superposition is

$$|\Psi(F)\rangle = \frac{1}{E[f]}([R_0|0\rangle + I_0|4\rangle] + [R_1|1\rangle + I_1|5\rangle] + [R_2|2\rangle + I_2|6\rangle] + [R_3|3\rangle + I_3|7\rangle]).$$

This superposition represents the 8D vector $\vec{\Psi}(F) = (R_0, R_1, R_2, R_3, I_0, I_1, I_2, I_3)'/2$. After permutation $P_1 = (1,2,4)(3,6,5)$, we obtain the vector $P_1\left[\vec{\Psi}(F)\right] = (R_0, I_0, R_1, I_1, R_2, I_2, R_3, I_3)'/2$. In the matrix form, the linear filtration can be written as follows. First, the Kronecker sum of four matrices is applied

$$K\left[\vec{\Psi}(F)\right] = \left(\begin{bmatrix} A_0 & -B_0 \\ B_0 & A_0 \end{bmatrix} \oplus \begin{bmatrix} A_1 & -B_1 \\ B_1 & A_1 \end{bmatrix} \oplus \begin{bmatrix} A_2 & -B_2 \\ B_2 & A_2 \end{bmatrix} \oplus \begin{bmatrix} A_3 & -B_3 \\ B_3 & A_3 \end{bmatrix}\right) P_1\left[\vec{\Psi}(F)\right]. \tag{9.38}$$

Four matrices of rotation R_{ϑ_0}, R_{ϑ_1}, R_{ϑ_2}, and R_{ϑ_3} are used,

$$R_{\vartheta_k} = \begin{bmatrix} A_k & -B_k \\ B_k & A_k \end{bmatrix} = \begin{bmatrix} \cos\vartheta_k & -\sin\vartheta_k \\ \sin\vartheta_k & \cos\vartheta_k \end{bmatrix}, \qquad k = 0, 1, 2, 3.$$

Here, $\vartheta_2 = 0$ and therefore $A_k = 1$, $B_k = 0$, and $R_{\vartheta_2} = I$ is the identity 2×2 matrix. Eq. 9.38 can be written as

$$K\left[\vec{\Psi}(G)\right] = \begin{bmatrix} \mathrm{Re}(G_0) \\ \mathrm{Im}(G_0) \\ \mathrm{Re}(G_1) \\ \mathrm{Im}(G_1) \\ R_2 \\ I_2 \\ \mathrm{Re}(G_3) \\ \mathrm{Im}(G_3) \end{bmatrix} = \frac{1}{2}\begin{bmatrix} R_{\vartheta_0} & & & \\ & R_{\vartheta_1} & & \\ & & I & \\ & & & R_{\vartheta_3} \end{bmatrix} \begin{bmatrix} R_0 \\ I_0 \\ R_1 \\ I_1 \\ R_2 \\ I_2 \\ R_3 \\ I_3 \end{bmatrix}. \tag{9.39}$$

On the next step, a zero qubit $|0\rangle$ is added, so that the composed 4-qubit will be in superposition that corresponds to the 16D vector

$$(1,0)' \otimes K\left[\vec{\Psi}(F)\right] = \left(K\left[\vec{\Psi}(F)\right]', \underbrace{0,0,0,0,0,0,0,0}\right)'.$$

Then, for example, the following permutation can be used:

$$P_2 = (4,8)(5,9) = \begin{pmatrix} 0\,1\,2\,3\,4\,5\,6\,7\,8\,9\,10\,11\,12\,13\,14\,15 \\ 0\,1\,2\,3\,8\,9\,6\,7\,4\,5\,10\,11\,12\,13\,14\,15 \end{pmatrix}.$$

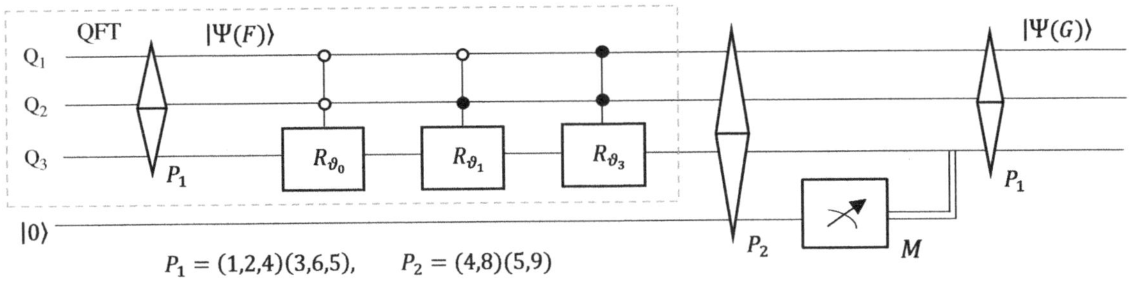

Fig. 9.2 The logic scheme for processing the low-pass filter in Example 9.5.

This permutation moves the component R_2 and I_2 into the second part of the vector. The result is the vector

$$\Phi = P_2\left(K\left[\vec{\Psi}(F)\right]', \underbrace{0,0,0,0,0,0,0,0}\right)' = \left(\tilde{R}_0, \tilde{I}_0, \tilde{R}_1, \tilde{I}_1, 0, 0, \tilde{R}_3, \tilde{I}_3, \underbrace{R_2, I_2, 0, 0, 0, 0, 0, 0}\right)'.$$

Here, we use the notations $\tilde{R}_p = \mathrm{Re}(G_p)$ and $\tilde{I}_p = \mathrm{Im}(G_p)$, for $p = 0,1,3$.
In the computational basis $B_4 = \{|0\rangle, |1\rangle, ..., |15\rangle\}$, this vector describes the 4-qubit superposition

$$|\Phi\rangle = \underbrace{[\tilde{R}_0|0\rangle + \tilde{I}_0|1\rangle] + [\tilde{R}_1|2\rangle + \tilde{I}_1|3\rangle] + [\tilde{R}_3|6\rangle + \tilde{I}_3|7\rangle]} + \underbrace{[R_2|8\rangle + I_2|9\rangle]}.$$

If we separate (measure) this superposition with the last qubit in 0, we obtain the following 3-qubit superposition in the computational basis $B_3 = \{|0\rangle, |1\rangle, ..., |7\rangle\}$:

$$|\Phi_1\rangle = [\tilde{R}_0|0\rangle + \tilde{I}_0|1\rangle] + [\tilde{R}_1|2\rangle + \tilde{I}_1|3\rangle] + [\tilde{R}_3|6\rangle + \tilde{I}_3|7\rangle].$$

Thus, $|\Phi\rangle = |0\rangle|\Phi_1\rangle + |1\rangle(R_2|0\rangle + I_2|1\rangle)$. After permutation P_1, the 3-qubit superposition $|\Phi_1\rangle$ as the 8D vector $\vec{\Phi}_1$ will be the vector with quantum representation $|\Psi(G)\rangle$, that is,

$$\left[\vec{\Psi}(G)\right] = P_1\left[\vec{\Phi}_1\right] = \frac{1}{2}\begin{bmatrix}\tilde{R}_0\\ \tilde{R}_1\\ 0\\ \tilde{R}_3\\ \tilde{I}_0\\ \tilde{I}_1\\ 0\\ \tilde{I}_3\end{bmatrix} = \frac{1}{2}\begin{bmatrix}\mathrm{Re}(G_0)\\ \mathrm{Re}(G_1)\\ 0\\ \mathrm{Re}(G_3)\\ \mathrm{Im}(G_0)\\ \mathrm{Im}(G_1)\\ 0\\ \mathrm{Im}(G_3)\end{bmatrix}. \tag{9.40}$$

The corresponding logic scheme for filtering the 4-point signal by the 3-qubit superposition of the Fourier transform is given in Fig. 9.2. The measurement operation M is placed after the permutation P_2.

Example 9.6 4-Point Ideal Filters
For the general case, the scheme for filtering the 4-point signal by the 3-qubit superposition is shown in Fig. 9.3. For the above example, the third rotation is the identity operation and can be removed from this general circuit together with its logical element. That will result in the scheme in Fig. 9.2. This circuit can be used for other ideal filters, for instance, when $\{Y_p; p = 0, 1, 2, 3\} = \{e^{i\vartheta_0}, 0, 0, e^{i\vartheta_2}\}$ or $\{0, e^{i\vartheta_1}, e^{i\vartheta_2}, 0\}$. In these cases, the permutation P_2 in the scheme should be changed correspondingly. For instance, for the first case, this permutation is $P_2 = (2, 10)(3, 11)(4, 8)(5, 9)$. The quantum circuits for the QFT are not included in these two schemes. A similar circuit can be

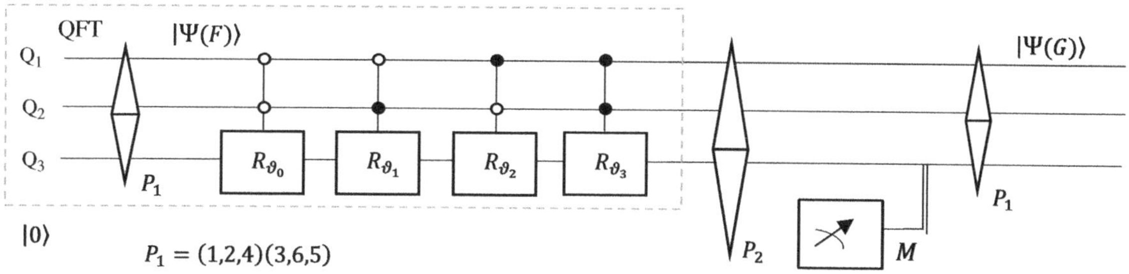

Fig. 9.3 The general logic scheme for processing the 3-qubit superposition of the Fourier transform, $|\Psi(F)\rangle \to |\Psi(G)\rangle$.

used for filtering 2^r-point signals by other ideal filters with amplitudes equal 1, such as low-pass and high-pass filters. For this, the $(r+1)$-qubit superposition of the 2^{r+1}-point DFT, $r \geq 2$, is used.

9.3.1 General Method of Filtration by Ideal Filters

We call a filter ideal if its spectral characteristic Y_p has only magnitudes equal to 1 and 0, when $p = 0{:}\,(N-1)$, where $N = 2^r$, $r > 1$. Namely, there exists such a subset A of the integer interval $X_N = \{0, 1, 2, \ldots, N-1\}$, that

$$|\,Y_p\,| = \begin{cases} 1, & \text{if } \; p \in A; \\ 0, & \text{if } \; p \in X_N \setminus A. \end{cases} \tag{9.41}$$

Here, $X_N\ A$ denotes the set complement of A. In Example 9.5, the case was described when $N = 4$, $X_4 = \{0, 1, 2, 3\}$, and $A = \{0, 1, 3\}$. Note that for a real filter, to preserve the property $Y_{N-p} = \overline{Y_p}$, we consider that if $p \in A$, then $N - p \in A$.

Algorithm to calculate the superposition of the DFT of the filtered signal $G_p = Y_p F_p$.

1) Perform the permutation P_1 of the $(r+1)$-qubit right-sided superposition $|\Psi(F)\rangle$, or the vector

$$\vec{\Psi}(F) = (R_0, R_1, R_2, \ldots, R_{N-1}, I_0, I_1, I_2, \ldots, I_{N-1})'$$

to the vector $P_1\left[\vec{\Psi}(F)\right] = (R_0, I_0, R_1, I_1, R_0, I_2, \ldots, R_{N-1}, I_{N-1})$.
2) Compose the unitary matrix from the 2×2 matrices of Givens rotations $M = \oplus_{p \in \{0,1,\ldots,N-1\}} R_{\vartheta_p}$, where $Y_p = e^{i\vartheta_p}$ and $\vartheta_p = 0$, if $p \notin A$.
3) Apply this matrix to the new permuted superposition $P_1|\Psi(F)\rangle$.
4) Add a zero qubit.
5) Use the permutation P_2 of the amplitudes of the $(r+2)$-qubit superposition, which is defined as $p \leftrightarrow p + 2N$, for all $p \in X_N\ A$.
6) Measure the resulting superposition in a state where the last added qubit is $|0\rangle$.
7) Apply the permutation P_1.The result is the $(r+1)$-qubit superposition $|\Psi(G)\rangle$.

As an example, Fig. 9.4 illustrates the low-pass filtration of 1D speech signal of length $854128 < 2^{20}$ on the classical computer. On a quantum computer, the above algorithm will require $(r+2) = 22$ qubits.

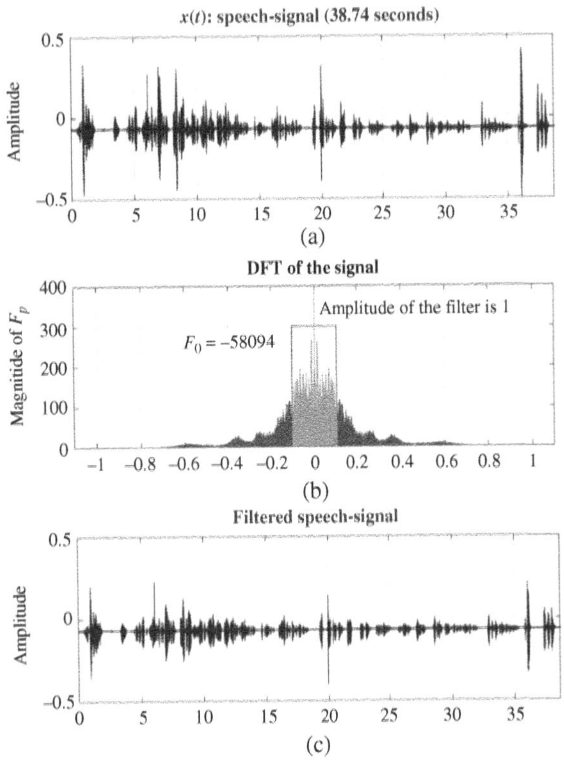

Fig. 9.4 1D speech signal (a) before and (c) after processing by (b) the ideal low-pass filter in the frequency domain.

9.3.2 Application: Linear Convolution of Signals

Let us consider the Fourier transform qubit representation (FTQR) of the image or signal, which is described in Section 8. A signal of length $N = 2^r$, $r > 1$, is presented by the r-qubit equal probability superposition (see Eq. 8.2)

$$|\psi(f)\rangle = \frac{1}{\sqrt{N}} \sum_{n=0}^{N-1} e^{i\alpha f_n} \mid n\rangle. \tag{9.42}$$

At the state $|n\rangle$, the complex amplitude $e^{i\alpha f_n}$ is the 2D vector $(\cos\alpha f_n, \sin\alpha f_n)'$ which in quantum computation represents the qubit $|q_n\rangle = \cos\alpha f_n|0\rangle + \sin\alpha f_n|1\rangle$. It means that this model already includes the model FRQI, wherein the signal f_n is the image written into the 1D signal,

$$|\psi(f)\rangle = \frac{1}{\sqrt{N}} \sum_{n=0}^{N-1} e^{i\alpha f_n} \mid n\rangle = \frac{1}{\sqrt{N}} \sum_{n=0}^{N-1} (\cos\alpha f_n|0\rangle + \sin\alpha f_n|1\rangle) \mid n\rangle. \tag{9.43}$$

The operation of multiplication of superpositions in the FTQR model has the following properties:

1) Let f_n and g_n be two signals of length N each. Them the multiplication of superpositions equals to

$$|\psi(f)\rangle \cdot |\psi(g)\rangle = \frac{1}{\sqrt{N}} \sum_{n=0}^{2^r-1} e^{i\alpha(f_n + g_n)}|n\rangle. \tag{9.44}$$

Indeed,

$$\begin{aligned}|\psi(f)\rangle\cdot|\psi(g)\rangle &= \frac{1}{\sqrt{N}}\sum_{n=0}^{N-1}(\cos\alpha f_n|0\rangle+\sin\alpha f_n|1\rangle)|n\rangle\cdot\frac{1}{\sqrt{N}}\sum_{n=0}^{N-1}(\cos\alpha g_n|0\rangle+\sin\alpha g_n|1\rangle)|n\rangle\\ &= \frac{1}{\sqrt{N}}\sum_{n=0}^{N-1}[(\cos\alpha f_n|0\rangle+\sin\alpha f_n|1\rangle)\cdot(\cos\alpha g_n|0\rangle+\sin\alpha g_n|1\rangle)]|n\rangle\\ &= \frac{1}{\sqrt{N}}\sum_{n=0}^{N-1}(\cos\alpha(f_n+g_n)|0\rangle+\sin\alpha(f_n+g_n)|1\rangle)|n\rangle.\end{aligned}$$

Thus, the sum of signals is described by the multiplication of the superpositions.

2) Consider the linear (aperiodic) convolution of the signal $f=\{f_n\}$ by the filter $h=\{h_{-1}, h_0, h_1\}$,

$$y_n=(f*h)_n=\sum_{k=n-1}^{n+1}f_k h_{n-k}=f_{n-1}h_1+f_n h_0+f_{n+1}h_1,\quad n=0:(N-1). \tag{9.45}$$

For example, we can consider the mean filter, $h=\{1,1,1\}/3$. The signal is assumed to have been padded with zeros, $f\to(0,f,0)$. The FTQR of the convolution can be calculated (defined) as follows:

$$\begin{aligned}|\psi(y)\rangle &= \frac{1}{\sqrt{N}}\sum_{n=0}^{N-1}e^{i\alpha y_n}\,|n\rangle=\frac{1}{\sqrt{N}}\sum_{n=0}^{N-1}\left(\prod_{k=n-1}^{n+1}e^{i\alpha(f_k h_{n-k})}\right)\\ |n\rangle &= \frac{1}{\sqrt{N}}\sum_{n=0}^{N-1}\underbrace{\prod_{k=n-1}^{n+1}(\cos\alpha(f_k h_{n-k})|0\rangle+\sin\alpha(f_k h_{n-k})|1\rangle)}\\ |n\rangle &= \frac{1}{\sqrt{N}}\sum_{n=0}^{N-1}\underbrace{\prod_{k=-1}^{1}(\cos\alpha(f_{n+k}h_{-k})|0\rangle+\sin\alpha(f_{n+k}h_{-k})|1\rangle)\,|n\rangle}.\end{aligned}$$

The large operation, $\prod$, of the product in the last two lines in this equation is the product of qubits. Thus, the above quantum representation of the convolution is calculated by

$$|\psi(f*h)\rangle=\frac{1}{\sqrt{N}}\sum_{n=0}^{N-1}\underbrace{\prod_{k=-1}^{1}(\cos\alpha(f_{n+k}h_{-k})|0\rangle+\sin\alpha(f_{n+k}h_{-k})|1\rangle)}\,|n\rangle. \tag{9.46}$$

In the case when $h=\{1,1,1\}$, this superposition equals to

$$|\psi(f*h)\rangle=\frac{1}{\sqrt{N}}\sum_{n=0}^{N-1}\underbrace{\prod_{k=-1}^{1}(\cos\alpha(f_{n+k})|0\rangle+\sin\alpha(f_{n+k})|1\rangle)}\,|n\rangle$$

This definition of convolution can be also used in the general case. For example, we take a filter of length 5, $h=\{h_{-2},h_{-1},h_0,h_1,h_2\}$. It can be considered normalized, $h=h/||h||$. Then, the quantum convolution is defined as the superposition calculated by

$$|\psi(f*h)\rangle=\frac{1}{\sqrt{N}}\sum_{n=0}^{N-1}\underbrace{\prod_{k=-2}^{2}(\cos\alpha(f_{n+k}h_{-k})|0\rangle+\sin\alpha(f_{n+k}h_{-k})|1\rangle)}\,|n\rangle, \tag{9.46'}$$

or in the above case

$$|\psi(f*h)\rangle = \frac{1}{\sqrt{N}}\sum_{n=0}^{N-1}\underbrace{\prod_{k=-2}^{2}(\cos\alpha_1(f_{n+k})|0\rangle + \sin\alpha_1(f_{n+k})|1\rangle)}\,|\,n\rangle, \quad \alpha_1 = \alpha/3.$$

References

1 Grigoryan, A.M. and Agaian, S.S. (2023) 'Multiplicative group of quantum representations of signals,' *Quantum Information Processing*, 22, 82, p. 23. https://doi.org/10.1007/s11128-022-03765-6

2 Grigoryan, A.M. and Agaian, S.S. (2024) 'Quaternion-based arithmetic in quantum information processing: A Promising approach for efficient color quantum imaging [Hypercomplex Signal and Image Processing],' *IEEE Signal Processing Magazine*, p. 41 (2) 64–74. doi: https://doi.org/10.1109/MSP.2023.3327627

10

Quaternion-Based Arithmetic in Quantum Image Processing

Classical color image processing, image recognition, and machine learning introduce nonlinearity, causing the collapse of the quantum state into classical probability perceptrons after measurements due to the inherent linearity of quantum computing. To address this challenge, quaternion-based arithmetic offers a promising approach [1, 2]. By treating the primary color components (red, green, blue, and gray) as a single unit using quaternion algebra, nonlinear relationships can be implemented, effectively manipulating higher-dimensional color data. This section aims to achieve efficient and accurate color quantum image processing (QIP) by introducing new quaternion-quantum-based color imaging tools based on multiplicative arithmetic on 2-qubits and quantum superpositions [3, 4]. The approach includes the concept of quaternion Fourier transforms in 2-qubit-based color image representation. To end, we discuss possible applications of the proposed methods in color quantum imaging.

QIP is a research branch of quantum information and quantum computing. QIP presents novel perspectives and approaches to tasks such as image compression, denoising, feature extraction, and optimal filtration` [5–10]. Although the field is still in its early stages and faces challenges, researchers are actively working to overcome these obstacles and fully unleash the potential of QIP. Several methods have been developed to represent quantum images in recent years, offering the advantage of simultaneously encoding positions and colors in a normalized quantum state. A few such methods are described in Section 7.2. These methods allow for performing image operations on all pixels to leverage the superposition phenomenon of quantum mechanics.

Hypercomplex algebra, specifically quaternion algebra, has emerged as a powerful tool in color image processing. Quaternions are hypercomplex numbers that extend the concept of complex numbers in four dimensions [11–14]. Quaternion arithmetic allows for processing primary color components as a single unit, enabling the implementation of nonlinear relationships and facilitating more efficient color image operations. Quaternion-based arithmetic's in the (1, 3)- and (2, 2)-models [1, 2, 15] are powerful tools that unlock new possibilities for manipulating and analyzing color or quaternion data, paving the way for advancements in quantum signal and image processing. Manipulating images directly on a quantum computer presents challenges, such as the requirement for reversible operations, the normalization of quantum qubit superpositions after each operation, and the proper representation and definition of classic operations like convolution operators. By incorporating quaternion arithmetic into quantum computing, new operations can be developed for 2-qubits and superpositions and new quantum circuits with applications in color image processing.

This section aims to introduce a multiplicative group of 2-qubits-based model inspired by the widely used (1, 3)-model of quaternion algebra and demonstrate its potential in quantum signal/image processing applications. The proposed arithmetic offers novel capabilities and advantages for manipulating quantum signals and images. Through a series of applications and experimental results, this paper showcases the effectiveness of the new arithmetic model, showcasing its ability to enhance signal processing tasks and improve image analysis, including

- New operations include the polar representation of 2-qubits, quaternion exponentiation, and the computation of powers for 2-qubits and 2-qubit-based superpositions.

Quantum Image Processing in Practice: A Mathematical Toolbox, First Edition. Artyom M. Grigoryan and Sos S. Agaian.

Companion website: www.wiley.com/go/grigoryan/quantumimageprocessing

- New models for 2-qubit image representation based on quaternions and quantum quaternion Fourier transforms (QQFT).
- Comprehensive arithmetic on qubits as a specific case, encompassing multiplication, inverse operations, powering, exponentiation, and division of qubits and quantum qubit superpositions [16]. Additionally, the extension of low-pass and high-pass filtering methods for quantum signals and images in the frequency domain through the QQFT.

In this section, two arithmetic of quaternions are described, namely the (1, 3)- and (2, 2)-models. The multiplicative group of 2-qubits-based model with new concepts is introduced. The quaternion qubit image representation with examples is given and the concept of the quantum quaternion Fourier transform is presented.

10.1 Noncommutative Quaternion Arithmetic

This section presents a quaternion arithmetic mathematical framework that extends the concept of complex numbers to four dimensions. The need for quaternion arithmetic arises from the limitations of traditional complex numbers when dealing with three-dimensional data, such as color.

A quaternion is a two-component number q representing a vector $\boldsymbol{q} = (a, b, c, d)$ in the 4D space. The first component is a real number a and is called the real part of q. The second component, or the imaginary part, of the quaternion is the 3-component number $q' = bi + cj + dk$.This number represents a 3D vector $\boldsymbol{q}' = (b, c, d)$. In this (1,3)-model, the quaternion is the number $q = a + q'$. All coefficients c, b, and d are real. The numbers i, j, and k are three different imaginary units, such that $i^2 = j^2 = k^2 = -1$. For these units, the following multiplication laws hold [12]:

$$ij = -ji = k, \quad jk = -kj = i, \quad ki = -ik = j. \tag{10.1}$$

Table 10.1 shows the multiplications of these imaginary units and real number 1 in part (a).

Because of these laws, the multiplication $q = q_1q_2$ of quaternions $q_n = a_n + q_n' = a_n + (b_ni + c_nj + d_nk)$, $n = 1$, 2, is calculated as follows:

$$q = \underbrace{a_1a_2 - [b_1b_2 + c_1c_2 + d_1d_2]}_{\text{real part}} + \underbrace{[a_1q_2' + a_2q_1'] + \begin{vmatrix} i & j & k \\ b_1 & c_1 & d_1 \\ b_2 & c_2 & d_2 \end{vmatrix}}_{\text{imaginary part}}. \tag{10.2}$$

Here, the formal determinant is used for the cross product $\boldsymbol{q}_1' \times \boldsymbol{q}_2'$ of two vectors $\boldsymbol{q}_n' = (b_n, c_n, d_n)'$, $n = 1, 2$. The quaternion $\overline{q} = a - (bi + cj + dk)$ is called the conjugate of q. As a point in the 4D space, the quaternion lies on the sphere with radius $R = |q| = \sqrt{a^2 + b^2 + c^2 + d^2}$ which is called the modulus of q. It directly follows from Eq. 10.1 that the operation of multiplication is not commutative. In the matrix form, the product $q = q_1q_2$ is the 4D vector calculated by

$$\boldsymbol{q} = \begin{bmatrix} a \\ b \\ c \\ d \end{bmatrix} = \boldsymbol{A}_{q_1} \begin{bmatrix} a_2 \\ b_2 \\ c_2 \\ d_2 \end{bmatrix} = \begin{bmatrix} a_1 & -b_1 & -c_1 & -d_1 \\ b_1 & a_1 & -d_1 & c_1 \\ c_1 & d_1 & a_1 & -b_1 \\ d_1 & -c_1 & b_1 & a_1 \end{bmatrix} \begin{bmatrix} a_2 \\ b_2 \\ c_2 \\ d_2 \end{bmatrix}. \tag{10.3}$$

Table 10.1 (a) Table $T(1, i, j, k)$ of multiplications of the four quaternion unit numbers. (b) Table $T(E, I, J, K)$ of multiplications of the four 4 × 4 basis matrices.

	1	***i***	***j***	***k***
1	1	i	j	k
i	i	−1	k	$-j$
j	j	$-k$	−1	i
k	k	j	$-i$	−1
	E	***I***	***J***	***K***
$\boldsymbol{E}$	$\boldsymbol{E}$	$\boldsymbol{I}$	$\boldsymbol{J}$	$\boldsymbol{K}$
$\boldsymbol{I}$	$\boldsymbol{I}$	$-\boldsymbol{E}$	$\boldsymbol{K}$	$-\boldsymbol{J}$
$\boldsymbol{J}$	$\boldsymbol{J}$	$-\boldsymbol{K}$	$-\boldsymbol{E}$	$\boldsymbol{I}$
$\boldsymbol{K}$	$\boldsymbol{K}$	$\boldsymbol{J}$	$-\boldsymbol{I}$	$-\boldsymbol{E}$

The determinant of this matrix $\det \boldsymbol{A}_{q_1} = |q_1|^4$, the matrix is orthogonal, and its inverse is defined by the transpose matrix as $\boldsymbol{A}_{q_1}^{-1} = \boldsymbol{A}_{q_1}^T/|q_1|^2$ [1]. The matrix has a block structure

$$\boldsymbol{A}_{q_1} = \begin{bmatrix} \boldsymbol{A}_2 & -\boldsymbol{B}_2 \\ \boldsymbol{B}_2 & \boldsymbol{A}_2 \end{bmatrix} = \begin{bmatrix} a_1 & -b_1 & -c_1 & -d_1 \\ b_1 & a_1 & -d_1 & c_1 \\ c_1 & d_1 & a_1 & -b_1 \\ d_1 & -c_1 & b_1 & a_1 \end{bmatrix}, \quad \boldsymbol{A}_2 = \begin{bmatrix} a_1 & -b_1 \\ b_1 & a_1 \end{bmatrix}, \boldsymbol{B}_2 = \begin{bmatrix} c_1 & d_1 \\ d_1 & -c_1 \end{bmatrix}. \tag{10.4}$$

The 4 × 4 matrix above is generated by the first quaternion, the operation is performed on the second quaternion. The same product $q = q_1q_2$ can be calculated by the following operation on the first quaternion:

$$\boldsymbol{q} = \begin{bmatrix} a \\ b \\ c \\ d \end{bmatrix} = \boldsymbol{A}_{q_2} \begin{bmatrix} a_1 \\ b_1 \\ c_1 \\ d_1 \end{bmatrix} = \begin{bmatrix} a_2 & -b_2 & -c_2 & -d_2 \\ b_2 & a_2 & d_2 & -c_2 \\ c_2 & -d_2 & a_2 & b_2 \\ d_2 & c_2 & -b_2 & a_2 \end{bmatrix} \begin{bmatrix} a_1 \\ b_1 \\ c_1 \\ d_1 \end{bmatrix}.$$

Here, the 4 × 4 matrix is generated by the second quaternion and this matrix does not have simple block structure as the matrix $\boldsymbol{A}_{q_1}$. Therefore, we will stand on the case with the matrix $\boldsymbol{A}_{q_1}$.

The arithmetic operations of inverse and division can also be described in the matrix form. The division of quaternions

$$q = q_1/q_2 = q_1\frac{\overline{q}_2}{|q_2|^2}$$

is the quaternion calculated by

$$\boldsymbol{q} = \begin{bmatrix} a \\ b \\ c \\ d \end{bmatrix} = \begin{bmatrix} a_1 & -b_1 & -c_1 & -d_1 \\ b_1 & a_1 & -d_1 & c_1 \\ c_1 & d_1 & a_1 & -b_1 \\ d_1 & -c_1 & b_1 & a_1 \end{bmatrix} \times \frac{1}{|q_2|^2} \begin{bmatrix} a_2 \\ -b_2 \\ -c_2 \\ -d_2 \end{bmatrix} = \frac{1}{|q_2|^2} \begin{bmatrix} a_1 & b_1 & c_1 & d_1 \\ b_1 & -a_1 & d_1 & -c_1 \\ c_1 & -d_1 & -a_1 & b_1 \\ d_1 & c_1 & -b_1 & -a_1 \end{bmatrix} \begin{bmatrix} a_2 \\ b_2 \\ c_2 \\ d_2 \end{bmatrix}.$$

In the case with the unit quaternions q_1 and q_2, the matrix of division is equal to

$$B_{q_1} = \begin{bmatrix} a_1 & b_1 & c_1 & d_1 \\ b_1 & -a_1 & d_1 & -c_1 \\ c_1 & -d_1 & -a_1 & b_1 \\ d_1 & c_1 & -b_1 & -a_1 \end{bmatrix}, \quad \det B_{q_1} = -1.$$

Therefore, in the (1,3)-model, we can work with quaternions like 4D vectors, without calling the units i, j, and k imaginary. They are real vectors, not imaginary ones; the unit vectors $\boldsymbol{i} = (0, 1, 0, 0)'$, $\boldsymbol{j} = (0, 0, 1, 0)'$, and $\boldsymbol{k} = (0, 0, 0, 1)'$. For instance, the multiplication $ij = k$ is described as

$$\boldsymbol{ij} = \boldsymbol{A_i}\begin{bmatrix} 0 \\ 0 \\ 1 \\ 0 \end{bmatrix} = \begin{bmatrix} 0 & -1 & 0 & 0 \\ 1 & 0 & 0 & 0 \\ 0 & 0 & 0 & -1 \\ 0 & 0 & 1 & 0 \end{bmatrix}\begin{bmatrix} 0 \\ 0 \\ 1 \\ 0 \end{bmatrix} = \begin{bmatrix} 0 \\ 0 \\ 0 \\ 1 \end{bmatrix} = \boldsymbol{k}, \quad (\det\boldsymbol{A_i} = 1), \tag{10.5}$$

and the product $ji = -k$ as

$$\boldsymbol{ji} = \boldsymbol{A_j}\begin{bmatrix} 0 \\ 1 \\ 0 \\ 0 \end{bmatrix} = \begin{bmatrix} 0 & 0 & -1 & 0 \\ 0 & 0 & 0 & 1 \\ 1 & 0 & 0 & 0 \\ 0 & -1 & 0 & 0 \end{bmatrix}\begin{bmatrix} 0 \\ 1 \\ 0 \\ 0 \end{bmatrix} = \begin{bmatrix} 0 \\ 0 \\ 0 \\ -1 \end{bmatrix} = -\begin{bmatrix} 0 \\ 0 \\ 0 \\ 1 \end{bmatrix} = -\boldsymbol{k}, \quad (\det\boldsymbol{A_j} = 1). \tag{10.6}$$

We can also write the divisions of these vectors. For example, $\dfrac{\boldsymbol{i}}{\boldsymbol{j}} = \boldsymbol{i}\bar{\boldsymbol{j}} = -\boldsymbol{ij} = -\boldsymbol{k}$, and $\dfrac{\boldsymbol{k}}{\boldsymbol{j}} = \boldsymbol{k}\bar{\boldsymbol{j}} = -\boldsymbol{kj} = \boldsymbol{i}$.

10.2 Commutative Quaternion Arithmetic

In this section, we consider another, namely the (2, 2)-model of representation of quaternions [2, 17]. In this model, the quaternion is a pair of two complex numbers a_1 and a_2,

$$q = [a_1, a_2] = a_1 + ja_2 = (a + ib) + j(c + id) = a + (bi + cj + dk), \tag{10.7}$$

where $a_1 = a + ib$ and $a_2 = c + id$. Here, we add only one imaginary unit $j \neq i$, and new unit k is the product of the primary units, that is, $k = ji$. The conjugate operation over the quaternion is defined by complex conjugates as $\bar{q} = [\bar{a}_1, \bar{a}_2] = [(a, -b), (c, -d)]$. The sum of quaternions is accomplished component wise, as in the (1,3)-model. However, the multiplication of quaternions $q_1 = [a_1, a_2]$ and $q_2 = [b_1, b_2]$ is defined as

$$q = q_1q_2 = [a_1, a_2][b_1, b_2] \triangleq [a_1b_1 - a_2b_2, a_1b_2 + a_2b_1]. \tag{10.8}$$

The operations in square brackets are calculated by the traditional multiplication of the complex numbers a_1, a_2, b_1, and b_2. This model uses four basic unit vectors,

1) The real quaternion $\boldsymbol{e}_0 = [(1, 0), (0, 0)]$, or shortly $\boldsymbol{e}_0 = (1, 0) = 1$.
2) Three imaginary units, $\boldsymbol{e}_1 = [(0, 1), (0, 0)]$, $\boldsymbol{e}_2 = [(0, 0), (1, 0)]$, and $\boldsymbol{e}_3 = [(0, 0), (0, 1)]$.

Table 10.2 Table $T(1, \boldsymbol{e}_1, \boldsymbol{e}_2, \boldsymbol{e}_3)$ of multiplications of the four basic unit numbers.

	$\boldsymbol{e}_0$	$\boldsymbol{e}_1$	$\boldsymbol{e}_2$	$\boldsymbol{e}_3$
$\boldsymbol{e}_0$	1	$\boldsymbol{e}_1$	$\boldsymbol{e}_2$	$\boldsymbol{e}_3$
$\boldsymbol{e}_1$	$\boldsymbol{e}_1$	-1	$\boldsymbol{e}_3$	$-\boldsymbol{e}_2$
$\boldsymbol{e}_2$	$\boldsymbol{e}_2$	$\boldsymbol{e}_3$	-1	$\boldsymbol{e}_1$
$\boldsymbol{e}_3$	$\boldsymbol{e}_3$	$-\boldsymbol{e}_2$	$-\boldsymbol{e}_1$	1

The multiplication laws of these units are given in Table 10.2. There is one imaginary unit q such that $q^2 = 1$, namely, $\boldsymbol{e}_3^2 = 1$.

The multiplication $q = q_1q_2$ of two quaternions in Eq. 10.8 can be written in the matrix form as

$$q = \begin{bmatrix} a \\ b \\ c \\ d \end{bmatrix} = \boldsymbol{M}_{q_1} \begin{bmatrix} a_2 \\ b_2 \\ c_2 \\ d_2 \end{bmatrix} = \begin{bmatrix} a_1 & -b_1 & -c_1 & d_1 \\ b_1 & a_1 & -d_1 & -c_1 \\ c_1 & -d_1 & a_1 & -b_1 \\ d_1 & c_1 & b_1 & a_1 \end{bmatrix} \begin{bmatrix} a_2 \\ b_2 \\ c_2 \\ d_2 \end{bmatrix}. \tag{10.9}$$

This matrix has a block structure, as the matrix $\boldsymbol{A}_{q_1}$ in Eq. 10.4,

$$\boldsymbol{M}_{q_1} = \begin{bmatrix} \boldsymbol{A}_2 & -\boldsymbol{C}_2 \\ \boldsymbol{C}_2 & \boldsymbol{A}_2 \end{bmatrix} = \begin{bmatrix} a_1 & -b_1 & -c_1 & d_1 \\ b_1 & a_1 & -d_1 & -c_1 \\ c_1 & -d_1 & a_1 & -b_1 \\ d_1 & c_1 & b_1 & a_1 \end{bmatrix}, \ \boldsymbol{A}_2 = \begin{bmatrix} a_1 & -b_1 \\ b_1 & a_1 \end{bmatrix}, \boldsymbol{C}_2 = \begin{bmatrix} c_1 & -d_1 \\ d_1 & c_1 \end{bmatrix}. \tag{10.10}$$

The difference only is the submatrix $\boldsymbol{C}_2 \neq \boldsymbol{B}_2$, wherein

$$\boldsymbol{C}_2 = \begin{bmatrix} c_1 & -d_1 \\ d_1 & c_1 \end{bmatrix} = \begin{bmatrix} c_1 & d_1 \\ d_1 & -c_1 \end{bmatrix} \begin{bmatrix} 1 & 0 \\ 0 & -1 \end{bmatrix} = \boldsymbol{B}_2 Z.$$

Example 10.1 Multiplication in Two Models

Consider two unit 4D vectors $\boldsymbol{q}_1 = (1, -2, 1, -1)/\sqrt{7}$ and $\boldsymbol{q}_2 = (1, -1, 3, 2)/\sqrt{15}$. In the (1, 3)-model, these vectors describe the quaternions $q_1 = (1 + (-i + j - k))/\sqrt{7}$ and $q_2 = (1 + (-i + 3j + 2k))/\sqrt{15}$, respectively. The product of these quaternions is calculated by

$$q = \boldsymbol{A}_{q_1} \frac{1}{\sqrt{15}} \begin{bmatrix} 1 \\ -1 \\ 3 \\ 2 \end{bmatrix} = \frac{1}{\sqrt{7}} \begin{bmatrix} 1 & 2 & -1 & 1 \\ -2 & 1 & 1 & 1 \\ 1 & -1 & 1 & 2 \\ -1 & -1 & -2 & 1 \end{bmatrix} \times \frac{1}{\sqrt{15}} \begin{bmatrix} 1 \\ -1 \\ 3 \\ 2 \end{bmatrix} = \frac{1}{\sqrt{7}\sqrt{15}} \begin{bmatrix} -2 \\ 2 \\ 9 \\ -4 \end{bmatrix}, \tag{10.11}$$

and $\det \boldsymbol{A}_{q_1} = 1$; therefore, $|q| = 1$.

In the (2, 2)-model, the vectors $\mathbf{q}_1$ and $\mathbf{q}_2$ correspond to the quaternions $q_1 = [a_1, a_2] = [(1, -2), (1, -1)]/\sqrt{7}$ and $q_2 = [b_1, b_2] = [(1, -1), (3, 2)]/\sqrt{15}$, respectively. The product q_1q_2 is calculated by

$$q = \boldsymbol{M}_{q_1}\frac{1}{\sqrt{15}}\begin{bmatrix} 1 \\ -1 \\ 3 \\ 2 \end{bmatrix} = \frac{1}{\sqrt{7}}\begin{bmatrix} 1 & 2 & -1 & -1 \\ -2 & 1 & 1 & -1 \\ 1 & 1 & 1 & 2 \\ -1 & 1 & -2 & 1 \end{bmatrix} \times \frac{1}{\sqrt{15}}\begin{bmatrix} 1 \\ -1 \\ 3 \\ 2 \end{bmatrix} = \frac{1}{\sqrt{7}\sqrt{15}}\begin{bmatrix} -6 \\ -2 \\ 7 \\ -6 \end{bmatrix}. \tag{10.12}$$

For the matrix $\boldsymbol{M}_{q_1}$, the determinant $\det\boldsymbol{M}_{q_1} = 0.9184$ and $|q| = 1{,}0911$. The inverse matrix is

$$\boldsymbol{M}_{q_1}^{-1} = k\frac{1}{\sqrt{35}}\begin{bmatrix} 3 & -4 & 1 & -3 \\ 4 & 3 & 3 & 1 \\ -1 & 3 & 3 & -4 \\ -3 & -1 & 4 & 3 \end{bmatrix}, \quad k = 1.0435.$$

The main difference between the (1, 3)- and (2, 2)-models is that the multiplication in the (2, 2)-model is commutative. However, in the general case, the matrix $\boldsymbol{M}_{q_1}$ is not orthogonal, while the matrix of multiplication $\boldsymbol{A}_{q_1}$ is orthogonal. For a unit quaternion q_1, $\det\boldsymbol{A}_{q_1} = |q_1|^4 = 1$, the matrix is unitary and its inverse is the transpose matrix, $\boldsymbol{A}_{q_1}^{-1} = \boldsymbol{A}_{q_1}^T$. This is the main argument why we prefer to use the noncommutative quaternion (1, 3) model in quantum computation, rather than the (2,2)-model.

10.3 Geometry of the Quaternions

Since we cannot plot 4D vectors, a unit quaternion $q = (a, q')$ can be illustrated in the following way. Imagine the real line R with the point a on it and the sphere S_r of radius $r = \sqrt{1-a^2}$ with the center at point a, as shown in Fig. 10.1. The 3D point $q' = (b, c, d)$ is a point on this sphere. All unit quaternions with real part equals a are represented by points on this sphere

$$S_r = \{q' = (b, c, d) \in R^3;\ |q'|^2 = b^2 + c^2 + d^2 = 1\}.$$

All numbers a, b, c, and d are real, the point q' on the sphere is defined by the position vector $\boldsymbol{q}'$ from the original point a.

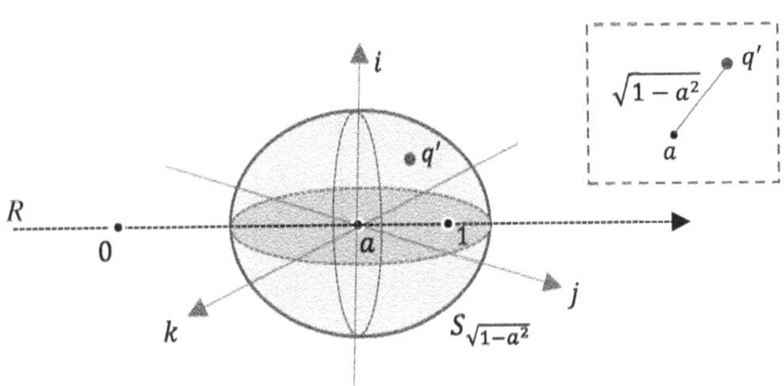

Fig. 10.1 The sphere with a point presents a quaternion $q = (a, q')$.

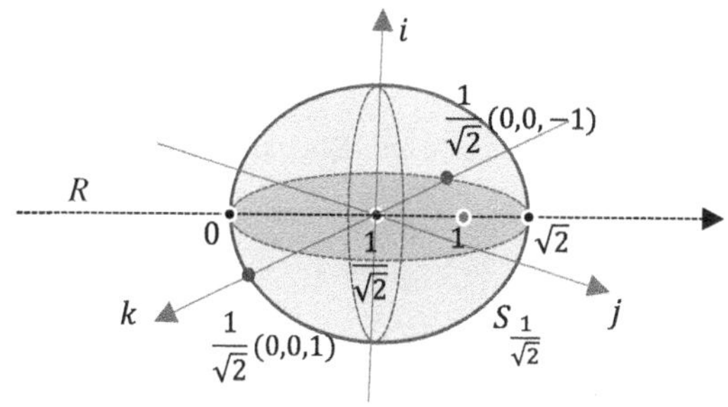

Fig. 10.2 The sphere with the points presenting the Bell states $|\beta_1\rangle$ and $|\beta_2\rangle$.

Such spheres can be considered as locus for 2-qubits. For instance, the Bell states $|\beta_{1,2}\rangle = \frac{1}{\sqrt{2}}(|00\rangle \pm |11\rangle)$ are presented by two points $\frac{1}{\sqrt{2}}(0,0,\pm 1)$ on the same sphere $S_{1/\sqrt{2}}$ with the center at point $1/\sqrt{2}$, as shown in Fig. 10.2.

Example 10.2 A Point on the Sphere
Consider the quaternion

$$q = \frac{4 + (5i + 2j + 6k)}{9} = \frac{4}{9} + \frac{(5i + 2j + 6k)}{9}.$$

As illustrated in Fig. 10.3, this quaternion is represented by the point q'=(5/9,2/9,6/9)≈(0.55,0.22,0.66) on the sphere $S_{4/9}$ with radius $r = \sqrt{1-(4/9)^2} \approx 0.8958$.

Figure 10.4 shows the sphere $S_{0.7}$ in part (a) and two spheres $S_{0.9}$ and $S_{-0.9}$ in part (b).

The spheres $S_{-0.95}$, $S_{0.65}$, and S_0 are shown together in Fig. 10.5 in part (a). The sphere S_0 has the largest radius $r = 1$. There is no intersection of these spheres here, as it seems, since the spheres are given in different coordinate systems. This is the moving coordinate system; the motion parameter is a, the real part of the quaternions. If all spheres are drawn side by side, we see an ellipsoidal body. The plane sections of 21 such spheres S_a are shown in part (b). Here, the numbers a are taken from the interval [−1, 1] with the step 0.1.

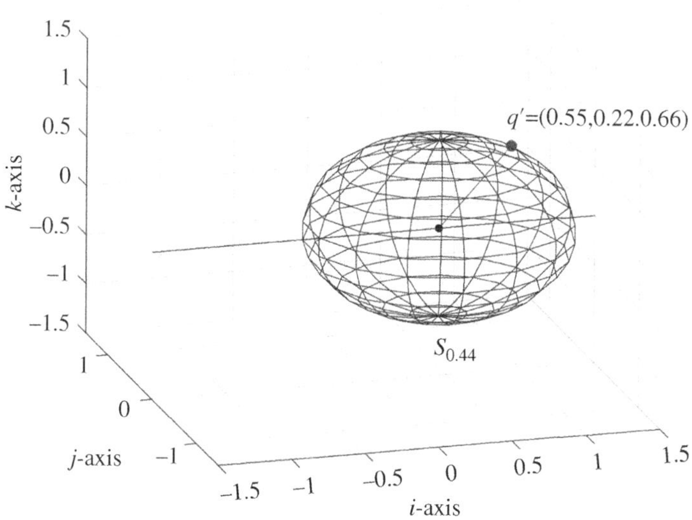

Fig. 10.3 The sphere $S_{4/9}$ with the point (5/9,2/9,6/9).

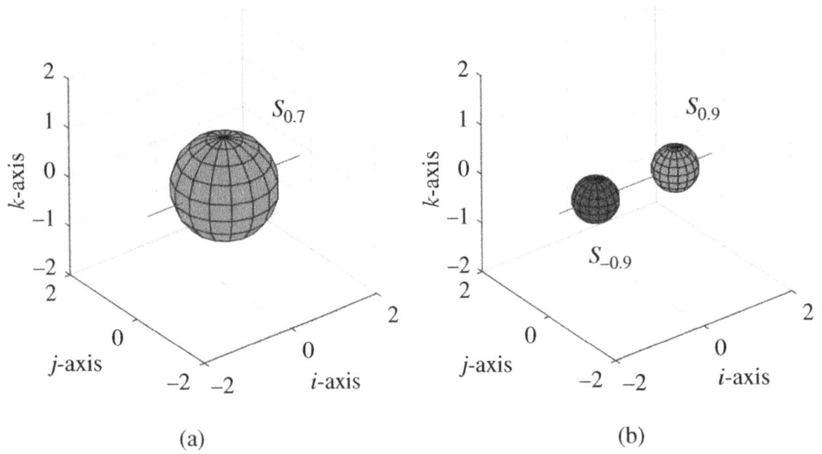

Fig. 10.4 The spheres (a) $S_{0.7}$ and (b) $S_{-0.9}$ and $S_{0.9}$.

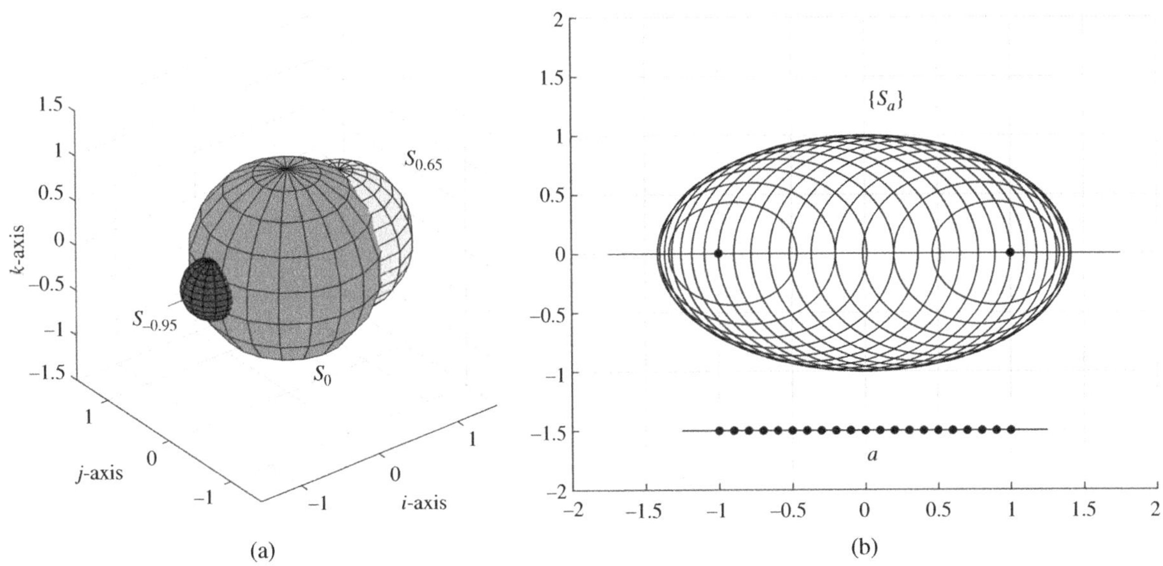

Fig. 10.5 (a) Three spheres $S_{-0.95}$, S_0, and $S_{0.65}$ and (b) the plane sections of 21 spheres.

10.4 Multiplicative Group on 2-Qubits

In this section, we describe how the concept of quaternion multiplication in the (1,3)-model can be used on 2-qubits [3]. The main properties of the multiplication are presented. Consider two 2-qubits $|\varphi_1\rangle = a_1|0\rangle + b_1|1\rangle + c_1|2\rangle + d_1|3\rangle$ and $|\varphi_2\rangle = a_2|0\rangle + b_2|1\rangle + c_2|2\rangle + d_2|3\rangle$.

Definition 10.1
The multiplication of 2-qubits $|\varphi_1\rangle$ and $|\varphi_2\rangle$ is calculated by

$$|\varphi\rangle \triangleq |\varphi_1\rangle \cdot |\varphi_2\rangle = a|0\rangle + b|1\rangle + c|2\rangle + d|3\rangle \tag{10.13}$$

which is the 2-qubit in the superposition

Table 10.3 Rules for multiplying the four basis states.

	$\|0\rangle$	$\|1\rangle$	$\|2\rangle$	$\|3\rangle$
$\|0\rangle$	$\|0\rangle$	$\|1\rangle$	$\|2\rangle$	$\|3\rangle$
$\|1\rangle$	$\|1\rangle$	$-\|0\rangle$	$\|3\rangle$	$-\|2\rangle$
$\|2\rangle$	$\|2\rangle$	$-\|3\rangle$	$-\|0\rangle$	$\|1\rangle$
$\|3\rangle$	$\|3\rangle$	$\|2\rangle$	$-\|1\rangle$	$-\|0\rangle$
	$\|00\rangle$	$\|01\rangle$	$\|10\rangle$	$\|11\rangle$
$\|00\rangle$	$\|00\rangle$	$\|01\rangle$	$\|10\rangle$	$\|11\rangle$
$\|01\rangle$	$\|01\rangle$	$-\|00\rangle$	$\|11\rangle$	$-\|10\rangle$
$\|10\rangle$	$\|10\rangle$	$-\|11\rangle$	$-\|00\rangle$	$\|01\rangle$
$\|11\rangle$	$\|11\rangle$	$\|10\rangle$	$-\|01\rangle$	$-\|00\rangle$

$$|\varphi\rangle = (a_1a_2 - b_1b_2 - c_1c_2 - d_1d_2)|0\rangle + (b_1a_2 + a_1b_2 - d_1c_2 + c_1d_2)|1\rangle + (c_1a_2 + d_1b_2 + a_1c_2 - b_1d_2)|2\rangle + (d_1a_2 - c_1b_2 + b_1c_2 + a_1d_2)|3\rangle.$$

Thus, the amplitudes of the states in the 2-qubit $|\varphi\rangle$ are calculated by Eq. 10.13,

$$a = a_1a_2 - b_1b_2 - c_1c_2 - d_1d_2, \quad b = b_1a_2 + a_1b_2 - d_1c_2 + c_1d_2, \tag{10.14}$$

$$c = c_1a_2 + d_1b_2 + a_1c_2 - b_1d_2, \quad d = d_1a_2 - c_1b_2 + b_1c_2 + a_1d_2. \tag{10.15}$$

The multiplication rules of basis states are given in the Table 10.3.

Example 10.3 Product of 2-Qubit Superpositions
Consider the following two 2-qubits:

$$|\varphi_1\rangle = \frac{1}{\sqrt{15}}(3|0\rangle + 2|1\rangle + |2\rangle - |3\rangle) \text{ and } |\varphi_2\rangle = \frac{1}{5}(2|0\rangle - 4|1\rangle + 2|2\rangle + |3\rangle).$$

Then, the products of these 2-qubits are

$$|\varphi_1\rangle \cdot |\varphi_2\rangle = \frac{1}{5\sqrt{15}}(13|0\rangle - 5|1\rangle + 10|2\rangle + 9|3\rangle),$$

$$|\varphi_2\rangle \cdot |\varphi_1\rangle = \frac{1}{5\sqrt{15}}(13|0\rangle - 11|1\rangle + 6|2\rangle - 7|3\rangle).$$

Indeed, the non-normalized amplitudes of these 2-qubits can be calculated in the matrix form as

$$\begin{bmatrix} 3 & -2 & -1 & 1 \\ 2 & 3 & 1 & 1 \\ 1 & -1 & 3 & -2 \\ -1 & -1 & 2 & 3 \end{bmatrix} \begin{bmatrix} 2 \\ -4 \\ 2 \\ 1 \end{bmatrix} = \begin{bmatrix} 13 \\ -5 \\ 10 \\ 9 \end{bmatrix} \text{ and } \begin{bmatrix} 2 & 4 & -2 & -1 \\ -4 & 2 & -1 & 2 \\ 2 & 1 & 2 & 4 \\ 1 & -2 & -4 & 2 \end{bmatrix} \begin{bmatrix} 3 \\ 2 \\ 1 \\ -1 \end{bmatrix} = \begin{bmatrix} 13 \\ -11 \\ 6 \\ -7 \end{bmatrix}.$$

Also, we can calculate the squares of these qubits,

$$|\varphi_1\rangle^2 = \frac{1}{15}(3|0\rangle + 12|1\rangle + 6|2\rangle - 6|3\rangle), \quad |\varphi_2\rangle^2 = \frac{1}{25}(-17|0\rangle - 16|1\rangle + 8|2\rangle + 4|3\rangle).$$

Multiplication Properties:

1) $|0\rangle$ is a unit, that is, $|0\rangle\cdot|\varphi\rangle = |\varphi\rangle$.
2) (Noncommutativity) $|\varphi_1\rangle \cdot |\varphi_2\rangle \neq |\varphi_2\rangle \cdot |\varphi_1\rangle$, if $|\varphi_2\rangle \neq |\varphi_1\rangle$.
3) (Distributivity) $(|\varphi_1\rangle \cdot |\varphi_2\rangle) \cdot |\varphi_3\rangle = |\varphi_1\rangle \cdot (|\varphi_2\rangle \cdot |\varphi_3\rangle)$.
4) (Associativity)
   a) $(|\varphi_1\rangle + |\varphi_2\rangle) \cdot |\varphi_3\rangle = |\varphi_1\rangle \cdot |\varphi_3\rangle + |\varphi_2\rangle \cdot |\varphi_3\rangle$,
   b) $|\varphi_3\rangle \cdot (|\varphi_1\rangle + |\varphi_2\rangle) = |\varphi_3\rangle \cdot |\varphi_1\rangle + |\varphi_3\rangle \cdot |\varphi_2\rangle$.
5) $|0\rangle\cdot|\varphi\rangle = |\varphi\rangle \cdot |0\rangle = |\varphi\rangle$.
6) $|1\rangle\cdot|\varphi\rangle = -b|0\rangle + a|1\rangle - d|2\rangle + c|3\rangle$; $|\varphi\rangle\cdot|1\rangle = -b|0\rangle + a|1\rangle + d|2\rangle - c|3\rangle$.
7) $|2\rangle\cdot|\varphi\rangle = -c|0\rangle + d|1\rangle + a|2\rangle - b|3\rangle$; $|\varphi\rangle\cdot|2\rangle = -c|0\rangle - d|1\rangle + a|2\rangle + b|3\rangle$.
8) $|3\rangle\cdot|\varphi\rangle = -d|0\rangle - c|1\rangle + b|2\rangle + a|3\rangle$; $|\varphi\rangle\cdot|3\rangle = -d|0\rangle + c|1\rangle - b|2\rangle + a|3\rangle$.

The properties in (6)–(8) describe the permutation of the amplitudes of the 2-qubit with the sign change of two of them. (For more properties, see [3])

A few operations on the 2-qubit $|\varphi\rangle = a|0\rangle + b|1\rangle + c|2\rangle + d|3\rangle$:

1) The quaternion square is calculated by

$$q^2 = \left(a^2 - |q'|^2, 2aq'\right) = (2a^2 - 1) + 2a(ib + jc + kd).$$

The square of the 2-qubit is defined as

$$|\varphi\rangle^2 = |\varphi\rangle \cdot |\varphi\rangle = (2a^2 - 1)|0\rangle + 2a(b|1\rangle + c|2\rangle + d|3\rangle). \tag{10.16}$$

In particular, if $a = 0$, then $|\varphi\rangle^2 = -|0\rangle$.

2) Using the square of the quaternion $q_1^2 = (2a_1^2 - 1) + 2a_1(ib_1 + jc_1 + kd_1)$, one can obtain the square root of the quaternion q, by solving the equation $q_1^2 = q = a + (ib + jc + dk)$. The solution is described by the system

$$2a_1^2 - 1 = a, \quad 2a_1b_1 = b, \quad 2a_1c_1 = c, \quad 2a_1d_1 = d.$$

Thus,

$$a_1 = \pm\sqrt{\frac{1+a}{2}}, \quad b_1 = \frac{b}{2a_1}, \quad c_1 = \frac{c}{2a_1}, \quad d_1 = \frac{d}{2a_1}.$$

The square root of the 2-qubit $|\varphi\rangle$ is defined as

$$\sqrt{|\varphi\rangle} = \sqrt{\frac{1+a}{2}}|0\rangle + \frac{1}{\sqrt{2(1+a)}}(b|1\rangle + c|2\rangle + d|3\rangle), \quad a \neq -1. \tag{10.17}$$

In the $a = -1$ case, that is, when $|\varphi\rangle = -|0\rangle$, this state is equivalent to $|0\rangle$ with the square root $|0\rangle$.

Example 10.4 Powers of Several 2-Qubits
The following multiplications of 2-qubits hold when $n = 1$, 2, and 3:

$$\left(\frac{|0\rangle + |n\rangle}{\sqrt{2}}\right)^2 = \frac{|0\rangle + |n\rangle + |n\rangle + |n\rangle^2}{2} = \frac{|0\rangle + |n\rangle + |n\rangle - |0\rangle}{2} = |n\rangle,$$

$$\left(\frac{|0\rangle - |n\rangle}{\sqrt{2}}\right)^2 = \frac{|0\rangle - |n\rangle - |n\rangle + |n\rangle^2}{2} = \frac{|0\rangle - |n\rangle - |n\rangle - |0\rangle}{2} = -|n\rangle,$$

$$\left(\frac{|0\rangle + |n\rangle}{\sqrt{2}}\right) \cdot \left(\frac{|0\rangle - |n\rangle}{\sqrt{2}}\right) = \frac{|0\rangle - |n\rangle^2}{2} = \frac{|0\rangle + |0\rangle}{2} = |0\rangle.$$

The Hadamard states are the square roots of $\pm|1\rangle$,

$$\left(|\varphi_{H,1}\rangle\right)^2 = \left(\frac{|0\rangle + |1\rangle}{\sqrt{2}}\right)^2 = |1\rangle, \qquad \left(|\varphi_{H,2}\rangle\right)^2 = \left(\frac{|0\rangle - |1\rangle}{\sqrt{2}}\right)^2 = -|1\rangle.$$

The Bell states are roots of $\pm|3\rangle$,

$$\left(|\varphi_{B,1}\rangle\right)^2 = \left(\frac{|00\rangle + |11\rangle}{\sqrt{2}}\right)^2 = |11\rangle, \quad \left(|\varphi_{B,2}\rangle\right)^2 = \left(\frac{|00\rangle - |11\rangle}{\sqrt{2}}\right)^2 = -|11\rangle.$$

The multiplication of two Bell states

$$|\varphi_{B,1}\rangle \cdot |\varphi_{B,3}\rangle = \left(\frac{|00\rangle + |11\rangle}{\sqrt{2}}\right) \cdot \left(\frac{|01\rangle + |10\rangle}{\sqrt{2}}\right) = |10\rangle \quad \text{and} \quad |\varphi_{B,3}\rangle \cdot |\varphi_{B,1}\rangle = |01\rangle.$$

The Hadamard states and Bell states are the divisors of the zero state, that is,

$$|\varphi_{H,1}\rangle \cdot |\varphi_{H,2}\rangle = |0\rangle, \qquad |\varphi_{B,1}\rangle \cdot |\varphi_{B,2}\rangle = |0\rangle. \tag{10.18}$$

The multiplication of the Hadamard and Bell states is calculated by

$$|\varphi_{H,1}\rangle \cdot |\varphi_{B,1}\rangle = \frac{|0\rangle + |1\rangle - |2\rangle + |3\rangle}{2}, \quad |\varphi_{H,1}\rangle \cdot |\varphi_{B,2}\rangle = \frac{|0\rangle + |1\rangle + |2\rangle - |3\rangle}{2},$$

$$|\varphi_{H,2}\rangle \cdot |\varphi_{B,1}\rangle = \frac{|0\rangle - |1\rangle + |2\rangle + |3\rangle}{2}, \quad |\varphi_{H,1}\rangle \cdot |\varphi_{B,2}\rangle = \frac{|0\rangle - |1\rangle - |2\rangle - |3\rangle}{2}.$$

The square roots of the Hadamard states are equal to

$$\sqrt{|\varphi_{H,1}\rangle} = \frac{1}{2}\sqrt{\sqrt{2}\left(\sqrt{2}+1\right)}|00\rangle + \frac{1}{\sqrt{2\sqrt{2}(\sqrt{2}+1)}}|01\rangle$$

$$\sqrt{|\varphi_{H,2}\rangle} = \frac{1}{2}\sqrt{\sqrt{2}\left(\sqrt{2}+1\right)}|00\rangle - \frac{1}{\sqrt{2\sqrt{2}(\sqrt{2}+1)}}|01\rangle.$$

The square roots of the Bell states are equal to

$$\sqrt{|\varphi_{B,1}\rangle} = \frac{1}{2}\sqrt{\sqrt{2}\left(\sqrt{2}+1\right)}|00\rangle + \frac{1}{\sqrt{2\sqrt{2}(\sqrt{2}+1)}}|11\rangle, \sqrt{|\varphi_{B,2}\rangle} = \frac{1}{2}\sqrt{\sqrt{2}\left(\sqrt{2}+1\right)}|00\rangle - \frac{1}{\sqrt{2\sqrt{2}(\sqrt{2}+1)}}|11\rangle.$$

3) The conjugate of the 2-qubit is defined as

$$|\overline{\varphi}\rangle = a|0\rangle - b|1\rangle - c|2\rangle - d|3\rangle. \tag{10.19}$$

Therefore, $\left|\overline{(\overline{\varphi})}\right\rangle = |\varphi\rangle$.

Example 10.5 Bell State Conjugate
The first two Bell superpositions are conjugate to each other,

$$|\varphi_{B,1}\rangle = (|00\rangle + |11\rangle)/\sqrt{2} \quad \text{and} \quad |\overline{\varphi}_{B,1}\rangle = |\varphi_{B,2}\rangle = (|00\rangle - |11\rangle)/\sqrt{2}.$$

For the superpositions $|\varphi_{B,3}\rangle = (|01\rangle + |10\rangle)/\sqrt{2}$ and $|\varphi_{B,4}\rangle = (|01\rangle - |10\rangle)/\sqrt{2}$, the conjugates $|\overline{\varphi}_{B,3}\rangle = -|\varphi_{B,3}\rangle$ and $|\overline{\varphi}_{B,4}\rangle = -|\varphi_{B,4}\rangle$.

4) Polar form of the 2-qubit:

$$|\varphi\rangle = \cos\vartheta|0\rangle + \sin\vartheta|\phi\rangle, \tag{10.20}$$

where $\vartheta = \arccos\alpha$ and $|\phi\rangle$ is the quantum superposition

$$|\phi\rangle = \frac{b|1\rangle + c|2\rangle + d|3\rangle}{\sqrt{b^2 + c^2 + d^2}}. \tag{10.21}$$

For the quaternion, this polar form is defined as follows [1]:

$$q = a + q' = a + \frac{q'}{|q'|}|q'| = e^{\mu\vartheta} \triangleq \cos\vartheta + \mu\sin\vartheta. \tag{10.22}$$

Here, $\mu = q'/|q'|$ is a pure unit quaternion, $\mu^2 = -1$. The superposition $|\phi\rangle$ plays the role of μ.

Example 10.6 Bell State in Polar Form
Consider the following 2-qubit: $|\varphi\rangle = (|0\rangle - |1\rangle + |2\rangle + |3\rangle)/2$ that corresponds to the 4D vector $[1, -1, 1, 1]/2$, or the quaternion

$$q = \frac{1 + (-i + j + k)}{2} = \frac{1}{2} + \frac{(-i + j + k)}{\sqrt{3}} \cdot \frac{\sqrt{3}}{2} = \cos(60^\circ) + \frac{(-i + j + k)}{\sqrt{3}}\sin(60^\circ).$$

Here, $\vartheta = 60^\circ$, or $\pi/3$, and $\mu = (-i + j + k)/\sqrt{3}$. The above 2-qubit can be written as

$$|\varphi\rangle = \cos(60^\circ)|0\rangle + \sin(60^\circ)|\phi\rangle, \quad |\phi\rangle = \frac{-|1\rangle + |2\rangle + |3\rangle}{\sqrt{3}}. \tag{10.23}$$

For the Bell state $|\varphi_{B,1}\rangle$, the polar form is

$$|\varphi_{B,1}\rangle = \frac{|00\rangle}{\sqrt{2}} + \frac{|11\rangle}{\sqrt{2}} = \cos(45^\circ)|00\rangle + \sin(45^\circ)|\phi\rangle, \qquad |\phi\rangle = |11\rangle.$$

10.4.1 2-Qubit to the Power

By using the polar form of the 2-qubits, we can define different powers and roots of 2-qubits. For that, we consider the polar form of the 2-qubit in Eq. 10.20, $|\varphi\rangle = \cos\vartheta|0\rangle + \sin\vartheta|\phi\rangle$. By using the polar form of the quaternion, the quaternion in power x is defined as [1]

$$q^x = \left(e^{\mu\vartheta}\right)^x = e^{\mu(\vartheta x)} = \cos(x\vartheta) + \mu\sin(x\vartheta). \tag{10.24}$$

Therefore, it is possible to calculate powers of the 2-qubits.

Definition 10.2
The power of x of the 2-qubit $|\varphi\rangle$ is defined as

$$|\varphi\rangle^x = \cos(x\vartheta)|0\rangle + \sin(x\vartheta)|\phi\rangle. \tag{10.25}$$

Since $|\phi\rangle^2 = -|0\rangle$, then $|\varphi\rangle^x \cdot |\varphi\rangle^y = |\varphi\rangle^{x+y}$, for real any numbers x and y.

It is not difficult to verify that this equation for $x = 2$ and 0.5 results in Eqs. 10.16 and 10.17, for the square and square root of the 2-qubit $|\varphi\rangle$, respectively.

Example 10.7 Five Powers of 2-Qubit

Consider the 2-qubit in Eq. 10.23, $|\varphi\rangle = \cos(60^\circ)|0\rangle + \sin(60^\circ)|\phi\rangle$, where $|\phi\rangle = (-|1\rangle + |2\rangle + |3\rangle)/\sqrt{3}$. The following powers of this 2-qubit are held below:

$$\begin{aligned}
|\varphi\rangle^2 &= \cos(120^\circ)|0\rangle + \sin(120^\circ)|\phi\rangle = -0.5|0\rangle + 0.0866|\phi\rangle = -|\overline{\varphi}\rangle,\\
|\varphi\rangle^3 &= \cos(180^\circ)|0\rangle + \sin(180^\circ)|\phi\rangle = -|0\rangle,\\
|\varphi\rangle^4 &= \cos(240^\circ)|0\rangle + \sin(240^\circ)|\phi\rangle = 0.5|0\rangle - 0.0866|\phi\rangle = |\overline{\varphi}\rangle,\\
|\varphi\rangle^{1/2} &= \cos(30^\circ)|0\rangle + \sin(30^\circ)|\phi\rangle = 0.0866|0\rangle + 0.5|\phi\rangle,\\
|\varphi\rangle^{1/3} &= \cos(20^\circ)|0\rangle + \sin(20^\circ)|\phi\rangle = 0.9397|0\rangle + 0.3420|\phi\rangle.
\end{aligned}$$

The 2-qubit $|\varphi_1\rangle$ inverse to $|\varphi\rangle$ is defined from the equation $|\varphi_1\rangle \cdot |\varphi\rangle = |0\rangle$. For instance, if $|\varphi\rangle = (|0\rangle + |1\rangle + |2\rangle + |3\rangle)/2$, then from the equation

$$A_{|\varphi\rangle}|\varphi_1\rangle = \frac{1}{2}\begin{bmatrix} 1 & -1 & -1 & -1\\ 1 & 1 & 1 & -1\\ 1 & -1 & 1 & 1\\ 1 & 1 & -1 & 1\end{bmatrix}\begin{bmatrix} a_1\\ b_1\\ c_1\\ d_1\end{bmatrix} = \begin{bmatrix} 1\\ 0\\ 0\\ 0\end{bmatrix}, \quad or \quad \frac{1}{2}\begin{bmatrix} 1 & 1 & 1 & 1\\ -1 & 1 & -1 & 1\\ -1 & 1 & 1 & -1\\ -1 & -1 & 1 & 1\end{bmatrix}\begin{bmatrix} 1\\ 0\\ 0\\ 0\end{bmatrix} = \begin{bmatrix} a_1\\ b_1\\ c_1\\ d_1\end{bmatrix},$$

we obtain $|\varphi_1\rangle = (|0\rangle - |1\rangle - |2\rangle - |3\rangle)/2 = |\overline{\varphi}\rangle$. Thus, the inverse is the 2-qubit conjugate $|\overline{\varphi}\rangle$. The same is true in the general case. Indeed,

$$\underline{|\varphi_1\rangle \cdot |\varphi\rangle} \cdot |\overline{\varphi}\rangle = \underline{|0\rangle} \cdot |\overline{\varphi}\rangle \quad \rightarrow \quad |\varphi_1\rangle \cdot |0\rangle = |\overline{\varphi}\rangle \quad \rightarrow |\varphi_1\rangle = |\varphi\rangle^{-1} = |\overline{\varphi}\rangle.$$

The inverse from the right, that is, 2-qubit $|\varphi_2\rangle$ defined from the equation $|\varphi\rangle \cdot |\varphi_2\rangle = |0\rangle$ is also equal to $|\varphi_1\rangle$. Indeed, from $|\varphi\rangle \cdot |\varphi_2\rangle = |0\rangle$, it follows that $|\varphi\rangle = |\overline{\varphi}_2\rangle$, or $|\varphi_2\rangle = |\overline{\varphi}\rangle$.

Using the multiplication operation, we can define the division operation of 2-qubits.

Definition 10.3 The division of 2-qubits $|\varphi_1\rangle$ and $|\varphi_2\rangle$ is defined as

$$|\varphi\rangle \triangleq \frac{|\varphi_1\rangle}{|\varphi_2\rangle} \triangleq |\varphi_1\rangle \cdot |\varphi_2\rangle^{-1} = |\varphi_1\rangle \cdot |\overline{\varphi_2}\rangle. \tag{10.26}$$

The division of 2-qubits is the multiplication of the first 2-qubit by the conjugate of the second 2-qubit. It follows from the definition that $|\varphi_1\rangle/|\varphi_1\rangle = |\overline{\varphi}_1\rangle \cdot |\varphi_1\rangle = |00\rangle$, for any 2-qubit $|\varphi_1\rangle$.

Example 10.8 Bell State as the Division of 2-Qubits

Consider two quantum superpositions in the basis $\mathcal{B}_2 = \{|0\rangle, |1\rangle, |2\rangle, |3\rangle\}$,

$$|\varphi_1\rangle = \frac{|0\rangle - |1\rangle - |2\rangle + |3\rangle}{2} \text{ and } |\varphi_2\rangle = |\varphi_{H,2}\rangle = \frac{|0\rangle - |1\rangle}{\sqrt{2}}. \tag{10.27}$$

Dividing these two superpositions, we obtain the Bell superposition $|\varphi_{B,1}\rangle$,

$$\frac{|\varphi_1\rangle}{|\varphi_2\rangle} = \frac{|0\rangle - |1\rangle - |2\rangle + |3\rangle}{2} \cdot \frac{|0\rangle + |1\rangle}{\sqrt{2}} = \frac{|0\rangle + |3\rangle}{\sqrt{2}} = \frac{|00\rangle + |11\rangle}{\sqrt{2}}. \tag{10.28}$$

In the matrix form, the calculation of this superposition is described as

$$\frac{1}{2}\begin{bmatrix} 1 & 1 & 1 & -1 \\ -1 & 1 & -1 & -1 \\ -1 & 1 & 1 & 1 \\ 1 & 1 & -1 & 1 \end{bmatrix} \times \frac{1}{\sqrt{2}}\begin{bmatrix} 1 \\ 1 \\ 0 \\ 0 \end{bmatrix} = \frac{1}{2\sqrt{2}}\begin{bmatrix} 2 \\ 0 \\ 0 \\ 2 \end{bmatrix} = \frac{1}{\sqrt{2}}\begin{bmatrix} 1 \\ 0 \\ 0 \\ 1 \end{bmatrix} = \frac{|0\rangle + |3\rangle}{\sqrt{2}}. \tag{10.29}$$

Thus, we have operations of multiplication, inverse, and division on 2-qubits and with these operations, the set of 2-qubits is described as a multiplicative group. These operations can also be extended to 2-qubit superpositions, which we will cover in the next section.

Comments: Note that the operation of multiplication by 2-qubits described above can be considered as an extension of the operation of multiplication by single qubits [16], by analogy, as the multiplication of quaternions, the concept of complex multiplication was extended. In the basis $\mathcal{B}_1 = \{|0\rangle, |1\rangle\}$, the multiplication of two single qubits $|\zeta_1\rangle = a_1|0\rangle + b_1|1\rangle$ and $|\zeta_2\rangle = a_2|0\rangle + b_2|1\rangle$ is defined as the single qubit in the superposition

$$|\zeta\rangle = |\zeta_1\rangle \cdot |\zeta_2\rangle = a|0\rangle + b|1\rangle, \tag{10.30}$$

where the amplitudes of the qubit are calculated in the matrix form as

$$\begin{bmatrix} a \\ b \end{bmatrix} = \begin{bmatrix} a_1 & -b_1 \\ b_1 & a_1 \end{bmatrix}\begin{bmatrix} a_2 \\ b_2 \end{bmatrix}.$$

This is operation of complex multiplication of numbers $a_1 + ib_1$ and $a_2 + ib_2$. In other words, the amplitudes of the qubit $|\zeta\rangle$ are $a = a_1a_2 - b_1b_2$ and $b = b_1a_2 + a_1b_2$.

If directly follows from Eq. 10.13 that in the $c_1 = c_2 = d_1 = d_2 = 0$ case, that is, when multiplying two superpositions $|\varphi_1\rangle = a_1|00\rangle + b_1|01\rangle$ and $|\varphi_2\rangle = (a_2|00\rangle + b_2|01\rangle)$, we obtain

$$|\varphi\rangle = |\varphi_1\rangle \cdot |\varphi_2\rangle = |0\rangle[(a_1a_2 - b_1b_2)|0\rangle + (b_1a_2 + a_1b_2)|1\rangle] = |0\rangle|\zeta\rangle. \tag{10.31}$$

Square brackets are the multiplication of the single qubits $|\zeta_1\rangle$ and $|\zeta_2\rangle$. This is the case, when $|\varphi_1\rangle = |0\rangle \mid \zeta_1\rangle$ and $|\varphi_2\rangle = |0\rangle \mid \zeta_2\rangle$, the multiplication is defined by the matrix

$$\boldsymbol{A}_{\zeta_1} = \begin{bmatrix} a_1 & -b_1 & 0 & 0 \\ b_1 & a_1 & 0 & 0 \\ 0 & 0 & a_1 & -b_1 \\ 0 & 0 & b_1 & a_1 \end{bmatrix} = \begin{bmatrix} 1 & 0 \\ 0 & 1 \end{bmatrix} \otimes \begin{bmatrix} a_1 & -b_1 \\ b_1 & a_1 \end{bmatrix}. \tag{10.32}$$

In this particular case, the multiplication of 2-qubits is the tensor product of the basis state $|0\rangle$ with the product of single qubits.

10.4.2 Second Model of Quaternion and 2-Qubits

The 2-qubit superposition $|\varphi\rangle = a|00\rangle + b|01\rangle + c|10\rangle + d|11\rangle$ with the quaternion $q = a + (bi + cj + dk)$ is referred to as the vector $q = (a, b, c, d)$ in the 4D real space R^4 with the basic vectors $e = (1, 0, 0, 0)$, $i = (0, 1, 0, 0)$, $j = (0, 0, 1, 0)$, and $k = (0, 0, 0, 1)$. We cannot plot vectors in the 4D space. This space does not appear to us in reality. Therefore, we consider the geometry of quaternions, which is different from the geometry given in Section 10.3. The quaternion q can be represented as a 3D ball at center (b, c, d) and radius defined as $r = |a|$. Thus, in 3D space, we can represent a quaternion as a ball with a position vector at its center. Such representation is not unique, since the same ball will represent two quaternions with $\pm a$. Therefore, we consider drawing balls with an arrow, as shown in

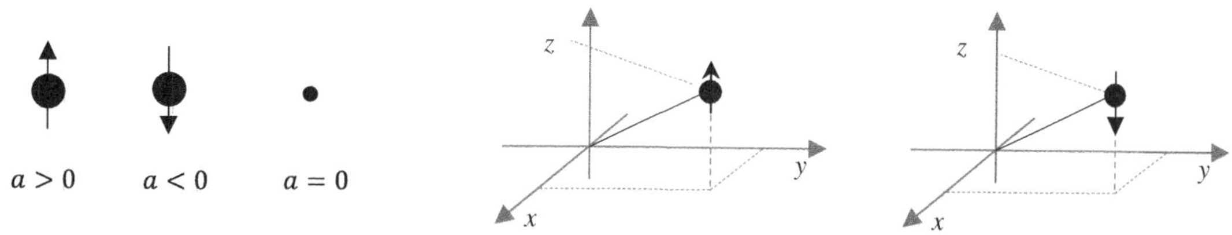

Fig. 10.6 The graphical symbol of the quaternion in the 3D space.

Fig. 10.6. Positive direction (arrow vertically up) is considered if $a > 0$, and negative direction (arrow horizontally down) if $a < 0$. If $a = 0$, then the pure quaternion q is a point in the 3D space with coordinates (b, c, d) without arrow in it.

Thus, the notation is used $\dot{q} = (q_0, \boldsymbol{q}')$, for the quaternion $q = (a, b, c, d)$. Here, q_0 is the radius and $\boldsymbol{q}'$ is the 3D vector. The component $\boldsymbol{q}'$ of the quaternion refers to the space. The first component q_0 has some kind of special meaning. This is not a time or a coordinate; in real models, it describes some kind of vital substance, which we represent as a ball. The length of the vector $\boldsymbol{q}'$ equals to $\sqrt{b^2 + c^2 + d^2} = \sqrt{1 - a^2}$. If we consider the radius of this ball be

$$r = 1 - |\boldsymbol{q}'| = 1 - \sqrt{1 - a^2} = 1 - \sqrt{b^2 + c^2 + d^2}, \tag{10.33}$$

then, not only the center but entire ball will be inside the unit sphere (see Fig. 10.7). The length $|q| = 1$ is the energy of the quaternion. The number r is the remainder of the energy of the imaginary part. The closer the number a is to 1, the larger the radius of the ball, and vice versa. The notation $B(\dot{q})$ will be used for such a ball. If $a = 0$, then the pure quaternion $q = (0, b, c, d)$ is the point in the 3D unit sphere with coordinates (b, c, d) and without arrow in it. In this case, $r = 1 - \sqrt{1 - a^2} = 0$.

As an example, Fig. 10.8 in part (a) illustrates the geometry of the quaternion $q = (1, 1, 1, 1)$ for the 2-qubit $|\varphi_1\rangle = (|00\rangle + |01\rangle + |10\rangle + |11\rangle)/2$. All amplitudes of the 2-qubit are equal to 0.5. The center of the ball is in the point (0.5,0.5,0.5) and the radius is $r = 1 - \sqrt{1 - 1/4} = 0.1340$. Other three 2-qubits are also shown. The balls $B(1, 1, 0, 0)$ and $B(1, -1, 0, 0)$ represent two quaternions $q = (1, \pm 1, 0, 0)$, or two Hadamard states $(|00\rangle \pm |01\rangle)/\sqrt{2}$, respectively. The Bell state $(|10\rangle - |01\rangle)/\sqrt{2}$ that corresponds to the quaternion $q = (1, 0, 0, -1)$ is shown as the ball $B(1, 0, 0, -1)$. The same four balls inside the unit sphere are shown in part (b). All 2-qubits, or quaternions with length 1, can be viewed as balls inside the unit sphere, as shown in part (b). This sphere corresponds to the basis state $|00\rangle$, or the quaternion 1; it is the ball $B(1, 0, 0, 0)$.

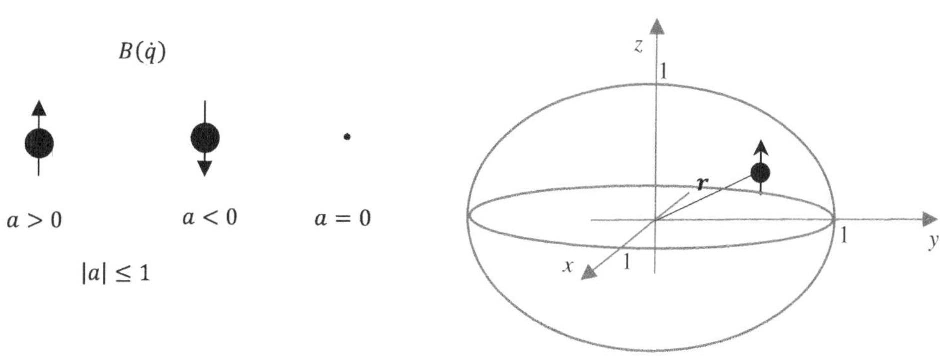

Fig. 10.7 The graphical symbol of a quaternion in 3D space.

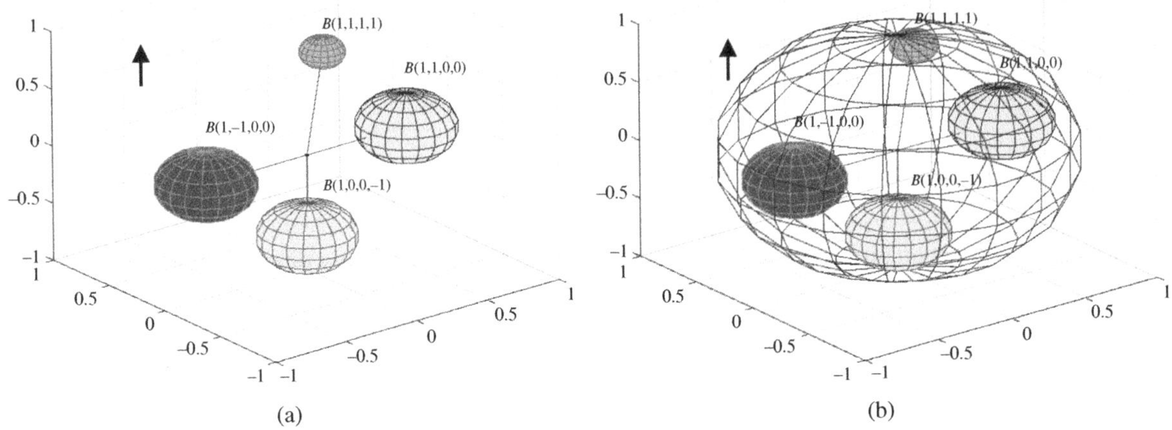

Fig. 10.8 The graphical representation of four different 2-qubits in 3-D space (a) without the unit sphere and (b) with the unit sphere (all balls are positive).

Figure 10.9 shows the ball $B(1, 1, 0.2, 0.8)$ that represents the normalized quaternion $q_1 = (1, 1, 0.2, 0.8)$, or the 2-qubit in the superposition $|q_1\rangle = (|00\rangle + |01\rangle + 0.2|10\rangle + 0.8|11\rangle)/1.6371$. Also, the ball is shown for the quaternion $q_2 = (1, -1, -1, 0.4)$ representing the following 2-qubit: $|q_2\rangle = (|00\rangle - |01\rangle - |10\rangle + 0.4|11\rangle)/1.7776$. The superpositions of the quaternion multiplications q_1q_2 and q_2q_1 (which are different numbers since the quaternion multiplication is not commutative) are also illustrated in this figure. In addition, we added the *quaternion ball* for the superposition $|q_1 + q_2\rangle$, which is described by the normalized sum of quaternions $q_1 + q_2$.

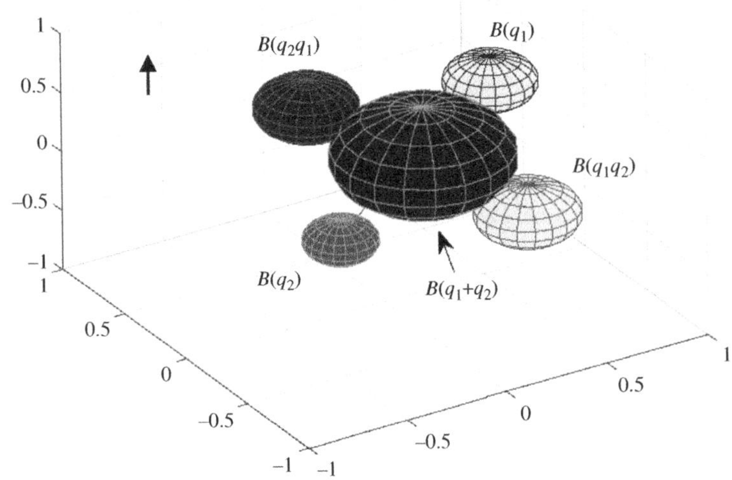

Fig. 10.9 The graphical representation of two 2-qubits, their multiplications, and sum (all balls are positive).

References

1 Grigoryan, A.M. and Agaian, S.S. (2018). Quaternion and Octonion Color Image Processing with MATLAB. *SPIE* PM279: https://doi.org/10.1117/3.2278810.

2 Grigoryan, A.M. and Agaian, S.S. (2022) 'Commutative quaternion algebra and DSP fundamental properties: Quaternion convolution and Fourier transform,' *Signal Processing*, 196, 108533, p. 14. https://doi.org/10.1016/j.sigpro.2022.108533

3 Grigoryan, A.M. and Agaian, S.S. (2023). Introduction to multiplicative group on 2-qubits with quantum color image processing. *Quantum Information Processing* 22 (351): 34. https://doi.org/10.1007/s11128-023-04091-1.

4 Grigoryan, A.M. and Agaian, S.S. (2024) 'Quaternion-based arithmetic in quantum information processing: a promising approach for efficient color quantum imaging [Hypercomplex Signal and Image Processing],' *IEEE Signal Processing Magazine*, 41 (2) p. 64–74. doi: https://doi.org/10.1109/MSP.2023.3327627

5 Mastriani, M. (2017). Quantum image processing? *Quantum Information Processing* 16 (1): 27.

6 Yan, F. and Venegas-Andraca, S.E. (2020). *Quantum Image Processing*. Springer Nature Singapore Pte Ltd.

7 Grigoryan, A.M. and Agaian, S.S. (2018). Color facial image representation with new quaternion gradients. *Proceedings of IS&T International Symposium, Electronic Imaging: Algorithms and Systems*, p. 7, 28 January–2 February, Burlingame, CA. https://doi.org/10.2352/ISSN.2470-1173.2018.13.IPAS-383

8 Grigoryan, A.M. John, A., and Agaian, S.S. (2018). Enhancement of underwater color images by two-side 2-D quaternion discrete Fourier transform. *Proceedings of IS&T International Symposium, Electronic Imaging: Algorithms and Systems*, p. 6, 28 January–2 February, Burlingame, CA. https://doi.org/10.2352/ISSN.2470-1173.2018.13.IPAS-441

9 Grigoryan, A.M. and Agaian, S.S.(2020). Optimal restoration of multiple signals in quaternion algebra. *Processing of SPIE 11399, Mobile Multimedia/Image Processing, Security, and Applications*. 1139909, p. 13. https://doi.org/10.1117/12.2558209

10 Grigoryan, A.M., Jeninson, J., and Agaian, S.S. (2015). Quaternion Fourier transform based alpha-rooting method for color image measurement and enhancement. *Signal Processing* 109: 269–289. https://doi.org/10.1016/j.sigpro.2014.11.019.

11 Graves, J. (1845). On a connection between the general theory of normal couples and the theory of complete quadratic functions of two variables. *Philosophical Magazine* 26: 315–320. https://doi.org/10.1080/14786444508645136.

12 Hamilton, W.R. (1866). *Elements of Quaternions*. London: Logmans, Green and Co.

13 Kantor, I.L. and Solodovnikov, A.S. (1973). *Hypercomplex Numbers*. Moscow: Nauka.

14 Dixon, G.M. (1994). *Division Algebras: Octonions, Quaternions, Complex Numbers, and the Algebraic Design of Physics*. Berlin: Kluwer Academic Publishers.

15 Grigoryan, A.M. and Agaian, S.S. (2014). Retooling of color imaging in the quaternion algebra. *Applied Mathematics and Sciences: An International Journal* 1 (3): 23–39.

16 Grigoryan, A.M. and Agaian, S.S. (2023). Multiplicative group of quantum representations of signals. *Quantum Information Processing* 22 (82): 23. https://doi.org/10.1007/s11128-022-03765-6.

17 Davenport, C. (2000). Commutative hypercomplex mathematics, p. 11, http://home.usit.net/cmdaven/cmdaven1.htm.

11

Quantum Schemes for Multiplication of 2-Qubits

In this section, we describe a few quantum schemes for the multiplication matrix of 2-qubits. The matrix will be diagonalized by using the DsiHT with Givens rotations [1, 2]. The method of QD-decomposition is described in Section 6.2 in detail and will be used here. Some of the rotation matrices in these two sections are denoted differently. These sections were written independently and we decided to keep the original texts. Thus, this section can be read without referring to Sections 6.2 and 6.3. The matrix of multiplication will be written, by using different systems of coordinates. Namely, we consider three such systems and for each of them the corresponding quantum scheme(s) for multiplication of 2-qubits will be described.

The 2-qubit is the quantum superposition

$$|\varphi\rangle = a_0|00\rangle + a_1|01\rangle + a_2|10\rangle + a_3|11\rangle, \tag{11.1}$$

($|a_0|^2 + |a_1|^2 + |a_2|^2 + |a_3|^2 = 1$). The basis states can be presented by the 4D unit vectors $|0\rangle = [1, 0, 0, 0]'$, $|1\rangle = [0, 1, 0, 0]'$, $|2\rangle = [0, 0, 1, 0]'$, and $|3\rangle = [0, 0, 0, 1]'$. We need quantum schemes on 2-qubits, to calculate the multiplication of two qubits. Consider two 2-qubits $|\varphi_1\rangle = a_1|0\rangle + b_1|1\rangle + c_1|2\rangle + d_1|3\rangle$ and $|\varphi_2\rangle = a_2|0\rangle + b_2|1\rangle + c_2|2\rangle + d_2|3\rangle$. The multiplication of 2-qubits $|\varphi_1\rangle$ and $|\varphi_2\rangle$ is calculated by

$$|\varphi\rangle \triangleq |\varphi_1\rangle \cdot |\varphi_2\rangle = a|0\rangle + b|1\rangle + c|2\rangle + d|3\rangle, \tag{11.2}$$

where the amplitudes are calculated by

$$\begin{bmatrix} a \\ b \\ c \\ d \end{bmatrix} = A_{|\varphi_1\rangle} \begin{bmatrix} a_2 \\ b_2 \\ c_2 \\ d_2 \end{bmatrix} = \begin{bmatrix} a_1 & -b_1 & -c_1 & -d_1 \\ b_1 & a_1 & -d_1 & c_1 \\ c_1 & d_1 & a_1 & -b_1 \\ d_1 & -c_1 & b_1 & a_1 \end{bmatrix} \begin{bmatrix} a_2 \\ b_2 \\ c_2 \\ d_2 \end{bmatrix}. \tag{11.3}$$

The 4×4 matrix is defined by the first 2-qubit and it is unitary [3]. Thus, we can consider the multiplication operation as a unitary transform. Note that the result of this multiplication can also be obtained by using another matrix,

$$\begin{bmatrix} a \\ b \\ c \\ d \end{bmatrix} = A_{|\varphi_2\rangle} \begin{bmatrix} a_1 \\ b_1 \\ c_1 \\ d_1 \end{bmatrix} = \begin{bmatrix} a_2 & -b_2 & -c_2 & -d_2 \\ b_2 & a_2 & d_2 & -c_2 \\ c_2 & -d_2 & a_2 & b_2 \\ d_2 & c_2 & -b_2 & a_2 \end{bmatrix} \begin{bmatrix} a_1 \\ b_1 \\ c_1 \\ d_1 \end{bmatrix}. \tag{11.4}$$

Quantum Image Processing in Practice: A Mathematical Toolbox, First Edition. Artyom M. Grigoryan and Sos S. Agaian.

Companion website: www.wiley.com/go/grigoryan/quantumimageprocessing

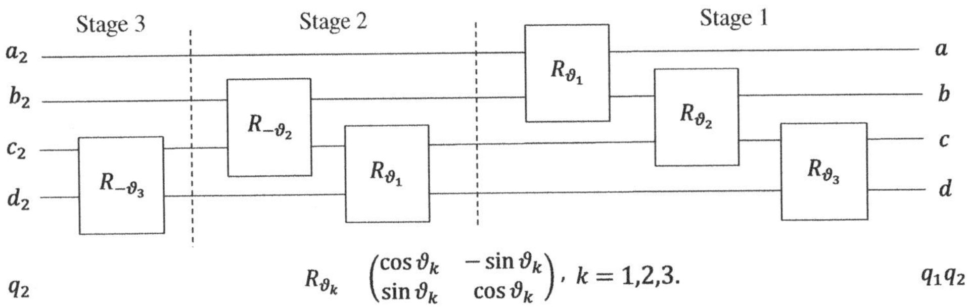

Fig. 11.1 The traditional block diagram of multiplication of quaternions.

Here, the unitary 4×4 matrix is defined by the second 2-qubit. This matrix is not composed by two 2×2 submatrices, as for the $\boldsymbol{A}_{|\varphi_1\rangle}$. Thus, the multiplication of 2-qubits associates with two different unitary transforms. We will use both transforms and describe in detail the corresponding quantum circuits. First, we will consider methods for constructing a quantum circuit for the transform with the matrix $\boldsymbol{A}_{|\varphi_1\rangle}$.

With 2-qubits $|\varphi_1\rangle$ and $|\varphi_2\rangle$, the corresponding amplitude vectors are treated as the unit quaternions $q_1 = (a_1, b_1, c_1, d_1)$ and $q_2 = (a_2, b_2, c_2, d_2)$. Therefore, we denote the above multiplication matrix by $\boldsymbol{A}_{q_1}$ and also by $\boldsymbol{A}_{|\varphi_1\rangle}$. The amplitudes of the 2-qubit $|\varphi_1\rangle$ can be written in spherical system of coordinates:

$$a_1 = \cos(\vartheta_1), \quad b_1 = \sin(\vartheta_1)\cos(\vartheta_2), \quad c_1 = \sin(\vartheta_1)\sin(\vartheta_2)\cos(\vartheta_3), \quad d_1 = \sin(\vartheta_1)\sin(\vartheta_2)\sin(\vartheta_3).$$

The angles are calculated by

$$\vartheta_1 = \operatorname{acos}(a_1), \quad \vartheta_2 = \operatorname{acos}\left(\frac{b_1}{\sqrt{1-a_1^2}}\right), \qquad \vartheta_3 = \operatorname{sign}(d_1)\operatorname{acos}\left(\frac{c_1}{\sqrt{1-a_1^2-b_1^2}}\right). \tag{11.5}$$

First, we will describe the block diagram shown in Fig. 11.1. It is the main block diagram of multiplication $q = q_1 q_2$ of the quaternion $q_2 = (a_2, b_2, c_2, d_2)$ by the quaternion $q_1 = (a_1, b_1, c_1, d_1)$. Here, the input is the quaternion q_2, or the 2-qubit $|\varphi_2\rangle$, and the product is the quaternion q, or $|\varphi_1\rangle \cdot |\varphi_2\rangle$.

11.1 Schemes for the 4 × 4 Gate A_{q_1}

The factorization of the 4×4 matrix of quaternion multiplication $\boldsymbol{A}_{q_1}$ can be done by the heap transforms with the weak and strong carriages on each stage of the QD-factorization [2]. We stand on the path with the strong carriage. The matrix diagonalization completed by three stages and with six matrices of rotation is described as follows:

$$\underbrace{Z_1} \cdot \underbrace{(Y_1 Y_2)} \cdot \underbrace{(X_1 X_2 X_3)} \boldsymbol{A}_{q_1} = I_4. \tag{11.6}$$

Here, I_4 is the 4×4 identity matrix and other six unitary matrices are

$$X_3 = \begin{bmatrix} 1 & 0 & 0 & 0 \\ 0 & 1 & 0 & 0 \\ 0 & 0 & \cos\vartheta_3 & \sin\vartheta_3 \\ 0 & 0 & -\sin\vartheta_3 & \cos\vartheta_3 \end{bmatrix}, \quad X_2 = \begin{bmatrix} 1 & 0 & 0 & 0 \\ 0 & \cos\vartheta_2 & \sin\vartheta_2 & 0 \\ 0 & -\sin\vartheta_2 & \cos\vartheta_2 & 0 \\ 0 & 0 & 0 & 1 \end{bmatrix}, \quad X_1 = \begin{bmatrix} \cos\vartheta_1 & \sin\vartheta_1 & 0 & 0 \\ -\sin\vartheta_1 & \cos\vartheta_1 & 0 & 0 \\ 0 & 0 & 1 & 0 \\ 0 & 0 & 0 & 1 \end{bmatrix},$$

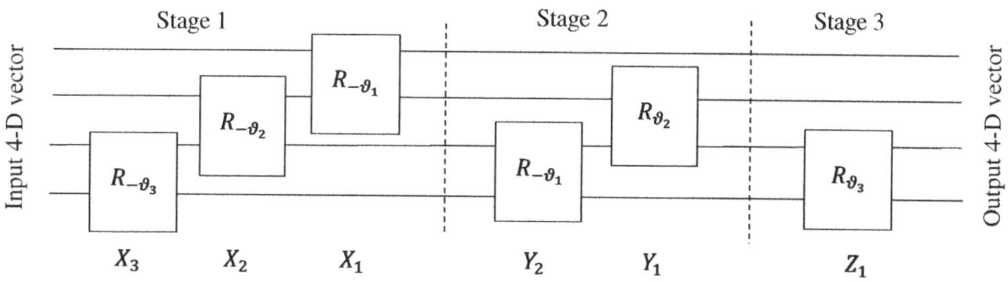

Fig. 11.2 The diagram of application of three strong DsiHTs for $\boldsymbol{A}_{q_1}$ matrix factorization.

$$Y_2 = \begin{bmatrix} 1 & 0 & 0 & 0 \\ 0 & 1 & 0 & 0 \\ 0 & 0 & \cos\vartheta_1 & \sin\vartheta_1 \\ 0 & 0 & -\sin\vartheta_1 & \cos\vartheta_1 \end{bmatrix}, \quad Y_1 = \begin{bmatrix} 1 & 0 & 0 & 0 \\ 0 & \cos\vartheta_2 & -\sin\vartheta_2 & 0 \\ 0 & \sin\vartheta_2 & \cos\vartheta_2 & 0 \\ 0 & 0 & 0 & 1 \end{bmatrix}, \quad Z_1 = \begin{bmatrix} 1 & 0 & 0 & 0 \\ 0 & 1 & 0 & 0 \\ 0 & 0 & \cos\vartheta_3 & -\sin\vartheta_3 \\ 0 & 0 & \sin\vartheta_3 & \cos\vartheta_3 \end{bmatrix}.$$

The corresponding diagram of application of DsiHTs in the above matrix factorization is shown in Fig. 11.2. Note that this diagonalization has an interesting property. The same three angles ϑ_1, ϑ_2, and ϑ_3 are used in the matrix factorization; each angle in two rotations.

After transposing all six rotation matrices in the Eq. 11.6, we obtain the following matrix $\boldsymbol{A}_{q_1}$ representation:

$$\boldsymbol{A}_{q_1} = \begin{bmatrix} a_1 & -b_1 & -c_1 & -d_1 \\ b_1 & a_1 & -d_1 & c_1 \\ c_1 & d_1 & a_1 & -b_1 \\ d_1 & -c_1 & b_1 & a_1 \end{bmatrix} = \underbrace{X'_3X'_2X'_1}\cdot\underbrace{Y'_2Y'_1}\cdot\underbrace{Z'_1}. \tag{11.7}$$

Here, the notation X' means the transpoe of the matrix X. The block diagram given in Fig. 11.1 corresponds to this representation. Considering that the matrix $Y'_1 = X_2$ and the matrix $Z'_1 = X_3$, the complete factorization of the $\boldsymbol{A}_{q_1}$ matrix can be written as

$$\boldsymbol{A}_{q_1} = \underbrace{X'_3X'_2X'_1}\cdot\underbrace{Y'_2X_2}\cdot\underbrace{X_3} \tag{11.8}$$

$$= \underbrace{\begin{bmatrix} 1 & 0 & 0 & 0 \\ 0 & 1 & 0 & 0 \\ 0 & 0 & \cos\vartheta_3 & -\sin\vartheta_3 \\ 0 & 0 & \sin\vartheta_3 & \cos\vartheta_3 \end{bmatrix} \begin{bmatrix} 1 & 0 & 0 & 0 \\ 0 & \cos\vartheta_2 & -\sin\vartheta_2 & 0 \\ 0 & \sin\vartheta_2 & \cos\vartheta_2 & 0 \\ 0 & 0 & 0 & 1 \end{bmatrix} \begin{bmatrix} \cos\vartheta_1 & -\sin\vartheta_1 & 0 & 0 \\ \sin\vartheta_1 & \cos\vartheta_1 & 0 & 0 \\ 0 & 0 & 1 & 0 \\ 0 & 0 & 0 & 1 \end{bmatrix}}$$

$$\times \underbrace{\begin{bmatrix} 1 & 0 & 0 & 0 \\ 0 & 1 & 0 & 0 \\ 0 & 0 & \cos\vartheta_1 & -\sin\vartheta_1 \\ 0 & 0 & \sin\vartheta_1 & \cos\vartheta_1 \end{bmatrix} \begin{bmatrix} 1 & 0 & 0 & 0 \\ 0 & \cos\vartheta_2 & \sin\vartheta_2 & 0 \\ 0 & -\sin\vartheta_2 & \cos\vartheta_2 & 0 \\ 0 & 0 & 0 & 1 \end{bmatrix}} \underbrace{\begin{bmatrix} 1 & 0 & 0 & 0 \\ 0 & 1 & 0 & 0 \\ 0 & 0 & \cos\vartheta_3 & \sin\vartheta_3 \\ 0 & 0 & -\sin\vartheta_3 & \cos\vartheta_3 \end{bmatrix}}.$$

The matrices X_3, X'_3, and Y'_2 are the controlled gates (or 1-controlled gates)

$$X_3 = I_2 \oplus R_{-\vartheta_3}, \qquad X'_3 = I_2 \oplus R_{\vartheta_3}, \qquad Y'_2 = I_2 \oplus R_{\vartheta_1}. \tag{11.9}$$

Here, the Givens rotations, or gates, are

$$R_{\vartheta_k} = W(\vartheta_k) = \begin{pmatrix} \cos\vartheta_k & -\sin\vartheta_k \\ \sin\vartheta_k & \cos\vartheta_k \end{pmatrix}, \quad k = 1, 2, 3.$$

Also, the product of two matrices $X'_1 Y'_1$ can be written as

$$X'_1 Y'_2 = \begin{bmatrix} \cos\vartheta_1 & -\sin\vartheta_1 & 0 & 0 \\ \sin\vartheta_1 & \cos\vartheta_1 & 0 & 0 \\ 0 & 0 & 1 & 0 \\ 0 & 0 & 0 & 1 \end{bmatrix} \begin{bmatrix} 1 & 0 & 0 & 0 \\ 0 & 1 & 0 & 0 \\ 0 & 0 & \cos\vartheta_1 & -\sin\vartheta_1 \\ 0 & 0 & \sin\vartheta_1 & \cos\vartheta_1 \end{bmatrix} = I_2 \otimes R_{\vartheta_1}.$$

A) Note that the matrix X_1 is the 0-controlled gate $R_{-\vartheta_1} \oplus I_2$ and can be calculated by the 1-controlled gate as follows:

$$\begin{bmatrix} \cos\vartheta_1 & \sin\vartheta_1 & 0 & 0 \\ -\sin\vartheta_1 & \cos\vartheta_1 & 0 & 0 \\ 0 & 0 & 1 & 0 \\ 0 & 0 & 0 & 1 \end{bmatrix} = \begin{bmatrix} 0 & 0 & 1 & 0 \\ 0 & 0 & 0 & 1 \\ 1 & 0 & 0 & 0 \\ 0 & 1 & 0 & 0 \end{bmatrix} \begin{bmatrix} 1 & 0 & 0 & 0 \\ 0 & 1 & 0 & 0 \\ 0 & 0 & \cos\vartheta_1 & \sin\vartheta_1 \\ 0 & 0 & -\sin\vartheta_1 & \cos\vartheta_1 \end{bmatrix} \begin{bmatrix} 0 & 0 & 1 & 0 \\ 0 & 0 & 0 & 1 \\ 1 & 0 & 0 & 0 \\ 0 & 1 & 0 & 0 \end{bmatrix}.$$

Thus,

$$X_1 = R_{-\vartheta_1} \oplus I_2 = P_1 (I_2 \oplus R_{-\vartheta_1}) P_1 \quad \text{and} \quad X'_1 = R_{\vartheta_1} \oplus I_2 = P_1 (I_2 \oplus R_{\vartheta_1}) P_1. \tag{11.10}$$

The permutation $P_1 = (0, 2)(1, 3)$ is used. Its matrix equals to

$$P_1 = \begin{bmatrix} 0 & 0 & 1 & 0 \\ 0 & 0 & 0 & 1 \\ 1 & 0 & 0 & 0 \\ 0 & 1 & 0 & 0 \end{bmatrix} = \begin{bmatrix} 0 & 1 \\ 1 & 0 \end{bmatrix} \otimes \begin{bmatrix} 1 & 0 \\ 0 & 1 \end{bmatrix} = X \otimes I_2. \tag{11.11}$$

One can see that P_1 is the operation $(x, y) \rightarrow (x \oplus 1, y)$ on two-bit plane, $x, y \in \{0, 1\}$. The quantum circuit element for the 4×4 gate X'_1 is shown in Fig. 11.3. The flowchart "sort" is used for permutations in quantum schemes. The permutation P_1 is also shown as the local Pauli NOT gate.

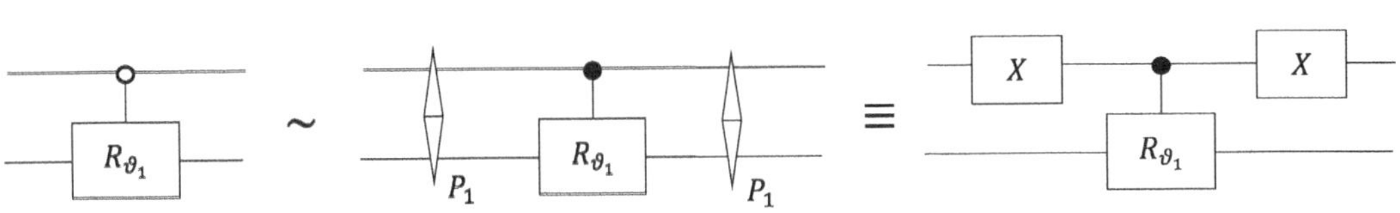

Fig. 11.3 The 0-controlled rotation gate X'_1 and its equivalent circuits with 1-controlled gate.

Note that the matrix X_2 can be described by the controlled gate as follows:

$$X_2 = \begin{bmatrix} 1 & 0 & 0 & 0 \\ 0 & \cos\vartheta_2 & \sin\vartheta_2 & 0 \\ 0 & -\sin\vartheta_2 & \cos\vartheta_2 & 0 \\ 0 & 0 & 0 & 1 \end{bmatrix} = \begin{bmatrix} 0 & 1 & 0 & 0 \\ 0 & 0 & 1 & 0 \\ 0 & 0 & 0 & 1 \\ 1 & 0 & 0 & 0 \end{bmatrix} \begin{bmatrix} 1 & 0 & 0 & 0 \\ 0 & 1 & 0 & 0 \\ 0 & 0 & \cos\vartheta_2 & \sin\vartheta_2 \\ 0 & 0 & -\sin\vartheta_2 & \cos\vartheta_2 \end{bmatrix} \begin{bmatrix} 0 & 0 & 0 & 1 \\ 1 & 0 & 0 & 0 \\ 0 & 1 & 0 & 0 \\ 0 & 0 & 1 & 0 \end{bmatrix},$$

or

$$X_2 = P_2(I_2 \oplus R_{-\vartheta_2})P_2' \quad \text{and} \quad X_2' = P_2(I_2 \oplus R_{\vartheta_2})P_2'. \tag{11.12}$$

The permutation $P_2 = (0, 3, 2, 1)$ and its inverse are used. Their matrices are

$$P_2 = \begin{bmatrix} 0 & 1 & 0 & 0 \\ 0 & 0 & 1 & 0 \\ 0 & 0 & 0 & 1 \\ 1 & 0 & 0 & 0 \end{bmatrix} \quad \text{and} \quad P_2' = \begin{bmatrix} 0 & 0 & 0 & 1 \\ 1 & 0 & 0 & 0 \\ 0 & 1 & 0 & 0 \\ 0 & 0 & 1 & 0 \end{bmatrix}.$$

The circuit elements for the gates X_2 and X_2' are shown in Fig. 11.4 in parts (a) and (b) respectively. Thus, the matrix of multiplication can be written as

$$\mathbf{A}_{|\varphi_1\rangle} = \underbrace{(I_2 \oplus R_{\vartheta_3})}_{X_3'}\underbrace{P_2(I_2 \oplus R_{\vartheta_2})P_2'}_{X_2'}\underbrace{(I_2 \otimes R_{\vartheta_1})}_{X_1'Y_2'}\underbrace{P_2(I_2 \oplus R_{-\vartheta_2})P_2'}_{X_2}\underbrace{(I_2 \oplus R_{-\vartheta_3})}_{X_3}, \tag{11.13}$$

or with only 1-controlled gates as

$$\mathbf{A}_{|\varphi_1\rangle} = \underbrace{(I_2 \oplus R_{\vartheta_3})}_{X_3'}\underbrace{P_2(I_2 \oplus R_{\vartheta_2})P_2'}_{X_2'}\underbrace{P_1(I_2 \oplus R_{\vartheta_1})P_1}_{X_1'}\underbrace{(I_2 \oplus R_{\vartheta_1})}_{Y_2'}\underbrace{P_2(I_2 \oplus R_{-\vartheta_2})P_2'}_{X_2}\underbrace{(I_2 \oplus R_{-\vartheta_3})}_{X_3}. \tag{11.14}$$

Here, the multiplication $P_2'P_1$ of two permutations can be reduced to one, considering that $P_1 = P_2^2$ or $P_2'P_1 = P_2$. Indeed, the following matrix multiplication holds:

$$P_2'P_1 = \begin{bmatrix} 0 & 0 & 0 & 1 \\ 1 & 0 & 0 & 0 \\ 0 & 1 & 0 & 0 \\ 0 & 0 & 1 & 0 \end{bmatrix} \begin{bmatrix} 0 & 0 & 1 & 0 \\ 0 & 0 & 0 & 1 \\ 1 & 0 & 0 & 0 \\ 0 & 1 & 0 & 0 \end{bmatrix} = \begin{bmatrix} 0 & 1 & 0 & 0 \\ 0 & 0 & 1 & 0 \\ 0 & 0 & 0 & 1 \\ 1 & 0 & 0 & 0 \end{bmatrix} = P_2. \tag{11.15}$$

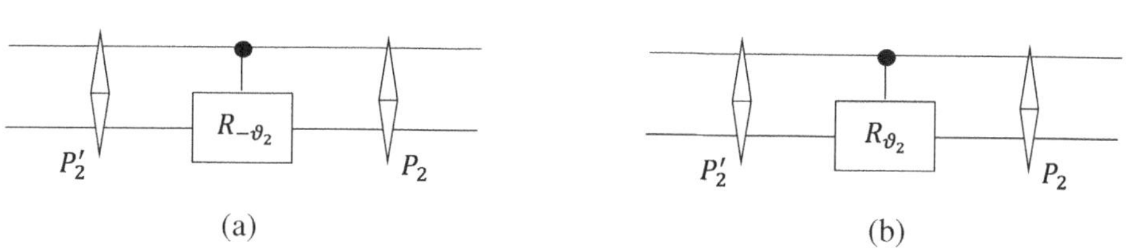

Fig. 11.4 The circuits for the gates (a) X_2 and (b) X_2' with 1-controlled gate and two permutations for each.

Therefore, the matrix of multiplication can be written as

$$A_{|\varphi_1\rangle} = \underbrace{(I_2 \oplus R_{\vartheta_3})P_2(I_2 \oplus R_{\vartheta_2})P_2(I_2 \oplus R_{\vartheta_1})P_1} \cdot \underbrace{(I_2 \oplus R_{\vartheta_1})P_2(I_2 \oplus R_{-\vartheta_2})P_2'} \cdot \underbrace{I_2 \oplus R_{-\vartheta_3}} =$$

$$\underbrace{\begin{bmatrix} 1 & 0 & 0 & 0 \\ 0 & 1 & 0 & 0 \\ 0 & 0 & \cos\vartheta_3 & -\sin\vartheta_3 \\ 0 & 0 & \sin\vartheta_3 & \cos\vartheta_3 \end{bmatrix} \begin{bmatrix} 0 & 1 & 0 & 0 \\ 0 & 0 & 1 & 0 \\ 0 & 0 & 0 & 1 \\ 1 & 0 & 0 & 0 \end{bmatrix} \begin{bmatrix} 1 & 0 & 0 & 0 \\ 0 & 1 & 0 & 0 \\ 0 & 0 & \cos\vartheta_2 & -\sin\vartheta_2 \\ 0 & 0 & \sin\vartheta_2 & \cos\vartheta_2 \end{bmatrix} \begin{bmatrix} 0 & 1 & 0 & 0 \\ 0 & 0 & 1 & 0 \\ 0 & 0 & 0 & 1 \\ 1 & 0 & 0 & 0 \end{bmatrix} \begin{bmatrix} 1 & 0 & 0 & 0 \\ 0 & 1 & 0 & 0 \\ 0 & 0 & \cos\vartheta_1 & -\sin\vartheta_1 \\ 0 & 0 & \sin\vartheta_1 & \cos\vartheta_1 \end{bmatrix} \begin{bmatrix} 0 & 0 & 1 & 0 \\ 0 & 0 & 0 & 1 \\ 1 & 0 & 0 & 0 \\ 0 & 1 & 0 & 0 \end{bmatrix}}$$

$$\underbrace{\begin{bmatrix} 1 & 0 & 0 & 0 \\ 0 & 1 & 0 & 0 \\ 0 & 0 & \cos\vartheta_1 & -\sin\vartheta_1 \\ 0 & 0 & \sin\vartheta_1 & \cos\vartheta_1 \end{bmatrix} \begin{bmatrix} 0 & 1 & 0 & 0 \\ 0 & 0 & 1 & 0 \\ 0 & 0 & 0 & 1 \\ 1 & 0 & 0 & 0 \end{bmatrix} \begin{bmatrix} 1 & 0 & 0 & 0 \\ 0 & 1 & 0 & 0 \\ 0 & 0 & \cos\vartheta_2 & \sin\vartheta_2 \\ 0 & 0 & -\sin\vartheta_2 & \cos\vartheta_2 \end{bmatrix} \begin{bmatrix} 0 & 0 & 0 & 1 \\ 1 & 0 & 0 & 0 \\ 0 & 1 & 0 & 0 \\ 0 & 0 & 1 & 0 \end{bmatrix}}$$

$$\underbrace{\begin{bmatrix} 1 & 0 & 0 & 0 \\ 0 & 1 & 0 & 0 \\ 0 & 0 & \cos\vartheta_3 & \sin\vartheta_3 \\ 0 & 0 & -\sin\vartheta_3 & \cos\vartheta_3 \end{bmatrix}}$$

The quantum scheme of the gate $A_{|\varphi_1\rangle}$ for multiplication of the 2-qubit $|\varphi_2\rangle$ by $|\varphi_1\rangle$ is given in Fig. 11.5. Six controlled gates of rotations and five permutations are used.

As follows from Eq. 11.13, one local gate of rotation can be used with four controlled gates and the above scheme can be drawn, as shown in Fig. 11.6. Four permutations are used.

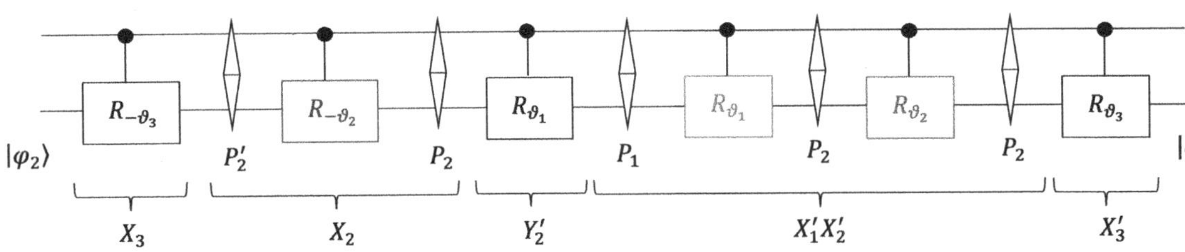

Fig. 11.5 The quantum scheme of multiplication of the 2-qubits $|\varphi_2\rangle$ and $|\varphi_1\rangle$, by using six controlled rotations.

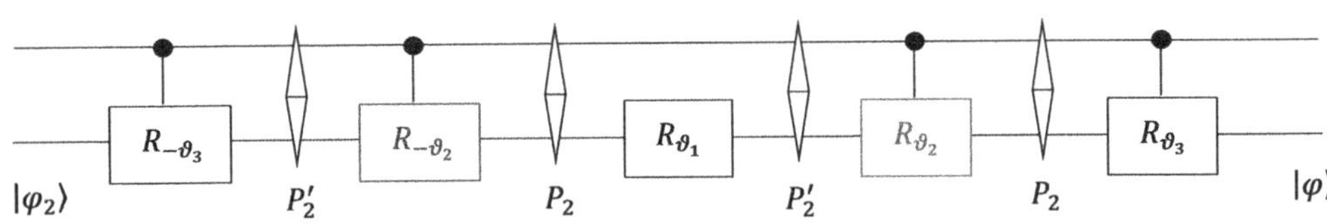

Fig. 11.6 The quantum scheme of multiplication of the 2-qubits $|\varphi_2\rangle$ and $|\varphi_1\rangle$.

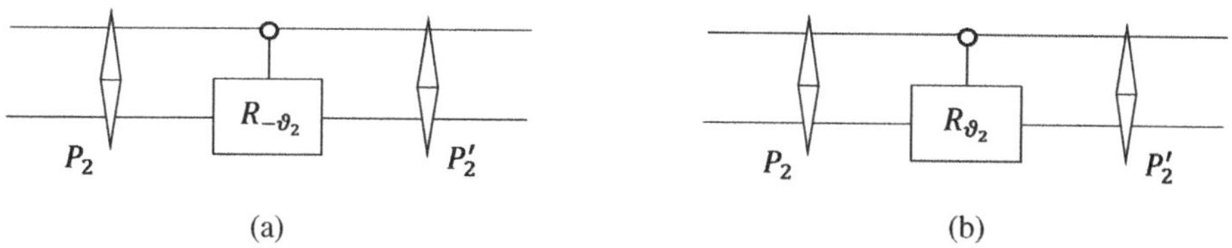

Fig. 11.7 The circuit for the gates (a) X_2 and (b) X_2' with 0-controlled gate and two permutations for each.

B) The above diagram can also be written with three 0-contolled gates. Indeed, the following representation of the matrix X_2 holds:

$$X_2 = \begin{bmatrix} 1 & 0 & 0 & 0 \\ 0 & \cos\vartheta_2 & \sin\vartheta_2 & 0 \\ 0 & -\sin\vartheta_2 & \cos\vartheta_2 & 0 \\ 0 & 0 & 0 & 1 \end{bmatrix} = \begin{bmatrix} 0 & 0 & 0 & 1 \\ 1 & 0 & 0 & 0 \\ 0 & 1 & 0 & 0 \\ 0 & 0 & 1 & 0 \end{bmatrix} \begin{bmatrix} \cos\vartheta_2 & \sin\vartheta_2 & 0 & 0 \\ -\sin\vartheta_2 & \cos\vartheta_2 & 0 & 0 \\ 0 & 0 & 1 & 0 \\ 0 & 0 & 0 & 1 \end{bmatrix} \begin{bmatrix} 0 & 1 & 0 & 0 \\ 0 & 0 & 1 & 0 \\ 0 & 0 & 0 & 1 \\ 1 & 0 & 0 & 0 \end{bmatrix},$$

or

$$X_2 = P_2'(R_{-\vartheta_2} \oplus I_2)P_2 \quad \text{and} \quad X_2' = P_2'(R_{\vartheta_2} \oplus I_2)P_2.$$

The circuit elements for these two inverse gates are shown in Fig. 11.7.
Thus,

$$\boldsymbol{A}_{|\varphi_1\rangle} = \underbrace{X_3'X_2'X_1'}\cdot\underbrace{Y_2'X_2}\cdot\underbrace{X_3} =$$

$$\underbrace{\begin{bmatrix} 1 & 0 & 0 & 0 \\ 0 & 1 & 0 & 0 \\ 0 & 0 & \cos\vartheta_3 & -\sin\vartheta_3 \\ 0 & 0 & \sin\vartheta_3 & \cos\vartheta_3 \end{bmatrix} \begin{bmatrix} 0 & 0 & 0 & 1 \\ 1 & 0 & 0 & 0 \\ 0 & 1 & 0 & 0 \\ 0 & 0 & 1 & 0 \end{bmatrix} \begin{bmatrix} \cos\vartheta_2 & -\sin\vartheta_2 & 0 & 0 \\ \sin\vartheta_2 & \cos\vartheta_2 & 0 & 0 \\ 0 & 0 & 1 & 0 \\ 0 & 0 & 0 & 1 \end{bmatrix} \begin{bmatrix} 0 & 1 & 0 & 0 \\ 0 & 0 & 1 & 0 \\ 0 & 0 & 0 & 1 \\ 1 & 0 & 0 & 0 \end{bmatrix} \begin{bmatrix} \cos\vartheta_1 & -\sin\vartheta_1 & 0 & 0 \\ \sin\vartheta_1 & \cos\vartheta_1 & 0 & 0 \\ 0 & 0 & 1 & 0 \\ 0 & 0 & 0 & 1 \end{bmatrix}}$$

$$\underbrace{\begin{bmatrix} 1 & 0 & 0 & 0 \\ 0 & 1 & 0 & 0 \\ 0 & 0 & \cos\vartheta_1 & -\sin\vartheta_1 \\ 0 & 0 & \sin\vartheta_1 & \cos\vartheta_1 \end{bmatrix} \begin{bmatrix} 0 & 0 & 0 & 1 \\ 1 & 0 & 0 & 0 \\ 0 & 1 & 0 & 0 \\ 0 & 0 & 1 & 0 \end{bmatrix} \begin{bmatrix} \cos\vartheta_2 & \sin\vartheta_2 & 0 & 0 \\ -\sin\vartheta_2 & \cos\vartheta_2 & 0 & 0 \\ 0 & 0 & 1 & 0 \\ 0 & 0 & 0 & 1 \end{bmatrix} \begin{bmatrix} 0 & 1 & 0 & 0 \\ 0 & 0 & 1 & 0 \\ 0 & 0 & 0 & 1 \\ 1 & 0 & 0 & 0 \end{bmatrix}} \underbrace{\begin{bmatrix} 1 & 0 & 0 & 0 \\ 0 & 1 & 0 & 0 \\ 0 & 0 & \cos\vartheta_3 & \sin\vartheta_3 \\ 0 & 0 & -\sin\vartheta_3 & \cos\vartheta_3 \end{bmatrix}}$$

The corresponding quantum circuit for the 4×4 gate $\boldsymbol{A}_{|\varphi_1\rangle}$ with six rotations and four permutations is shown in Fig. 11.8. Note that sequential execution of the 1- and 0-controlled R_{ϑ_1} gates is equal to two parallel local R_{ϑ_1} gates, $X_1'Y_2' = R_{\vartheta_1} \oplus R_{\vartheta_1} = I_2 \otimes R_{\vartheta_1}$.

Figure 11.9 shows the simple symbol that can be used for the 4×4 gate $\boldsymbol{A}_{|\varphi_1\rangle}$ of multiplication of two 2-qubits in quantum circuits.

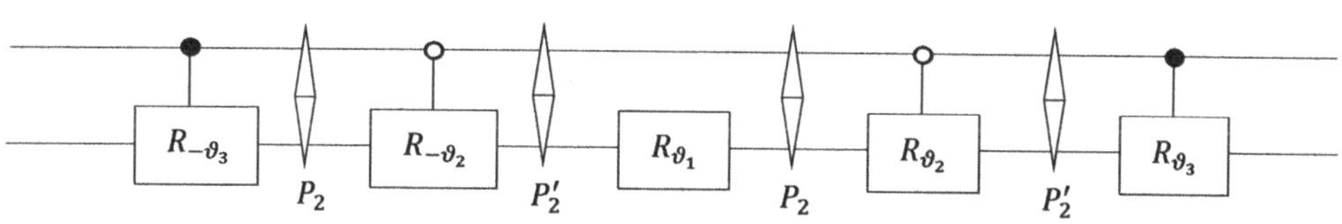

Fig. 11.8 The quantum scheme of calculation of 4 × 4 matrix by six rotations and four permutations.

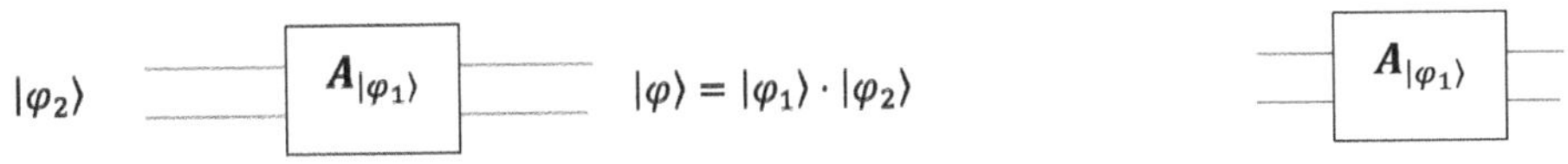

Fig. 11.9 The 2-qubit gate of multiplication of the 2-qubit $|\varphi_2\rangle$ by $|\varphi_1\rangle$.

11.2 The 4 × 4 Gate with 4 Rotations

In this section, we present a system of coordinates in the 4-D space, which allows for factorization of the matrix $\boldsymbol{A}_{|\varphi_1\rangle}$ with only four gates of rotations. Examples of such matrices which relate to different 4-point Hadamard transforms are also described.

In the subspace of unentangled 2-qubits, which can be presented as the tensor product of single qubits, we consider the following system of coordinates. Let $|\phi_1\rangle = a_{1,0}|0\rangle + a_{1,1}|1\rangle$ and $|\phi_2\rangle = a_{2,0}|0\rangle + a_{2,1}|1\rangle$ be single qubits with real amplitudes and let $|\varphi_1\rangle$ be the 2-qubit $|\varphi\rangle = a_1|00\rangle + b_1|01\rangle + c_1|10\rangle + d_1|11\rangle$ calculated as the tensor product

$$|\varphi_1\rangle = |\phi_1\rangle \otimes |\phi_2\rangle = a_{1,0}a_{2,0}|00\rangle + a_{1,0}a_{2,1}|01\rangle + a_{1,1}a_{2,0}|10\rangle + a_{1,1}a_{2,1}|11\rangle. \tag{11.16}$$

We consider the polar representation of single qubits $|\phi_1\rangle = \cos\vartheta_1|0\rangle + \sin\vartheta_1|1\rangle$ and $|\phi_2\rangle = \cos\vartheta_2|0\rangle + \sin\vartheta_2|1\rangle$. The angles in this representation are calculated by

$$\vartheta_1 = \text{sign}(b_1)\text{acos}(a_1), \quad \vartheta_2 = \text{sign}(d_1)\text{acos}(c_1). \tag{11.17}$$

Thus,

$$|\varphi_1\rangle = \cos(\vartheta_1)\cos(\vartheta_2)|00\rangle + \cos(\vartheta_1)\sin(\vartheta_2)|01\rangle + \sin(\vartheta_1)\cos(\vartheta_2)|10\rangle + \sin(\vartheta_1)\sin(\vartheta_2)|11\rangle.$$

Now, we introduce the following "Kronecker system of coordinates" in the subspace of the 4-D space:

$$a_1 = \cos(\vartheta_1)\cos(\vartheta_2), \quad b_1 = \cos(\vartheta_1)\sin(\vartheta_2), \quad c_1 = \sin(\vartheta_1)\cos(\vartheta_2), \quad d_1 = \sin(\vartheta_1)\sin(\vartheta_2).$$

It is not difficult to show that for $\boldsymbol{A}_{|\varphi_1\rangle}$ matrix factorization, we can use the DsiHT with the path shown in Fig. 11.10.

According to this block diagram, the matrix factorization is completed by four matrices of rotation as follows:

$$X_1X_2X_3X_4\boldsymbol{A}_{|\varphi_1\rangle} = I_4. \tag{11.18}$$

Here, four unitary matrices are

$$X_4 = C(R_{-\vartheta_2}) = \begin{bmatrix} 1 & 0 & 0 & 0 \\ 0 & 1 & 0 & 0 \\ 0 & 0 & \cos\vartheta_2 & \sin\vartheta_2 \\ 0 & 0 & -\sin\vartheta_2 & \cos\vartheta_2 \end{bmatrix}, \quad X_3 = C_0(R_{-\vartheta_2}) = \begin{bmatrix} \cos\vartheta_2 & \sin\vartheta_2 & 0 & 0 \\ -\sin\vartheta_2 & \cos\vartheta_2 & 0 & 0 \\ 0 & 0 & 1 & 0 \\ 0 & 0 & 0 & 1 \end{bmatrix},$$

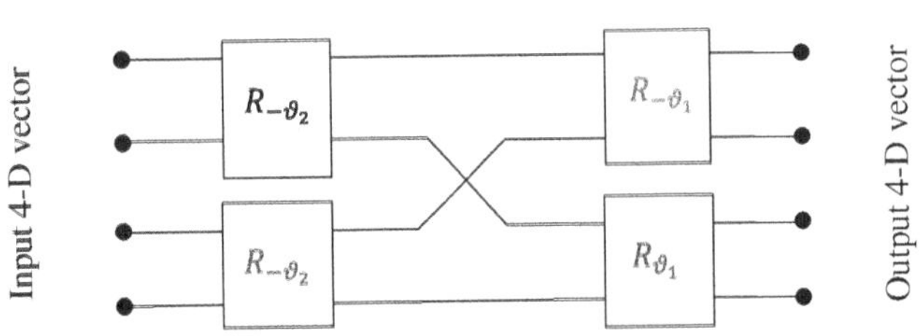

Fig. 11.10 The Transform for factorization of the matrix of multiplication of quaternions.

$$X_2 = \begin{bmatrix} 1 & 0 & 0 & 0 \\ 0 & \cos\vartheta_1 & 0 & -\sin\vartheta_1 \\ 0 & 0 & 1 & 0 \\ 0 & \sin\vartheta_1 & 0 & \cos\vartheta_1 \end{bmatrix}, \quad X_1 = \begin{bmatrix} \cos\vartheta_1 & 0 & \sin\vartheta_1 & 0 \\ 0 & 1 & 0 & 0 \\ -\sin\vartheta_1 & 0 & \cos\vartheta_1 & 0 \\ 0 & 0 & 0 & 1 \end{bmatrix}.$$

Therefore, the 4×4 matrix $\boldsymbol{A}_{|\varphi_1\rangle}$ can be written as the following:

$$\boldsymbol{A}_{|\varphi_1\rangle} = \begin{bmatrix} a_1 & -b_1 & -c_1 & -d_1 \\ b_1 & a_1 & -d_1 & c_1 \\ c_1 & d_1 & a_1 & -b_1 \\ d_1 & -c_1 & b_1 & a_1 \end{bmatrix} = X_4'X_3'X_2'X_1' = \tag{11.19}$$

$$\begin{bmatrix} 1 & 0 & 0 & 0 \\ 0 & 1 & 0 & 0 \\ 0 & 0 & \cos\vartheta_2 & -\sin\vartheta_2 \\ 0 & 0 & \sin\vartheta_2 & \cos\vartheta_2 \end{bmatrix} \begin{bmatrix} \cos\vartheta_2 & -\sin\vartheta_2 & 0 & 0 \\ \sin\vartheta_2 & \cos\vartheta_2 & 0 & 0 \\ 0 & 0 & 1 & 0 \\ 0 & 0 & 0 & 1 \end{bmatrix} \begin{bmatrix} 1 & 0 & 0 & 0 \\ 0 & \cos\vartheta_1 & 0 & \sin\vartheta_1 \\ 0 & 0 & 1 & 0 \\ 0 & -\sin\vartheta_1 & 0 & \cos\vartheta_1 \end{bmatrix} \begin{bmatrix} \cos\vartheta_1 & 0 & -\sin\vartheta_1 & 0 \\ 0 & 1 & 0 & 0 \\ \sin\vartheta_1 & 0 & \cos\vartheta_1 & 0 \\ 0 & 0 & 0 & 1 \end{bmatrix}.$$

Thus, for the case when the 2-qubit is the tensor product of single qubits, the above decomposition requires only four rotations. The corresponding scheme for multiplying the quaternion $q_2 = (a_2, b_2, c_2, d_2)$ by this matrix is given in Fig. 11.11.

The traditional symbol of a small block with an entry inside is not very convenient to use for non-adjacent inputs. Therefore, in this diagram and henceforth, the following new graphic symbol is used for the rotation gate R_{ϑ_k}:

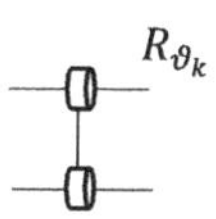

With these symbols, the diagram in Fig. 11.1 can be drawn as shown in Fig. 11.12.

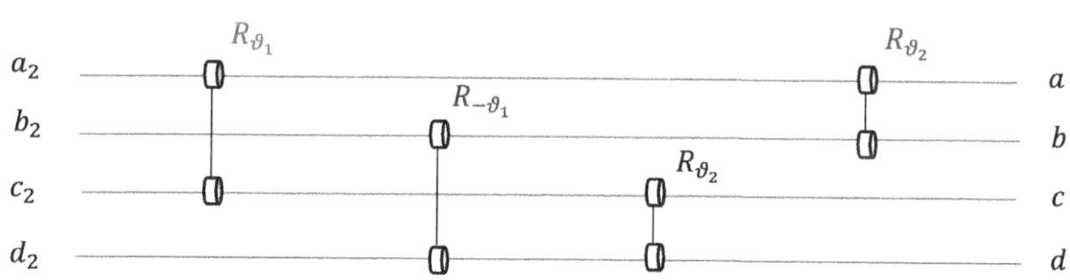

Fig. 11.11 The traditional diagram of multiplication of quaternion q_2 by q_1.

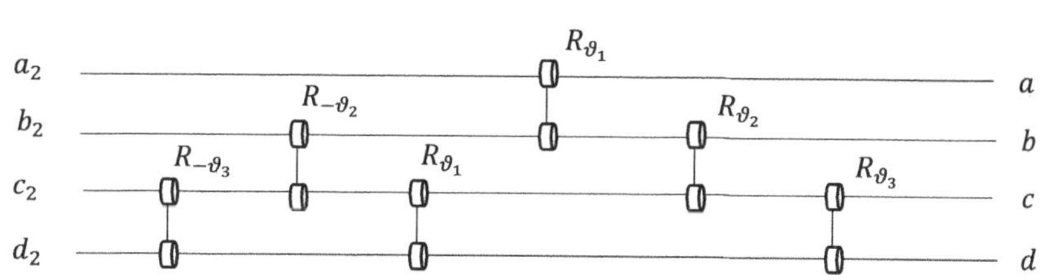

Fig. 11.12 The traditional diagram of multiplication of quaternions by six rotations.

In Eq. 11.19, the matrices X_2' and X_1' can be calculated by 1-controlled gates as follows:

$$X_2' = \begin{bmatrix} 1 & 0 & 0 & 0 \\ 0 & \cos\vartheta_1 & 0 & \sin\vartheta_1 \\ 0 & 0 & 1 & 0 \\ 0 & -\sin\vartheta_1 & 0 & \cos\vartheta_1 \end{bmatrix} = \begin{bmatrix} 1 & 0 & 0 & 0 \\ 0 & 0 & 1 & 0 \\ 0 & 1 & 0 & 0 \\ 0 & 0 & 0 & 1 \end{bmatrix} \begin{bmatrix} 1 & 0 & 0 & 0 \\ 0 & 1 & 0 & 0 \\ 0 & 0 & \cos\vartheta_1 & \sin\vartheta_1 \\ 0 & 0 & -\sin\vartheta_1 & \cos\vartheta_1 \end{bmatrix} \begin{bmatrix} 1 & 0 & 0 & 0 \\ 0 & 0 & 1 & 0 \\ 0 & 1 & 0 & 0 \\ 0 & 0 & 0 & 1 \end{bmatrix},$$

$$X_1' = \begin{bmatrix} \cos\vartheta_1 & 0 & -\sin\vartheta_1 & 0 \\ 0 & 1 & 0 & 0 \\ \sin\vartheta_1 & 0 & \cos\vartheta_1 & 0 \\ 0 & 0 & 0 & 1 \end{bmatrix} = \begin{bmatrix} 0 & 0 & 1 & 0 \\ 0 & 1 & 0 & 0 \\ 0 & 0 & 0 & 1 \\ 1 & 0 & 0 & 0 \end{bmatrix} \begin{bmatrix} 1 & 0 & 0 & 0 \\ 0 & 1 & 0 & 0 \\ 0 & 0 & \cos\vartheta_1 & -\sin\vartheta_1 \\ 0 & 0 & \sin\vartheta_1 & \cos\vartheta_1 \end{bmatrix} \begin{bmatrix} 0 & 0 & 0 & 1 \\ 0 & 1 & 0 & 0 \\ 1 & 0 & 0 & 0 \\ 0 & 0 & 1 & 0 \end{bmatrix}.$$

Thus,

$$X_2' = P_3(I_2 \oplus R_{-\vartheta_1})P_3 \quad \text{and} \quad X_1' = P_4(I_2 \oplus R_{\vartheta_1})P_4'. \tag{11.20}$$

Here, the following permutations are used: $P_4 = (0, 3, 2)$ and $P_3 = (1, 2)$, or in the matrix form

$$P_4 = \begin{bmatrix} 0 & 0 & 1 & 0 \\ 0 & 1 & 0 & 0 \\ 0 & 0 & 0 & 1 \\ 1 & 0 & 0 & 0 \end{bmatrix} \quad \text{and} \quad P_3 = \begin{bmatrix} 1 & 0 & 0 & 0 \\ 0 & 0 & 1 & 0 \\ 0 & 1 & 0 & 0 \\ 0 & 0 & 0 & 1 \end{bmatrix}.$$

Summarized the above matrix presentations, we can write the 4×4 matrix of multiplication with four controlled gates as follows:

$$A_{|\varphi_1\rangle} = \underbrace{(I_2 \oplus R_{\vartheta_2})}\underbrace{(R_{\vartheta_2} \oplus I_2)}\,\underbrace{P_3(I_2 \oplus R_{-\vartheta_1})P_3}\,\underbrace{P_4(I_2 \oplus R_{\vartheta_1})P_4'}$$

or

$$A_{|\varphi_1\rangle} = (R_{\vartheta_2} \oplus R_{\vartheta_2})P_3\underbrace{(I_2 \oplus R_{-\vartheta_1})}\,P_5\underbrace{(I_2 \oplus R_{\vartheta_1})}P_4', \tag{11.21}$$

where $R_{\vartheta_2} \oplus R_{\vartheta_2} = I_2 \otimes R_{\vartheta_2}$ and the permutation $P_5 = (0, 3, 1, 2)$ is calculated as

$$P_5 = P_3P_4 = \begin{bmatrix} 1 & 0 & 0 & 0 \\ 0 & 0 & 1 & 0 \\ 0 & 1 & 0 & 0 \\ 0 & 0 & 0 & 1 \end{bmatrix} \begin{bmatrix} 0 & 0 & 1 & 0 \\ 0 & 1 & 0 & 0 \\ 0 & 0 & 0 & 1 \\ 1 & 0 & 0 & 0 \end{bmatrix} = \begin{bmatrix} 0 & 0 & 1 & 0 \\ 0 & 0 & 0 & 1 \\ 0 & 1 & 0 & 0 \\ 1 & 0 & 0 & 0 \end{bmatrix}.$$

The corresponding quantum scheme of multiplication of 2-qubits is given in Fig. 11.13. Four controlled gates of rotations and three permutations are used.

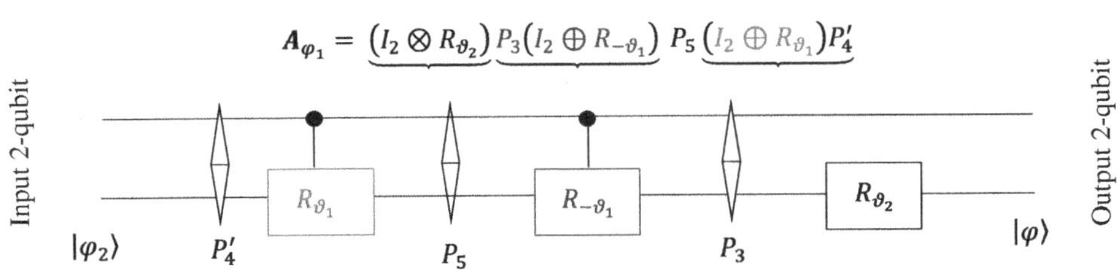

Fig. 11.13 The quantum scheme of multiplication of the 2-qubit $|\varphi_2\rangle$ by $|\varphi_1\rangle$, by using four rotations.

The 4×4 unitary matrix $A_{|\varphi_1\rangle}$ can be written in terms of angles as

$$A_{|\varphi_1\rangle}=A(\vartheta_1,\vartheta_2)=\begin{bmatrix}\cos(\vartheta_1)\cos(\vartheta_2) & -\cos(\vartheta_1)\sin(\vartheta_2) & -\sin(\vartheta_1)\cos(\vartheta_2) & -\sin(\vartheta_1)\sin(\vartheta_2)\\ \cos(\vartheta_1)\sin(\vartheta_2) & \cos(\vartheta_1)\cos(\vartheta_2) & -\sin(\vartheta_1)\sin(\vartheta_2) & \sin(\vartheta_1)\cos(\vartheta_2)\\ \sin(\vartheta_1)\cos(\vartheta_2) & \sin(\vartheta_1)\sin(\vartheta_2) & \cos(\vartheta_1)\cos(\vartheta_2) & -\cos(\vartheta_1)\sin(\vartheta_2)\\ \sin(\vartheta_1)\sin(\vartheta_2) & -\sin(\vartheta_1)\cos(\vartheta_2) & \cos(\vartheta_1)\sin(\vartheta_2) & \cos(\vartheta_1)\cos(\vartheta_2)\end{bmatrix}. \tag{11.22}$$

11.3 Examples of 12 Hadamard Matrices

In this section, a few examples of the parameterized matrix of multiplication $A_{|\varphi_1\rangle}$ are considered.

Example 11.1 Four Hadamard Matrices 4 × 4
We consider four different unitary matrices with coefficients ± 1, that is Hadamard matrices.

1) Consider the case when the angles $\vartheta_1=\vartheta_2=\pi/4$. Then, the 4×4 matrix A_{φ_1} can be written with four rotations by the angle of $\pi/4$ each as follows:

$$A(\pi/4,\pi/4)=\frac{1}{2}\begin{bmatrix}1 & -1 & -1 & -1\\ 1 & 1 & -1 & 1\\ 1 & 1 & 1 & -1\\ 1 & -1 & 1 & 1\end{bmatrix}=$$

$$\begin{bmatrix}\frac{1}{\sqrt{2}} & -\frac{1}{\sqrt{2}} & 0 & 0\\ \frac{1}{\sqrt{2}} & \frac{1}{\sqrt{2}} & 0 & 0\\ 0 & 0 & 1 & 0\\ 0 & 0 & 0 & 1\end{bmatrix}\begin{bmatrix}1 & 0 & 0 & 0\\ 0 & 1 & 0 & 0\\ 0 & 0 & \frac{1}{\sqrt{2}} & -\frac{1}{\sqrt{2}}\\ 0 & 0 & \frac{1}{\sqrt{2}} & \frac{1}{\sqrt{2}}\end{bmatrix}\begin{bmatrix}1 & 0 & 0 & 0\\ 0 & \frac{1}{\sqrt{2}} & 0 & \frac{1}{\sqrt{2}}\\ 0 & 0 & 1 & 0\\ 0 & -\frac{1}{\sqrt{2}} & 0 & \frac{1}{\sqrt{2}}\end{bmatrix}\begin{bmatrix}\frac{1}{\sqrt{2}} & 0 & -\frac{1}{\sqrt{2}} & 0\\ 0 & 1 & 0 & 0\\ \frac{1}{\sqrt{2}} & 0 & \frac{1}{\sqrt{2}} & 0\\ 0 & 0 & 0 & 1\end{bmatrix}.$$

This matrix as a 4×4 gate can be described by 4 controlled gates as follows:

$$A(\pi/4,\pi/4)=(I_2\otimes R_{\pi/4})\underbrace{P_3(I_2\oplus R_{-\pi/4})}\,P_5\underbrace{(I_2\oplus R_{\pi/4})}P'_4. \tag{11.23}$$

This matrix as the gate on 2-qubits is presented by the circuit shown in Fig. 11.14.

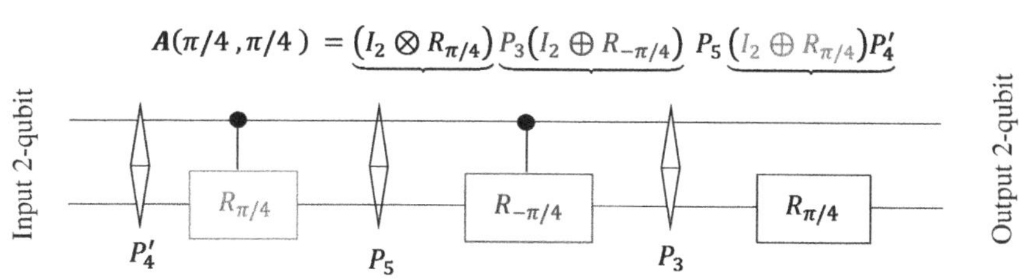

Fig. 11.14 The quantum scheme of the gate $A(\pi/4,\pi/4)$ with four rotations.

2) For another matrix with angles $\vartheta_1 = -\vartheta_2 = \pi/4$,

$$\boldsymbol{A}(\pi/4, -\pi/4) = \frac{1}{2}\begin{bmatrix} 1 & 1 & -1 & 1 \\ -1 & 1 & 1 & 1 \\ 1 & -1 & 1 & 1 \\ -1 & -1 & -1 & 1 \end{bmatrix} =$$

$$\begin{bmatrix} \frac{1}{\sqrt{2}} & \frac{1}{\sqrt{2}} & 0 & 0 \\ -\frac{1}{\sqrt{2}} & \frac{1}{\sqrt{2}} & 0 & 0 \\ 0 & 0 & 1 & 0 \\ 0 & 0 & 0 & 1 \end{bmatrix} \begin{bmatrix} 1 & 0 & 0 & 0 \\ 0 & 1 & 0 & 0 \\ 0 & 0 & \frac{1}{\sqrt{2}} & \frac{1}{\sqrt{2}} \\ 0 & 0 & -\frac{1}{\sqrt{2}} & \frac{1}{\sqrt{2}} \end{bmatrix} \begin{bmatrix} 1 & 0 & 0 & 0 \\ 0 & \frac{1}{\sqrt{2}} & 0 & \frac{1}{\sqrt{2}} \\ 0 & 0 & 1 & 0 \\ 0 & -\frac{1}{\sqrt{2}} & 0 & \frac{1}{\sqrt{2}} \end{bmatrix} \begin{bmatrix} \frac{1}{\sqrt{2}} & 0 & -\frac{1}{\sqrt{2}} & 0 \\ 0 & 1 & 0 & 0 \\ \frac{1}{\sqrt{2}} & 0 & \frac{1}{\sqrt{2}} & 0 \\ 0 & 0 & 0 & 1 \end{bmatrix}.$$

With four controlled gates of rotation, it can be written as

$$\boldsymbol{A}(\pi/4, -\pi/4) = \underbrace{(I_2 \otimes R_{-\pi/4})P_3(I_2 \oplus R_{-\pi/4})}\, P_5\underbrace{(I_2 \oplus R_{\pi/4})P_4'}. \tag{11.24}$$

and its quantum circuit is given in Fig. 11.15.

3) The Hadamard matrix with the angles $\vartheta_1 = -\vartheta_2 = -\pi/4$ is described as

$$\boldsymbol{A}(-\pi/4, \pi/4) = \frac{1}{2}\begin{bmatrix} 1 & -1 & 1 & 1 \\ 1 & 1 & 1 & -1 \\ -1 & -1 & 1 & -1 \\ -1 & 1 & 1 & 1 \end{bmatrix} =$$

$$\begin{bmatrix} \frac{1}{\sqrt{2}} & -\frac{1}{\sqrt{2}} & 0 & 0 \\ \frac{1}{\sqrt{2}} & \frac{1}{\sqrt{2}} & 0 & 0 \\ 0 & 0 & 1 & 0 \\ 0 & 0 & 0 & 1 \end{bmatrix} \begin{bmatrix} 1 & 0 & 0 & 0 \\ 0 & 1 & 0 & 0 \\ 0 & 0 & \frac{1}{\sqrt{2}} & -\frac{1}{\sqrt{2}} \\ 0 & 0 & \frac{1}{\sqrt{2}} & \frac{1}{\sqrt{2}} \end{bmatrix} \begin{bmatrix} 1 & 0 & 0 & 0 \\ 0 & \frac{1}{\sqrt{2}} & 0 & -\frac{1}{\sqrt{2}} \\ 0 & 0 & 1 & 0 \\ 0 & \frac{1}{\sqrt{2}} & 0 & \frac{1}{\sqrt{2}} \end{bmatrix} \begin{bmatrix} \frac{1}{\sqrt{2}} & 0 & \frac{1}{\sqrt{2}} & 0 \\ 0 & 1 & 0 & 0 \\ -\frac{1}{\sqrt{2}} & 0 & \frac{1}{\sqrt{2}} & 0 \\ 0 & 0 & 0 & 1 \end{bmatrix}.$$

Using four controlled gates, this matrix can be written as

$$\boldsymbol{A}(-\pi/4, \pi/4) = \underbrace{(I_2 \otimes R_{\pi/4})}\underbrace{P_3(I_2 \oplus R_{\pi/4})}\, P_5\underbrace{(I_2 \oplus R_{-\pi/4})P_4'}. \tag{11.25}$$

The quantum circuit for this Hadamard gate is given in Fig. 11.16.

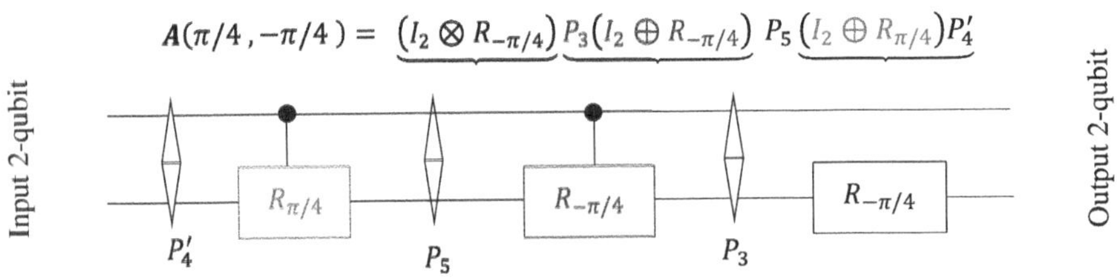

Fig. 11.15 The quantum scheme of the gate $\boldsymbol{A}(\pi/4, -\pi/4)$ with four rotations.

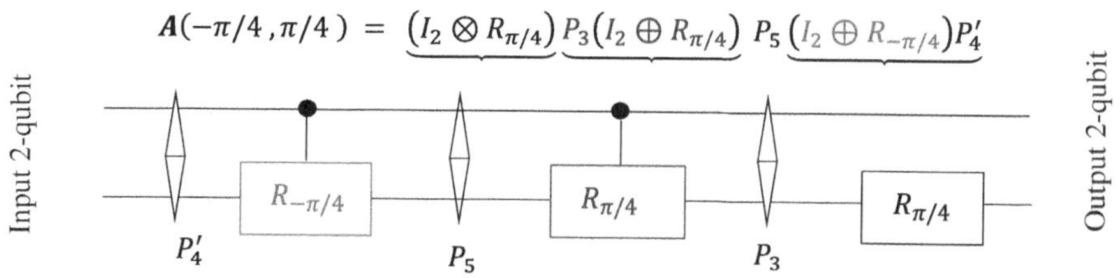

Fig. 11.16 The quantum scheme of the gate $\boldsymbol{A}(-\pi/4, \pi/4)$ with four rotations.

4) The Hadamard matrix with the angles $\vartheta_1 = \vartheta_2 = -\pi/4$ is

$$\boldsymbol{A}(-\pi/4, -\pi/4) = \frac{1}{2}\begin{bmatrix} 1 & 1 & 1 & -1 \\ -1 & 1 & -1 & -1 \\ -1 & 1 & 1 & 1 \\ 1 & 1 & -1 & 1 \end{bmatrix}$$

$$= \begin{bmatrix} \frac{1}{\sqrt{2}} & \frac{1}{\sqrt{2}} & 0 & 0 \\ -\frac{1}{\sqrt{2}} & \frac{1}{\sqrt{2}} & 0 & 0 \\ 0 & 0 & 1 & 0 \\ 0 & 0 & 0 & 1 \end{bmatrix}\begin{bmatrix} 1 & 0 & 0 & 0 \\ 0 & 1 & 0 & 0 \\ 0 & 0 & \frac{1}{\sqrt{2}} & \frac{1}{\sqrt{2}} \\ 0 & 0 & -\frac{1}{\sqrt{2}} & \frac{1}{\sqrt{2}} \end{bmatrix}\begin{bmatrix} 1 & 0 & 0 & 0 \\ 0 & \frac{1}{\sqrt{2}} & 0 & -\frac{1}{\sqrt{2}} \\ 0 & 0 & 1 & 0 \\ 0 & \frac{1}{\sqrt{2}} & 0 & \frac{1}{\sqrt{2}} \end{bmatrix}\begin{bmatrix} \frac{1}{\sqrt{2}} & 0 & \frac{1}{\sqrt{2}} & 0 \\ 0 & 1 & 0 & 0 \\ -\frac{1}{\sqrt{2}} & 0 & \frac{1}{\sqrt{2}} & 0 \\ 0 & 0 & 0 & 1 \end{bmatrix}.$$

Figure 11.17 shows the quantum circuit for the Hadamard gate with this matrix factorization

$$\boldsymbol{A}(-\pi/4, -\pi/4) = \underbrace{(I_2 \otimes R_{-\pi/4})P_3(I_2 \oplus R_{\pi/4})}\, P_5\underbrace{(I_2 \oplus R_{-\pi/4})P_4'}. \tag{11.26}$$

Here also, two controlled gates and one local rotation gate are used and the same permutations P_4', P_5, and P_3. Thus, we obtain four examples of 4×4 Hadamard matrices $\boldsymbol{A}_{|\varphi_1\rangle}$,

$$\frac{1}{2}\begin{bmatrix} 1 & -1 & -1 & -1 \\ 1 & 1 & -1 & 1 \\ 1 & 1 & 1 & -1 \\ 1 & -1 & 1 & 1 \end{bmatrix}, \quad \frac{1}{2}\begin{bmatrix} 1 & 1 & -1 & 1 \\ -1 & 1 & 1 & 1 \\ 1 & -1 & 1 & 1 \\ -1 & -1 & -1 & 1 \end{bmatrix}, \quad \frac{1}{2}\begin{bmatrix} 1 & -1 & 1 & 1 \\ 1 & 1 & 1 & -1 \\ -1 & -1 & 1 & -1 \\ -1 & 1 & 1 & 1 \end{bmatrix}, \quad \frac{1}{2}\begin{bmatrix} 1 & 1 & 1 & -1 \\ -1 & 1 & -1 & -1 \\ -1 & 1 & 1 & 1 \\ 1 & 1 & -1 & 1 \end{bmatrix}.$$

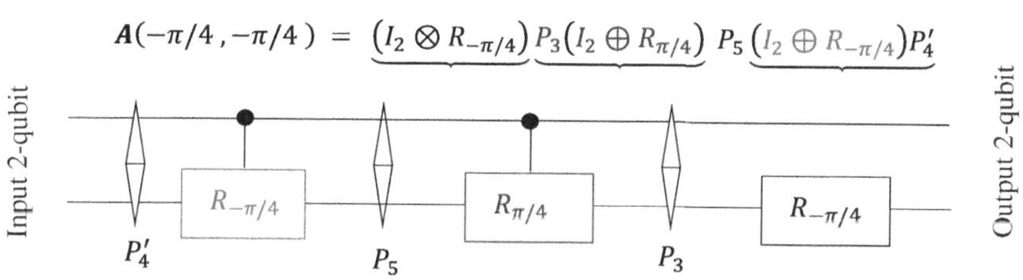

Fig. 11.17 The quantum scheme of the gate $\boldsymbol{A}(\pi/4, \pi/4)$ with four rotations.

These matrices are matrices of multiplications of 2-qubits by the following unentangled 2-qubits $|\varphi_1\rangle$:

$$|\varphi_{1;+,+}\rangle = \frac{1}{\sqrt{2}}(|0\rangle + |1\rangle)\otimes\frac{1}{\sqrt{2}}(|0\rangle + |1\rangle) = \frac{1}{2}(|00\rangle + |01\rangle + |10\rangle + |11\rangle),$$

$$|\varphi_{1;+,-}\rangle = \frac{1}{\sqrt{2}}(|0\rangle + |1\rangle)\otimes\frac{1}{\sqrt{2}}(|0\rangle - |1\rangle) = \frac{1}{2}(|00\rangle - |01\rangle + |10\rangle - |11\rangle),$$

$$|\varphi_{1,-,+}\rangle = \frac{1}{\sqrt{2}}(|0\rangle - |1\rangle)\otimes\frac{1}{\sqrt{2}}(|0\rangle + |1\rangle) = \frac{1}{2}(|00\rangle + |01\rangle - |10\rangle - |11\rangle),$$

$$|\varphi_{1,-,-}\rangle = \frac{1}{\sqrt{2}}(|0\rangle - |1\rangle)\otimes\frac{1}{\sqrt{2}}(|0\rangle - |1\rangle) = \frac{1}{2}(|00\rangle - |01\rangle - |10\rangle + |11\rangle).$$

The composition of all these unitary Hadamard matrices $\boldsymbol{A} = \boldsymbol{A}_{|\varphi_1\rangle}$ differs from the well-known Sylvester construction using the Kronecker product, or tensor product,

$$\boldsymbol{H}_4 = \boldsymbol{H}_2 \otimes \boldsymbol{H}_2 = \frac{1}{2}\begin{bmatrix} 1 & 1 & 1 & 1 \\ 1 & -1 & 1 & -1 \\ 1 & 1 & -1 & -1 \\ 1 & -1 & -1 & 1 \end{bmatrix}, \quad \det\boldsymbol{H}_4 = 1, \quad \boldsymbol{H}_4^2 = I_4.$$

This Hadamard matrix $\boldsymbol{H}_4$ is also called the matrix of the 4-point discrete Walsh-Hadamard transform. For all above 4×4 matrices $\boldsymbol{A}$, we have $\boldsymbol{AA}' = \boldsymbol{I}_4$, but $\boldsymbol{A}^2 \neq I_4$.

The Walsh-Hadamard matrix can be seen in the amplitudes of the four 2-qubits $|\varphi_{1,\pm,\pm}\rangle$ above. In other words, this matrix is composed by the first columns of four Hadamard matrices above. We write it as $\boldsymbol{H}_4 = \boldsymbol{H}_{c1}$. In addition, if we compose 4×4 matrices by using only columns number 2, 3, and 4, we obtain the following three new matrices:

$$\boldsymbol{H}_{c2} = \frac{1}{2}\begin{bmatrix} -1 & 1 & -1 & 1 \\ 1 & 1 & 1 & 1 \\ 1 & -1 & -1 & 1 \\ -1 & -1 & 1 & 1 \end{bmatrix}, \quad \boldsymbol{H}_{c3} = \frac{1}{2}\begin{bmatrix} -1 & -1 & 1 & 1 \\ -1 & 1 & 1 & -1 \\ 1 & 1 & 1 & 1 \\ 1 & -1 & 1 & -1 \end{bmatrix}, \quad \boldsymbol{H}_{c4} = \frac{1}{2}\begin{bmatrix} -1 & 1 & 1 & -1 \\ 1 & 1 & -1 & -1 \\ -1 & 1 & -1 & 1 \\ 1 & 1 & 1 & 1 \end{bmatrix}.$$

All these matrices are unitary with determinant 1. Thus, we obtain four Hadamard matrices that present the gates of multiplication of 2-qubits. Other four Hadamard matrices, including the matrix of the Walsh-Hadamard transform, are constructed from the columns of the first 4 multiplication matrices.

Comment 1

If transform the matrix $\boldsymbol{A}_{|\varphi_1\rangle}$ in Eq. 11.3 to the matrix of type $\boldsymbol{A}_{|\varphi_2\rangle}$ in Eq. 11.4,

$$\boldsymbol{T}: \begin{bmatrix} a_1 & -b_1 & -c_1 & -d_1 \\ b_1 & a_1 & -d_1 & c_1 \\ c_1 & d_1 & a_1 & -b_1 \\ d_1 & -c_1 & b_1 & a_1 \end{bmatrix} \rightarrow \begin{bmatrix} a_1 & -b_1 & -c_1 & -d_1 \\ b_1 & a_1 & d_1 & -c_1 \\ c_1 & -d_1 & a_1 & b_1 \\ d_1 & c_1 & -b_1 & a_1 \end{bmatrix}, \tag{11.27}$$

then from four 4×4 Hadamard matrices $\boldsymbol{A}(\pm\pi,\pm\pi)$ we obtain the following matrices $\boldsymbol{TA}(\pm\pi,\pm\pi)$:

$$\frac{1}{2}\begin{bmatrix} 1 & -1 & -1 & -1 \\ 1 & 1 & 1 & -1 \\ 1 & -1 & 1 & 1 \\ 1 & 1 & -1 & 1 \end{bmatrix}, \frac{1}{2}\begin{bmatrix} 1 & 1 & -1 & 1 \\ -1 & 1 & -1 & -1 \\ 1 & 1 & 1 & -1 \\ -1 & 1 & 1 & 1 \end{bmatrix}, \frac{1}{2}\begin{bmatrix} 1 & -1 & 1 & 1 \\ 1 & 1 & -1 & 1 \\ -1 & 1 & 1 & 1 \\ -1 & -1 & -1 & 1 \end{bmatrix}, \frac{1}{2}\begin{bmatrix} 1 & 1 & 1 & -1 \\ -1 & 1 & 1 & 1 \\ -1 & -1 & 1 & -1 \\ 1 & -1 & 1 & 1 \end{bmatrix}.$$

All these matrices have determinant equal 1 and they are unitary. The quantum circuits for these matrices will be described in the next section. Thus, we obtain 12 matrices of the 4-point DHdT. Eight matrices are matrices of multiplication of 2-qubits.

Comment 2

The QD-decomposition of the Walsh-Hadamard matrix by the strong DsiHTs results in the following factorization:

$$\underbrace{Z_1}\cdot\underbrace{(Y_1Y_2)}\cdot\underbrace{(X_1X_2X_3)}\boldsymbol{H}_4 = R. \tag{11.28}$$

Here, R is the 4×4 diagonal matrix diag$\{1,-1,-1,1\}$ and other six unitary matrices are

$$X_3=\begin{bmatrix}1&0&0&0\\0&1&0&0\\0&0&\cos\vartheta_3&\sin\vartheta_3\\0&0&-\sin\vartheta_3&\cos\vartheta_3\end{bmatrix},\quad X_2=\begin{bmatrix}1&0&0&0\\0&\cos\vartheta_2&\sin\vartheta_2&0\\0&-\sin\vartheta_2&\cos\vartheta_2&0\\0&0&0&1\end{bmatrix},\quad X_1=\begin{bmatrix}\cos\vartheta_1&\sin\vartheta_1&0&0\\-\sin\vartheta_1&\cos\vartheta_1&0&0\\0&0&1&0\\0&0&0&1\end{bmatrix},$$

$$Y_2=\begin{bmatrix}1&0&0&0\\0&1&0&0\\0&0&\cos\vartheta_1&-\sin\vartheta_1\\0&0&\sin\vartheta_1&\cos\vartheta_1\end{bmatrix},\quad Y_1=\begin{bmatrix}1&0&0&0\\0&\cos\vartheta_2&-\sin\vartheta_2&0\\0&\sin\vartheta_2&\cos\vartheta_2&0\\0&0&0&1\end{bmatrix},\quad Z_1=\begin{bmatrix}1&0&0&0\\0&1&0&0\\0&0&\cos\vartheta_3&\sin\vartheta_3\\0&0&-\sin\vartheta_3&\cos\vartheta_3\end{bmatrix}.$$

The angles are $\vartheta_3=-45°$, $\vartheta_2=-54.7356°$, and $\vartheta_1=-60°$. The matrix R is similar to X_2 and Y_1, but for the angle 180°. The corresponding diagram of QD-decomposition by three DsiHTs is shown in Fig. 11.18.

Therefore, the following matrix representation with seven rotation gates holds:

$$\boldsymbol{H}_4=\frac{1}{2}\begin{bmatrix}1&1&1&1\\1&-1&1&-1\\1&1&-1&-1\\1&-1&-1&1\end{bmatrix}=\underbrace{X_3'X_2'X_1'}\cdot\underbrace{Y_2'Y_1'}\cdot\underbrace{Z_1'}\cdot R. \tag{11.29}$$

The corresponding diagram of calculation by DsiHTs in the above matrix factorization is shown in Fig. 11.19.

It is important to note that Here, $Y_2X_1\neq I_2\otimes R_{\vartheta_1}$, but can be written as $Y_2X_1=C_0(Z)(I_2\otimes R_{\vartheta_1})C_0(Z)$ (see Eq. 3.24). Thus, it turned out a very complicated scheme with seven angles for $\boldsymbol{H}_4$. The corresponding quantum scheme of the Walsh-Hadamard gate is similar to the scheme given in Fig. 11.13. Only one additional gate for the diagonal matrix R should be added.

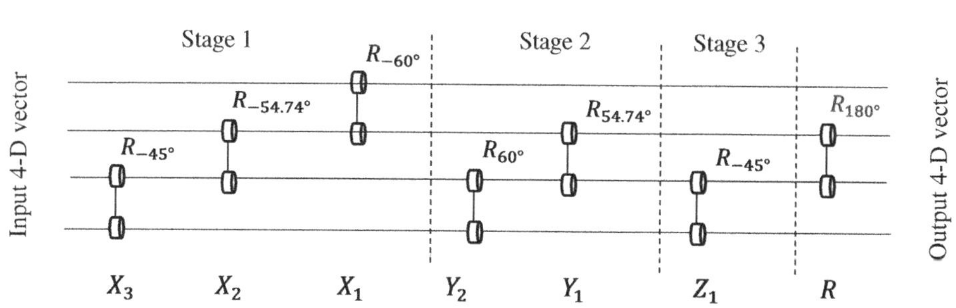

Fig. 11.18 The traditional diagram of QD-decomposition of the Walsh–Hadamard matrix.

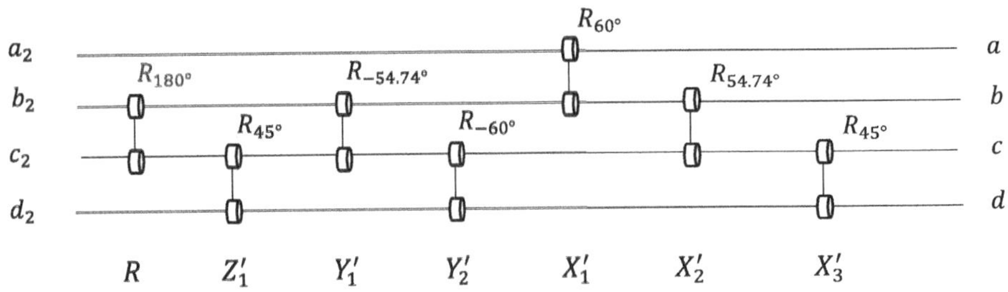

Fig. 11.19 The diagram of the Walsh–Hadamard gate H_4 by seven rotations.

11.4 The General Case: 4 × 4 Gate with 5 Rotations

In this section, we describe the circuit of the 4×4 gate of multiplication of 2-qubits in the general case, when the 2-qubit $|\varphi_1\rangle$ is not necessary be unentangled. It is shown that this gate can be calculated by maximum five gates of rotations. For that, we describe a new system of spherical coordinates in the 4-D space.

Consider the 4×4 matrix of multiplication $\boldsymbol{A} = \boldsymbol{A}_{|\varphi_1\rangle}$, or $\boldsymbol{A}_{q_1}$. It has the following block structure:

$$\boldsymbol{A} = \begin{bmatrix} A_2 & -B_2 \\ B_2 & A_2 \end{bmatrix} = \begin{bmatrix} a_1 & -b_1 & -c_1 & -d_1 \\ b_1 & a_1 & -d_1 & c_1 \\ c_1 & d_1 & a_1 & -b_1 \\ d_1 & -c_1 & b_1 & a_1 \end{bmatrix},$$

where the matrices A_2 and B_2 are similar to the 2×2 matrices of rotation,

$$A_2 = \begin{bmatrix} a_1 & -b_1 \\ b_1 & a_1 \end{bmatrix}, \qquad B_2 = \begin{bmatrix} c_1 & d_1 \\ d_1 & -c_1 \end{bmatrix} = \begin{bmatrix} c_1 & -d_1 \\ d_1 & c_1 \end{bmatrix} \begin{bmatrix} 1 & 0 \\ 0 & -1 \end{bmatrix}. \tag{11.30}$$

The following properties hold for these matrices:

1) $A_2A_2' + B_2B_2' = (a_1^2 + b_1^2 + c_1^2 + d_1^2)I_2 = I_2$,
2) $A_2B_2' - B_2A_2' = 0 \rightarrow A_2B_2' = B_2A_2' = (A_2B_2')'$,
3) $\boldsymbol{AA}' = \begin{pmatrix} A_2 & -B_2 \\ B_2 & A_2 \end{pmatrix} \begin{pmatrix} A_2' & B_2' \\ -B_2' & A_2' \end{pmatrix} = \begin{pmatrix} A_2A_2' + B_2B_2' & A_2B_2' - B_2A_2' \\ B_2A_2' - A_2B_2' & A_2A_2' + B_2B_2' \end{pmatrix} = I_4.$

These two 2×2 matrices can be written as follows:

$$\boldsymbol{A}_2 = \sqrt{a_1^2 + b_1^2} \begin{bmatrix} \dfrac{a_1}{\sqrt{a_1^2 + b_1^2}} & -\dfrac{b_1}{\sqrt{a_1^2 + b_1^2}} \\ \dfrac{b_1}{\sqrt{a_1^2 + b_1^2}} & \dfrac{a_1}{\sqrt{a_1^2 + b_1^2}} \end{bmatrix} = \sqrt{a_1^2 + b_1^2} \begin{bmatrix} \cos\vartheta_1 & -\sin\vartheta_1 \\ \sin\vartheta_1 & \cos\vartheta_1 \end{bmatrix},$$

$$\boldsymbol{B}_2 = \sqrt{c_1^2 + d_1^2} \begin{bmatrix} \dfrac{c_1}{\sqrt{c_1^2 + d_1^2}} & \dfrac{d_1}{\sqrt{c_1^2 + d_1^2}} \\ \dfrac{d_1}{\sqrt{c_1^2 + d_1^2}} & -\dfrac{c_1}{\sqrt{c_1^2 + d_1^2}} \end{bmatrix} = \sqrt{c_1^2 + d_1^2} \begin{bmatrix} \cos\vartheta_2 & \sin\vartheta_2 \\ \sin\vartheta_2 & -\cos\vartheta_2 \end{bmatrix}.$$

The angles ϑ_1 and ϑ_2 are calculated from equations

$$\cos\vartheta_1 = \frac{a_1}{\sqrt{a_1^2+b_1^2}}, \quad \sin\vartheta_1 = \frac{b_1}{\sqrt{a_1^2+b_1^2}}, \qquad \vartheta_1 \in [0, 2\pi),$$

$$\cos\vartheta_2 = \frac{c_1}{\sqrt{c_1^2+d_1^2}}, \quad \sin\vartheta_2 = \frac{d_1}{\sqrt{c_1^2+d_1^2}}, \qquad \vartheta_2 \in [0, 2\pi).$$

Because of condition $a_1^2 + b_1^2 + c_1^2 + d_1^2 = 1$, we can introduce the new angle ϑ_3,

$$\cos\vartheta_3 = \sqrt{a_1^2+b_1^2} > 0, \quad \sin\vartheta_3 = \sqrt{c_1^2+d_1^2} > 0, \quad \vartheta_3 \in (0, \pi/2).$$

Then, the above two matrices can be written as

$$\boldsymbol{A}_2 = \cos\vartheta_3 \begin{bmatrix} \cos\vartheta_1 & -\sin\vartheta_1 \\ \sin\vartheta_1 & \cos\vartheta_1 \end{bmatrix} \quad \text{and} \quad \boldsymbol{B}_2 = \sin\vartheta_3 \begin{bmatrix} \cos\vartheta_2 & \sin\vartheta_2 \\ \sin\vartheta_2 & -\cos\vartheta_2 \end{bmatrix}. \tag{11.31}$$

The matrix $\boldsymbol{A}_{|\varphi_1\rangle}$ is described by three angles ϑ_1, ϑ_2, and ϑ_3 and can be written as

$$\boldsymbol{A}_{|\varphi_1\rangle} = \boldsymbol{A}_{\alpha_1,\alpha_2,\alpha_3} = \begin{bmatrix} \cos\vartheta_3\cos\vartheta_1 & -\cos\vartheta_3\sin\vartheta_1 & -\sin\vartheta_3\cos\vartheta_2 & -\sin\vartheta_3\sin\vartheta_2 \\ \cos\vartheta_3\sin\vartheta_1 & \cos\vartheta_3\cos\vartheta_1 & -\sin\vartheta_3\sin\vartheta_2 & \sin\vartheta_3\cos\vartheta_2 \\ \sin\vartheta_3\cos\vartheta_2 & \sin\vartheta_3\sin\vartheta_2 & \cos\vartheta_3\cos\vartheta_1 & -\cos\vartheta_3\sin\vartheta_1 \\ \sin\vartheta_3\sin\vartheta_2 & -\sin\vartheta_3\cos\vartheta_2 & \cos\vartheta_3\sin\vartheta_1 & \cos\vartheta_3\cos\vartheta_1 \end{bmatrix}.$$

Therefore, we can use the following coordinate representation of amplitudes of the 2-qubit:

$$a_1 = \cos\vartheta_3\cos\vartheta_1, \quad b_1 = \cos\vartheta_3\sin\vartheta_1, \quad c_1 = \sin\vartheta_3\cos\vartheta_2, \quad d_1 = \sin\vartheta_3\sin\vartheta_2. \tag{11.32}$$

The following factorization holds for this matrix:

$$ZYX\boldsymbol{A}_{\alpha_1,\alpha_2,\alpha_3} = I_4, \tag{11.33}$$

where the matrices

$$X = X_{\vartheta_1,\vartheta_2} = \begin{bmatrix} \cos\vartheta_1 & \sin\vartheta_1 & 0 & 0 \\ -\sin\vartheta_1 & \cos\vartheta_1 & 0 & 0 \\ 0 & 0 & 1 & 0 \\ 0 & 0 & 0 & 1 \end{bmatrix} \begin{bmatrix} 1 & 0 & 0 & 0 \\ 0 & 1 & 0 & 0 \\ 0 & 0 & \cos\vartheta_2 & \sin\vartheta_2 \\ 0 & 0 & -\sin\vartheta_2 & \cos\vartheta_2 \end{bmatrix}, \tag{11.34}$$

$$Y = Y_{\vartheta_3} = \begin{bmatrix} 1 & 0 & 0 & 0 \\ 0 & \cos\vartheta_3 & 0 & -\sin\vartheta_3 \\ 0 & 0 & 1 & 0 \\ 0 & \sin\vartheta_3 & 0 & \cos\vartheta_3 \end{bmatrix} \begin{bmatrix} \cos\vartheta_3 & 0 & \sin\vartheta_3 & 0 \\ 0 & 1 & 0 & 0 \\ -\sin\vartheta_3 & 0 & \cos\vartheta_3 & 0 \\ 0 & 0 & 0 & 1 \end{bmatrix}, \tag{11.35}$$

$$Z = Z_{\vartheta_1-\vartheta_2} = \begin{bmatrix} 1 & 0 & 0 & 0 \\ 0 & 1 & 0 & 0 \\ 0 & 0 & \cos(\vartheta_2-\vartheta_1) & -\sin(\vartheta_2-\vartheta_1) \\ 0 & 0 & \sin(\vartheta_2-\vartheta_1) & \cos(\vartheta_2-\vartheta_1) \end{bmatrix}. \tag{11.36}$$

The block-diagram of the matrix factorization in Eq. 11.33 is shown in Fig. 11.20. Five rotations are used in this factorization. The angles of rotations are ϑ_1, ϑ_2, and $(\vartheta_2 - \vartheta_1)$.

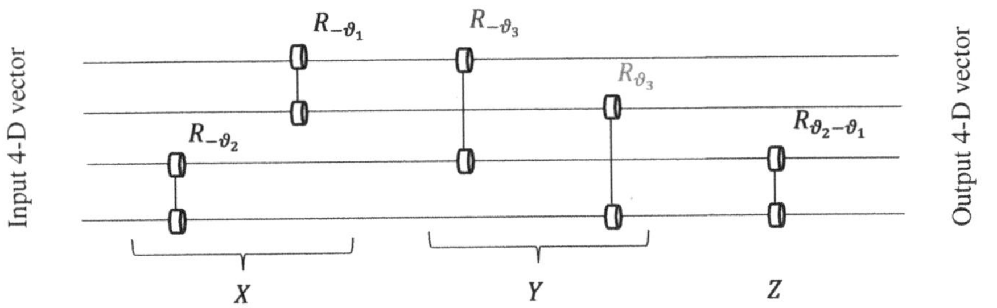

Fig. 11.20 Block diagram of factorization of the 4 × 4-multiplication matrix $\boldsymbol{A}$.

For the matrix $\boldsymbol{A}$, we obtain the following representation by five rotations:

$$\boldsymbol{A} = \boldsymbol{A}_{\alpha_1,\alpha_2,\alpha_3} = X'Y'Z' = \tag{11.37}$$

$$\begin{bmatrix} \cos\vartheta_1 & -\sin\vartheta_1 & 0 & 0 \\ \sin\vartheta_1 & \cos\vartheta_1 & 0 & 0 \\ 0 & 0 & \cos\vartheta_2 & -\sin\vartheta_2 \\ 0 & 0 & \sin\vartheta_2 & \cos\vartheta_2 \end{bmatrix} \begin{bmatrix} \cos\vartheta_3 & 0 & -\sin\vartheta_3 & 0 \\ 0 & \cos\vartheta_3 & 0 & \sin\vartheta_3 \\ \sin\vartheta_3 & 0 & \cos\vartheta_3 & 0 \\ 0 & -\sin\vartheta_3 & 0 & \cos\vartheta_3 \end{bmatrix}$$

$$\begin{bmatrix} 1 & 0 & 0 & 0 \\ 0 & 1 & 0 & 0 \\ 0 & 0 & \cos(\vartheta_1-\vartheta_2) & -\sin(\vartheta_1-\vartheta_2) \\ 0 & 0 & \sin(\vartheta_1-\vartheta_2) & \cos(\vartheta_1-\vartheta_2) \end{bmatrix}.$$

The block-diagram for multiplying the quaternion $q_2 = (a_2, b_2, c_2, d_2)$ by the matrix $\boldsymbol{A}$ is given in Fig. 11.21.

Note that the diagram in Fig. 11.11 is the particular case of the diagram above when the angles $\vartheta_2 = \vartheta_1$. In this case, the matrix $Z' = I_2 \oplus R_{\vartheta_1-\vartheta_2} = I_2 \oplus I_2 = I_4$. Only we need to write ϑ_3 as ϑ_1.

The matrix $Y' = X'_1X'_2$, where X'_2 and X'_1 are matrices similar to ones defined in Eq. 11.20 for the angle ϑ_1. With controlled gates, these matrices

$$X'_2 = \begin{bmatrix} 1 & 0 & 0 & 0 \\ 0 & \cos\vartheta_3 & 0 & \sin\vartheta_3 \\ 0 & 0 & 1 & 0 \\ 0 & -\sin\vartheta_3 & 0 & \cos\vartheta_3 \end{bmatrix} = P_3(I_2 \oplus R_{-\vartheta_3})P_3,$$

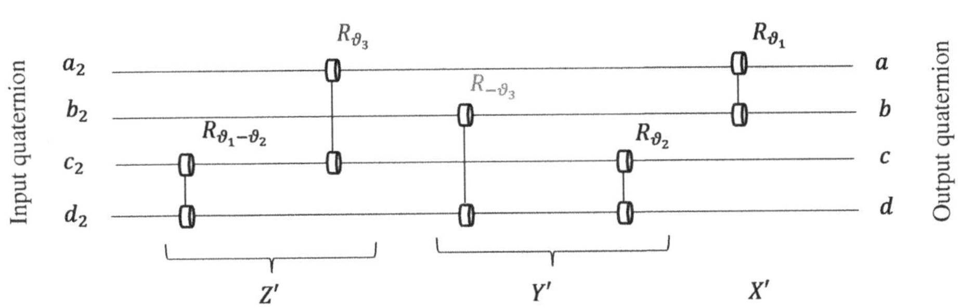

Fig. 11.21 The traditional diagram of multiplication of quaternion q_2 by q_1.

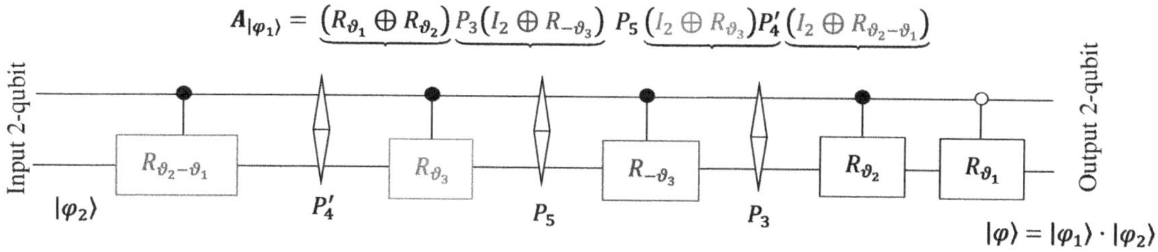

Fig. 11.22 The quantum scheme of multiplication of the 2-qubit $|\varphi_2\rangle$ by $|\varphi_1\rangle$, by using five rotations.

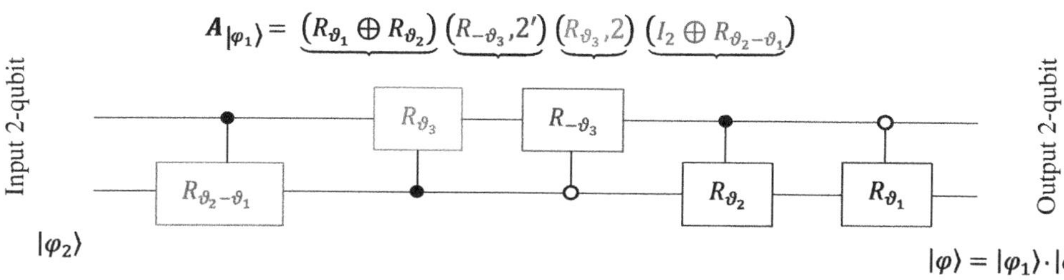

Fig. 11.23 The simplified quantum scheme of multiplication of the 2-qubit $|\varphi_2\rangle$ by $|\varphi_1\rangle$, by using five rotations.

$$X_1' = \begin{bmatrix} \cos\vartheta_3 & 0 & -\sin\vartheta_3 & 0 \\ 0 & 1 & 0 & 0 \\ \sin\vartheta_3 & 0 & \cos\vartheta_3 & 0 \\ 0 & 0 & 0 & 1 \end{bmatrix} = P_4(I_2 \oplus R_{\vartheta_3})P_4'.$$

Therefore, the 4×4 matrix $A_{|\varphi_1\rangle}$ of multiplication can be calculated as follows (considering that $P_3P_4 = P_5$):

$$A_{|\varphi_1\rangle} = \underbrace{(R_{\vartheta_1} \oplus R_{\vartheta_2})}\underbrace{P_3(I_2 \oplus R_{-\vartheta_3})}P_5\underbrace{(I_2 \oplus R_{\vartheta_3})P_4'}\underbrace{(I_2 \oplus R_{\vartheta_2-\vartheta_1})}.$$

The corresponding quantum scheme of multiplication $|\varphi\rangle = |\varphi_1\rangle|\varphi_2\rangle$ is given in Fig. 11.22. Five controlled gates of rotations and three permutations are used in calculation of this gate.

Comment 3

If we also use the second qubit as a control qubit for gates X_1' and Y', then this scheme can be simplified as shown in Fig. 11.23 (see Eq. 3.40). This scheme is without the permutations P_3, P_5, and P_4'.

11.5 Division of 2-Qubits

In this section, we describe quantum scheme for the 4×4 gate of division of 2-qubits. The division of one 2-qubit $|\varphi_1\rangle$ by another 2-qubit $|\varphi_2\rangle$ is calculated by

$$|\varphi\rangle = |\varphi_1\rangle / |\varphi_2\rangle = |\varphi_1\rangle \cdot |\overline{\varphi}_2\rangle, \tag{11.38}$$

where $|\overline{\varphi}_2\rangle$ is the conjugate 2-qubit $a_2|00\rangle - (b_2|01\rangle + c_2|10\rangle + d_2|11\rangle)$. The complex conjugate operation can be presented in matrix form as

$$T = \begin{bmatrix} 1 & & & \\ & -1 & & \\ & & -1 & \\ & & & -1 \end{bmatrix} = \begin{bmatrix} 1 & & & \\ & -1 & & \\ & & 1 & \\ & & & -1 \end{bmatrix}\begin{bmatrix} 1 & & & \\ & 1 & & \\ & & -1 & \\ & & & 1 \end{bmatrix} = (I_2 \otimes Z)(I_2 \oplus Z),$$

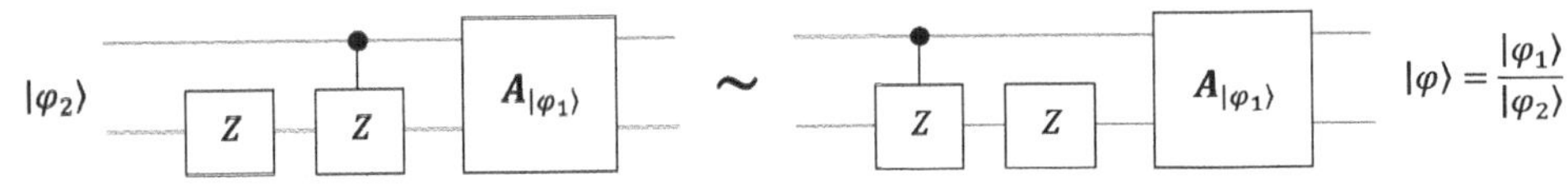

Fig. 11.24 The 2-qubit gate of multiplication of the 2-qubit $|\varphi_2\rangle$ by $|\varphi_1\rangle$.

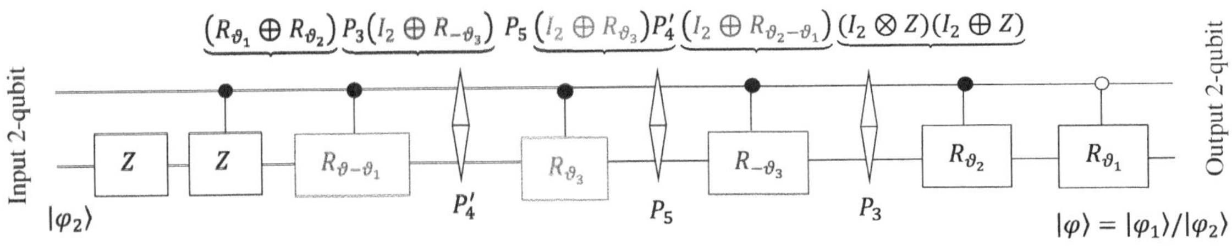

Fig. 11.25 The quantum scheme of division of the 2-qubit $|\varphi_1\rangle$ by $|\varphi_2\rangle$, by using five rotations.

or $(I_2 \oplus Z)(Z \otimes I_2)$. Therefore, one local and one controlled gates Z should be added in the beginning of each of the circuits described above, as shown in Fig. 11.24.

For example, when using the circuit with 5 rotations, which is given in Fig. 11.22, the division scheme of 2-qubits will look like that shown in Fig. 11.25. Two Z gates have been added at the beginning of the scheme.

11.6 Multiplication Circuit by 2nd 2-Qubit (A_{q_2})

In this section, new circuits are described for multiplication of 2-qubits, $|\varphi\rangle = |\varphi_1\rangle \cdot |\varphi_2\rangle$. These two 2-qubits are represented by two quaternions $q_1 = (a_1, b_1, c_1, d_1)$ and $q_2 = (a_2, b_2, c_2, d_2)$. In Section 11.1, factorization of the 4×4 multiplication matrix was accomplished by the matrix $\boldsymbol{A}_{q_1}$ generated by the first 2-qubit. Now we consider the same multiplication by the matrix $\boldsymbol{A}_{q_2}$ generated by the 2nd 2-qubit.

$$\begin{bmatrix} a \\ b \\ c \\ d \end{bmatrix} = \boldsymbol{A}_{q_2} \begin{bmatrix} a_1 \\ b_1 \\ c_1 \\ d_1 \end{bmatrix} = \begin{bmatrix} a_2 & -b_2 & -c_2 & -d_2 \\ b_2 & a_2 & d_2 & -c_2 \\ c_2 & -d_2 & a_2 & b_2 \\ d_2 & c_2 & -b_2 & a_2 \end{bmatrix} \begin{bmatrix} a_1 \\ b_1 \\ c_1 \\ d_1 \end{bmatrix} = \boldsymbol{A}_{q_1} \begin{bmatrix} a_2 \\ b_2 \\ c_2 \\ d_2 \end{bmatrix}. \tag{11.39}$$

A) Multiplication with 6 rotations

First, we consider the QD-decomposition of this matrix by 6 rotations. We will use results of the QD-decomposition of the matrix $\boldsymbol{A}_{q_1}$. The amplitudes of the 2-qubit $|\varphi_2\rangle$ can be written, as was done for the 2-qubit $|\varphi_1\rangle$, when factorizing the matrix in Eq. 11.6, in spherical system of coordinates:

$$a_2 = \cos(\vartheta_1),\ \ b_2 = \sin(\vartheta_1)\cos(\vartheta_2), c_2 = \sin(\vartheta_1)\sin(\vartheta_2)\cos(\vartheta_3),\ \ d_2 = \sin(\vartheta_1)\sin(\vartheta_2)\sin(\vartheta_3). \tag{11.40}$$

The angles are calculated by

$$\vartheta_1 = \operatorname{acos}(a_2),\ \ \vartheta_2 = \operatorname{acos}\left(\frac{b_2}{\sqrt{1-a_2^2}}\right),\qquad \vartheta_3 = \operatorname{sign}(d_2)\operatorname{acos}\left(\frac{c_2}{\sqrt{1-a_2^2-b_2^2}}\right). \tag{11.41}$$

The factorization of the 4×4 matrix of quaternion multiplication $\boldsymbol{A}_{q_2}$ by the heap transforms with the strong carriage

$$\underbrace{Z_1}\cdot\underbrace{(Y_1Y_2)}\cdot\underbrace{(X_1X_2X_3)}\boldsymbol{A}_{q_2} = I_4. \tag{11.42}$$

results in the following six matrices (see `Test_for_matrix4x4q2.m`):

$$X_3 = \begin{bmatrix} 1 & 0 & 0 & 0 \\ 0 & 1 & 0 & 0 \\ 0 & 0 & \cos\vartheta_3 & \sin\vartheta_3 \\ 0 & 0 & -\sin\vartheta_3 & \cos\vartheta_3 \end{bmatrix},\ X_2 = \begin{bmatrix} 1 & 0 & 0 & 0 \\ 0 & \cos\vartheta_2 & \sin\vartheta_2 & 0 \\ 0 & -\sin\vartheta_2 & \cos\vartheta_2 & 0 \\ 0 & 0 & 0 & 1 \end{bmatrix},\ X_1 = \begin{bmatrix} \cos\vartheta_1 & \sin\vartheta_1 & 0 & 0 \\ -\sin\vartheta_1 & \cos\vartheta_1 & 0 & 0 \\ 0 & 0 & 1 & 0 \\ 0 & 0 & 0 & 1 \end{bmatrix},$$

$$Y_2 = \begin{bmatrix} 1 & 0 & 0 & 0 \\ 0 & 1 & 0 & 0 \\ 0 & 0 & \cos\vartheta_1 & -\sin\vartheta_1 \\ 0 & 0 & \sin\vartheta_1 & \cos\vartheta_1 \end{bmatrix},\ Y_1 = \begin{bmatrix} 1 & 0 & 0 & 0 \\ 0 & \cos\vartheta_2 & -\sin\vartheta_2 & 0 \\ 0 & \sin\vartheta_2 & \cos\vartheta_2 & 0 \\ 0 & 0 & 0 & 1 \end{bmatrix},\ Z_1 = \begin{bmatrix} 1 & 0 & 0 & 0 \\ 0 & 1 & 0 & 0 \\ 0 & 0 & \cos\vartheta_3 & -\sin\vartheta_3 \\ 0 & 0 & \sin\vartheta_3 & \cos\vartheta_3 \end{bmatrix}.$$

Here, $Z_1 = X_3'$ and $Y_1 = X_2'$. The matrices in this QD-decomposition are the same as in QD-decomposition of the matrix $\boldsymbol{A}_{q_1}$, only with one difference. The rotations in the matrices Y_2 are accomplished in opposite directions, i.e., the angle ϑ_1 changes as $-\vartheta_1$. Therefore, the diagram in Fig. 11.2 must be changed, as shown in Fig. 11.26.

The matrix factorization is $\boldsymbol{A}_{q_2} = \underbrace{(X_3'X_2'X_1')}\underbrace{(Y_2'Y_1')}Z_1' = \underbrace{(X_3'X_2'X_1')}\underbrace{(Y_2'X_2)}X_3$. Therefore, the multiplication of quaternions can be described by the block-diagram shown Fig. 11.27. The input is the quaternion q_1.

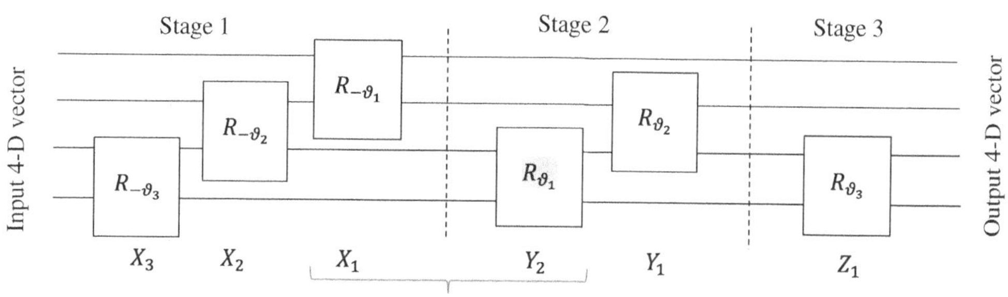

Fig. 11.26 The block diagram of $\boldsymbol{A}_{q_2}$ matrix factorization by 3 strong DsiHTs.

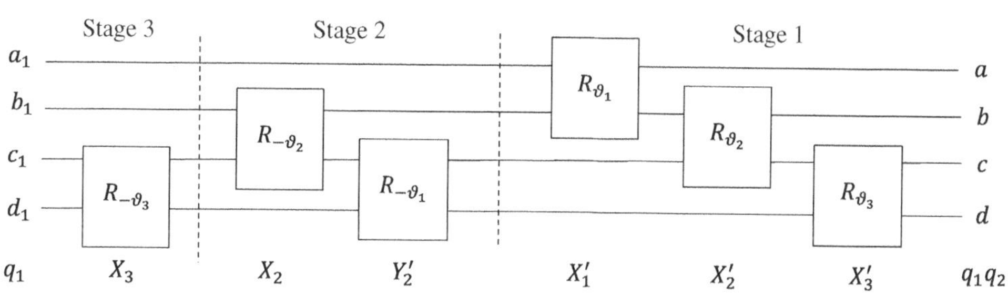

Fig. 11.27 The block diagram of multiplication of quaternions.

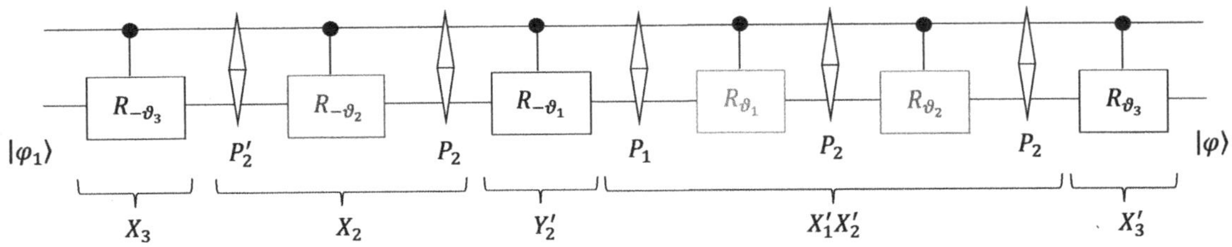

Fig. 11.28 The quantum scheme of multiplication of 2-qubits $|\varphi_2\rangle$ and $|\varphi_1\rangle$, by using 6 controlled rotations.

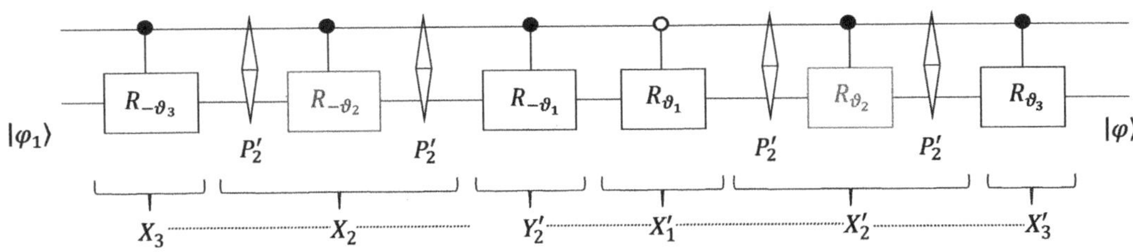

Fig. 11.29 The simplified quantum scheme of multiplication of the 2-qubits $|\varphi_2\rangle$ and $|\varphi_1\rangle$, by using 6 gates of rotation.

Considering only one change in the sign of the angle ϑ_1 (in the matrix Y_2), we can use the circuits described in Section 11.1 for the matrix $\boldsymbol{A}_{q_1}$. As an example, we use the circuit with six controlled rotations given in Fig. 11.5 and obtain the circuit for matrix $\boldsymbol{A}_{q_2}$, which is shown in Fig. 11.28. Input for this circuit is the 2-qubit $|\varphi_1\rangle$.

The scheme with six controlled gates of rotation also can be simplified, as shown in Fig. 11.29. Here, the gate X_1' is presented by the gate R_{ϑ_1} with the 0-control qubit. These two quantum schemes are considered equivalent.

B) Multiplication with 5 rotations (for A_{q_2})

The QD-decomposition of the matrix $\boldsymbol{A}_{q_2}$ can also be accomplished with only five rotations, as for the matrix $\boldsymbol{A}_{q_1}$ considered in Section 11.4. For that, we consider the 3-angle representation of the second 2-qubit

$$a_2 = \cos\vartheta_3\cos\vartheta_1, \quad b_2 = \cos\vartheta_3\sin\vartheta_1, \quad c_2 = \sin\vartheta_3\cos\vartheta_2, \quad d_2 = \sin\vartheta_3\sin\vartheta_2. \tag{11.43}$$

Here, the angles are calculated by

$$\vartheta_1 = \text{sign}(b_2)\arccos\left(\frac{a_2}{\sqrt{a_2^2+b_2^2}}\right), \vartheta_2 = \text{sign}(d_2)\arccos\left(\frac{c_2}{\sqrt{c_2^2+d_2^2}}\right), \quad \vartheta_3 = \arccos\sqrt{a_2^2+b_2^2}.$$

In this system, the matrix $\boldsymbol{A}_{q_2}$ is parameterized by 3 angles. The QD-diagonalization $ZYX\boldsymbol{A}_{q_2} = I_4$ of the matrix is described by the block-diagram in Fig. 11.30. This block-diagram is similar to one used for the matrix $\boldsymbol{A}_{q_1}$, given in Fig. 11.20. Five rotations are used in this factorization. The angles of rotations are ϑ_1, ϑ_2, and $(-\vartheta_2 - \vartheta_1)$, for the last rotation. For the matrix $\boldsymbol{A}_{q_1}$, the last rotation is performing by the angle $(\vartheta_2 - \vartheta_1)$. The difference only in the last rotation.

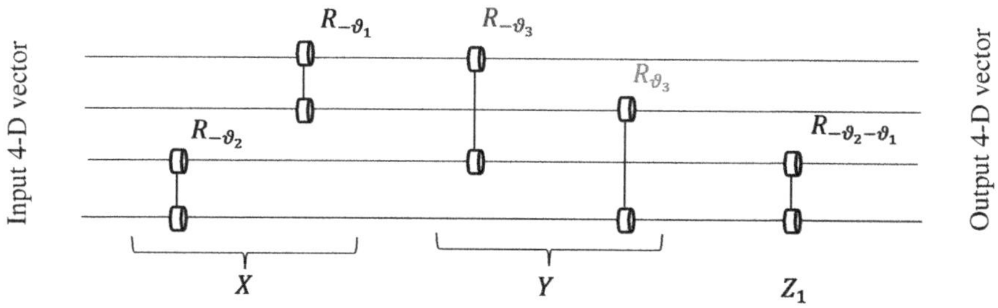

Fig. 11.30 Block diagram of factorization of the 4 × 4-multiplication matrix A_{q_2}.

Therefore, we obtain the following representation by five rotations:

$$A_{q_2} = X'Y'Z_1' = \begin{bmatrix} \cos\vartheta_1 & -\sin\vartheta_1 & 0 & 0 \\ \sin\vartheta_1 & \cos\vartheta_1 & 0 & 0 \\ 0 & 0 & \cos\vartheta_2 & -\sin\vartheta_2 \\ 0 & 0 & \sin\vartheta_2 & \cos\vartheta_2 \end{bmatrix}$$

$$\begin{bmatrix} \cos\vartheta_3 & 0 & -\sin\vartheta_3 & 0 \\ 0 & \cos\vartheta_3 & 0 & \sin\vartheta_3 \\ \sin\vartheta_3 & 0 & \cos\vartheta_3 & 0 \\ 0 & -\sin\vartheta_3 & 0 & \cos\vartheta_3 \end{bmatrix} \begin{bmatrix} 1 & 0 & 0 & 0 \\ 0 & 1 & 0 & 0 \\ 0 & 0 & \cos(\vartheta_1+\vartheta_2) & \sin(\vartheta_1+\vartheta_2) \\ 0 & 0 & -\sin(\vartheta_1+\vartheta_2) & \cos(\vartheta_1+\vartheta_2) \end{bmatrix}. \tag{11.44}$$

The block-diagram for multiplying the quaternion $q_1 = (a_1, b_1, c_1, d_1)$ by the matrix A_{q_2} is given in Fig. 11.31. Therefore, the 4×4 matrix $A_{|\varphi_2\rangle}$ of multiplication can be calculated as follows:

$$A_{|\varphi_2\rangle} = \underbrace{(R_{\vartheta_1} \oplus R_{\vartheta_2})P_3}\underbrace{(I_2 \oplus R_{-\vartheta_3})P_5}\underbrace{(I_2 \oplus R_{\vartheta_3})P_4'}\underbrace{(I_2 \oplus R_{-\vartheta_2-\vartheta_1})}.$$

The corresponding quantum scheme of multiplication $|\varphi\rangle = |\varphi_1\rangle \cdot |\varphi_2\rangle$ is given in Fig. 11.32.

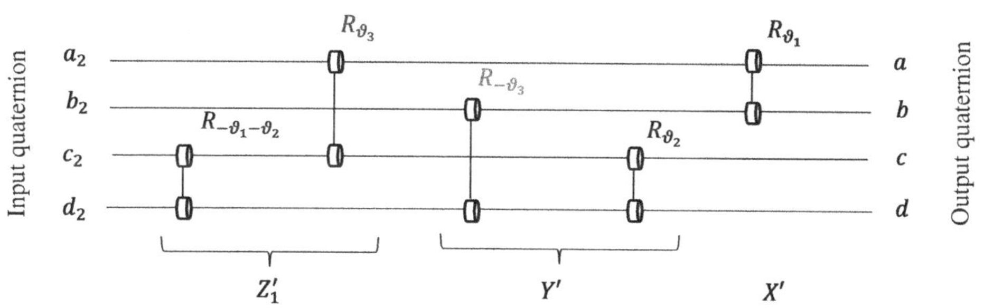

Fig. 11.31 The traditional diagram of multiplication of quaternion q_2 by q_1.

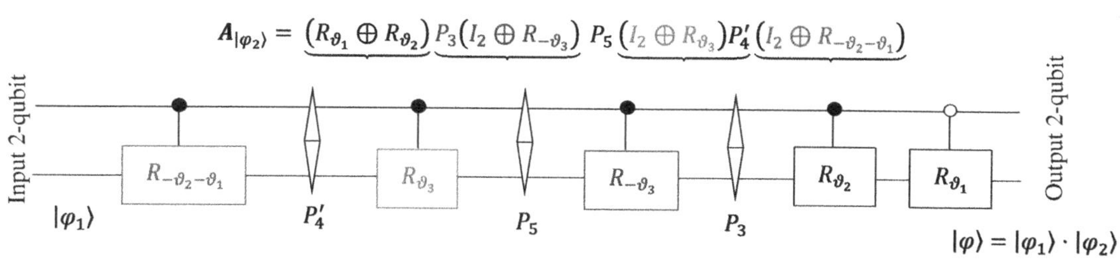

Fig. 11.32 The quantum scheme of multiplication of the 2-qubit $|\varphi_2\rangle$ by $|\varphi_1\rangle$, by using five rotations.

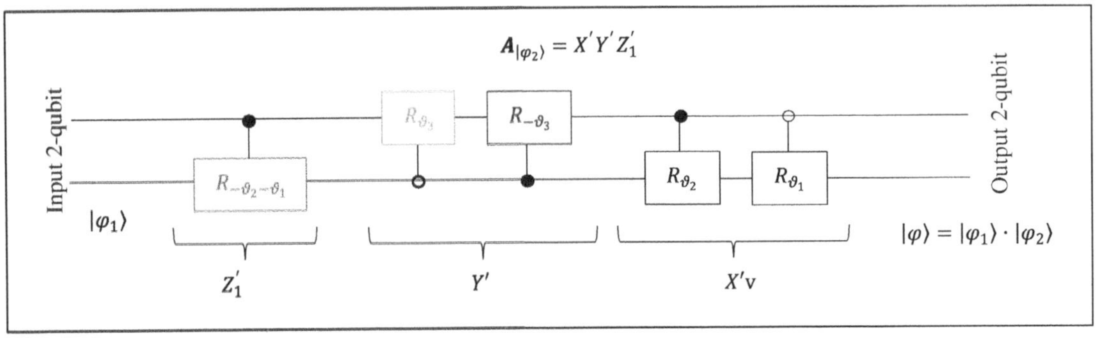

Fig. 11.33 The quantum scheme of multiplication of the 2-qubit $|\varphi_2\rangle$ by $|\varphi_1\rangle$, by using five controlled rotations.

It should be noted that for all quantum circuits in this chapter the first qubit is considered only as the control qubit. If we consider the second qubit also as a control one, or both as control ones, then the circuits can be simplified. For example, the circuit in Fig. 11.32 can be written without permutations, as shown in Fig. 11.33.

References

1 Grigoryan, A.M. and Grigoryan, M.M. (2009). *Brief Notes in Advanced DSP: Fourier Analysis with MATLAB.* Taylor and Francis Group: CRC Press.
2 Grigoryan, A.M. (2014). New method of Givens rotations for triangularization of square matrices. *Journal of Advances in Linear Algebra & Matrix Theory (ALAMT)* 4 (2): 65–78. https://doi.org/10.4236/alamt.2014.42004.
3 Grigoryan, A.M. and Agaian, S.S. (2018). Quaternion and octonion color image processing with MATLAB. *SPIE* PM279: https://doi.org/10.1117/3.2278810.
4 Agaian, S.S. (1985). Hadamard matrices and their applications. In: *Lecture Notes in Mathematics 1168*. New York: Springer - Verlag.
5 Grigoryan, A.M. and Agaian, S.S. (2003). *Multidimensional Discrete Unitary Transforms: Representation, Partitioning, and Algorithms.* New York: Marcel Dekker.

12

Quaternion Qubit Image Representation (QQIR)

Quaternion qubit image representation (QQIR) aims to provide an efficient approach for representing and manipulating color images using quaternion algebra and 2-qubits in the quantum computing paradigm. The aim is to overcome the limitations of traditional image representations by leveraging the unique properties of quaternions and quantum computing. In this section, we consider the RGB color model and let $f = \{f_{n_1,n_2}\}$ be the discrete color image of $N \times M$ pixels. Primary color components are r_{n_1,n_2}, g_{n_1,n_2}, and b_{n_1,n_2} for the red, green, and blue colors, respectively. It is assumed that the dimensions of the image are powers of 2, that is, $N = 2^r$, $r > 1$, and $M = 2^s$, $s > 1$. The color image together or without the grayscale component or intensity in the quaternion space is presented as the image $q = \{q_{n_1,n_2}\}$ defined as

$$q_{n_1,n_2} = a_{n_1,n_2} + \left(ir_{n_1,n_2} + jg_{n_1,n_2} + kb_{n_1,n_2}\right). \tag{12.1}$$

The real part of this quaternion image is the grayscale image, and color image is its imaginary part. The grayscale image is calculated as $a_{n_1,n_2} = \left(r_{n_1,n_2} + g_{n_1,n_2} + b_{n_1,n_2}\right)/3$. The brightness of the image, $\left(0.30r_{n_1,n_2} + 0.59g_{n_1,n_2} + 0.11b_{n_1,n_2}\right)$, can also be used for a_{n_1,n_2}. Such images are widely used in quaternion arithmetic and are called *quaternion images* [1, 2].

To simplify our next calculations, we consider the image as the 1D quaternion signal of length $L = NM = 2^{r+s}$, by using the mapping $q_{n_1,n_2} \to q_n = a_n + ir_n + jg_n + kb_n$, $n = n_2N + n_1$, where $n_2 = 0{:}\,(M-1)$ and $n_1 = 0{:}\,(N-1)$. At each point n, this image-signal is described by the 4D vector (a_n, r_n, g_n, b_n). With normalization, this vector represents a 2-qubit in superposition

$$|\varphi(q_n)\rangle = \frac{1}{|q_n|}(a_n|00\rangle + r_n|01\rangle + g_n|10\rangle + b_n|11\rangle). \tag{12.2}$$

Here, $|q_n| = \sqrt{a_n^2 + r_n^2 + g_n^2 + b_n^2}$, $n = 0{:}\,(L-1)$. In Section 7.2.7, the transformation 2×2 model is described, when the quaternion image is presented by the right-sided $(r+s+2)$-qubit superposition

$$|\varphi(q)\rangle = \frac{1}{\sqrt{L}}\sum_{n=0}^{L-1}\left(\frac{a_n|00\rangle + r_n|01\rangle + g_n|10\rangle + b_n|11\rangle}{\sqrt{a_n^2 + r_n^2 + g_n^2 + b_n^2}}\right)|n\rangle. \tag{12.3}$$

The left-sided superposition can also be used

$$|\psi(q)\rangle = \frac{1}{\sqrt{L}}\sum_{n=0}^{L-1}|n\rangle\left(\frac{a_n|00\rangle + r_n|01\rangle + g_n|10\rangle + b_n|11\rangle}{\sqrt{a_n^2 + r_n^2 + g_n^2 + b_n^2}}\right).$$

Quantum Image Processing in Practice: A Mathematical Toolbox, First Edition. Artyom M. Grigoryan and Sos S. Agaian.

Companion website: www.wiley.com/go/grigoryan/quantumimageprocessing

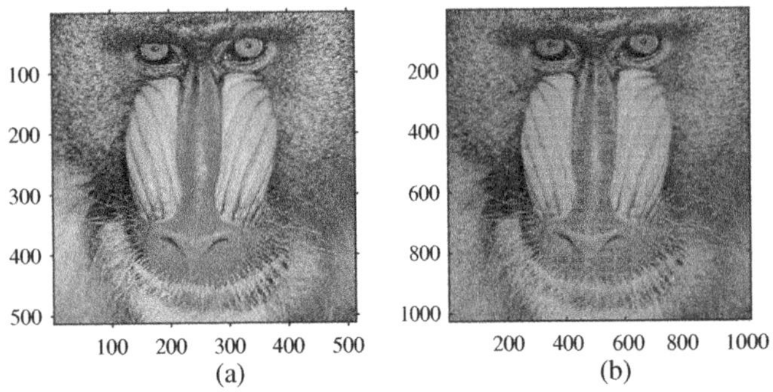

Fig. 12.1 (a) Color image "mandrill.tiff" and (b) its quaternion-in-grayscale image.

The superposition $|\varphi(q)\rangle$ is by 2-qubits and differs from the $(r+s+2)$-qubit superposition

$$|\,\psi(q)\rangle = \frac{1}{A}\sum_{n=0}^{NM-1}(a_n|00\rangle + r_n|01\rangle + g_n|10\rangle + b_n|11\rangle)|n\rangle, \tag{12.4}$$

where $A^2 = E(f)$ is the energy of the quaternion image. We can use the above representations in two ways.

1) The first is the 2×2 model with grayscale image (2×2 MGI). The quaternion image of $N\times M$ pixels can be mapped into the grayscale image of $2N\times 2M$ pixels [3, 4]. For instance, the following 2×2 model can be used:

$$q_{n,m} \to [q]_{n,m} = \begin{bmatrix} a_{n,m} & r_{n,m} \\ g_{n,m} & b_{n,m} \end{bmatrix}.$$

   As an example, Fig. 12.1 shows the color image in part (a) and its quaternion-in-grayscale image in part (b). In this model, many well-known methods of grayscale image processing can be used to obtain the analogous methods on color and quaternion images.
2) The second way is in developing new methods of quantum computation in the quaternion space. For instance, the operations of multiplication and division of 2-qubits can be used [5]. Examples of processing quaternion images are given in Section 12.2.

12.1 Model 2 for Quaternion Images

We describe a simple model for quaternion images, wherein the four colors of the quaternion image are coded in one qubit at each pixel. It should be noted that the quaternion numbers can be written as a coupled complex number

$$q = a + ib + jc + kd = (a + jc) + i(b + jd).$$

Each of these complex numbers can be written in the polar form, $(a+jc) = z_0e^{j\varphi_0}$, $(b+jd) = z_1e^{j\varphi_1}$, where z_1 and z_2 are positive real numbers. Therefore, the value $q = q_{n,m}$ of the image at pixel (n, m) can be written in the form of single qubit as

$$|\,q\rangle = \frac{z_0e^{j\varphi_0}|0\rangle + z_1e^{j\varphi_2}|1\rangle}{|q|}.$$

Here, $|z_0|^2 + |z_1|^2 = |q|^2$. Thus, the quaternion image $f_{n,m}$ can be presented by the following $(r + s + 1)$-qubit state:

$$\left|\check{f}\right\rangle = \frac{1}{\sqrt{NM}} \sum_{n=0}^{N-1} \sum_{m=0}^{M-1} \left(\frac{z_{n,m;0} e^{j\varphi_1(n,m)}|0\rangle + z_{n,m;1} e^{j\varphi_2(n,m)}|1\rangle}{|\, q_{n,m} \,|} \right) \otimes |n, m\rangle. \tag{12.5}$$

The angles $\varphi_1(n, m)$ and $\varphi_2(n, m)$ are calculated from the components of the images as

$$\left(a_{n,m} + jr_{n,m}\right) = z_{n,m;0} e^{j\varphi_1(n,m)}, \quad \left(g_{n,m} + jb_{n,m}\right) = z_{n,m;1} e^{j\varphi_1(n,m)},$$

The quaternion image requires $(r+s+1)$ qubits. The above models for representing quaternion images in quantum computing can also be used for color images in other than RGB color models, such as the CMYK and XYZ models.

12.1.1 Comments: Abstract Models with Quaternion Exponential Function

A quaternion is a number $q = a + q' = a + (bi + cj + dk)$, where a is the real part and $q' = (bi + cj + dk)$ is the three-component imaginary part. The coefficients a, b, c, and d are real. The numbers i, j, and k are imaginary units with the following multiplication laws:

$$ij = -ji = k, \quad ki = -ik = j, \quad jk = -kj = i, \quad i^2 = j^2 = k^2 = ijk = -1. \tag{12.6}$$

A quaternion number can be written as

$$q = a + q' = a + \frac{q'}{|\, q' \,|}|q'| = a + \mu\vartheta,$$

where the pure unit quaternion $\mu = q'/|q'|, \mu^2 = -1$, and the real number $\vartheta = |q'| = \sqrt{b^2 + c^2 + d^2}$ is considered as an angle. The quaternion exponent is defined as $e^q = e^a(\cos\vartheta + \mu \sin \vartheta)$.

We consider the discrete RGB color image $f = \{f_{n,m}\}$ of size $N \times N$, where $N = 2^r$, $r > 1$, as the quaternion image $q = \{q_k = a_k + ir_k + jg_k + kb_k; k = 0{:}\,(N^2 - 1)\}$. Note that for only primary colors, when $a_k = 0$, $q_k = q_k'$, and $\mu_k = q_k/|\, q_k|$, the exponent is $e^{q_k} = e^{\mu_k \vartheta_k} = \cos\vartheta_k + \mu_k \sin\vartheta_k$. At each pixel k, the image is represented by the single qubit

$$|\check{q}_k\rangle = e^{ia_k}(\cos\vartheta_k|0\rangle + \mu_k \sin\vartheta_k|1\rangle). \tag{12.7}$$

The quantum representation of the quaternion image is defined by the $(2r + 1)$-qubit superposition

$$|\check{q}\rangle = \frac{1}{N}\sum_{n=0}^{N^2-1} e^{ia_k}(\cos\vartheta_k|0\rangle + \mu_k \sin\vartheta_k|1\rangle)|k\rangle. \tag{12.8}$$

A quaternion $q = a + q'$ in the polar form $q = |\, q \,| e^{\mu\vartheta}$, where the angle $\vartheta \in [0, \pi]$. Here, $\mu = q'/|\, q' \,|, \mu^2 = -1$, and $\cos\vartheta = a/|\, q|$. The module of the quaternion is $|q| = \sqrt{a^2 + (q')^2} = \sqrt{a^2 + b^2 + c^2 + d^2}$. Therefore, any pixel value q_k of the image can be written as

$$q_k = |q_k|(\cos\vartheta_k + \mu_k \cdot \sin\vartheta_k), \qquad \vartheta_k \in [0, \pi],$$

or as the single qubit

$$|\varphi(q_k)\rangle = |q_k|(\cos\vartheta_k|0\rangle + \mu_k \cdot \sin\vartheta_k|1\rangle) = \cos\vartheta_k|0\rangle + \mu_k \cdot \sin\vartheta_k|1\rangle. \tag{12.9}$$

The quantum representation of the image is defined by

$$|\varphi(q)\rangle = \frac{1}{A}\sum_{n=0}^{N^2-1} |q_k|(\cos\vartheta_k|0\rangle + \mu_k \cdot \sin\vartheta_k|1\rangle)|k\rangle \tag{12.10}$$

with $(2r+1)$ qubits. The normalized coefficient is calculated as in Eq. 12.4.

Thus, two concepts of the quaternion and quantum representations of color images can be united to obtain the so-called quaternion-in-quantum representation of color images. The images are considered in the RGB color models, but other color models, such as the CMY(K) and XYZ, can also be used with similar quantum representations.

12.1.2 Multiplication of Colors

At each time-point n, the quaternion signal $q_n = (a_n, r_n, g_n, b_n)$ of the RGB image can be considered as the 4×4 matrix of colors. This matrix is also the 4×4 matrix of multiplication A_{q_n} described in Section 10.1. Indeed, we consider the following four matrices:

$$\boldsymbol{E} = \begin{bmatrix} 1 & 0 & 0 & 0 \\ 0 & 1 & 0 & 0 \\ 0 & 0 & 1 & 0 \\ 0 & 0 & 0 & 1 \end{bmatrix}, \boldsymbol{I} = \begin{bmatrix} 0 & -1 & 0 & 0 \\ 1 & 0 & 0 & 0 \\ 0 & 0 & 0 & -1 \\ 0 & 0 & 1 & 0 \end{bmatrix}, \boldsymbol{J} = \begin{bmatrix} 0 & 0 & -1 & 0 \\ 0 & 0 & 0 & 1 \\ 1 & 0 & 0 & 0 \\ 0 & -1 & 0 & 0 \end{bmatrix}, \boldsymbol{K} = \begin{bmatrix} 0 & 0 & 0 & -1 \\ 0 & 0 & -1 & 0 \\ 0 & 1 & 0 & 0 \\ 1 & 0 & 0 & 0 \end{bmatrix}.$$

In other words, $\boldsymbol{E} = \boldsymbol{A}_e$, $\boldsymbol{I} = \boldsymbol{A}_i$, $\boldsymbol{J} = \boldsymbol{A}_j$, and $\boldsymbol{K} = \boldsymbol{A}_k$ (see Eq. 10.4). All these matrices have determinant 1 and the multiplication laws of them are given in Table 10.1 in part (b). Then, the quaternion signal at the point n can be written as a matrix

$$\boldsymbol{q}_n = a_n\boldsymbol{E} + r_n\boldsymbol{I} + g_n\boldsymbol{J} + b_n\boldsymbol{K} = \begin{bmatrix} a_n & -r_n & -g_n & -b_n \\ r_n & a_n & b_n & -g_n \\ g_n & -b_n & a_n & r_n \\ b_n & g_n & -r_n & a_n \end{bmatrix} = \boldsymbol{A}_{q_n}.$$

In 2-qubit representation $|\varphi(q_n)\rangle$, the values of colors are normalized and we consider therefore that $|q_n| = 1$. In this case, the determinant of this matrix equals to 1 and the matrix is unitary. Thus, the 2-qubit $|\varphi(q_n)\rangle$ is represented by this color matrix at each pixel, which we call $\boldsymbol{A}_{RGB}(q_n)$. It means that in quaternion arithmetic and in the space of 2-qubits representing quaternions, the multiplication of colors is possible.

12.1.3 2-Qubit Superposition of Quaternion Images

The following $(r+s+2)$-qubit representation of the quaternion image is considered:

$$|\varphi(q)\rangle = \frac{1}{\sqrt{L}}\sum_{n=0}^{L-1} |\varphi(q_n)\rangle|n\rangle. \tag{12.11}$$

Here, $|n\rangle$ is the basis state in the space of $(r+s)$-qubit superpositions. We call this representation the quantum quaternion image representation (QQIR). The quaternion algebra allows for processing all colors at each pixel as one unit. The traditional methods process each color separately, which leads to unwanted effects in the colors, after uniting all components into a new color image. Therefore, we consider the above quantum representation of the quaternion image.

The operations of multiplication and division on quantum superpositions composed by 2-qubits are described in the following way [5]. Let $|\varphi(q_1)\rangle$ and $|\varphi(q_2)\rangle$ be two 2-qubit-based superpositions representing the quaternion image-signals q_1 and q_2 of length $L = 2^{(r+s)}$ each,

$$|\varphi(q_1)\rangle = \frac{1}{\sqrt{L}}\sum_{n=0}^{L-1}|q_{1,n}\rangle|n\rangle \quad \text{and} \quad |\varphi(q_2)\rangle = \frac{1}{\sqrt{L}}\sum_{n=0}^{L-1}|q_{2,n}\rangle|n\rangle. \tag{12.12}$$

Definition 12.1

The product of the superpositions $|\varphi(q_1)\rangle$ and $|\varphi(q_2)\rangle$ is the following $(r+s)$-2-qubit superposition:

$$|\varphi(q)\rangle = |\varphi(q_1)\rangle \cdot |\varphi(q_2)\rangle \triangleq \frac{1}{\sqrt{L}}\sum_{n=0}^{L-1}\left(|q_{1,n}\rangle \cdot |q_{2,n}\rangle\right)|n\rangle. \tag{12.13}$$

We can define, for example, the square of the quantum image as

$$|\varphi(q_1)\rangle^2 = \frac{1}{\sqrt{L}}\sum_{n=0}^{L-1}\left|q_{1,n}\right\rangle^2|n\rangle \tag{12.14}$$

and similarly define different powers and roots of this image.

Definition 12.2

The division of the superpositions $|\varphi(q_1)\rangle$ by $|\varphi(q_2)\rangle$ is the following $(r+s)$-2-qubit superposition:

$$|\varphi(q)\rangle = \frac{|\varphi(q_2)\rangle}{|\varphi(q_1)\rangle} \triangleq \frac{1}{\sqrt{L}}\sum_{n=0}^{L-1}\left(|q_{2,n}\rangle / |q_{1,n}\rangle\right)|n\rangle. \tag{12.15}$$

Note that if $q_2 = q_1$, this definition results in the QQIR in the form

$$|E\rangle = \frac{|\varphi(q_1)\rangle}{|\varphi(q_1)\rangle} = \frac{1}{\sqrt{L}}\sum_{n=0}^{L-1}\left(|q_{1,n}\rangle \cdot |\bar{q}_{1,n}\rangle\right)|n\rangle = \frac{1}{\sqrt{L}}\sum_{n=0}^{L-1}|00\rangle|n\rangle = \frac{1}{\sqrt{L}}\sum_{n=0}^{L-1}|n\rangle.$$

In the last sum, $|n\rangle$ is the state in the computation basis $\mathcal{B}_{r+s+2} = \{|0\rangle, |1\rangle, |2\rangle, \ldots, |4L-1\rangle\}$ in the space of $(r+s+2)$-qubit superpositions. This superposition is the unit with respect to the above multiplication. This means that $|\varphi(q)\rangle \cdot |E\rangle = |E\rangle \cdot |\varphi(q)\rangle = |\varphi(q)\rangle$, for any QQIR $|\varphi(q)\rangle$.

Note that for the 2-qubits, the equality $|\varphi\rangle/|\varphi\rangle = |00\rangle$ is valid. However, for superpositions with more than one 2-qubit, the equation

$$|\varphi(f_1)\rangle/|\varphi(f_2)\rangle = |0\rangle = \left|\underbrace{00\ldots0}_{r+2}\right\rangle$$

does not make sense because it is unsolvable. The role of the unit here is played by the superposition $|E\rangle$, not the basis state $|0\rangle$. Indeed, the above equation leads to the system of equations $|q_{1,0}\rangle|\bar{q}_{2,0}\rangle = |00\rangle, |q_{1,1}\rangle|\bar{q}_{2,1}\rangle = |\emptyset\rangle, |q_{1,2}\rangle|\bar{q}_{2,2}\rangle = |\emptyset\rangle, \ldots$. Except the first equality, we obtain states of impossibility for 2-qubits.

Although we describe below examples with the above operations on right-sided superpositions of quaternion images, left-sided superpositions of images can also be considered. This will change the multiplicative arithmetic on images.

Definition 12.3

The product of the left-sided superpositions $|\varphi(q_1)\rangle$ and $|\varphi(q_2)\rangle$ is the following $(r+s)$-2-qubit superposition:

$$|\varphi(q)\rangle = |\varphi(q_1)\rangle \cdot |\varphi(q_2)\rangle \triangleq \frac{1}{\sqrt{L}}\sum_{n=0}^{L-1}|n\rangle\left(|q_{1,n}\rangle \cdot |q_{2,n}\rangle\right). \tag{12.16}$$

The square of the quantum image as

$$|\varphi(q_1)\rangle^2 = \frac{1}{\sqrt{L}}\sum_{n=0}^{L-1}|n\rangle\left|q_{1,n}\right\rangle^2. \tag{12.17}$$

Definition 12.4
The division of the left-sided superpositions $|\varphi(q_1)\rangle$ by $|\varphi(q_2)\rangle$ is the following $(r+s)$-2-qubit superposition:

$$|\varphi(q)\rangle = \frac{|\varphi(q_2)\rangle}{|\varphi(q_1)\rangle} \triangleq \frac{1}{\sqrt{L}}\sum_{n=0}^{L-1}|n\rangle\left(|q_{2,n}\rangle/|q_{1,n}\rangle\right). \tag{12.18}$$

Note that if $q_2 = q_1$, this definition results in the QQIR in the form

$$|E\rangle = \frac{|\varphi(q_1)\rangle}{|\varphi(q_1)\rangle} = \frac{1}{\sqrt{L}}\sum_{n=0}^{L-1}|n\rangle\left(|q_{1,n}\rangle \cdot |\bar{q}_{1,n}\rangle\right) = \frac{1}{\sqrt{L}}\sum_{n=0}^{L-1}|n\rangle|00\rangle = \frac{1}{\sqrt{L}}\sum_{n=0}^{L-1}|4n\rangle.$$

This superposition is the unit with respect to the above multiplication.

12.2 Examples in Color Image Processing

In this section, we present a few examples of processing quaternion and color images with the above operation of multiplication of 2-qubits. First, we illustrate the concept of the square root of the image in quantum quaternion image representation.

Example 12.1 Square Root of Quantum Superposition of Color Image
Consider the quaternion image $q_n = q_{n_1,n_2}$ composed by the RGB color image f_{n_1,n_2} of size 763×524 pixels together with its grayscale image a_{n_1,n_2}. The color image is “picasso252.jpg” and its grayscale component as one quaternion image is shown in part (a) in Fig. 12.2. The dimensions of the image are not the powers of 2. Therefore, we consider the space of 1024×1024-pixel images and the 22-qubit quantum representation for this image,

$$|\varphi(q)\rangle = \frac{1}{2^{10}}\sum_{n=0}^{2^{20}-1}|\varphi(q_n)\rangle|n\rangle. \tag{12.19}$$

The square root superposition is calculated by

$$\left|\varphi\left(q_{\sqrt{2}}\right)\right\rangle \triangleq \sqrt{|\varphi(q)\rangle} = \frac{1}{2^{10}}\sum_{n=0}^{2^{20}-1}\left|q_n\right\rangle^{1/2}|n\rangle. \tag{12.20}$$

The quaternion image $q_{\sqrt{2}}$ with its gray and color components, that should correspond to this 22-qubit representation, is shown in part (b). For comparison, the quaternion image composed by the square root transform of each color component with its gray color is shown in part (c).

Figure 12.3 illustrates another example. The quaternion image of the 512×512-pixel “pepper” is shown in part (a). On this image with 20-qubit superposition $|\varphi(q)\rangle$, we applied the square operation $|\varphi(q)\rangle^2$. The quaternion image that corresponds to this square superposition is shown in part (b).

12.2.1 Grayscale-2-Quaternion Image Model

The quaternion arithmetic is also used for processing grayscale images [1]. For example, we can consider the model when grayscale images are mapped into quaternion images and then use the quantum representation of such images by 2-qubits as described above. Indeed, let us consider, for example, $N \times M$-pixel grayscale image presented as $N/2 \times M/2$-pixel quaternion image by following 2×2 model:

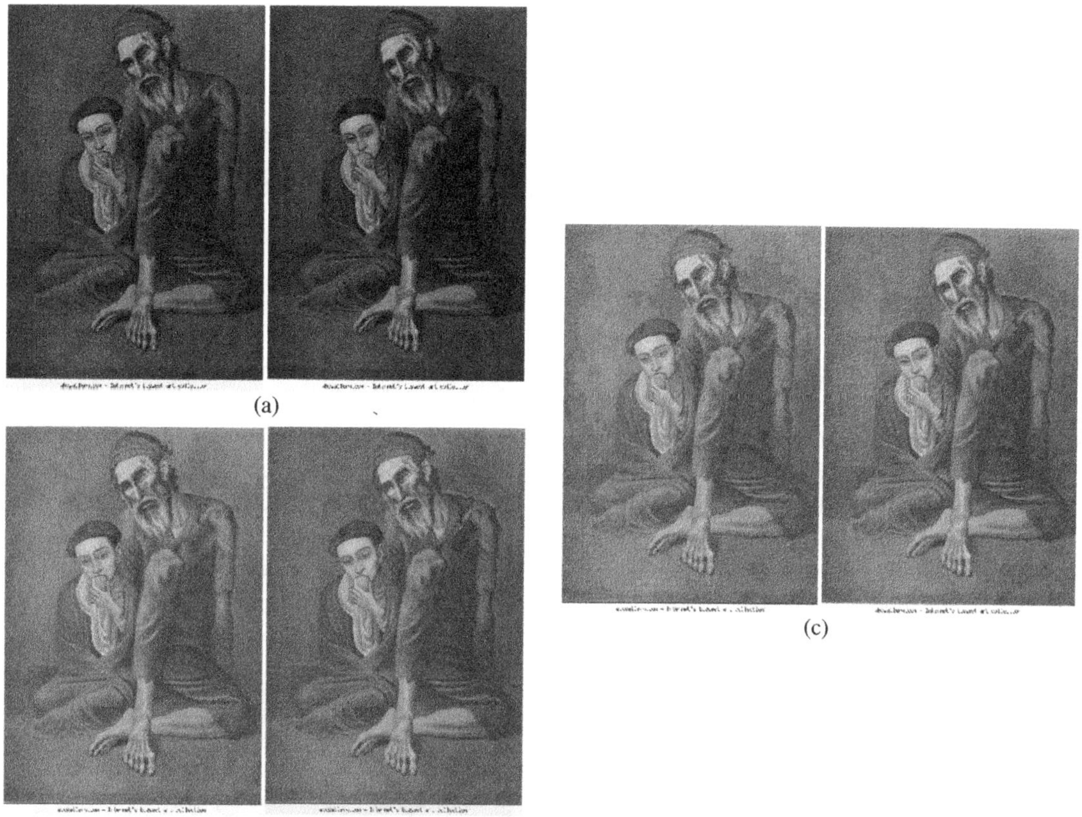

(a) (b) (c)

Fig. 12.2 The real and color components of (a) the original quaternion image, (b) the quaternion square root image, and (c) the traditional square root transform on each color component (after scaling to the same range of 256). *Source:* Grigoryan & Agaian, 2020 / IEEE.

(a) (b)

Fig. 12.3 (a) The real and color components of the quaternion image "pepper" and (b) the quaternion image of the quantum square superposition.

$$\begin{bmatrix} f_{2n,2m} & f_{2n,2m+1} \\ f_{2n+1,2m} & f_{2n+1,2m+1} \end{bmatrix} \rightarrow q_t = q_{nN/2+m} = f_{2n,2m} + \left(if_{2n,2m+1} + jf_{2n+1,2m} + kf_{2n+1,2m+1}\right), \tag{12.21}$$

where $n = 0{:}\left(\frac{N}{2}-1\right), m = 0{:}\left(\frac{M}{2}-1\right)$. Here, we assume that N and M are even numbers. The 2-qubit at point t, or state $|t\rangle$, is $|q_t\rangle = (f_{2n,2m}|0\rangle + f_{2n,2m+1}|1\rangle + f_{2n+1,\,2m}|2\rangle + f_{2n+1,\,2m+1}|3\rangle)/A(t)$. Here, $A(t)$ is the normalized coefficient. Therefore, the above quantum representation of this new quaternion image can be written as

$$|\varphi(q)\rangle = \frac{4}{NM} \sum_{t=0}^{MN/4-1} \frac{1}{A(t)} |q_t\rangle|t\rangle. \tag{12.22}$$

This quantum representation requires $((r-1)+(s-1)+2) = (r+s)$ qubits. For comparison, the flexible representation of quantum images (FRQI) uses $2r+1$ qubits, when $s = r$.

To illustrate the effectiveness of the 2×2 model, we consider the 440×750-pixel grayscale image shown in Fig. 12.4 in part (a). In the 2×2 model, this image is presented as the 220×375-pixel quaternion image; its imaginary part is shown in part (b). With powers of 2, this image can be placed in the 256×512-pixel image, which will require $(9 + 10) = 19$ qubits in the superposition $|\varphi(q)\rangle$. If at each pixel t, perform the square root of the 2-qubit $|q_t\rangle$, we obtain the superposition of the image whose imaginary part is in part (c). When back-mapping a new quaternion image from a 2×2 model, the enhanced grayscale image will be as shown in part (d).

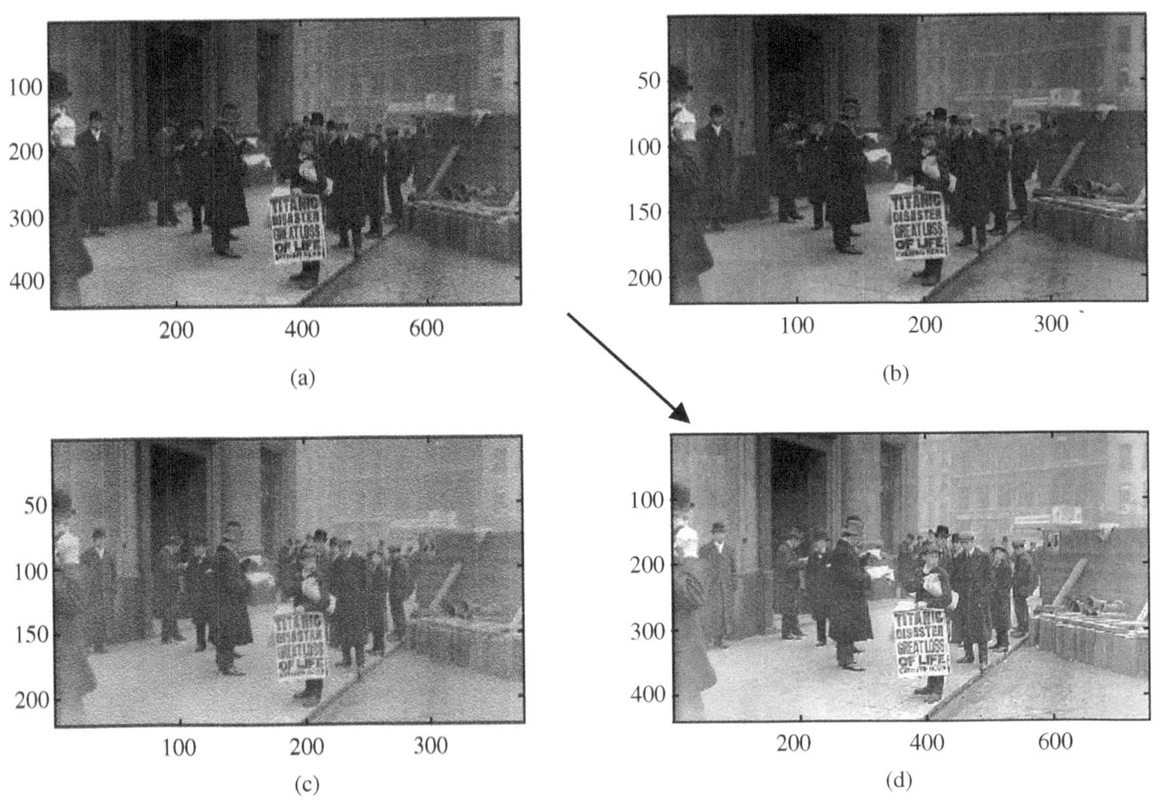

Fig. 12.4 (a) "The original 'Titanic_paperboy.jpg' grayscale image (https://en.wikipedia.org/wiki/File:Titanic_paperboy.jpg), the imaginary, or color" part of the quaternion image of half the size (b) before and (c) after processing by the quantum square root operation in the 2 × 2 model, and (d) new gray image in its original size (after multiplying intensity by 12). *Source:* Fine Art America / Wikimedia Commons / Public domain.

12.3 Quantum Quaternion Fourier Transform

In this section, we consider the concept of the 1D quantum quaternion Fourier transform (QQFT). Let $q_n = (a_n, r_n, g_n, b_n)$ be a quaternion image-signal of length $N = 2^{r+s}$. The N-point right-sided discrete quaternion Fourier transform (DQFT) of a quaternion signal q_n is calculated by [6–8]

$$Q_p = (Q_e)_p + \left[(Q_i)_p i + (Q_j)_p j + (Q_k)_p k\right] = \frac{1}{\sqrt{N}} \sum_{n=0}^{N-1} q_n W_N^{np\mu}, \quad p = 0 : (N-1). \tag{12.23}$$

The exponential kernel of the quaternion transform is

$$W_N^{np\mu} = (W_N^{\mu})^{np} = \exp\left(-\mu \frac{2\pi}{N} np\right) = \cos\left(\frac{2\pi}{N} np\right) - \mu \sin\left(\frac{2\pi}{N} np\right). \tag{12.24}$$

All coefficients $(Q_e)_p$, $(Q_i)_p$, $(Q_j)_p$, and $(Q_k)_p$ are real and μ is a pure unit quaternion $\mu = m_1 i + m_2 j + m_3 k$, $|\mu| = 1$. For instance, we can consider $\mu = (i + j + k)/\sqrt{3}$ or $\mu = j$. The DQFT is parameterized by the pure unit μ. The N-point DQFT is calculated by the corresponding 1D DFTs A_p, R_p, G_p, and B_p of the components a_n, r_n, g_n, and b_n as follows [1]:

$$\begin{pmatrix} (Q_e)_p \\ (Q_i)_p \\ (Q_j)_p \\ (Q_k)_p \end{pmatrix} = \mathrm{Re}\begin{pmatrix} A_p \\ R_p \\ G_p \\ B_p \end{pmatrix} - \begin{pmatrix} 0 & -m_1 & -m_2 & -m_3 \\ m_1 & 0 & m_3 & -m_2 \\ m_2 & -m_3 & 0 & m_1 \\ m_3 & m_2 & -m_1 & 0 \end{pmatrix} \times \mathrm{Im}\begin{pmatrix} A_p \\ R_p \\ G_p \\ B_p \end{pmatrix}. \tag{12.25}$$

Here, the 4×4-matrix $\boldsymbol{M}_\mu$ in the above equation is skew-symmetric and unitary. Re(z) and Im(z) denote the real and imaginary parts of the complex numbers z, respectively.

It is considered that signal q_n has been normalized at each point, that is, $|q_n| = 1$. The energy of the quaternion signal is preserved after transforming to the frequency domain, that is,

$$\frac{1}{N} \sum_{n=0}^{N-1} |q_n|^2 = \sum_{p=0}^{N-1} |Q_p|^2 = 1.$$

The quantum signal that corresponds to q_n is the $(r + s + 2)$-qubit superposition

$$|\varphi(q)\rangle = \frac{1}{\sqrt{N}} \sum_{n=0}^{N-1} |\varphi(q_n)\rangle |n\rangle.$$

Definition 12.5 The QQFT is defined as the $(r + s + 2)$-qubit quantum superposition with amplitudes which are values Q_p of the quaternion Fourier transform (QFT),

$$|\Phi(Q)\rangle = \frac{1}{\sqrt{N}} (Q_0|0\rangle + Q_1|1\rangle + Q_2|2\rangle + Q_3|3\rangle + \ldots + Q_{N-1}|N-1\rangle). \tag{12.26}$$

Unlike the traditional N-point QFT, the components of Q_p are not complex, but quaternion numbers. We call them amplitudes but mean the following 4D vector (or 2-qubit):

$$Q_p = \begin{pmatrix} (Q_e)_p \\ (Q_i)_p \\ (Q_j)_p \\ (Q_k)_p \end{pmatrix} \cong |\Phi(Q_p)\rangle = \frac{1}{|Q_p|} \left[(Q_e)_p|0\rangle + (Q_i)_p|1\rangle + (Q_j)_p|2\rangle + (Q_k)_p|3\rangle\right]. \tag{12.27}$$

12.4 Ideal Filters on QQIR

In this section, we analyze examples of quaternion image processing by ideal filters, such as low-pass, band-pass, and high-pass filters [5]. The processing is reduced to multiplication of the QQFT of the signal/image by the quaternion exponents at a given subset of frequency-points, and zeroing components of the transforms in the remaining frequency-points.

Unlike complex arithmetic, in the non-commutative quaternion arithmetic, linear filters as convolutions cannot be reduced to the multiplication of the QDFT of the signal Q_p and the frequency characteristic Y_p of the filter. Thus, Y_pQ_p-type filtration does not correspond to the convolution, but to another operation (see also [24]). This is why we call this operation *frequency domain filtration*. In quantum computation, to perform this operation on a 2-qubit superposition of $|\varphi(Q)\rangle$, normalization is required at each stage of computation. Therefore, we consider the case of ideal filters, when at each frequency-point p, the magnitude $|Y_p| = 1$ or 0. Non-zero values of the filter can be written as $e^{\mu\theta(p)}$, where μ is a pure unit quaternion, and $\theta = \theta(p)$ is the angle, or the phase, of the filter.

Definition 12.6 The filtered value of the QFT at frequency-point p is calculated by

$$G_p = \begin{cases} e^{\mu\theta(p)}Q_p, & \text{if } |Y_p| = 1; \\ 0, & \text{if } |Y_p| = 0, \end{cases} \tag{12.28}$$

and, in the quantum superposition, it is the 2-qubit

$$|\Phi(G_p)\rangle = \begin{cases} (\cos\theta(p)|0\rangle + \sin\theta(p)|\mu\rangle)\cdot|\Phi(Q_p)\rangle, & \text{if } |Y_p| = 1; \\ |0>, & \text{if } |Y_p| = 0, \end{cases} \tag{12.29}$$

where $|\mu\rangle = m_1|1\rangle + m_2|2\rangle + m_3|3\rangle$.

At a frequency point p at which $Y_p \neq 0$, new amplitudes of the filtered 2-qubits

$$|\Phi(G_p)\rangle = (Q_e)'_p|0\rangle + (Q_e)'_p|1\rangle + (Q_e)'_p|2\rangle + (Q_e)'_p|3\rangle = |e^{\mu\theta}\rangle\cdot|\Phi(Q_p)\rangle$$

are calculated in the matrix form as (see Eq. 12.27)

$$\begin{bmatrix}(Q_e)'_p\\(Q_i)'_p\\(Q_j)'_p\\(Q_k)'_p\end{bmatrix} = \begin{bmatrix} \cos\theta(p) & -m_1\sin\theta(p) & -m_2\sin\theta(p) & -m_3\sin\theta(p)\\ m_1\sin\theta(p) & \cos\theta(p) & -m_3\sin\theta(p) & m_2\sin\theta(p)\\ m_2\sin\theta(p) & m_3\sin\theta(p) & \cos\theta(p) & m_1\sin\theta(p)\\ m_3\sin\theta(p) & -m_2\sin\theta(p) & m_1\sin\theta(p) & \cos\theta(p)\end{bmatrix}\frac{1}{|Q_p|}\begin{bmatrix}(Q_e)_p\\(Q_i)_p\\(Q_j)_p\\(Q_k)_p\end{bmatrix}. \tag{12.30}$$

This 4×4 matrix $\boldsymbol{M}_{\theta(p)}$ is unitary and a quantum scheme for this 2-qubit operation can be developed. The QQFT of the processed quantum signal can be written as

$$|\Phi(G)\rangle = \frac{1}{\sqrt{K}}\sum_{\{p;\,Y_p\neq 0\}}|\Phi(G_p)\rangle|p\rangle. \tag{12.31}$$

The summation in this equation is only for frequency-points wherein the filter is not zero (K is the number of these points). To zero all other 2-qubits in the superposition with states $|p\rangle$, one additional qubit can be added to the superposition $|\Phi(F)\rangle$. Then, a permutation can be used to move these states $|p\rangle$ to the second part of the new superposition and then to measure the last qubit (for more detail, see the ideal filters described of single qubit-based superpositions in [9]).

12.4.1 Algorithm of Filtration $G_p = Y_pF_p$ by Ideal Filters

The characteristic of the quaternion filter Y_p has magnitudes 1 or 0. Let A be a subset of the integer interval $X_N = \{0, 1, 2, ..., N-1\}$ and $|Y_p| = 1$, if $p \in A$, and $|Y_p| = 0$, if $p \in X_N \backslash A$. Consider for simplicity of calculation that $E[q] = 1$. The N-point QFT, or the $(r+s+2)$-qubit superposition $|\Phi(Q)\rangle$ is presented by the $4N$-D vector

$$\vec{\Phi}(Q) = \Big(\underbrace{(Q_e)_0, (Q_e)_1, (Q_e)_2, ..., (Q_e)_{N-1}}, \underbrace{(Q_i)_0, (Q_i)_1, (Q_i)_2, ..., (Q_i)_{N-1}},$$

$$\underbrace{(Q_j)_0, (Q_j)_1, (Q_j)_2, ..., (Q_j)_{N-1}}, \underbrace{(Q_k)_0, (Q_k)_1, (Q_k)_2, ..., (Q_k)_{N-1}} \Big)'.$$

1) Perform the permutation $\boldsymbol{P}_1$ of the $(r+s+2)$-qubit superposition $|\Psi(Q)\rangle$, or the vector $\vec{\Phi}(Q)$ to the vector

$$\boldsymbol{P}_1\left[\vec{\Phi}(Q)\right] = \Big(\underbrace{(Q_e)_0, (Q_i)_0, (Q_j)_0, (Q_k)_0}, \underbrace{(Q_e)_1, (Q_i)_1, (Q_j)_1, (Q_k)_1},$$

$$\underbrace{(Q_e)_2, (Q_i)_2, (Q_j)_2, (Q_k)_2}, ..., \underbrace{(Q_e)_{N-1}, (Q_i)_{N-1}, (Q_j)_{N-1}, (Q_k)_{N-1}} \Big)'.$$

   In fact, this vector presents the left-sided superposition of the N-point QFT.
2) Compose the $4N \times 4N$ matrix as the Kronecker sum of the 4×4 matrices $\boldsymbol{M}_{\theta(p)}$ of multiplications

$$\boldsymbol{M}_\vartheta = \oplus_{p \in \{0,1,...,N-1\}} \boldsymbol{M}_{\vartheta(p)}.$$

   Here, $Y_p = e^{\mu\vartheta(p)}$, if $p \in A$, and $\vartheta_p = 0$, if $p \notin A$.
3) Apply this matrix on the new permuted superposition $\boldsymbol{P}_1[|\Phi(Q)\rangle]$.
4) Add one zero qubit and apply the permutation $\boldsymbol{P}_2$ of the amplitudes of the new $(r+s+3)$-qubit superposition $|0\rangle \otimes \boldsymbol{P}_1[|\Phi(Q)\rangle]$, which is defined as $p \leftrightarrow p+4N$, for all $p \notin A$.
5) Measure (M) the obtained superposition at the state when the last qubit is $|0\rangle$ and apply the inverse permutation $\boldsymbol{P}_1^{-1}$.

The result is the qubit superposition $|\Phi(G)\rangle$. The scheme for filtering the $(r+s+2)$-qubit superposition $|\Phi(Q)\rangle$ is shown in Fig. 12.5. As follows from Eq. 12.25, the quaternion QFT can be computed from the QFT, for which the quantum circuits are described in Section 9 (see also known [10–12]). This transform and its inverse are included in this scheme.

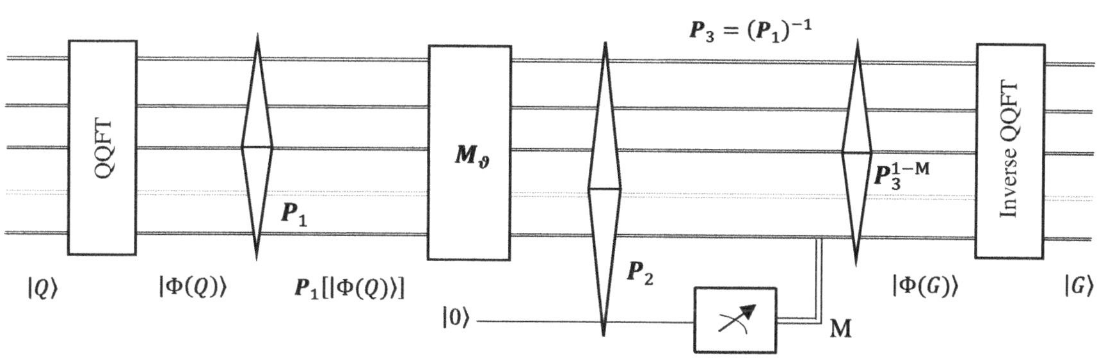

Fig. 12.5 The abstract logic scheme for filtering the quantum quaternion Fourier transform, $|\Phi(Q)\rangle \rightarrow |\Phi(G)\rangle$.

12.5 Cyclic Convolution of 2-Qubit Superpositions

Now we consider the realization of the aperiodic convolution of the quaternion image signal, by using the traditional Fourier transforms. Together with a quaternion signal $q_n = (a_n, r_n, g_n, b_n)$ of the color image, we consider a quaternion sequence h_n which is the impulse response characteristic of a linear system. This sequence is denoted by $h_n = ((h_e)_n, (h_i)_n, (h_j)_n, (h_k)_n)$. The concept of the circular linear convolution is defined as

$$\tilde{q}_n = (q * h)_n = \sum_{m=0}^{N-1} q_{n-m} h_m \triangleq \sum_{m=0}^{N-1} q_{(n-m)\mathrm{mod}N} h_m. \tag{12.32}$$

The arithmetic of the numbers of quaternions is not commutative in the (1,3)-model, $(q * h)_n \neq (h * q)_n$. Note that the quaternion convolution $(q * h)_n$ is not reduced to multiplication of QDFT, that is, $(q * h)_n \nrightarrow Q_p H_p$. It was shown in [13] that the components of the convolution $\tilde{q}_n$ can be calculated at each point n by 16 traditional aperiodic convolutions (the subscript n is omitted from the equations)

$$\left.\begin{array}{ll} \tilde{a} = a * h_e - r * h_i - g * h_j - b * h_k, & \tilde{r} = a * h_i + r * h_e - g * h_k + b * h_j \\ \tilde{g} = a * h_j + r * h_k + g * h_e - b * h_i, & \tilde{b} = a * h_k - r * h_J + g * h_i + b * h_e \end{array}\right\}. \tag{12.33}$$

Using the capital letter $\mathcal{F}$ for the 1D discrete Fourier transformation, the above system of equations in the frequency domain can be written in the matrix form as follows:

$$\begin{bmatrix} \tilde{A}_p \\ \tilde{R}_p \\ \tilde{G}_p \\ \tilde{B}_p \end{bmatrix} = \begin{bmatrix} \mathcal{F}(h_e) & -\mathcal{F}(h_i) & -\mathcal{F}(h_j) & -\mathcal{F}(h_k) \\ \mathcal{F}(h_i) & \mathcal{F}(h_e) & -\mathcal{F}(h_k) & \mathcal{F}(h_j) \\ \mathcal{F}(h_j) & \mathcal{F}(h_k) & \mathcal{F}(h_e) & -\mathcal{F}(h_i) \\ \mathcal{F}(h_k) & -\mathcal{F}(h_j) & \mathcal{F}(h_i) & \mathcal{F}(h_e) \end{bmatrix} \begin{bmatrix} A_p \\ R_p \\ G_p \\ B_p \end{bmatrix}. \tag{12.34}$$

The frequency-point $p \in \{0, 1, 2, ..., N-1\}$ was omitted from the notations. For instance, $\tilde{A}_p = \mathcal{F}(\tilde{a})_p, A_p = \mathcal{F}(a)_p$, $\mathcal{F}(r) = \mathcal{F}(r)_p$, and $\mathcal{F}(h_j) = \mathcal{F}(h_j)_p$ at the frequency-point p. It is important to note that all numbers in this matrix equation are traditional 1D DFTs and not the components of the 1D discrete quaternion FT. In this equation, the 4×4 matrix $\boldsymbol{H}$ is orthogonal. This matrix is similar to the matrix of multiplication of quaternions, only coefficients of the matrix are complex. This matrix is not unitary, but the matrix $\boldsymbol{H}\overline{\boldsymbol{H}}' = c\boldsymbol{I}$, where c is the complex number $\mathcal{F}(h_e)^2 + \mathcal{F}(h_i)^2 + \mathcal{F}(h_j)^2 + \mathcal{F}(h_k)^2$.

We have a fundamental formula for the cyclic convolution in quaternion algebra. If we assumed that the frequency characteristic of the 4-component filter H_p are known or given, then only computation of four traditional 1-D DFTs A_p, R_p, G_p, and B_p is required. It is assumed that the realization of multiplication in Eq. 12.34 in quantum computation is possible. The matrix in this equation is similar to the matrix of 2-qubit multiplication given in Eq. 10.3, for which the quantum circuits have been described in Section 11. Then, the convoluted quaternion signal $\tilde{q}_n = \tilde{a}_n + \tilde{r}_n i + \tilde{g}_n j + \tilde{b}_n k$ can be calculated as $\tilde{a}_n = \mathcal{F}^{-1}(\tilde{A}_p)$, $\tilde{r}_n = \mathcal{F}^{-1}(\tilde{R}_p)$, $\tilde{g}_n = \mathcal{F}^{-1}(\tilde{G}_p)$, and $\tilde{b}_n = \mathcal{F}^{-1}(\tilde{B}_p)$. Here, $\mathcal{F}^{-1}$ stands for the inverse N-point DFT.

12.6 Windowed Convolution

In this section, the concept of linear convolution (LC) is described in more detail. Examples of these operations are image and signal smoothing and gradient operations. An implementation of 1D linear convolution in a quantum computer is presented on several examples. For that, a new paired transform-based quantum representation and

Fig. 12.6 The image convoluted with the matrix.

computation of 1D and 2D convolutions and gradients are described. The 1D convolutions and gradient operators with short-length masks such as [1 −2 1], [1 2 2 1], [1 2 0 −2 −1], [1 2 −6 2 1], and [1 1 −4 1 1] are considered. These operators are widely used in image processing, such as edge detection by different gradient operators [14–16]. In these examples, we show how to choose the quantum representation of the signal for computing the convolution; each case is unique and considered separately.

Many methods in image processing systems $\{f_{n,m}\} \rightarrow$ [Image Processing System] $\rightarrow \{g_{n,m}\}$ can be described by the linear operation which is called the 2D linear convolution. This is an operation of the image with the characteristic of the system that completely describes the system. This characteristic can be represented as a matrix h. The convolution operation is denoted by "$*$" and $g_{n,m} = (f * h)_{n,m}$. We also say that the image g is the image f convoluted with h, as shown in Fig. 12.6. The knowledge of the matrix h allows for calculating responses of the system.

In general, the convoluted image g is of a size larger than f, but in many applications, this operation is calculated only for the pixels of the original image f.

The windowed convolution is the basic operation in imaging, including image enhancement, filtration, edge detection, and face recognition. The convolution is defined by a given matrix h of coefficients $h_{k,l}$ with a specified location of nonzero coefficients, which represents a window W. The window W may be of the form of a cross, square, rectangle, and circle of small and large sizes. If this window has the form of the cross, then nonzero coefficients are only in the middle column and row. When convoluting the image, this window is translated across the image pixel by pixel. One of the elements of the window is considered to be the center and the corresponding coefficient is denoted by $h_{0,\,0}$; usually it is the central coefficient in the matrix. In the convolution, this matrix first is rotated (by 180°), as $h_{k,l} \rightarrow \hat{h}_{k,l} = h_{-k,-l}$. Then, the matrix is put on the image or we can say translated at the pixel (n, m), and the sum of products of the matrix coefficients $h_{k,l}$ and the corresponding values of the image $f_{n-k,\,m-l}$ in the translated window is calculated. This sum is the value of the convolution at the pixel (n, m). Now, we consider a simple example of the convolution.

Example 12.2 Convolution by Window 2 × 3

Let us consider the following image as the 3 × 5 matrix:

$$f = \begin{bmatrix} \underline{f_{0,0}} & f_{0,1} & f_{0,2} & f_{0,3} & f_{0,4} \\ f_{1,0} & f_{1,1} & f_{1,2} & f_{1,3} & f_{1,4} \\ f_{20} & f_{2,1} & f_{2,2} & f_{2,3} & f_{2,4} \end{bmatrix} = \begin{bmatrix} \underline{2} & 1 & 2 & 3 & 4 \\ 1 & 4 & 9 & 5 & 5 \\ 2 & 6 & 7 & 8 & 4 \end{bmatrix}$$

and

$$h = \frac{1}{9}\begin{bmatrix} h_{-1,0} & h_{-1,1} & h_{-1,2} \\ \underline{h_{0,0}} & h_{0,1} & h_{0,2} \end{bmatrix} = \frac{1}{9}\begin{bmatrix} 2 & 1 & 2 \\ \underline{1} & 2 & 1 \end{bmatrix}.$$

Here, the underline coefficient shows the position of the original (0,0).

First, the 2×3 matrix h is rotated (by 180°), as

$$h = \frac{1}{9}\begin{bmatrix} 2 & 1 & 2 \\ \underline{1} & 2 & 1 \end{bmatrix} \rightarrow \hat{h} = \frac{1}{9}\begin{bmatrix} 1 & 2 & \underline{1} \\ 2 & 1 & 2 \end{bmatrix}. \tag{12.35}$$

To calculate the convolution at pixel (1, 2), the calculation is performed as

$$\begin{bmatrix} \underline{2} & 1 & 2 & 3 & 4 \\ 1 \cdot 1 & 4 \cdot 2 & 9 \cdot 1 & 5 & 5 \\ 2 \cdot 2 & 6 \cdot 1 & 7 \cdot 2 & 8 & 4 \end{bmatrix} \rightarrow \frac{1}{9}(1 + 8 + 9 + 4 + 6 + 14) = 4.7 = g_{1,2}.$$

Here, the matrix $\hat{h}$ is translated at pixel (1, 2) at which the image intensity is 9. In this translation, the center of the matrix is moved to the pixel (1, 2) in the image. Then, the six products are added and divided by 9; the result 4.7 is the value of the convolution at pixel (1, 2). At the last pixel (2, 4), the translated matrix covers only three pixels of the image and the calculation of the convolution is performed as

$$\begin{bmatrix} \underline{2} & 1 & 2 & 3 & 4 \\ 1 & 4 & 9 & 5 & 5 \\ 2 & 6 & 7 \cdot 1 & 8 \cdot 2 & 4 \cdot 1 \\ & & 0 \cdot 2 & 0 \cdot 1 & 0 \cdot 2 \end{bmatrix} \rightarrow \frac{1}{9}(7 + 16 + 4) = 3 = g_{2,4}.$$

Continuing the calculation at other pixels of the image, we obtain the convoluted image,

$$\begin{bmatrix} \underline{2} & 1 & 2 & 3 & 4 \\ 1 & 4 & 9 & 5 & 5 \\ 2 & 6 & 7 & 8 & 4 \end{bmatrix} * \frac{1}{9}\begin{bmatrix} 2 & 1 & 2 \\ \underline{1} & 2 & 1 \end{bmatrix} = \frac{1}{9}\begin{bmatrix} \underline{4} & 14 & 30 & 35 & 45 \\ 5 & 20 & 42 & 62 & 54 \\ 2 & 10 & 21 & 28 & 27 \end{bmatrix}.$$

The windowing near the frame boundary of the image requires extending the image to the large image, by adding the required number of pixels with zero intensity, as shown below

$$f \rightarrow \hat{f} = \begin{bmatrix} \underline{0} & 0 & 2 & 1 & 2 & 3 & 4 \\ 0 & 0 & 1 & 4 & 9 & 5 & 5 \\ 0 & 0 & 2 & 6 & 7 & 8 & 4 \\ 0 & 0 & 0 & 0 & 0 & 0 & 0 \end{bmatrix}. \tag{12.36}$$

This operation is called *the zero padding*. Two zero columns and one zero row were added to the image, and the new image is processed only at pixels of the original image. The zero-padded image $\hat{f}$ is of size 4×7, the original image f is of size 3×5, and the matrix h has the size 2×3. One can note that along the first dimension, $4 = (3 + 2) - 1$, and along the second dimension, $7 = (5 + 3) - 1$.

As we can see, in general, the convolution is a rather complex operation, and it will be much more difficult to perform it on a quantum computer. Therefore, we will use simple symmetrical masks, or windows W, of size 3×3, 5×5, ... This will facilitate the task of constructing algorithms for calculating the convolution on a quantum computer.

Example 12.3 Convolution by the Cross Window

Let us consider the 3×5 image

$$f = \begin{bmatrix} \underline{f_{0,0}} & f_{0,1} & f_{0,2} & f_{0,3} & f_{0,4} \\ f_{1,0} & f_{1,1} & f_{1,2} & f_{1,3} & f_{1,4} \\ f_{20} & f_{2,1} & f_{2,2} & f_{2,3} & f_{2,4} \end{bmatrix} = \begin{bmatrix} \underline{2} & 1 & 2 & 3 & 2 \\ 1 & 2 & 10 & 1 & 2 \\ 2 & 3 & 2 & 2 & 1 \end{bmatrix}$$

and the 3×3 matrix with the cross-window W,

$$h = \frac{1}{5}\begin{bmatrix} h_{-1,-1} & h_{-1,0} & h_{-1,1} \\ h_{0,-1} & \underline{h_{0,0}} & h_{0,1} \\ h_{1,-1} & h_{1,0} & h_{1,1} \end{bmatrix} = \frac{1}{5}\begin{bmatrix} & 1 & \\ 1 & \underline{1} & 1 \\ & 1 & \end{bmatrix}.$$

It can be assumed that the noisy pixel with intensity 10 occurs in the pixel (1, 2) in the image f, and our goal is to suppress this noise from the image. First, we perform the zero padding of the image

$$f \rightarrow \hat{f} = \begin{bmatrix} 0 & 0 & 0 & 0 & 0 & 0 & 0 \\ 0 & \underline{2} & 1 & 2 & 3 & 2 & 0 \\ 0 & 1 & 2 & 10 & 1 & 2 & 0 \\ 0 & 2 & 3 & 2 & 2 & 1 & 0 \\ 0 & 0 & 0 & 0 & 0 & 0 & 0 \end{bmatrix}.$$

The indexes in the matrix h run in both directions from -1 to 1, therefore, one row will be added on the top and one on the bottom, and one column from the right and one column from the left for the zero-padded image $\hat{f}$.

The matrix h is symmetric and does not require the rotation by 180°, that is, the rotated matrix $\hat{h} = h$. The value of the convolution at the first pixel (0, 0) is calculated as shown below,

$$\begin{bmatrix} 0 & 0 & 0 & 0 & 0 & 0 & 0 \\ 0 & \underline{2} & 1 & 2 & 3 & 2 & 0 \\ 0 & 1 & 2 & 10 & 1 & 2 & 0 \\ 0 & 2 & 3 & 2 & 2 & 1 & 0 \\ 0 & 0 & 0 & 0 & 0 & 0 & 0 \end{bmatrix} \rightarrow \frac{1}{5}(2 + 1 + 1) = 0.8 = g_{0,0}.$$

The calculation of convolution results in the following image:

$$\begin{bmatrix} \underline{2} & 1 & 2 & 3 & 2 \\ 1 & 2 & 10 & 1 & 2 \\ 2 & 3 & 2 & 2 & 1 \end{bmatrix} * \frac{1}{5}\begin{bmatrix} & 1 & \\ 1 & \underline{1} & 1 \\ & 1 & \end{bmatrix} = \frac{1}{5}\begin{bmatrix} \underline{4} & 7 & 16 & 8 & 7 \\ 7 & 17 & 17 & 18 & 6 \\ 6 & 9 & 17 & 6 & 5 \end{bmatrix}.$$

and if we round this convoluted image, we obtain the image

$$[g] = \begin{bmatrix} \underline{1} & 1 & 3 & 2 & 1 \\ 1 & 3 & 3 & 4 & 1 \\ 1 & 2 & 3 & 1 & 1 \end{bmatrix}.$$

Comparing this convoluted image with the image f, one can note that the noise at pixel (1, 2) has been suppressed.

(a) image (b) convoluted image

Fig. 12.7 (a) The image and (b) the convolution by the 3 × 3 matrix.

Figure 12.7 shows the "clock" image of size 256×256 in part (a) and the convoluted image by the 3×3 cross-window in part (b). The result of the windowed convolution is a smooth or a blur image; at each pixel, the average of the image together with four neighbors was calculated.

Such convolutions smooth the edges of the image and to see better this effect of blurring, we consider the binary image of size 512×512 that is shown in Fig. 12.8(a). The image convoluted with the 5×5 matrix

$$h = \frac{1}{13}\begin{bmatrix} 0 & 0 & 1 & 0 & 0 \\ 0 & 1 & 1 & 1 & 0 \\ 1 & 1 & \underline{1} & 1 & 1 \\ 0 & 1 & 1 & 1 & 0 \\ 0 & 0 & 1 & 0 & 0 \end{bmatrix}$$

is shown in part (b).

(a) image (b) convoluted image

Fig. 12.8 (a) The binary image and (b) the convolution by the 5 × 5 matrix.

12.6.1 Edges and Contours of Images

The human visual system is very sensitive to gradients. An image gradient is a directional change in the intensity of the image. Image gradients have been used to create new images with "visualized edges." Moreover, image gradients are very important (or they are central building blocks) in many graphics, computer vision, robotics, imaging systems, and visual surveillance applications. For instance, gradients can be used for extracting useful information (structure, feature, and properties of objects) from images [13, 17, 18].

12.6.2 Gradients and Thresholding

The image is considered as a function $f_{n,m}$ of the intensity in points (n, m) that are on the Cartesian lattice. The difference of the image in the horizontal direction can be considered as $f_{n+1,m} - f_{n,m}$ and in the vertical direction as $f_{n,m+1} - f_{n,m}$. They are very simple examples of the operations which are called *the gradients* of the image. Other operations can also be considered, as well as different directions along which these operations will be performed [14, 19]. The gradient operations result in differences in the intensity, which may be positive, zero, or negative. Therefore, when displaying the gradients, these gradients are calculated in the absolute mode. The simple gradient operations in the horizontal and vertical directions are the following differencing operators, respectively:

$$G_x\!: f_{n,m} \to G_x(f)_{n,m} = f_{n+1,m} - f_{n,m}, \quad G_y\!: f_{n,m} \to G_y(f)_{n,m} = f_{n,m+1} - f_{n,m} \tag{12.37}$$

The images $G_x(f)$ and $G_y(f)$ are called *the horizontal* (or *row*) *gradient* and *the vertical* (or *column*) *gradient* of the image, respectively. As an example, Fig. 12.9(a) shows the grayscale image that shows the line at the angle $\pi/6$ to the vertical that separates two regions with the intensities 200 and 10. The intensity of the line is 50. This image also has a small rectangle of size 80×60 with an intensity of 100 in the middle.

The magnitude of the horizontal gradient, $|G_x(f)|$, is shown in Fig. 12.9(b), and the magnitude of the vertical gradient, $|G_y(f)|$, is shown in part (c). One can notice that both gradients extract the points of the line. Two vertical sides of the rectangle are extracted by the gradient operator G_x and the horizontal sides by the gradient operator G_y.

The sum of the magnitudes of the gradients $G_x(f)$ and $G_y(f)$ is called *the magnitude gradient* of the image, $G_+(f) = |G_x(f)| + |G_y(f)|$. The diagram of the calculation of the magnitude gradient image is given in Fig. ~~12.10~~.

Figure 12.11 shows the magnitude gradient of the image in Fig. 12.9(a). This gradient image shows the line and rectangle in the image. The points of the line look brighter, because of the larger intensity increase which is equal to $200 - 50 = 150$. The points on the rectangle have a smaller change of intensity. The difference of the intensity in the

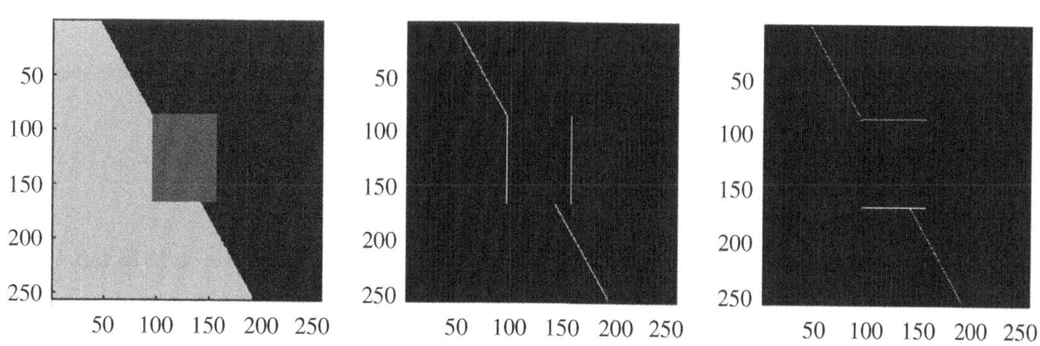

(a) f (b) $G_x(f)$ (c) $G_y(f)$

Fig. 12.9 (a) The modeled image and (b) the horizontal and (c) the vertical gradients of the image (in absolute scale).

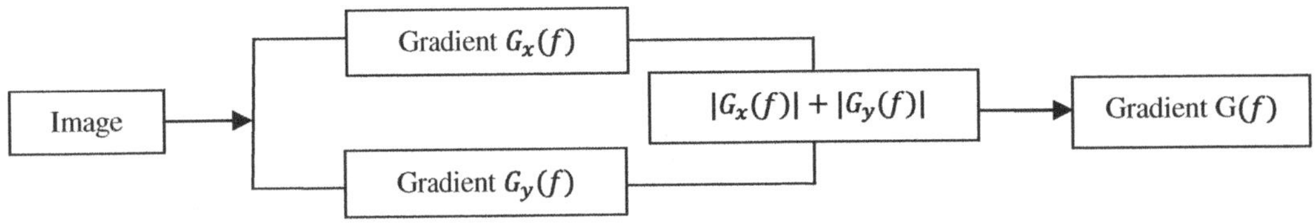

Fig. 12.10 The diagram of the calculation of the magnitude gradient image.

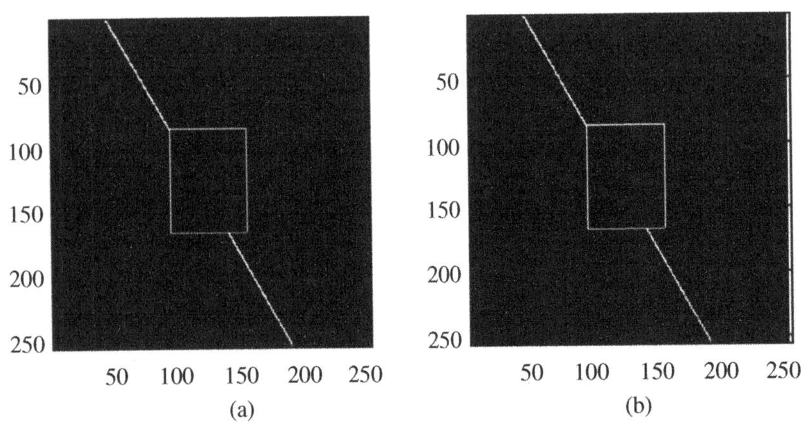

Fig. 12.11 (a) The gradient image $G(f)$ and (b) the threshold image.

points of the rectangle, that are in the bright area, equals $200 - 100 = 100$, and for the points in the dark area, it is equal to $100 - 10 = 90$.

After calculating the gradient image, a threshold T is selected and thresholding of the image is accomplished to calculate the counterimage, which is the binary image,

$$f_{n,m} \to (G(f)_T)_{n,m} = \begin{cases} 1, & \text{if } |G(f)_{n,m}| \geq T; \\ 0 & \text{otherwise.} \end{cases} \tag{12.38}$$

Figure 12.12 shows the diagram of the calculation of the counter image. The binary image $G(f)_1$ after thresholding by $T = 1$ is shown in Fig. 12.11(b).

Many gradient operators were developed and used in DIP [14]. For example, we consider the following second-order gradient operators in the horizontal and vertical directions:

$$G_x^2: f_{n,m} \to G_x^2(f)_{n,m} = f_{n,m-1} - 2f_{n,m} + f_{n,m+1},$$
$$G_y^2: f_{n,m} \to G_y^2(f)_{n,m} = f_{n-1,m} - 2f_{n,m} + f_{n+1,m}.$$

In the 3×3 window, these gradient operators are described by the 3×3 matrices (with the scale factors ½)

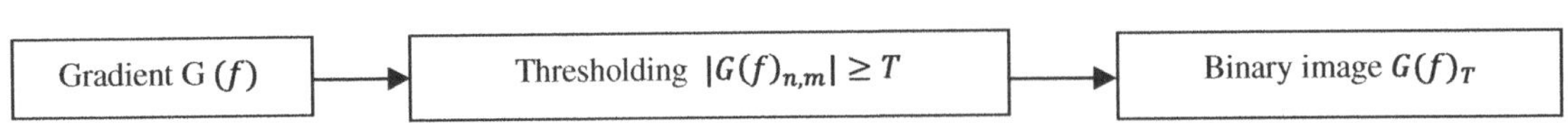

Fig. 12.12 The thresholding of the gradient image.

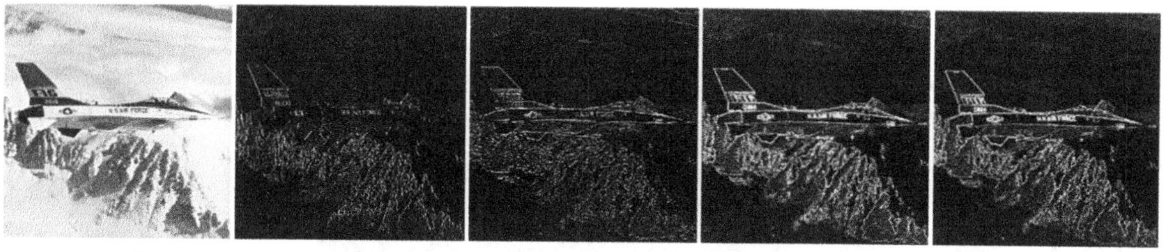

(a) f (b) $G_x^2(f)$ (c) $G_y^2(f)$ (d) $G^2(f)$ (e) $G_+^2(f)$

Fig. 12.13 (a) The "jetplane" image and the second gradient images in (b) the horizontal and (c) the vertical directions. (d) The magnitude gradient image and (e) the magnitude cross-gradient image.

$$[G_x^2] = \frac{1}{2}\begin{bmatrix} 0 & 0 & 0 \\ 1 & \underline{-2} & 1 \\ 0 & 0 & 0 \end{bmatrix} \quad \text{and} \quad [G_y^2] = \frac{1}{2}\begin{bmatrix} 0 & 1 & 0 \\ 0 & \underline{-2} & 0 \\ 0 & 1 & 0 \end{bmatrix}. \tag{12.39}$$

It is common to consider the second gradient as the transpose first matrix, $[G_y^2] = [G_x^2]'$, or $[G_y^2] = -[G_x^2]'$.

The magnitude gradient of the image is the sum of the magnitudes of the gradients, that is, $G^2(f) = |G_x^2(f)| + |G_y^2(f)|$. As an example, Fig 12.13 shows the "jetplane" image in part (a), the horizontal and vertical second-order gradients in parts (b) and (c), and the magnitude second gradient in part (d). Adding two matrices in Eq. 12.39, we obtain the following matrix of the second gradient, which we call *the 4-neighbor gradient* or *the cross-gradient*:

$$[G_+^2] = \frac{1}{2}\left([G_x^2] + [G_y^2]\right) = \frac{1}{4}\begin{bmatrix} 0 & 1 & 0 \\ 1 & \underline{-4} & 1 \\ 0 & 1 & 0 \end{bmatrix}. \tag{12.40}$$

The corresponding image $|G_+^2(f)|$ is called the magnitude cross-gradient image. For the "jetplane" image, such a gradient image is shown in Fig. 12.13(e).

We also consider the Prewitt 3-level gradient operators which are defined by the matrices

$$[P_x^2] = \frac{1}{3}\begin{bmatrix} 1 & 0 & -1 \\ 1 & \underline{0} & -1 \\ 1 & 0 & -1 \end{bmatrix} \quad \text{and} \quad [P_y^2] = \frac{1}{3}\begin{bmatrix} -1 & -1 & -1 \\ 0 & \underline{0} & 0 \\ 1 & 1 & 1 \end{bmatrix}. \tag{12.41}$$

Given image f, the gradient images in x- and y-directions are calculated by $P_x^2(f)$ and $P_y^2(f)$, respectively.

As an example, Fig. 12.14 shows the binary image in part (a), and the gradient images $P_x^2(f)$ and $P_y^2(f)$ in parts (b) and (c), respectively. One can notice that the gradient in the x-direction removes all horizontal lines from the image, and the gradient in the y-direction removes all vertical lines from the image.

The magnitude Prewitt gradient of the image is the sum of the magnitudes of the gradients, that is,

$$P^2(f) = \left(|P_x^2(f)| + |P_y^2(f)|\right)/2.$$

The *square-root Prewitt gradient* image is calculated by $P(f) = \sqrt{\left([P_x^2(f)]^2 + [P_y^2(f)]^2\right)/2}$.

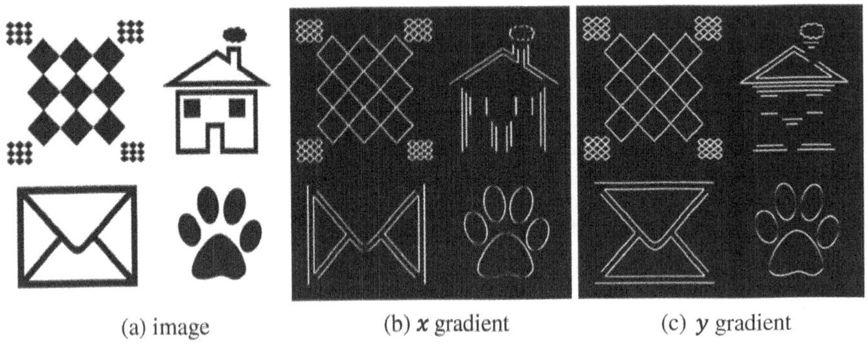

Fig. 12.14 (a) The binary image and (b) the horizontal and (c) the vertical differencing gradient images.

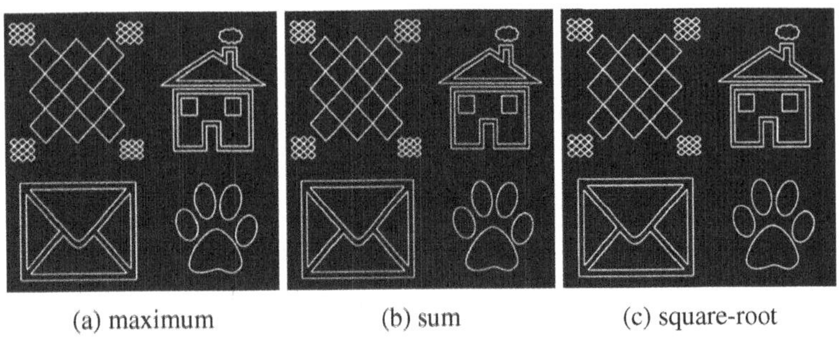

Fig. 12.15 The Prewitt gradient images: (a) maximum, (b) magnitude, and (c) square root.

Figure 12.15 shows the maximum gradient image $P_m(f) = \max\left\{|\,P_x^2(f)\,|\,,\,|\,P_y^2(f)\,|\right\}$ in part (a) and gradient images $P^2(f)$ and $P(f)$ in parts (b) and (c), respectively. These gradient operators work well on the above image.

12.7 Convolution Quantum Representation

Let us consider a signal $\{f_n;\ n = 0{:}\,(N-1)\}$ of length $=2^r$, $r > 2$, and the following mask for the convolution: $M = \left[1\ \underline{2}\ 2\ 1\right]$. The underlined number shows the position of the center of the mask. The signal can be represented by r qubits. We may also consider that the signal is periodic, to simplify the calculation of normalizing coefficients in the quantum representation of signals. The convolution of the signal with this mask is calculated by

$$y_n = (f * M)_n = f_{n-2} + 2f_{n-1} + 2f_n + f_{n+1}. \tag{12.42}$$

These components of the convolution at point n can be written into the vector $\mathbf{v}_n = (f_{n-2}, 2f_{n-1}, 2f_n, f_{n+1})'$. We assume that, for this vector, it is possible with the help of a classical computer to initiate the following superposition of 2-qubit states:

$$|v_n\rangle = \frac{1}{A_n}(f_{n-2}|00\rangle + 2f_{n-1}|01\rangle + 2f_n|10\rangle + f_{n+1}|11\rangle), \tag{12.43}$$

where the normalized coefficient $A_n = \sqrt{f_{n-2}^2 + 4f_{n-1}^2 + 4f_n^2 + f_{n+1}^2}$. Indeed, we do not need to come up with complex algorithms and schemes on shifted rows for a quantum computer. A classical computer will always be

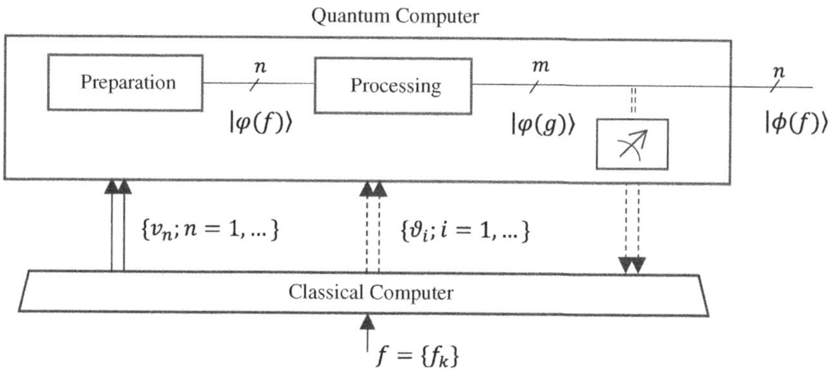

Fig. 12.16 The abstract interaction scheme of the quantum and classical computers.

connected to a quantum computer (see Fig. 12.16), and the task of preparing such data for the classical computer is not difficult.

For instance, in the $N = 8$ case, the preparation is described by the sequence $\mathbf{v} = \{\mathbf{v}_0, \mathbf{v}_1\ \mathbf{v}_2, \mathbf{v}_3\ \mathbf{v}_4, \mathbf{v}_5\ \mathbf{v}_6, \mathbf{v}_7\}$, or

$$\underbrace{0,0,2f_0,f_1}_{n=0},\underbrace{0,2f_0,2f_1,f_2}_{n=1},\underbrace{f_0,2f_1,2f_2,f_3}_{n=2},\underbrace{f_1,2f_2,2f_3,f_4}_{n=3},\underbrace{f_2,2f_3,2f_4,f_5}_{n=4},\underbrace{f_3,2f_4,2f_5,f_6}_{n=5},\underbrace{f_4,2f_5,2f_6,f_7}_{n=6},\underbrace{f_5,2f_6,2f_7,0}_{n=7}. \tag{12.44}$$

If the signal is periodic, then this sequence is composed as

$$\underbrace{f_6,2f_7,2f_0,f_1}_{n=0},\underbrace{f_7,2f_0,2f_1,f_2}_{n=1},\underbrace{f_0,2f_1,2f_2,f_3}_{n=2},\underbrace{f_1,2f_2,2f_3,f_4}_{n=3},\underbrace{f_2,2f_3,2f_4,f_5}_{n=4},\underbrace{f_3,2f_4,2f_5,f_6}_{n=5},\underbrace{f_4,2f_5,2f_6,f_7}_{n=6},\underbrace{f_5,2f_6,2f_7,f_0}_{n=7}.$$

The data register has increased four times, that is, only two more qubits are required for these data. It is also possible to prepare the input data in the classical computer as

$$\underbrace{0,0,f_0,f_1}_{n=0},\underbrace{0,f_0,f_1,f_2}_{n=1},\underbrace{f_0,f_1,f_2,f_3}_{n=2},\underbrace{f_1,f_2,f_3,f_4}_{n=3},\underbrace{f_2,f_3,f_4,f_5}_{n=4},\underbrace{f_3,f_4,f_5,f_6}_{n=5},\underbrace{f_4,f_5,f_6,f_7}_{n=6},\underbrace{f_5,f_6,f_7,0}_{n=7}.$$

These two data require different unitary matrices or gates, for subsequent processing.

The $(r+2)$-qubit state superposition of all convolution vectors y_n, that is, $\mathbf{v}$, is defined as

$$|\varphi\rangle = |\varphi(v)\rangle = \frac{1}{\sqrt{N}}\sum_{n=0}^{N-1}|n\rangle|v_n\rangle. \tag{12.45}$$

This is the right-sided superposition; the left-sided superposition can also be considered. We call this form of superposition with states of 2-qubits standard. This superposition requires two additional qubits. Now, we process $|v_n\rangle$ in Eq. 12.43 by the 2-qubit paired transform. The quantum circuit for the 2-qubit DPT is shown in Fig. 12.17. The bullet on the line denotes the control qubit.

The paired transform of the vector $\mathbf{v}_n$ at the point n is calculated by [20]

$$D_4\begin{bmatrix}1 & 0 & -1 & 0\\ 0 & 1 & 0 & -1\\ 1 & -1 & 1 & -1\\ 1 & 1 & 1 & 1\end{bmatrix}\begin{bmatrix}f_{n-2}\\ 2f_{n-1}\\ 2f_n\\ f_{n+1}\end{bmatrix} = D_4\begin{bmatrix}f_{n-2}-2f_n\\ 2f_{n-1}-f_{n+1}\\ f_{n-2}-2f_{n-1}+2f_n-f_{n+1}\\ f_{n-2}+2f_{n-1}+2f_n+f_{n+1}\end{bmatrix}.$$

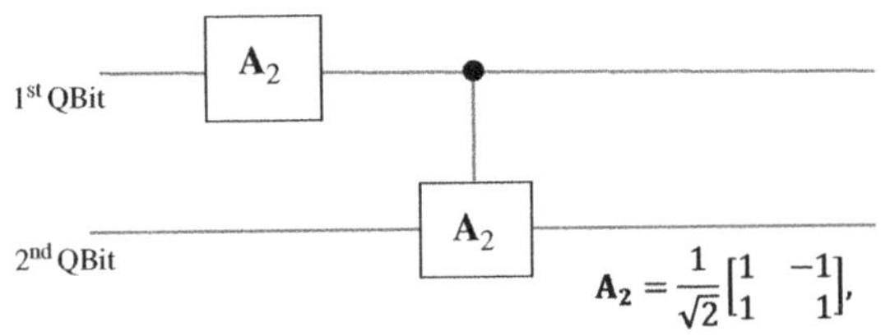

Fig. 12.17 The quantum circuit for the 2-qubit QPT.

The diagonal matrix $D_4 = \left\{ \frac{1}{\sqrt{2}}, \frac{1}{\sqrt{2}}, \frac{1}{2}, \frac{1}{2} \right\}$. The result of the transform is the 2-qubit superposition

$$|\chi_4'(y_n)\rangle = c_0|00\rangle + c_1|01\rangle + c_2|10\rangle + c_3\,|\,11\rangle$$

with the amplitudes: $c_0 = (f_{n-2} - 2f_n)/\sqrt{2},\ c_1 = (2f_{n-1} - f_{n+1})/\sqrt{2}$, and

$$c_2 = (f_{n-2} - 2f_{n-1} + 2f_n - f_{n+1})/2,\ c_3 = (f_{n-2} + 2f_{n-1} + 2f_n + f_{n+1})/2.$$

The paired transform is unitary. The amplitudes of this superposition are normalized by the coefficient $\sqrt{c_0^2 + c_1^2 + c_2^2 + c_3^2} = \sqrt{f_{n-2}^2 + 4f_{n-1}^2 + 4f_n^2 + f_{n+1}^2}$.

Up to the constant 3, the amplitude c_3 is the value of the convolution, y_n, with the mask $[1\ \underline{2}\ 2\ 1]/6$,

$$c_3 = 3y_n,\ \ y_n = \frac{1}{6}[1\ \ 2\ \ 2\ \ 1]\begin{bmatrix} f_{n-2} \\ f_{n-1} \\ f_n \\ f_{n+1} \end{bmatrix}. \tag{12.46}$$

Up to a constant 3/2, the amplitude c_2 is the value of the convolution which is described by the mask $[-1\ \underline{2}\ -2\ 1]/3$ and represents the 4-level gradient operator,

$$c_2 = \frac{3}{2}G_n(f),\ \ G_n(f) = \frac{1}{3}[1\ \ -2\ \ 2\ \ -1]\begin{bmatrix} f_{n-2} \\ f_{n-1} \\ f_n \\ f_{n+1} \end{bmatrix}. \tag{12.47}$$

Figure 12.18 shows the circuit element for processing 2-qubit $|y_n\rangle$, by the 2-qubit paired transform. Thus, the paired transform allows for parallel computing two different convolutions, one of which is the gradient operation.

It should be noted that the above reasoning is applicable to a superposition written in a nonstandard form

$$|\psi\rangle = \frac{1}{C}\sum_{n=0}^{N-1}|n\rangle|(f_{n-2}|00\rangle + 2f_{n-1}|01\rangle + 2f_n|10\rangle + f_{n+1}|11\rangle),$$

where the coefficient $C = \sqrt{6(f_0^2 + f_1^2 + \dots + f_{N-1}^2)}$.

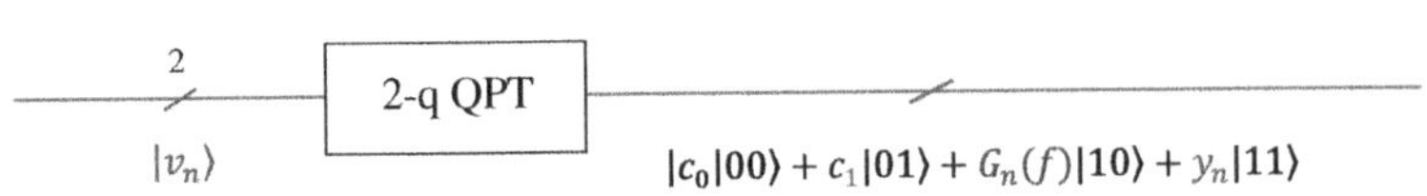

Fig. 12.18 The circuit element for processing the 2-qubit $|v_n\rangle$.

12.7.1 Gradient Operators and Numerical Simulations

In this section, we describe a few examples with images, to show that the presented above method works well.

Example 12.4 Gradient by Mask 1 × 3

We consider the quantum representation for the gradient with the mask $M = \left[1 - \underline{2}\ 1\right]/2$. The gradient of the signal, or the convolution of the signal with this mask, is calculated at each point n by

$$G_n(f) = (f * M)_n = [f_{n-1} - 2f_n + f_{n+1}]/2. \tag{12.48}$$

This operation can be written as $G_n(f) = [(f_{n-1} - f_n) + (f_{n+1} - f_n)]/2$. We define the following 2-qubit state superposition at the point-state $|n\rangle$:

$$|v_n\rangle = \frac{1}{A}(f_{n-1}|00\rangle - f_n|01\rangle + f_{n+1}|10\rangle - f_n|11\rangle) \tag{12.49}$$

and the vector $\boldsymbol{v}_n = (f_{n-2}, -f_n, f_{n+1}, f_n)'$. Here, the coefficient $A = A(n) = \sqrt{f_{n-1}^2 + 2f_n^2 + f_{n+1}^2}$. The $(r+2)$-qubit state superposition for the gradient of the signal is considered in standard form, as in Eq. 12.45.

Note that the following superposition also can be used:

$$|\psi\rangle = |\psi(v)\rangle = \frac{1}{C}\sum_{n=0}^{N-1}|n\rangle|v_n\rangle = \frac{1}{C}\sum_{n=0}^{N-1}|n\rangle|(f_{n-1}|00\rangle - f_n|01\rangle + f_{n+1}|10\rangle - f_n|11\rangle).$$

The coefficient $C = 2\sqrt{(f_0^2 + f_1^2 + \ldots + f_{N-1}^2)}$.

The 4-point discrete paired transform of the amplitudes of $|y_n\rangle$ is

$$\begin{bmatrix} 1 & 0 & -1 & 0 \\ 0 & 1 & 0 & -1 \\ 1 & -1 & 1 & -1 \\ 1 & 1 & 1 & 1 \end{bmatrix}\begin{bmatrix} f_{n-1} \\ -f_n \\ f_{n+1} \\ -f_n \end{bmatrix} = \begin{bmatrix} f_{n-1} - f_{n+1} \\ 0 \\ f_{n-1} + 2f_n + f_{n+1} \\ f_{n-1} - 2f_n + f_{n+1} \end{bmatrix}. \tag{12.50}$$

For simplicity of calculations, we omitted the diagonal matrix D_4 from the matrix of the paired transform. Thus, the 2-qubit paired transform on the input $|v_n\rangle$ is the 2-qubit state superposition

$$|\chi_4'(v_n)\rangle = c_0|00\rangle + c_1|01\rangle + c_2|10\rangle + c_3\,|\,11\rangle \tag{12.51}$$

with the following amplitudes of the states: $c_0 = f_{n-1} - f_{n+1}$, $c_1 = 0$, $c_2 = f_{n-1} + 2f_n + f_{n+1}$, and $c_3 = f_{n-1} - 2f_n + f_{n+1}$. These amplitudes of states should be normalized by the coefficient $A = \sqrt{c_0^2 + c_2^2 + c_3^2}$. The amplitude c_3 is the value of the gradient $G_n(f)$ at point n,

$$c_3 = c_3(n) = G_n(f) = \frac{1}{2}[1 \quad -2 \quad 1]\begin{bmatrix} f_{n-1} \\ f_n \\ f_{n+1} \end{bmatrix}. \tag{12.52}$$

Up to factor 4, the amplitude c_2 equals the convolution of the signal at point n, when the mask is $[1\ \underline{2}\ 1]/4$,

$$c_2 = c_2(n) = \frac{1}{4}[1 \quad 2 \quad 1]\begin{bmatrix} f_{n-1} \\ f_n \\ f_{n+1} \end{bmatrix}. \tag{12.53}$$

As an example, we consider the above operations over each row of the image "jetplane.jpg" of size 512 × 512 pixels in Fig. 12.19(a). The images composed by $c_0(n)$, $c_2(n)$, and $c_3(n)$ coefficients are shown in parts (b), (c),

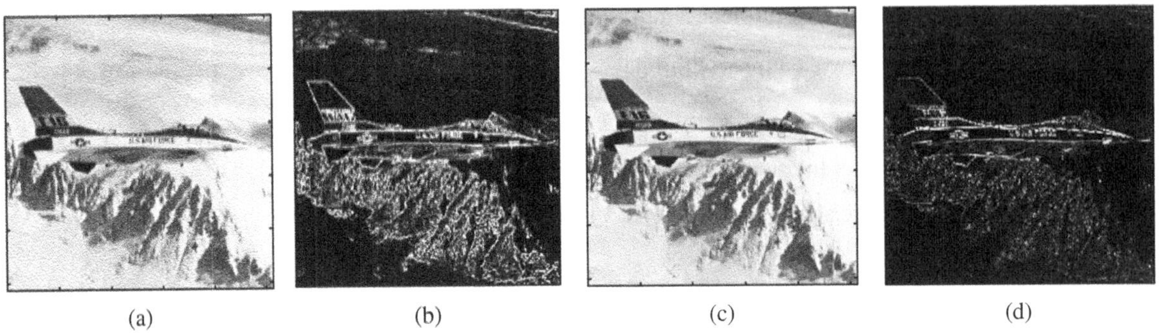

Fig. 12.19 (a) The original grayscale image, (b) the c_0-gradient image, (c) c_2-smooth image, and (d) c_3-gradient image.

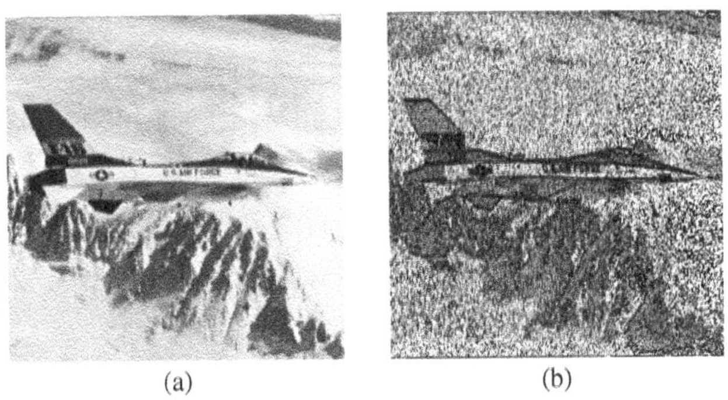

Fig. 12.20 (a) The "jetplane" image and (b) the simulated measured image with c_0, c_2, and c_3-images.

and (d), respectively. The images in parts (b) and (d) are gradient images, and, in part (c), the image is smoothed along the rows.

Next, we can model the process of measurements of all $(r+2)$ qubits $|\psi\rangle$ for this image. The probability of measurement of the 2-qubit $|y_n\rangle$ in the basis states $|00\rangle$, $|10\rangle$, and $|11\rangle$ are considered according to the coefficients $|c_0|^2$, $|c_2|^2$, and $|c_3|^2$. The result of such a simulation on the classical computer is shown in Fig. 12.20 in part (b). The image was processed by a row of 512 points each. For each row of the image, at each point $n \in \{0, 1, 2, ..., 511\}$, the value of the row-signal was taken randomly from the corresponding set of amplitudes $\{c_0(n), c_2(n), c_3(n)\}$. The edge points can be extracted by the threshold operation of the quantum bit sequence.

Example 12.5 Sobel Gradient Operators
We consider the quantum representation of the 5-level Sobel gradient operator with the mask $M=[-1\ -2\ \underline{0}\ 2\ 1]/3$. The gradient of the signal, or the convolution of the signal with this mask, is calculated at each point n by

$$G_5(n) = (f * M)_n = \frac{[f_{n-2} + 2f_{n-1} - 2f_{n+1} - f_{n+2}]}{3}. \tag{12.54}$$

This operation can be written as the sum of eight terms

$$\begin{aligned} G_5(n) &= \frac{1}{3}[(f_{n-2} - f_n) + 2(f_{n-1} - f_n) + 2(f_n - f_{n+1}) + (f_n - f_{n+2})] \\ &= \frac{1}{3}[(f_{n-2} - f_n) + 2(f_{n-1} - f_n) - 2(f_{n+1} - f_n) - (f_{n+2} - f_n)]. \end{aligned}$$

A. We consider the following quantum representation of the 3-qubit for the convolution at point $\boldsymbol{n}$:

$$|v_n\rangle = \frac{1}{A}\left(f_{n-2}|0\rangle - f_n|1\rangle + 2f_{n-1}|2\rangle - 2f_n|3\rangle + 2f_n|4\rangle - 2f_{n+1}|5\rangle + f_n|6\rangle - f_{n+2}|7\rangle\right).$$

Here, the coefficient $A = \sqrt{f_{n-2}^2 + 4f_{n-1}^2 + 10f_n^2 + 4f_{n+1}^2 + f_{n+2}^2}$. The corresponding 8D vector is

$$\boldsymbol{v}_n = \left(f_{n-2}, -f_n, 2f_{n-1}, -2f_n, 2f_n, -2f_{n+1}, f_n, -f_{n+2}\right)'.$$

The $(r+3)$-qubit state superposition for the gradient of the signal is considered in the standard form. We can also use the following superposition:

$$|\psi\rangle = \frac{1}{C}\sum_{n=0}^{N-1}|n\rangle\left|\left(f_{n-2}|0\rangle - f_n|1\rangle + 2f_{n-1}|2\rangle - 2f_n|3\rangle + \;+ 2f_n|4\rangle - 2f_{n+1}|5\rangle + f_n|6\rangle - f_{n+2}|7\rangle\right)\right.$$

with the coefficient $C = \sqrt{20\left(f_0^2 + f_1^2 + \ldots + f_{N-1}^2\right)}$.

The 8-point discrete paired transform, χ_8', over amplitudes of the 3-qubit $|v_n\rangle$ equals

$$\chi_8'[\boldsymbol{v}_n] = \chi_8'\begin{bmatrix} f_{n-2} \\ -f_n \\ 2f_{n-1} \\ -2f_n \\ 2f_n \\ -2f_{n+1} \\ f_n \\ -f_{n+2} \end{bmatrix} = \begin{bmatrix} -2f_n + f_{n-2} \\ -f_n + 2f_{n+1} \\ -f_n + 2f_{n-1} \\ -2f_n + f_{n+2} \\ f_{n-2} - 2f_{n-1} + f_n \\ f_n - 2f_{n+1} + f_{n+2} \\ f_{n-2} + 2f_{n-1} + 6f_n + 2f_{n+1} + f_{n+2} \\ f_{n-2} + 2f_{n-1} - 2f_{n+1} - f_{n+2} \end{bmatrix}. \tag{12.55}$$

The last four components of the paired transform are the convolutions of the signal with the masks $[\underline{1}\ -2\ 1]$, $[1\ -2\ \underline{1}]$, $[1\ 2\ \underline{6}\ 2\ 1]$, and $[-1\ -2\ \underline{0}\ 2\ 1]$. The first four outputs describe the signal plus gradients. Indeed,

$$c_0 = -2f_n + f_{n-2} = -\left[f_n + \left(f_n - f_{n-2}\right)\right],\quad c_1 = -f_n + 2f_{n+1} = f_{n+1} + \left(f_{n+1} - f_n\right),$$

$$c_2 = -f_n + 2f_{n-1} = f_{n-1} + \left(f_{n-1} - f_n\right),\quad c_3 = -2f_n + f_{n+2} = -\left[f_n + \left(f_n - f_{n+2}\right)\right].$$

When the first qubit is in state $|1\rangle$, the transform coefficients c_4, c_5, c_6, and c_7, or the amplitudes of the new superposition of states

$$\left|\chi_8'(v_n)\right\rangle_1 = \frac{1}{C_1}\left(c_4|00\rangle + c_5|01\rangle + c_6|10\rangle + c_7|11\rangle\right)$$

describe the different convolutions and gradients. Here, the normalizing coefficient $C_1 = \sqrt{c_4^2 + c_5^2 + c_6^2 + c_7^2}$. The amplitudes c_4 and c_5 are the values of the 2-level Sobel gradient

$$G_2(n) = \frac{1}{2}[1\quad -2\quad 1]\begin{bmatrix} f_{n-1} \\ f_n \\ f_{n+1} \end{bmatrix},$$

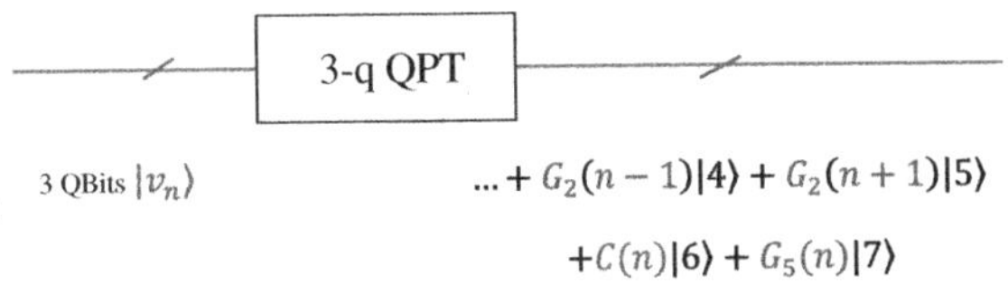

Fig. 12.21 The circuit element for processing the superposition $|y_n\rangle$.

which are calculated at two points, $(n-1)$ and $(n+1)$, that is,

$$c_4 = \frac{1}{2}[1 \quad -2 \quad 1]\begin{bmatrix} f_{n-2} \\ f_{n-1} \\ f_n \end{bmatrix} = G_2(n-1), c_5 = \frac{1}{2}[1 \quad -2 \quad 1]\begin{bmatrix} f_n \\ f_{n+1} \\ f_{n+2} \end{bmatrix} = G_2(n+1). \tag{12.56}$$

The amplitude c_6 describes the convolution of the signal at point n, when the mask is $[1\ 2\ \underline{6}\ 2\ 1]/12$,

$$c_6 = C(n) = \frac{1}{12}[1 \quad 2 \quad 6 \quad 2 \quad 1]\begin{bmatrix} f_{n-2} \\ f_{n-1} \\ f_n \\ f_{n+1} \\ f_{n+2} \end{bmatrix}. \tag{12.57}$$

The last amplitude c_7 corresponds to the 5-level Sobel gradient of the signal at point n,

$$c_7 = G_5(n) = \frac{1}{3}[1 \quad 2 \quad 0 \quad -2 \quad -1]\begin{bmatrix} f_{n-2} \\ f_{n-1} \\ f_n \\ f_{n+1} \\ f_{n+2} \end{bmatrix}. \tag{12.58}$$

Figure 12.21 shows the circuit element for processing the 2-qubit $|y_n\rangle$, by the 3-qubit paired transform over the input 3-qubit $|y_n\rangle$. This transform allows for parallel computing three different convolutions, one of which is the 5-level Sobel gradient transform. The 2-level gradient G_2 is calculated in two points, $(n-1)$ and $(n+1)$.

12.8 Other Gradient Operators

Consider the gradient operator with the mask $M = \left[1\,1 - \underline{4}\ 1\,1\right]/4$. The gradient of the signal f_n at each point n is calculated by

$$G_5(n) = (f * M)_n = [f_{n-2} + f_{n-1} - 4f_n + f_{n+1} + f_{n+2}]/4.$$

We consider the following quantum representation of the 3-qubit for the convolution at point n:

$$|v_n\rangle = \frac{1}{A}(f_{n-2}|0\rangle - f_n|1\rangle + f_{n-1}|2\rangle - f_n|3\rangle + f_{n+1}|4\rangle - f_n|5\rangle + f_{n+2}|6\rangle - f_n|7\rangle). \tag{12.59}$$

Here, the coefficient $A = A(n) = \sqrt{f_{n-2}^2 + f_{n-1}^2 + 4f_n^2 + f_{n+1}^2 + f_{n+2}^2}$. Therefore, the corresponding 8D vector at point n is defined as $\mathbf{v}_n = (f_{n-2}, -f_n, f_{n-1}, -f_n, f_{n+1}, -f_n, f_{n+2}, -f_n)'$.

The 8-point discrete paired transform of the convolution vector $\boldsymbol{v}_n$ equals

$$\chi_8'[\boldsymbol{v}_n] = \chi_8'\begin{bmatrix} f_{n-2} \\ -f_n \\ f_{n-1} \\ -f_n \\ f_{n+1} \\ -f_n \\ f_{n+2} \\ -f_n \end{bmatrix} = \begin{bmatrix} f_{n-2}-f_{n+1} \\ 0 \\ f_{n-1}-f_{n+2} \\ 0 \\ f_{n-2}-f_{n-1}+f_{n+1}-f_{n+2} \\ 0 \\ f_{n-2}+f_{n-1}+4f_n+f_{n+1}+f_{n+2} \\ f_{n-2}+f_{n-1}-4f_n+f_{n+1}+f_{n+2} \end{bmatrix}.$$

Therefore, the 3-qubit paired transform on the input $|v_n\rangle$ is the 3-qubit state superposition

$$\left|\chi_8'(v_n)\right\rangle = \frac{1}{C}\sum_{k=0}^{7} c_k \mid k\rangle$$

with the amplitudes $c_1 = c_3 = c_5 = 0$. Other five amplitudes represent a convolution and three gradient operations at point n,

$$c_4 = c_4(n) = \frac{1}{2}[1 \quad -1 \quad 0 \quad 1 \quad -1]\begin{bmatrix} f_{n-2} \\ f_{n-1} \\ f_n \\ f_{n+1} \\ f_{n+2} \end{bmatrix}, \quad c_6 = c_6(n) = \frac{1}{8}[1 \quad 1 \quad 4 \quad 1 \quad 1]\begin{bmatrix} f_{n-2} \\ f_{n-1} \\ f_n \\ f_{n+1} \\ f_{n+2} \end{bmatrix},$$

$$c_7 = c_7(n) = \frac{1}{4}[1 \quad 1 \quad -4 \quad 1 \quad 1]\begin{bmatrix} f_{n-2} \\ f_{n-1} \\ f_n \\ f_{n+1} \\ f_{n+2} \end{bmatrix}.$$

Thus, the 3-qubit paired transform allows for calculating three gradients and one convolution of the signal,

$$\left|\chi_8'(v_n)\right\rangle = \frac{1}{C}\left[\, c_0(n)|0\rangle + c_2(n)|2\rangle + c_4(n)|4\rangle + c_6(n)|6\rangle + c_7(n)|7\rangle \,\right]$$

where $C = \sqrt{c_0^2 + c_2^2 + c_4^2 + c_6^2 + c_7^2}$.

For the above "jetplane" image processed by rows, Figure 12.22 shows the images composed by $c_0(n)$, $c_4(n)$, $c_6(n)$, and $c_7(n)$ coefficients in parts (a), (b), (c), and (d), respectively.

(a) (b) (c) (d)

Fig. 12.22 (a) The c_0-gradient image, (b) the c_4-gradient image, (c) c_6-smooth image, and (d) c_7-gradient image.

Fig. 12.23 The simulated measurement of the "jetplane" image.

In this example, we also can model the process of measurements of all $(r + 2)$ qubits $|\psi\rangle$ for the "jetplane" image, and consider the probability of measurement of the 2-qubit $|y_n\rangle$ in five basis states $|000\rangle$, $|010\rangle$, $|100\rangle$, $|110\rangle$, and $|111\rangle$ according to the coefficients $|c_0|^2$, $|c_2|^2$, $|c_4|^2$, $|c_6|^2$, and $|c_7|^2$.

The result of such a simulation on the classical computer is shown in Fig. 12.23. For each row of the image, the values of the row-signal were taken randomly from the corresponding set of amplitudes $\{c_0(n),\ c_2(n),\ c_4(n),\ c_6(n),\ c_7(n)\}$.

Figure 12.24 shows the grayscale image of Leonardo Da Vinci painting "Lady with Ermine" in part (a) and the result of computer simulation of measurements of qubits $|\psi\rangle$ along each row of this image in part (b). The similar computer simulation of measurements for the grayscale image "pepper" of size 512×512 pixels is shown in parts (c) and (d).

It should be noted that in comparison with the method of the convolution by the fast Fourier transform, which is used in traditional computations, the paired transform is much faster. It is the core of the DFT [21, 22]. The processing signals and images do not require the inverse transforms, as in the method of DFT. All images in Figs.12.22–12.24 are results of the direct paired transforms calculated along the rows of these 2D signals. If we imagine that such a realization of the proposed convolution representation would be possible in a quantum computer, then (i) the computation of convolution with gradients would be very efficient, (ii) quantum computers have the potential to resolve other challenges of computer vision and image processing applications, including multiscale analysis, machine learning, segmentation, pattern recognition, and coding. The proposed method of representation and computation can be generalized and used for other unitary transforms used in image and signal processing, including the Hadamard transforms.

12.9 Gradient and Smooth Operators by Multiplication

The operation of multiplication of 2-qubits allows for simple implementation of gradient and convolution operations with small masks [5]. We consider a few of examples which are based on the following observation. The multiplication of two quaternions $q = q_1 q_2$ in matrix form is written as (see Eq. 10.3)

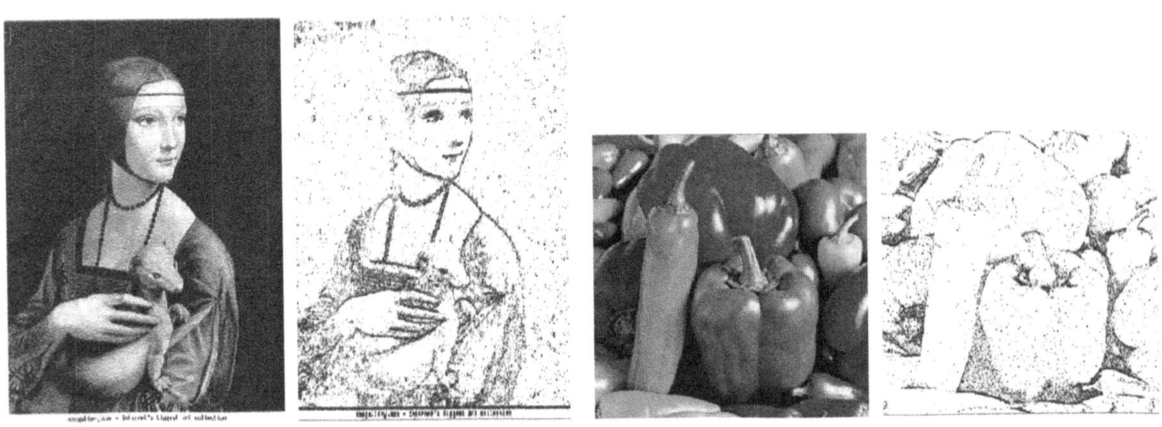

(a) (b) (c) (d)

Fig. 12.24 (a) The grayscale of [leonardo9.jpg] image of size 744 × 526 pixels (from http://www.abcgallery.com) and (b) the computer-simulated measured image with c_0, c_2, c_4, c_6, and c_7-amplitudes. (c) The "pepper" image and (d) the simulated random image. *Source:* National Museum / Wikimedia Commons / Public domain.

$$\boldsymbol{q} = \begin{bmatrix} a \\ b \\ c \\ d \end{bmatrix} = \boldsymbol{A}_{q_1} \begin{bmatrix} a_2 \\ b_2 \\ c_2 \\ d_2 \end{bmatrix} = \begin{bmatrix} a_1 & -b_1 & -c_1 & -d_1 \\ b_1 & a_1 & -d_1 & c_1 \\ c_1 & d_1 & a_1 & -b_1 \\ d_1 & -c_1 & b_1 & a_1 \end{bmatrix} \begin{bmatrix} a_2 \\ b_2 \\ c_2 \\ d_2 \end{bmatrix}. \tag{12.60}$$

Consider matrices $\boldsymbol{A}_{q_1}$ that will allow us to calculate the gradients and other convolutions. For example, the matrix of the quaternion $q_1 = \left(1, \underline{1}, 1, -1\right)/2$ can be considered,

$$\boldsymbol{A}_{q_1} = \frac{1}{2}\begin{bmatrix} 1 & -1 & -1 & 1 \\ 1 & 1 & 1 & 1 \\ 1 & -1 & 1 & -1 \\ -1 & -1 & 1 & 1 \end{bmatrix}, \qquad \det\left(\boldsymbol{A}_{q_1}\right) = 1.$$

The image is considered as a 1D signal composed by rows. As mentioned above, we consider that it is possible to prepare the input data of the image $\{f_n\}$ as

$$\underbrace{0,0,f_0,f_1}_{n=0},\ \underbrace{0,f_0,f_1,f_2}_{n=1},\ \underbrace{f_0,f_1,f_2,f_3}_{n=2},\ \underbrace{f_1,f_2,f_3,f_4}_{n=3},\ \underbrace{f_2,f_3,f_4,f_5}_{n=4},\ \ldots,\ \underbrace{f_4,f_5,f_6,f_7}_{n=6},\ \ldots,\ \underbrace{f_{N-3},f_{N-2},f_{N-1},0}_{n=N-1}.$$

The following left-sided superposition of the image is considered:

$$|\psi(f)\rangle = \frac{1}{C}\sum_{n=0}^{N-1}|n\rangle|v_n\rangle = \frac{1}{C}\sum_{n=0}^{N-1}|n\rangle\left|(f_{n-1}|00\rangle + f_n|01\rangle + f_n|10\rangle + f_{n+1}|11\rangle\right).$$

The coefficient $C = 2\sqrt{\left(f_0^2 + f_1^2 + \ldots + f_{N-1}^2\right)}$.

The 4-point discrete paired transform at the state $|n\rangle$ is the 2-qubit with the amplitudes calculated by

$$\frac{1}{2}\begin{bmatrix} 1 & -1 & -1 & 1 \\ 1 & 1 & 1 & 1 \\ 1 & -1 & 1 & -1 \\ -1 & -1 & 1 & 1 \end{bmatrix}\begin{bmatrix} f_{n-2} \\ f_{n-1} \\ f_n \\ f_{n+1} \end{bmatrix} = \frac{1}{2}\begin{bmatrix} f_{n-2} - f_{n-1} - f_n + f_{n+1} \\ f_{n-2} + f_{n-1} + f_n + f_{n+1} \\ f_{n-2} - f_{n-1} + f_n - f_{n+1} \\ -f_{n-2} - f_{n-1} + f_n + f_{n+1} \end{bmatrix}. \tag{12.61}$$

Thus, the 2-qubit paired transform on the input $|v_n\rangle$ is the 2-qubit state superposition

$$\left|\chi_4'(v_n)\right\rangle = c_0\left|00\right\rangle + c_1\left|01\right\rangle + c_2\left|10\right\rangle + c_3 \mid 11\rangle \tag{12.62}$$

with the following amplitudes of the states:

$$c_0 = \frac{1}{2}(f_{n-2} - f_{n-1} - f_n + f_{n+1}),\quad c_1 = \frac{1}{2}(f_{n-2} + f_{n-1} + f_n + f_{n+1}),$$
$$c_2 = \frac{1}{2}(f_{n-2} - f_{n-1} + f_n - f_{n+1}),\quad c_3 = \frac{1}{2}(-f_{n-2} - f_{n-1} + f_n + f_{n+1}).$$

The amplitudes c_0, c_2, and c_3 are the values of the gradients at point n, which are defined by masks

$$G_0 = \frac{1}{2}[1 - 1 - \underline{1}\ \ 1],\quad G_2 = \frac{1}{2}[1 - 1 + \underline{1}\ \ -1],\quad \text{and}\quad G_3 = \frac{1}{2}[-1 - 1 + \underline{1}\ \ 1].$$

The amplitude c_1 at point n is twice the mean value, that is, is defined by the mean operator $M_1 = \frac{1}{4}\left[1\ 1\ \underline{1}\ \ 1\right]$.

(a) (b) (c) (d)

Fig. 12.25 The images of the amplitudes (a) c_0, (b) c_1, (c) c_2, and (d) c_3 for the "jetplane" image.

As an example, Figure 12.25 shows the images of these four coefficients for the "jetplane.tif" image. To display the gradient images of coefficients c_0, c_2, and c_3 together with the average image in part (b), the gradient images were amplified by the coefficient 8.

12.9.1 Challenges

The utilization of quaternion algebra in quantum computing enables the development of operations on 2-qubits and superpositions, facilitating signal and image processing tasks, including color images. The introduction of quaternion-quantum-based color imaging tools, such as the quaternion Fourier transforms in 2-qubit-based representations, further enhances the capabilities of color quantum image processing. By combining quaternion technology with quantum computing, new algorithms can be developed that simultaneously process different image formats and yield practical color image processing outcomes. Future advancements in this area are promising for revolutionizing color image processing, contributing to advancements in various domains such as visual communications, multimedia systems, computer vision, and biomedical applications.

While quaternion-based arithmetic in quantum image processing holds great potential, several challenges need to be addressed for its effective implementation. First, one of the challenges is the development of efficient and optimized algorithms that can handle the complex computations involved in quaternion-based color image processing. As the size and complexity of image data increase, it is crucial to ensure scalability and computational efficiency to meet real-world applications' demands. Another challenge lies in integrating quaternion-based arithmetic with existing image processing frameworks and standards. Achieving interoperability with existing image processing techniques and tools is essential for the practical adoption and widespread use of quaternion-based approaches.

Finally, understanding and effectively leveraging the unique properties of quaternion-based arithmetic in quantum image processing requires further research and exploration. Investigating the theoretical foundations, developing new quaternion-based operations, and studying the performance characteristics and limitations of these approaches will contribute to optimizing their use in practical applications.

References

1 Grigoryan, A.M. and Agaian, S.S. (2018). Quaternion and Octonion color image processing with MATLAB. *SPIE*, PM279. https://doi.org/10.1117/3.2278810

2 Grigoryan, A.M. and Agaian, S.S. (2014). Retooling of color imaging in the quaternion algebra. *Applied Mathematics and Sciences: An International Journal* 1 (3): 23–39.

3 Grigoryan, A.M., John, A., and Agaian, S.S. (2017). A novel color image enhancement method by the transformation of color images to 2-D grayscale images. *International Journal of Signal Processing and Analysis* 2 (1): 18.
4 Grigoryan, A.M. John, A., and Agaian, S.S. (2018). A novel image enhancement method of 3-D medical images by transforming the 3-D images to 2-D grayscale images. *Proceedings of SPIE, Defense + Commercial Sensing, Mobile Multimedia/Image Processing, Security, and Applications*, 10668, 10. Orlando, Florida.
5 Grigoryan, A.M. and Agaian, S.S. (2023). Introduction to multiplicative group on 2-qubits with quantum color image processing. *Quantum Information Processing* 22 (351): 34. https://doi.org/10.1007/s11128-023-04091-1.
6 Sangwine, S.J. (1997). The discrete quaternion-Fourier transform. *IPA97, Conference Publication no. 443*, 790–793.
7 Sangwine, S.J., Ell, T.A., Blackledge, J.M., and Turner, M.J. (2000). The discrete Fourier transform of a color image. In: *Image Processing II Mathematical Methods, Algorithms and Applications* (ed. J.M. Blackledge and M.J. Turner), 430–441. Chichester: Horwood Publishing.
8 Yeh, M.H. (2008). Relationships among various 2-D quaternion Fourier transforms. *IEEE Signal Processing Letters* 15: 669–672.
9 Grigoryan, A.M. and Agaian, S.S. (2023). Multiplicative group of quantum representations of signals. *Quantum Information Processing* 22 (82): 23. https://doi.org/10.1007/s11128-022-03765.
10 Perez, L.R. and Garcia-Escartin, J.C. (2017). Quantum arithmetic with the quantum Fourier transform. *Quantum Information Processing* 16: 14.
11 Li, H.S., Fan, P., Xia, H. et al. (2018). The quantum Fourier transform based on quantum vision representation. *Quantum Information Processing* 17 (333): 25.
12 Grigoryan, A.M. and Agaian, S.S. (2019) 'Paired quantum Fourier transform with $\log_2 N$ Hadamard gates,' *Quantum Information Processing*, 18: 217, p. 26. https://doi.org/10.1007/s11128-019-2322-6
13 Grigoryan, A.M. and Agaian, S.S. (2020). Optimal restoration of multiple signals in quaternion algebra. *Processing of SPIE 11399, Mobile Multimedia/Image Processing, Security, and Applications 2020*. 1139909, 13. https://doi.org/10.1117/12.2558209
14 Pratt, W.K. (2001). *Digital Image Processing*, 3e. Wiley.
15 Robinson, G.S. (1977). Edge detection by compass gradient masks. *Computer Graphics and Image Processing* 6 (5): 492–501.
16 Grigoryan, A.M. Agaian, S.S., and Panetta, K. (2022). Quantum-inspired edge detection algorithms implementation using new dynamic visual data representation and short-length convolution computation. *arXiv*, 11. https://doi.org/10.48550/arXiv.2210.17490
17 Yuan, S., Venegas-Andraca, S.E., Wang, Y. et al. (2019). Quantum image edge detection algorithm. *International Journal of Theoretical Physics* 58: 2823–2833.
18 Xi-Wei, Y., Wang, H., Liao, Z. et al. (2017). Quantum image processing and its application to edge detection: theory and experiment. *Physical Review X* 7: 031041.
19 Grigoryan, A.M. and Agaian, S.S. (2022). Gradients and compass operators: method of rotations. *Proceedings of SPIE, Defense + Commercial Sensing, Multimodal Image Exploitation and Learning 2022*, 12, Orlando, Florida, United States. https://doi.org/10.1117/12.2618353.
20 Grigoryan, A.M. and Grigoryan, M.M. (2009). *Brief Notes in Advanced DSP: Fourier Analysis with MATLAB*. CRC Press, Taylor and Francis Group.
21 Grigoryan, A.M. (2001). 2-D and 1-D multi-paired transforms: frequency-time type wavelets. *IEEE Transactions on Signal Processing* 49 (2): 344–353. https://doi.org/10.1109/78.902116.
22 Grigoryan, A.M. (2020) 'Resolution map in quantum computing: signal representation by periodic patterns,' *Quantum Information Processing*, 19:177, p. 21. https://doi.org/10.1007/s11128-020-02685-7
23 Grigoryan, A.M. and Agaian, S.S. (2024). 3-Qubit circular quantum convolution computation using the Fourier transform with illustrative examples. *Journal of Quantum Computing* 6: 1–14. https://doi.org/10.32604/jqc.2023.026981.

13

Quantum Neural Networks: Harnessing Quantum Mechanics for Machine Learning

Quantum neural networks (QNNs) are a burgeoning field that merges the principles of quantum mechanics with machine learning [1–7]. These models aim to leverage the unique capabilities of quantum computers, such as superposition, entanglement, and interference, to overcome limitations faced by classical neural networks (NNs). Developed in the 1950s, NNs are inspired by the human brain's information processing and problem-solving abilities. Classical NNs often struggle with limitations

1) **Extensive training requirements:** Training NNs can be time consuming and data hungry.
2) **Noisy or incomplete data:** Handling noisy or incomplete datasets can be challenging for classical NNs.
3) **Scalability and generalization:** Scaling classical NNs to handle high-dimensional data and achieving good generalization can be difficult.

Quantum machine learning (QML), a promising intersection, has seen significant growth in recent years, fueled by a rise in impactful research papers. Both QML algorithms and quantum computers share characteristics that make them a promising match. QML algorithms often produce probabilistic results and contain correlated components. Quantum computers inherently provide probabilistic results upon measurement.

Classical machine learning algorithms struggle with the "curse of dimensionality," where computational complexity grows exponentially with data dimensions. Quantum computers, on the other hand, can perform parallel computations on superpositions of quantum states, potentially offering exponential speedups for specific tasks. While not a complete replacement for classical computing, quantum computers have the potential to be powerful tools within machine learning pipelines [8, 9]. Additionally, quantum computing offers significant speedups for tasks like factorization, search, and optimization compared to classical computers.

This chapter explores a general method for building NNs inspired by classical architectures on quantum computers. We will delve into the key differences and similarities between quantum and classical NNs and explore the potential of quantum-classical hybrid approaches. We will also discuss some potential applications and future trends in the exciting field of QNNs.

13.1 Introduction to Quantum Neural Networks: A New Frontier in Machine Learning

QNNs, also known as quantum ANNs or quantum circuit learning, represent a burgeoning field at the intersection of quantum mechanics and machine learning. These models aim to harness the unique power of quantum computing – quantum parallelism, interference, and entanglement – to surpass the capabilities of traditional NNs.

Quantum Image Processing in Practice: A Mathematical Toolbox, First Edition. Artyom M. Grigoryan and Sos S. Agaian.

Companion website: www.wiley.com/go/grigoryan/quantumimageprocessing

Artificial neural network (ANN), also known as NNs or neural nets, is a branch of machine learning and artificial intelligence (AI) inspired by the human brain's structure and function. ANNs are composed of interconnected processing units called artificial neurons, which can perform various operations on the input data and pass the output to the next layer of neurons. Arranged in layers, these networks leverage machine learning or deep learning techniques to adjust internal parameters and extract meaningful patterns from data. Quantum computing offers a compelling alternative. It boasts significant speedups for tasks like factorization, search, and optimization compared to classical computers. Additionally, quantum computing allows for the representation and manipulation of complex, high-dimensional data like quantum states and operators. ANNs can learn from data and adapt to new situations, making them powerful tools for solving complex problems in various domains, such as computer vision, natural language processing (NLP), speech recognition, and robotics. The biological NNs consist of billions of neurons, specialized cells that can receive, process, and transmit signals to other neurons through synapses. The neurons and synapses form a complex network that can store and process information parallel and distribute. The biological NNs can learn from experience and adapt to changing environments, enabling animals to perform various cognitive and behavioral tasks. ANNs try to mimic biological NNs by using mathematical models and computational algorithms.

Despite these advantages, directly applying classical NN architectures to quantum problems presents several challenges. Simulating complex, many-body quantum interactions on classical computers can be computationally expensive. Furthermore, quantum mechanics relies on linear operations for evolution, leading to probabilistic observations. This contrasts with the nonlinear activation functions (AFs) used in classical NNs. Consequently, adapting classical NN architectures to effectively handle quantum data requires innovative approaches that bridge these fundamental differences. Here, we mention the following.

1) **Classical ANNs: mimicking the brain for machine learning:** ANNs are a powerful tool in machine learning, inspired by the structure and function of the human brain. These networks consist of interconnected processing units, called artificial ncurons, that work together to solve complex problems.
2) **Building blocks of ANNs:** artificial neurons: First proposed in the 1950s, ANNs are built upon fundamental units called artificial neurons. Each neuron performs a simple mathematical operation involving three main steps: multiplication, summation, and activation.
3) **Weighted inputs:** Each input to a neuron is multiplied by a corresponding weight (see Fig. 13.1). These weights represent the strength of the connection between neurons and influence the overall output.
4) **Summation:** All weighted inputs are often summed together with a bias term. This bias term acts like a constant offset, allowing the neuron to shift its activation range.
5) **Activation:** The summed value is then passed through an AF, which introduces nonlinearity into the network. This nonlinearity allows ANNs to learn complex patterns in data that would not be possible with linear models.

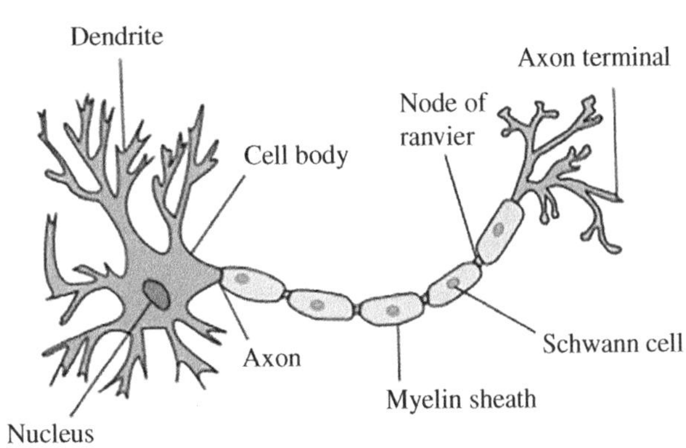

Fig. 13.1 Biologic neuron (Dhp1080 / Wikimedia / https://en.wikipedia.org/wiki/File:Neuron.svg, last accessed 17 July 20204 / CC BY 3.0.).

6) **Network structure and learning:** ANNs are parallel distributed information processors, meaning information is processed simultaneously across many interconnected neurons. These networks learn by adjusting the weights between neurons based on training data. The goal is to minimize the difference between the network's predicted output and the actual desired output. This process of weight adjustment is called learning, and various algorithms exist to optimize this process.

Applications of ANNs: ANNs are a versatile tool used in a wide range of applications, including:

ANNs are powerful and versatile computational tools widely applied across numerous domains due to their ability to learn complex patterns and relationships in data. Below, **we will explore** some of their key applications, **showcasing their adaptability and impact**.

One of the most prominent uses of ANNs is in **image classification and segmentation**, where they excel at identifying objects or regions of interest within images. This capability has revolutionized fields like medical imaging, enabling the detection of tumors, lesions, and other anomalies in radiographs, MRIs, and CT scans. Similarly, ANNs power advanced vision systems in autonomous driving that recognize pedestrians, traffic signs, road lanes, and other vehicles, ensuring safe and efficient navigation in dynamic environments.

In **speech processing**, ANNs have become integral to the functionality of virtual assistants such as Siri, Alexa, and Google Assistant. They enable these systems to understand spoken commands and respond intelligently. Additionally, ANNs drive automated transcription services, converting spoken language into written text, which finds applications in customer service, legal documentation, and accessibility tools for deaf people.

ANNs are also invaluable in **function approximation and regression analysis**, where they learn relationships between input and output data to make accurate predictions. In the financial sector, ANNs forecast stock prices, analyze market trends, and manage risks. In scientific research, they model complex phenomena like climate change, chemical reactions, or astrophysical events, providing researchers with predictive insights and aiding decision-making.

Another critical application lies in **pattern and sequence recognition**. ANNs can identify recurring patterns in data, such as financial market trends, consumer purchasing behavior, or DNA sequences. In bioinformatics, this ability aids in decoding genetic information, facilitating advancements in personalized medicine and genetic research. ANNs analyze market trends and consumer preferences in economics, enabling businesses to make data-driven strategic decisions.

ANNs are highly effective at **filtering and reducing noise**, improving the quality of audio recordings, images, and other data types. In telecommunications, they enhance signal clarity for better communication. In multimedia applications like photography and video production, ANNs reduce noise and artifacts, producing sharper images and higher-quality videos.

ANNs' **clustering capability** is particularly valuable in customer segmentation, market research, and data compression. By grouping similar data points, ANNs help businesses develop targeted marketing strategies and personalize customer experiences. In image compression, they optimize file sizes while preserving visual quality, making them indispensable in storage and transmission applications.

ANNs play a pivotal role in robotics by enabling robots to learn from their environment and adapt to new tasks. In manufacturing, ANNs improve robotic efficiency, ensuring precise and reliable operations. Beyond industrial applications, ANNs empower robots used in exploration – whether on space missions or deep-sea research – helping them navigate uncharted territories and perform complex tasks autonomously. This contributes to expanding human knowledge and capability in challenging environments.

ANNs transform **healthcare** through early disease detection, diagnosis, and treatment planning. For example, ANNs assist in identifying diseases like cancer, diabetes, and cardiovascular conditions by analyzing medical imaging, patient data, and genomic information. Additionally, they power personalized treatment recommendations and drug discovery by analyzing large datasets of patient responses and molecular structures.

In **NLP**, ANNs drive cutting-edge applications like **ChatGPT**, an advanced conversational AI model. ChatGPT is based on transformer architectures and leverages NNs to generate human-like text, making it highly effective for:

- **Customer support:** Automating responses, reducing wait times, and improving user experience.
- **Content creation:** Assisting in writing, editing, brainstorming ideas, and generating summaries.
- **Education:** Answering questions, explaining concepts, and providing tutoring assistance.
- **Programming:** Debugging code, providing examples, and explaining programming concepts.
- **Healthcare:** Offering preliminary guidance, scheduling appointments, and assisting with symptom analysis.

Other NLP applications powered by ANNs include **machine translation**, **sentiment analysis**, **text summarization**, and **speech-to-text** systems. ChatGPT and similar models have transformed how businesses interact with customers and how individuals engage with technology, enhancing productivity and accessibility.

In **cybersecurity**, ANNs are applied for anomaly detection, malware identification, and intrusion detection systems. They help identify suspicious activity, predict security threats, and enhance data protection. ANNs also improve fraud detection in banking and e-commerce by analyzing transaction patterns and identifying irregularities.

Gaming and entertainment have also benefited from ANNs, which create realistic characters, simulate physics-based interactions, and improve graphics rendering. ANNs enhance player experiences by powering adaptive game mechanics and personalized difficulty adjustments based on player behavior.

ANNs contribute significantly to **energy management** by optimizing power grids, predicting energy consumption patterns, and improving the efficiency of renewable energy systems. In solar and wind energy, ANNs help forecast energy generation based on weather conditions, ensuring better grid stability and resource management.

ANNs optimize route planning, traffic flow analysis, and supply chain management in transportation and logistics. Predicting traffic congestion and dynamically adjusting routes help reduce delivery times and fuel consumption. Additionally, ANNs are instrumental in fleet management and warehouse optimization, enhancing overall efficiency in logistics operations.

In **education**, ANNs support personalized learning platforms that adapt to individual student needs. They analyze student performance data to tailor learning materials, recommend resources, and identify improvement areas, making education more accessible and effective.

13.2 McCulloch–Pitts Processing Element

The artificial neurons are the key element or processing units (PN) of NNs that compute a weighted sum of their inputs and send the result through a nonlinear AF. The diagram of a biological neuron is shown in Fig. 13.1. The neuron contains a cell body that processes a signal receiving through the dendrite (a branch of a tree) connected to other neurons or from the environment. The neuron has also a single axon, from where the processed signal is transmitted to other neurons.

Figure 13.2 shows the model of an artificial neuron. A neuron receives the input $\boldsymbol{x} = (x_1, x_2, x_3, \ldots, x_n)$, $n \geq 2$, each input component x_j with its weight $w_j, j = 0{:}\ n$. The vector parameter of the neuron $\boldsymbol{w} = (w_1, w_2, \ldots, w_n)$. The neuron sums them up

$$s = \boldsymbol{w}\boldsymbol{x}^T - b = \sum_{j=1}^{n} w_j x_j - b \tag{13.1}$$

and adds a bias b which makes the neuron more effective. This shift of the weighted sum can be included in the sum, by adding the input $x_0 = -1$ with weight $w_0 = b$. The AF $F(s)$ processing the neuron's output using the weighted, that is, the output y of the neuron, is

$$y = F(s) = F\left(\sum_{j=1}^{n} w_j x_j - b\right) = F\left(\sum_{j=0}^{n} w_j x_j\right). \tag{13.2}$$

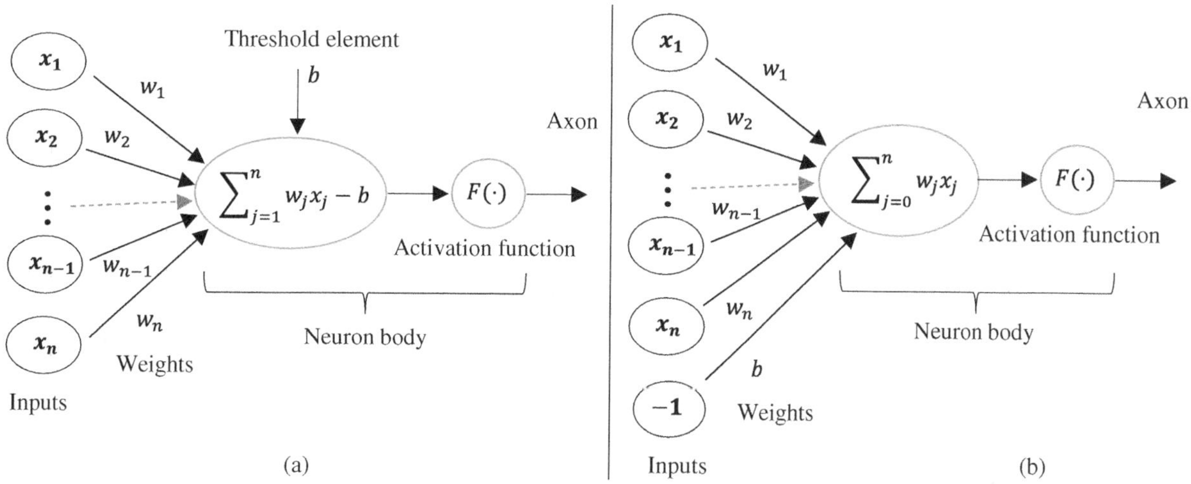

Fig. 13.2 Deterministic model of the artificial neuron (a) with n inputs and (b) with $n+1$ inputs.

The AF can be a step function, linear function, logistic sigmoid, or tanh function, which introduces nonlinearity into a neuron's behavior and maps the resulting output values into either the interval $[-1, 1]$ or $[0, 1]$. The AFs commonly used in many ANNs are the following:

1) **Threshold function:** Binary step function is a threshold-based AF that occurs after a specific threshold neuron is activated and below the threshold neuron is deactivated. The function is defined by

$$F(s) = \text{sign}(s) = \begin{cases} 1, & s > 0; \\ 0, & s \leq 0, \end{cases} \quad \text{or} \quad F(s) = \begin{cases} 1, & s > 0; \\ -1, & s \leq 0. \end{cases}$$

These two functions are shown in Fig. 13.3 in part (a). The binary step function is a binary classifier, where the decision function is the AF.

2) *Linear function* defined as (see Fig. 13.3 in part (b)),

$$F(s) = \begin{cases} 1, & s > 0.5; \\ s + 0.5, & s \in [-0.5, 0.5]; \\ 0, & s < -0.5. \end{cases} \tag{13.3}$$

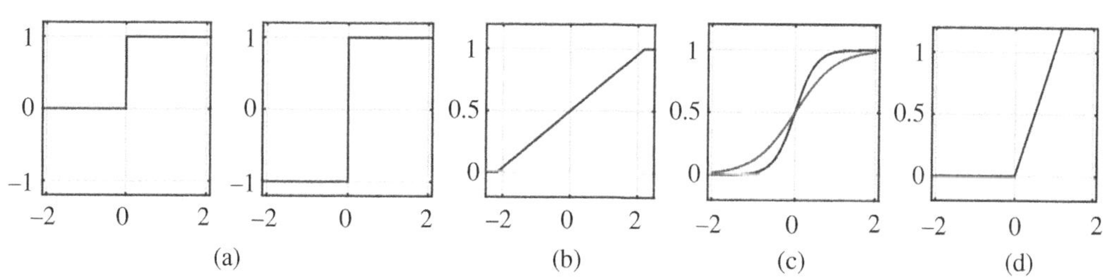

Fig. 13.3 Activation functions: (a) the threshold functions, (b) the linear, (c) logistic, and (d) rectified liner unit functions.

3) *Logistic (sigmoid) function* serves to transform an input into an output confined within the range of 0 to 1,

$$F(s) = \frac{1}{1 + e^{-cs}}, \tag{13.4}$$

where the coefficient $c > 0$ defines the width of the function along the s-axis. In Fig. 13.3, this function is shown for coefficients $c = 2$ and 4 in part (c). The sigmoid AF is particularly useful in classification tasks, as its output can be interpreted as a probability. Smooth and monotonic activation curves of the sigmoid function in combination with its everywhere differentiability facilitate the application of optimization techniques. Often employed in single-layer perceptron, the sigmoid AF models the activation of logistic units in NNs, particularly those with only one layer.

4) *Rectified linear unit function (ReLU)* is the most widely used AF in contemporary NNs. Defined as $F(s) = max(0, s)$, ReLU is a piecewise linear function that efficiently thresholds real-valued numbers at zero (see Fig. 13.3 part [d]). Its advantages include less computationally expensive operations and fast computation of both activation and derivatives. However, noting a discontinuity in the derivative at $x = 0$ is essential.

5) *Softmax function* emerges as the preferred choice for multiclass classification tasks, where probabilities for each class are computed, and the class with the highest probability is designated as the prediction. Mathematically expressed as

$$g(\boldsymbol{x}) = \text{softmax}(\boldsymbol{x}) = \left[\frac{e^{x_1}}{\sum_{k=1}^{K} e^{x_k}}, \ldots, \frac{e^{x_K}}{\sum_{k=1}^{K} e^{x_k}}\right]. \tag{13.5}$$

Here, x_j are the real values of the input vector $\boldsymbol{x} = (x_1, x_2, \ldots, x_K)$ on the last layer of the network. K is the number of classes in the multiclass classifier e_k^x.

The softmax function applies the standard exponential function to each element of the input vector. The denominator ensures normalization, guaranteeing that the output values form a valid probability distribution by summing to 1. Typically added as the last layer in NNs, the softmax AF aids in determining the most probable output among multiple classes. Its application is instrumental in converting network outputs into interpretable probability distributions, crucial for decision-making in multiclass scenarios.

Figure 13.4 shows the scheme of choosing the AF for the output layer of an ANN for different applications.

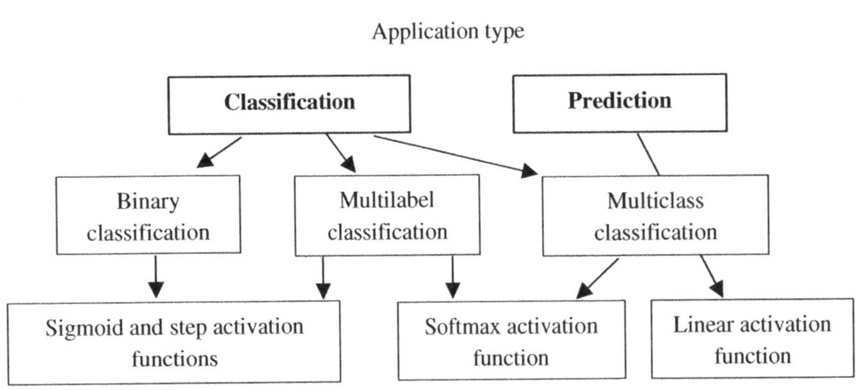

Fig. 13.4 AF suggestion for the ANN layers, depending on the application.

The main task when working with a NN is the selection of weights $w_j, j = 0, 1, \ldots, n$. There are different rules for training a NN. For example, we consider the simplest model of the network with one layer with one output layer and linear AF. It is so-called adaptive liner element (ADALINE). We consider the well-known delta rule which is a gradient descent learning rule for updating the weights. The training starts by initializing the weights randomly. Then, on each step t of training, the weight will be renewed, $w_{t,j} \to w_{t+1,j}$. Depending on the task, a training sample with a set of input and output data is selected, $\boldsymbol{X} = \{\boldsymbol{x}_t; t = 1{:}\,L\}$ and $\boldsymbol{D} = \{d_t; t = 1{:}\,L\}$, respectively. The inputs $\boldsymbol{x}_t$ are the n-D vectors $(x_{t,0}, x_{t,1}, \ldots, x_{t,n})$. For each pair $(\boldsymbol{x}_t, d_t)$ from the training sample, the mean-square error function (the loss function) is calculated

$$E(t) = \frac{1}{2}(y_t - d_t)^2 = \frac{1}{2}\left(F\left(\sum_{j=0}^{n} w_{t,j}x_{t,j}\right) - d_t\right)^2. \tag{13.6}$$

The training is based on the gradient descent method in the space of weights, and its training scheme is shown in Fig. 13.5. The process continues $L_1 \le L$ times until the error $E(t)$ decreases to the desired minimum error value $\varepsilon > 0$. If the condition $E(t) \le \varepsilon$ is met, this network is considered trained and the weights $w_j = w_{L_1,j}$ will be saved and used for this NN. It is clear that it is desirable to have a large training sample when working with such application, as image recognition.

The artificial neurons (perceptron) learning algorithm with a given learning rate, $a \in (0, 1)$, work as follows [10]:

Step-1: The weights $w_j = w_j(0)$, $j = 0{:}\,n$, are initialized to small random numbers.

Step-2: The pattern vectors $\boldsymbol{x}_t = (x_{t,0}, x_{t,1}, \ldots, x_{t,n})$ are alternately fed to the perceptron, and the rule in Eq. 13.2 produces the output y_t. The weights are updated according to the rule

$$w_j(t+1) = w_j(t) + ax_{tj}(y_t - d_t), \quad b(t+1) = b(t) - \alpha(y_t - d_t), \tag{13.7}$$

where t is the discrete time, and d_t is the desired output for training.

Step-3: The mean-square error function $E(t)$ is calculated. If $E(t) > \varepsilon$, go to Step 2, otherwise training stops and weighted coefficients are saved as $w_j = w_j(t+1)$, $j = 1{:}\,n$.

The learning rate can be calculated by $a = a(t) = 1/(1 + \|\boldsymbol{x}_t\|)$. In some work, it is considered $a = 0.0005$ [11].

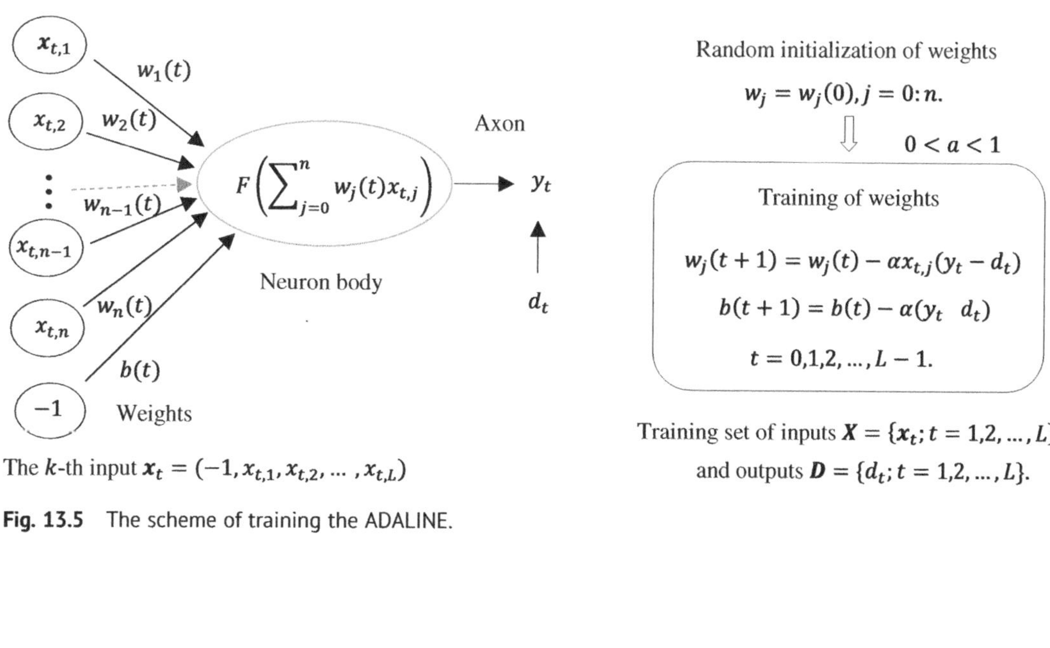

Fig. 13.5 The scheme of training the ADALINE.

13.3 Building Blocks: Layers and Architectures

Artificial neurons are the building blocks of ANNs. Let us explore how these neurons are organized to create functional networks. ANNs are structured into layers, where each layer contains groups of interconnected neurons performing similar operations. The simplest type of ANN is a single-layer network. It consists of just one layer of neurons directly receiving input data and producing an output. While these networks can handle basic tasks like linear regression or binary classification (predicting yes or no outcomes), they struggle with more complex problems that involve nonlinear relationships or require identifying multiple categories (multiclass classification).

To overcome this limitation, multilayer networks were introduced. These networks contain two or more layers of neurons, with the hidden layers being the key to unlocking greater capabilities. Hidden layers sit between the input and output layers, allowing the network to learn complex, nonlinear patterns in the data. Information flows from the input layer through the hidden layers, where it is progressively transformed and combined. Each hidden layer can have a different number of neurons, and stacking multiple hidden layers allows the network to extract increasingly intricate features from the data. This is what empowers multilayer networks to tackle complex tasks like image recognition or natural language processing. Networks with more than three hidden layers are often referred to as deep NNs. Deep learning architectures leverage the power of these hidden layers to achieve remarkable results in various fields. A multilayer network can perform more complex tasks, such as image recognition or natural language understanding, by extracting and combining features from the input data.

Example 13.1 Visualizing a Multi-layer Network

Let us take a closer look at a specific network architecture to understand how layers work together. Figure 13.6 depicts a 2-layer network with two outputs. This network has:

1) **Input layer:** Receives source data into the network. The number of neurons in this layer depends on the data format (for example, the number of pixels in an image).
2) **Hidden layers (2 in this example):** These layers are where the real "magic" happens. Each neuron in a hidden layer receives weighted inputs from the previous layer (either the input layer or another hidden layer). These weighted inputs are then processed by an AF, introducing nonlinearity, and allowing the network to learn complex relationships in the data. The number of neurons in hidden layers can significantly impact

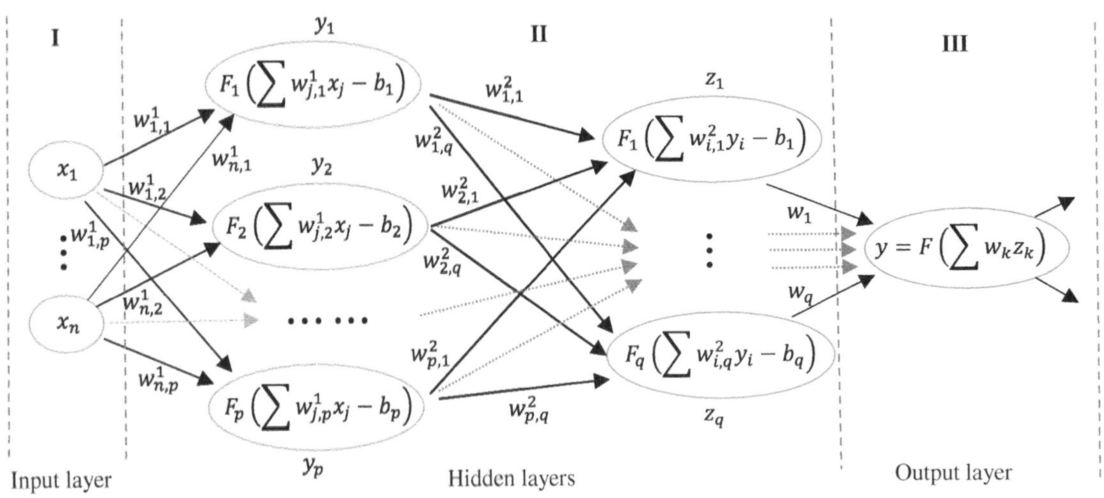

Fig. 13.6 Example of a three-layer network structure with two outputs of a feedforward ANN.

the network's capabilities. Too few neurons may limit its ability to learn complex patterns, while too many can lead to overfitting (poor performance on unseen data).

3) **Output layer:** Contains neurons that produce the results of the network. The number of neurons here depends on the task. For example, a network performing image classification might have one neuron per class it can identify.

As mentioned above, hidden layers are crucial for tackling complex tasks. Here is why:

a) **Modeling nonlinear relationships:** Real-world data often exhibits nonlinear relationships. Hidden layers, with their AFs, allow the network to model these nonlinearities, enabling it to learn intricate patterns that would not be possible with simpler linear models.
b) **Feature extraction:** Hidden layers can progressively extract and combine features from the data as it flows through the network. Imagine hidden layers acting like filters gradually refines the information to uncover the underlying patterns relevant to the task at hand. This feature of extraction capability is what empowers networks to excel in tasks like image recognition or natural language processing.
c) **Choosing network architecture:** The number of neurons in the input and output layers is determined by the specific problem to be solved. The input layer needs enough neurons to represent the data, while the output layer needs enough neurons to represent the possible outputs.

Let the input $\boldsymbol{x}$ be N-dimensional, $\boldsymbol{y}$ be the L-dimensional output (we can assume $L = N$), and M be the number of hidden layers. Assume that the single hidden layer number m in the network has K neurons. Each neuron in this layer is with N_1 inputs, that is, the input vector is $\boldsymbol{x} = (x_1, x_2, \ldots, x_{N_1})$, and N_1 outputs, or the vector $\boldsymbol{y} = (y_1, y_2, \ldots, y_{N_1})$. This processed unit can be described as $F(\boldsymbol{w}_k\boldsymbol{x}^T - b_k)$, $k = 1{:}\,K$. Here, $\boldsymbol{w}_k$ is the vector of weights and b_k is the corresponding bias. Then, calculations in the hidden layer can be described as input for the next hidden layer,

$$\boldsymbol{x}_{new} = \left(F\left(\boldsymbol{w}_1\boldsymbol{x}^T - b_1\right), F\left(\boldsymbol{w}_2\boldsymbol{x}^T - b_2\right), \ldots, F\left(\boldsymbol{w}_K\boldsymbol{x}^T - b_K\right)\right). \tag{13.8}$$

This operation in the matrix form can be written as $\boldsymbol{x}_{new} = F(\boldsymbol{W}_m\boldsymbol{x}^T - \boldsymbol{b}_m)$, where $\boldsymbol{W}_m$ is the matrix $K \times N_1$ composed by rows $\boldsymbol{w}_k$, $k = 1{:}\,K$, and $\boldsymbol{b}_m$ is a bias vector $(b_1, b_2, \ldots, b_K)$. Considering elementwise operation of the nonlinear function F to the vector input, the above equation can be written as

$$\boldsymbol{x}_{new} = F\left(\left(\boldsymbol{w}_1\boldsymbol{x}^T - b_1\right), \left(\boldsymbol{w}_2\boldsymbol{x}^T - b_2\right), \ldots, \left(\boldsymbol{w}_K\boldsymbol{x}^T - b_K\right)\right) = F\left(\boldsymbol{W}_m\boldsymbol{x}^T - \boldsymbol{b}_m\right).$$

This can be generalized to a NN with M hidden layers as:

$$y = N(\boldsymbol{x}) = \boldsymbol{W}_M \cdot F\left(\boldsymbol{W}_{M-1} \cdot F\left(\boldsymbol{W}_{M-2} \cdot \ldots \cdot F\left(\boldsymbol{W}_1 \cdot \boldsymbol{x}^T - \boldsymbol{b}_1\right)^T\right)^T \ldots\right)^T - \boldsymbol{b}_M. \tag{13.9}$$

13.4 Artificial Neural Network Architectures: From Simple to Complex

ANNs come in various architectures, each tailored for specific tasks. Here is a breakdown of some key types:

1) **Fully connected layers (FCLs):** In an FCL, every neuron in one layer connects to every neuron in the subsequent layer. This creates a dense network of connections for information exchange. FCLs are commonly used as building blocks in various ANN architectures. As an example. Figure 13.7 shows a four-layer fully connected network structure with a set of weights $\boldsymbol{W}_l = \left\{w^l_{j,k}; j, k = 1{:}\ n\right\}$, $l = 1, 2, 3$.

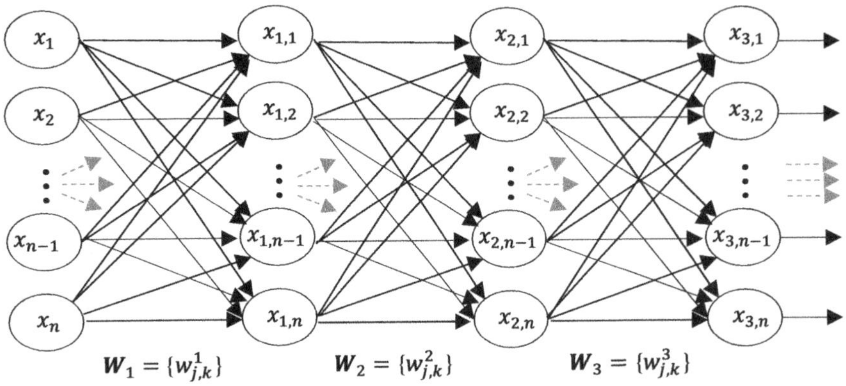

Fig. 13.7 The structure of a four-layer fully connected network.

2) **Deep learning (a multilayered approach)**: Deep learning refers to ANNs with more than three layers [2]. These deeper architectures allow for more complex feature extraction and learning, particularly beneficial for tackling intricate problems.
3) **Multilayer perceptrons (MLPs):** Stacking layers for learning. MLPs are feedforward networks with multiple hidden layers stacked between the input and output layers. Each layer in an MLP uses AFs like ReLU to introduce nonlinearity, enabling the network to learn complex patterns. MLPs excel at building mapping models, $f: x \rightarrow y$ and learning relationships between inputs and desired outputs. However, training them can be slow, and they may require substantial data.
4) **Feedforward networks, the workhorses of ANNs:** Feedforward networks represent the most common ANN architecture. Information flows unidirectionally, from the input layer through hidden layers (if present) to the output layer. These networks are well suited for supervised learning tasks like classification or regression. Feedforward networks encompass various sub-architectures, including:
   a) Fully connected networks (discussed earlier);
   b) **CNNs:** Specialized for image and video recognition tasks, leveraging filters to extract features;
   c) **Radial basis function networks (RBFNs):** Often used for function approximation or regression problems.
5) **Recurrent networks: remembering the past for sequential data:** Unlike feedforward networks, recurrent networks have feedback loops. Information can flow not just forward but also backward, allowing the network to consider past inputs when processing new information. This makes them ideal for sequential data like time series or natural language, where understanding the context of previous elements is crucial. Recurrent networks are used in tasks like:
   a) Unsupervised learning (e.g., clustering and anomaly detection)
   b) Supervised learning (e.g., sequence prediction)

   Common recurrent network architectures include:

   a) Simple recurrent networks (SRNs)
   b) Long short-term memory (LSTM) networks: Particularly adept at handling long-term dependencies in data.
   c) Gated recurrent unit (GRU) networks: Another variant for processing sequential data.
   d) Graph recurrent neural networks (GNNs): Designed for data structured as graphs, where nodes represent entities and edges represent relationships between them.

6) **Self-organizing networks (learning without a teacher):** Self-organizing networks (SONs) exhibit unsupervised learning behavior. They can learn from unlabeled or unstructured data without explicit guidance. This makes them suitable for tasks like:

- **Dimensionality reduction:** Compressing high-dimensional data into a lower-dimensional space while preserving essential information.
- **Feature extraction:** Identifying key characteristics within the data.
- **Data visualization:** Creating visual representations of complex data for easier interpretation.

Examples of SON architectures include: (i) Kohonen networks, (ii) Hopfield networks, and (iii) Boltzmann machines.

By understanding these different architectures, we can gain a deeper appreciation for the versatility and power of ANNs.

13.5 Key Properties and Operations of Artificial Neural Networks

1) **Learning from examples:** ANNs can learn from both labeled (with known outputs) and unlabeled data by adjusting connection weights through learning algorithms. These algorithms can be supervised (guided by a desired output) or unsupervised (learning from data patterns). They can also update weights after each example (online learning) or after processing batches of data (batch learning).
2) **Adaptability:** ANNs can adapt to new situations by modifying weights based on feedback (external or internal error signals). This adaptability improves their ability to generalize to unseen data, handle noise, and perform well in dynamic environments.
3) **Nonlinearity:** ANNs introduce nonlinearity using AFs. This complexity allows them to model intricate relationships and capture higher-order features within the data, leading to a more expressive and flexible network.
4) **Fault tolerance:** Parallel and distributed processing, along with redundancy, contribute to fault tolerance in ANNs. This parallelism improves efficiency and allows for handling large datasets. Redundancy enhances reliability by enabling the network to recover from failures or errors.

The main challenges and limitations of ANNs are:

a) ANNs can be difficult to interpret or explain, due to their black-box or opaque nature. ANNs are often criticized for being opaque or "black boxes." It can be difficult to understand how they arrive at their decisions, what features they learn from the data, or how they internally process information. This lack of interpretability can be problematic in sensitive applications like healthcare or finance, where understanding the reasoning behind decisions is crucial for trust and accountability.
b) Training large and complex ANNs often requires significant computational resources and memory. This can hinder their application on large datasets or in real-time scenarios with limited resources, such as on mobile devices, edge computing, or Internet of Things (IoT) applications.
c) **The high flexibility of ANNs can lead to two pitfalls:** overfitting and underfitting. Overfitting occurs when the network memorizes the training data too well, losing its ability to generalize to unseen data. Conversely, underfitting happens when the network fails to capture essential patterns from the data, resulting in poor performance. Both scenarios can significantly impact the validity and reliability of the model, especially when dealing with noisy or incomplete data.

Researchers are actively addressing these limitations, paving the way for a more robust and interpretable future. The current and future trends and developments in ANNs are:

a) **Explainable neural networks:** This area focuses on developing networks that can provide explanations for their outputs. Techniques like visualization, attribution methods, and model decomposition are being explored to bring transparency and interpretability to the decision-making process of these complex models. Explainable models can foster trust in ANNs, aid in debugging and improvement, and even lead to discoveries from the data.
b) **Federated learning: privacy-preserving collaboration:** Federated learning offers a groundbreaking approach to machine learning. It allows multiple devices to collaboratively train a shared model, all while keeping the data private on each device. This method is particularly valuable in scenarios where data privacy is critical. Traditional machine learning often aggregates data from various sources into a central location for training. This approach raises privacy concerns, as it involves collecting and storing sensitive user information on a central server. Federated learning tackles this issue by enabling model training directly on user devices. The key benefit is that sensitive data never leaves the devices. This significantly reduces privacy risks associated with centralized data storage. Additionally, federated learning circumvents the need to transfer large datasets, reducing network strain and computational demands. By leveraging the distributed processing power of participating devices, this approach not only safeguards user privacy but also fosters a more efficient learning process.
c) **Generative adversarial networks (GANs):** Creating new from existing: GANs are fascinating ANNs capable of generating entirely new and realistic data based on existing datasets. Imagine creating photorealistic images of people who do not exist or landscapes from entirely new worlds – that is the power of GANs.

GANs work as follows. Two NNs in a duel: GANs consist of two NNs locked in an adversarial competition.

- **Generator:** This network acts like a creative artist, continuously striving to produce novel, realistic data that resembles the training data.
- **Discriminator:** This network plays the role of a critical art expert, meticulously examining the generated data and attempting to distinguish it from real data.
- **A continuous learning process:** Through this ongoing competition, both networks progressively improve. The generator gets better at crafting realistic forgeries, while the discriminator hones its ability to detect them. Over time, this adversarial training allows the generator to produce increasingly convincing and high-quality data.

Applications of GANs:

- **Creating realistic images:** From generating new artistic styles to editing existing photos, GANs are pushing the boundaries of image creation.
- **Augmenting datasets:** In situations where real-world data is scarce, GANs can be used to generate synthetic data to supplement training datasets.
- **Developing new materials:** Researchers are exploring GANs to design and discover new materials with desired properties.

GANs are a rapidly evolving field with immense potential. As these networks continue to develop, they promise to revolutionize various creative and scientific domains.

13.5.1 Reinforcement Learning: Learning Through Trial and Reward

Reinforcement learning is a powerful branch of machine learning where agents learn to make optimal decisions through a feedback mechanism of rewards and penalties. Imagine an agent interacting with an environment, acting, and receiving feedback in the form of rewards for good choices and penalties for bad ones. This iterative process allows the agent to learn what actions lead to the most rewards over time, enabling it to adapt and perform well within a specific context.

Reinforcement learning has demonstrably versatile applications across various fields, including:

1) **Robotics:** Robots can leverage reinforcement learning to navigate complex environments, perform intricate tasks, and optimize their movements for efficiency.
2) **Video games:** AI game characters can be trained using reinforcement learning to exhibit strategic decision-making and human-like playing styles.
3) **Finance:** Reinforcement learning algorithms can be used to develop trading strategies that adapt to market fluctuations and aim to maximize returns.
4) **Healthcare:** Researchers are exploring reinforcement learning for drug discovery and optimizing treatment protocols based on patient responses.

Example 13.2 A Case Study: DeepMind's AlphaGo

A remarkable example of reinforcement learning's capabilities is DeepMind's AlphaGo program. By playing countless games against itself, AlphaGo learned from its experiences, refining its strategies and decision-making through trial and error. This ability to learn autonomously, guided by the principles of reinforcement learning, allowed AlphaGo to achieve superhuman performance in the complex game of Go, defeating some of the world's best human players.

The Power of Learning Through Interaction: A key strength of reinforcement learning lies in its ability for agents to learn from interacting with an environment. This makes it particularly valuable in situations where explicitly programming every possible scenario is impractical or infeasible. As a result, reinforcement learning remains a vibrant area of research and development, pushing the boundaries of AI by enabling machines to acquire complex behaviors through self-exploration and experience.

13.6 Quantum Neural Networks: A Computational Model Inspired by Quantum Mechanics

QNNs, also known as quantum nets, are a branch of machine learning and AI that are inspired by the principles of quantum mechanics. QNNs are composed of interconnected processing units called quantum neurons, which can perform various operations on the input data and pass the output to the next layer of quantum neurons (see Fig. 13.8 in part (b)). QNNs can learn from data and adapt to new situations, making them powerful tools for solving complex problems in various domains, such as computer vision, natural language processing, speech recognition, and robotics. The QNN combines the neuron structure with a quantum circuit.

Typically, the input vector data $\boldsymbol{x}$ is encoded component-wise into angles. For example, we consider the encoding

$$x_j \rightarrow \theta_j = 2\pi \frac{x_j - \min\limits_k \{x_k\}}{\max\limits_k \{x_k\} - \min\limits_k \{x_k\}}, \qquad j = 0 : (n-1). \tag{13.10}$$

Therefore, the qubits can be composed by $|x_j\rangle = \cos(\theta_j)\,|\,0\rangle + \sin(\theta_j)\,|\,1\rangle$. Then, the following $(n+1)$-qubit input superposition is considered:

$$|\,\varphi(x)\rangle = \sum_{j=0}^{n-1} |\,x_j\rangle\,|\,j\rangle = \cos(\theta_j)\,|0\rangle + \sin(\theta_j)\,|1\rangle$$

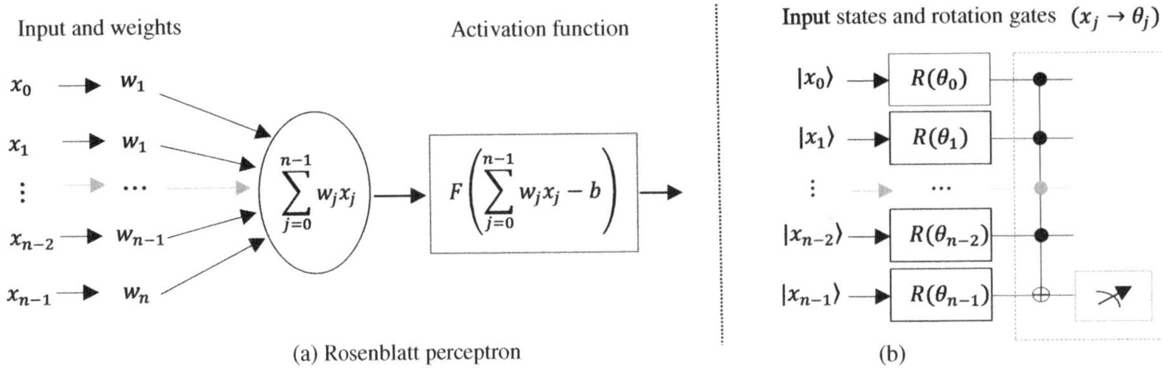

Fig. 13.8 (a) Conventionnel artificial neuron structure and (b) the quantum neuron structure [12].

or n-qubit superposition

$$|\varphi_1(x)\rangle = \frac{1}{\sqrt{N}}\sum_{j=0}^{n-1} x_j\,|j\rangle.$$

In Fig. 13.8 in part (b), $R(\theta_j), j = 0:(n-1)$ are rotation gates with encoded angles and all n qubits are rotated separately. A multi-qubit CNOT gate is used to read the output of the last qubit. What we can notice from this model of the quantum neuron is the absence of the weights in which the classical neuron uses in its body. Another structure of processing the input qubits is given in Fig. 13.9, for the 4-qubits. Here, too, at first all the qubits rotate separately. There is no quantum superposition here as such. At this stage, there is no connection between the qubits; they need to be entangled. The controlled parameterized rotation gates $R(\phi_j), j = 0{:}\ 3$, are added (instead of the controlled NOT gates) for the entanglement. The degree of entanglement is controlled by the parameters $\boldsymbol{\phi} = \{\phi_j; j = 0{:}\ 3\}$. Thus, instead of the summation $s = x_0w_0 + \ldots + x_3w_3$ in the neuron, the entanglement is used. This quantum circuit is shown to provide higher accuracy of image classification and is considered as an example of a perceptron with nonlinearity [11].

In general, the quantum perceptron model is shown in Fig. 13.10. Processing of the input superposition is carried out by a selected and pre-prepared parameterized or random quantum circuit (or a template), which is also called Ansatz. The parameters of this circuit are optimized using classical or quantum techniques. Such a circuit may contain multiple cores (or filters), and they may be so complex that the calculations may not be reproducible

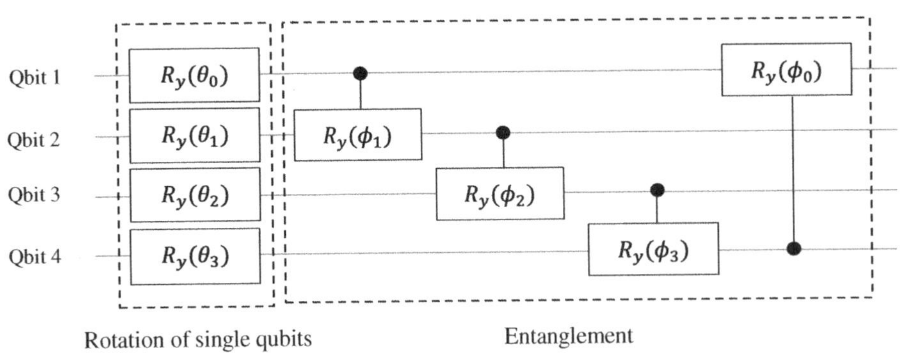

Fig. 13.9 The quantum circuit for 4-qubit filter in a quantum perceptron.

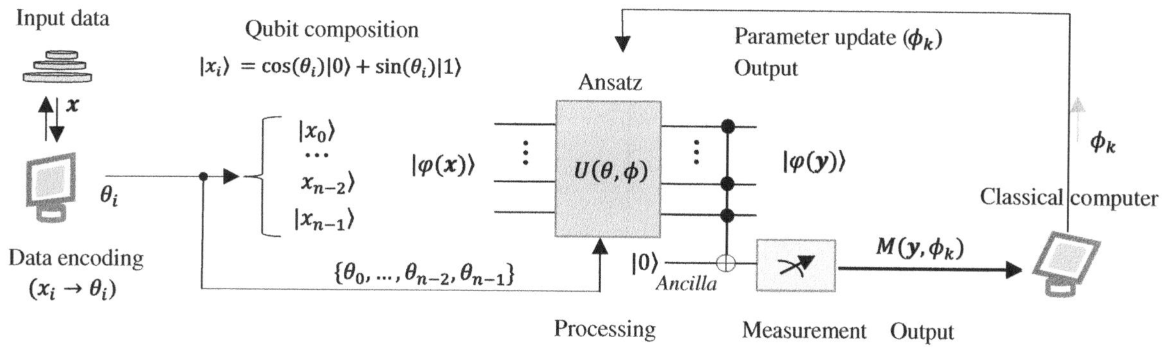

Fig. 13.10 Quantum-classical model of the perceptron with multi-CNOT gate on the ancilla qubit.

by classical computers. The required input superposition can be accomplished by the unitary DsiHT [13], as discussed in Section 6.1.2. The result, the quantum superposition $|\varphi(y)\rangle = U(\boldsymbol{\theta}, \boldsymbol{\phi}) \mid \varphi(x)\rangle$, written in an additional zero qubit (ancilla), will be measured and evaluated by a classical computer to further optimize the model parameters, $\boldsymbol{\phi}$. Other gates than CNOT can also be used on the ancilla qubit.

Let us assume the number of inputs equals $n = 2^r$, $r > 1$. The sum $s = x_0w_0 + \ldots + x_{n-1}w_{n-1}$ in the neuron can be presented by the quantum superposition

$$\mid \varphi(\mathbf{y})\rangle = (x_0w_0 + x_1w_1 + \ldots + x_{n-1}w_{n-1}) \mid 0\rangle + \cdots + \cdot \mid n\rangle. \tag{13.11}$$

Indeed, let us consider the DsiHT generated by the vector $w' = (w_0, w_1, \ldots w_{n-1})'$. The $(n \times n)$ matrix H_w of this transform is unitary with the first row composed by the weighting coefficients w_k. Then the multiplication

$$H_{\mathbf{w}}(|\varphi(x)\rangle) = H_{\mathbf{w}}\mathbf{x}' = \begin{bmatrix} w_0 & w_1 & \cdots & w_{n-1} \\ \cdot & \cdot & \cdots & \cdot \\ \cdot & \cdot & \cdots & \cdot \\ \cdot & \cdot & \cdots & \cdot \end{bmatrix} \begin{bmatrix} x_0 \\ x_1 \\ \cdot \\ x_{n-1} \end{bmatrix} = \begin{bmatrix} x_0w_0 + x_1w_1 + \ldots + x_{n-1}w_{n-1} \\ \cdot \\ \cdot \\ \cdot \end{bmatrix}$$
$$= (x_0w_0 + x_1w_1 + \ldots + x_{n-1}w_{n-1}) \mid 0\rangle + \cdot \mid 1\rangle + \cdots + \cdot \mid n-1\rangle.$$

It is assumed that both vectors of inputs and weights are normalized. Figure 13.11 illustrates the model of the perceptron, by using the DsiHT with an activation operator. The addition qubit (ancilla qubit) is added to measure the weighted sum s.

In general, the quantum perceptron model is shown in Fig. 13.12.

Unitary matrices other than DsiHT can also be considered (see, e.g., [14]) in the above model.

Example 13.3 Artificial Neuron with Two 2-Qubit QsiHTs
Consider the $n = 4$ case with the vector of weights $\boldsymbol{w} = (w_0, w_1, w_2, w_3) = (0.2, 0.3, 0.1, 0.4)$. Let the input vector be $\boldsymbol{x} = (x_0, x_1, x_2, x_3) = (1, 2, 4, 5)/\sqrt{46}$ with the 2-qubit superposition

$$\mid \varphi(\boldsymbol{x})\rangle = \frac{1}{\sqrt{30}}(|0\rangle + 2|1\rangle + 4|2\rangle + 5|3\rangle).$$

The normalized vector of weights $\boldsymbol{w} = (0.3651, 0.5477, 0.1826, 0.7303)$. The sum $s = (\boldsymbol{x}, \boldsymbol{w}) = (x_0w_0 + x_1w_1 + x_2w_2 + x_3w_3) = 0.8614$.

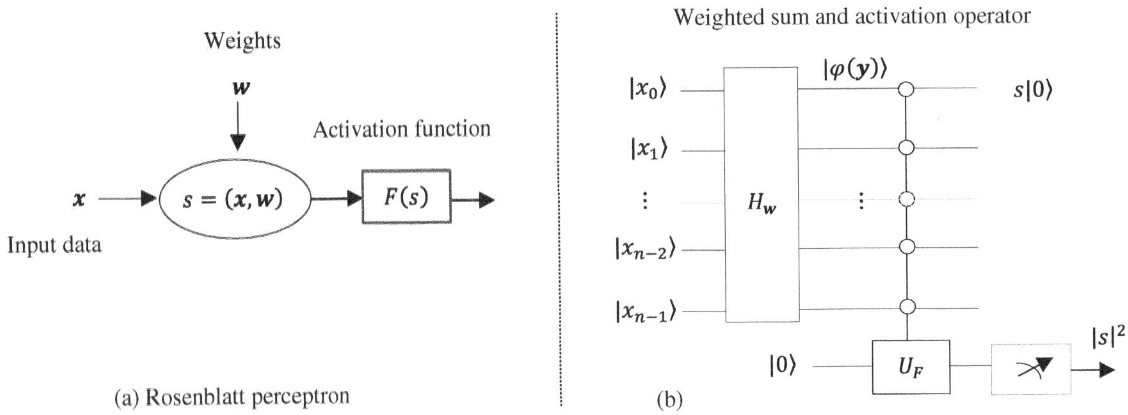

Fig. 13.11 (a) Conventionnel artificial neuron structure and (b) the quantum neurone structure.

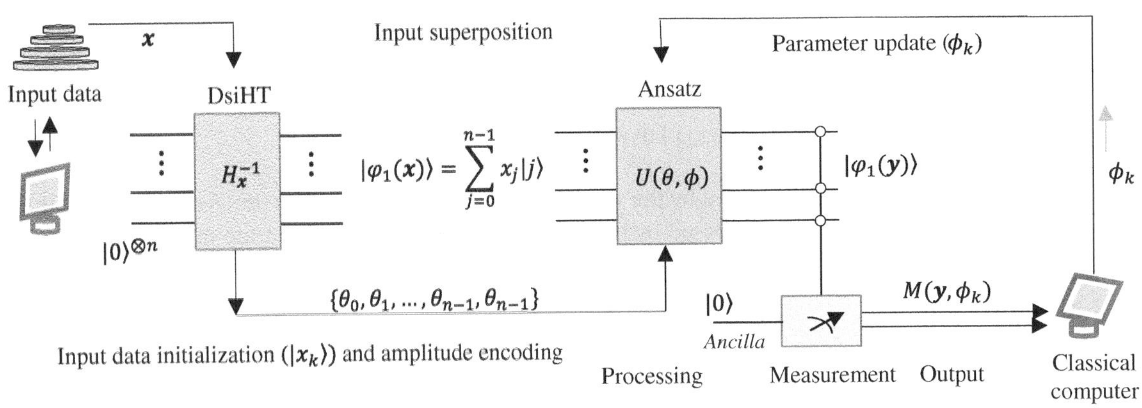

Fig. 13.12 Quantum-classical model of the perceptron with multi-CNOT gate on the ancilla qubit.

The generated by the vector $\boldsymbol{w}$ DsiHT with the strong wheel carriage [15] is

$$H_{\boldsymbol{w}} = \begin{bmatrix} 0.3651 & 0.5477 & 0.1826 & 0.7303 \\ -0.9309 & 0.2148 & 0.0716 & 0.2864 \\ 0 & -0.8086 & 0.1427 & 0.5708 \\ 0 & 0 & -0.9701 & 0.2425 \end{bmatrix} = DH_{\boldsymbol{w};i} = D\begin{bmatrix} 2 & 3 & 1 & 4 \\ -13 & 3 & 1 & 4 \\ 0 & -5.6667 & 1 & 4 \\ 0 & 0 & -16 & 4 \end{bmatrix},$$

where the diagonal matrix $D = \text{diag}\{0.1826, 0.0716, 0.1427, 0.0606\}$. Then,

$$H_{\boldsymbol{w}}|\,\varphi(\boldsymbol{x})\rangle = H_{\boldsymbol{w}}\frac{1}{\sqrt{46}}\begin{bmatrix} 1 \\ 2 \\ 4 \\ 5 \end{bmatrix} = \begin{bmatrix} 0.8614 \\ 0.1795 \\ 0.2665 \\ -0.3934 \end{bmatrix} = s|\,0\rangle + 0.1795|1\rangle + 0.2665|2\rangle - 0.3934|3\rangle.$$

This matrix is composed of three rotations by angles $\{-1.1970, -0.9418, -1.3258\}$, or in degrees $\{\vartheta_1, \vartheta_2, \vartheta_3\} = \{-68.5833°, -53.9601°, -75.9638°\}$,

$$H_w = \underbrace{\begin{bmatrix} \cos\vartheta_3 & -\sin\vartheta_3 & 0 & 0 \\ \sin\vartheta_3 & \cos\vartheta_3 & 0 & 0 \\ 0 & 0 & 1 & 0 \\ 0 & 0 & 0 & 1 \end{bmatrix}}_{\vartheta_3 = -68.5833°} \underbrace{\begin{bmatrix} 1 & 0 & 0 & 0 \\ 0 & \cos\vartheta_2 & -\sin\vartheta_2 & 0 \\ 0 & \sin\vartheta_2 & \cos\vartheta_2 & 0 \\ 0 & 0 & 0 & 1 \end{bmatrix}}_{\vartheta_2 = -53.9601°} \underbrace{\begin{bmatrix} 1 & 0 & 0 & 0 \\ 0 & 1 & 0 & 0 \\ 0 & 0 & \cos\vartheta_1 & -\sin\vartheta_1 \\ 0 & 0 & \sin\vartheta_1 & \cos\vartheta_1 \end{bmatrix}}_{\vartheta_1 = -75.9638°}. \tag{13.12}$$

The quantum circuit for this DsiHT with three rotation gates is given in Fig. 13.13. Two permutations $P_2 = (0, 3, 2, 1)$ and $P_2' = (0, 1, 2, 3)$ are used and the transform is described by

$$H_w = (R_{\vartheta_3} \oplus I_2)\underbrace{P_2(I_2 \oplus R_{\vartheta_2})P_2'}(I_2 \oplus R_{\vartheta_1}). \tag{13.13}$$

The initialization of the input vector $\boldsymbol{x}$ can be accomplished by the inverse DsiHT generated by this vector with the matrix

$$H_x = \begin{bmatrix} 0.1474 & 0.2949 & 0.5898 & 0.7372 \\ -0.9891 & 0.0440 & 0.0879 & 0.1099 \\ 0 & -0.9545 & 0.1862 & 0.2328 \\ 0 & 0 & -0.7809 & 0.6247 \end{bmatrix} = DH_{x;i} = D\begin{bmatrix} 1 & 2 & 4 & 5 \\ -45 & 2 & 4 & 5 \\ 0 & -20.5 & 4 & 5 \\ 0 & 0 & -6.25 & 5 \end{bmatrix}.$$

where the diagonal matrix $D = \mathrm{diag}\{0.1474, 0.0220, 0.0466, 0.1249\}$. One can notice that as for the matrix $H_{x;i}$, the first row and the main diagonal of the above matrix $H_{x;i}$ are the not normalized vector $\boldsymbol{x} = (x_0, x_1, x_2, x_3) = (1, 2, 4, 5)$. The graphs of the functions $v_k(n)$ which are the rows of this matrix are shown in Fig. 13.14. One

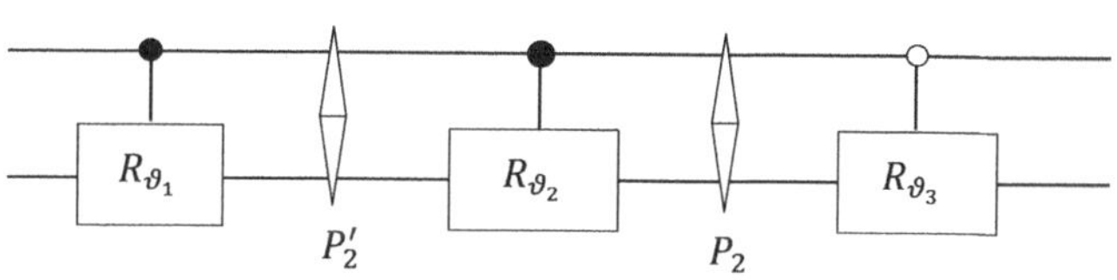

Fig. 13.13 The quantum scheme of the strong wheel-carriage 2-qubit QsiHT.

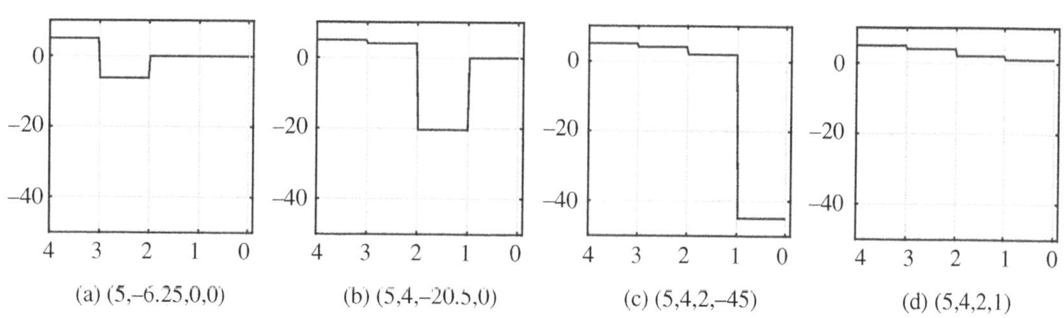

Fig. 13.14 Four row-waves of the matrix $H_{x;i}$: (a) $v_4(n)$, (b) $v_3(n)$, (c) $v_2(n)$, and (d) $v_1(n)$. All waves are drawn in reverse time order.

can notice the wave motion, when the two piece-wise function the wave becomes 3, and 4 piece-wise waves, and finally the wave $\boldsymbol{x}$. The motion of waves can be described by this diagram:

	n	3	2	1	0
	v_3	5	-6.25	0	0
v_2	Δ_1:	0	10.25	-20.5	0
v_1	Δ_2:	0	+	22.5	-45
v_0	Δ_3:	0	0	+	**46**

This diagram shows, for instance, that the wave $v_2 = v_3 + \Delta_1 = [5, -6.5, 0, 0] + [0, 10.25, -20.5, 0] = [5, 4, -20.5, 0]$ (written from the right to left). At the end of the movement, there is a strong splash in the wave $v_1(n)$, which is equal to the signal energy $\Delta_3 = 46 = \|\boldsymbol{x}\|^2$. The wave became the generatrix of the transform; $v_0 = v_1 + \Delta_3 = \boldsymbol{x}$. A detailed description of the motion of the DsiHT transform basis functions can be found in [16].

The input signal is calculated as $H_{\boldsymbol{x}}^T(1,0,0,0)' = \boldsymbol{x}$, or in terms of superpositions as

$$|\,\varphi(\boldsymbol{x})\rangle = \sum_{k=0}^{3} x_k|\,k\rangle = H_{\boldsymbol{x}}^T(|\,0\rangle^{\otimes 2}). \tag{13.14}$$

Three angles of rotations in the DsiHT, $H_{\boldsymbol{x}}$, are

$$\{\phi_1, \phi_2, \phi_3\} = \{-81.5213°, -72.6539°, -51.3402°\}.$$

The transpose matrix $H_{\boldsymbol{x}}^T = H_{\boldsymbol{x}}^{-1}$ can be described by the rotations as follows:

$$H_{\boldsymbol{x}}^T = (R_{-\phi_1} \oplus I_2)\underbrace{P_2(I_2 \oplus R_{-\phi_2})P_2'}(I_2 \oplus R_{-\phi_3}). \tag{13.15}$$

Thus, with two DsiHTs, the input superposition $|\varphi(\boldsymbol{x})\rangle$ is initiated by the transform $H_{\boldsymbol{x}}^T$ and then a new 2-qubit superposition $|\varphi(\boldsymbol{y})\rangle$ is calculated and measured. The probability of this superposition to be at state $|00\rangle$ equals to

$$\Pr(M|\varphi(\boldsymbol{y})\rangle = |00\rangle\,) = |\langle 00\,|\varphi(\boldsymbol{y})\rangle|^2 = |\,x_0w_0 + x_1w_1 + x_2w_2 + x_3w_3|^2 = |s|^2 = 0.8614^2.$$

The quantum model of calculation of the weighted sum by two DsiHTs is shown in Fig. 13.15. The zero state ancilla qubit is added to write on it and measure the outcome (the weighted sum $s = 0.8614$) as the probability s^2 of 2-qubit superposition at state $|0\rangle$.

The block scheme of the transform $H_{\boldsymbol{w}}\boldsymbol{x}$ with the following execution of the AF $F(s_k)$ is given in Fig. 13.16. All four outputs of the DsiHT are shown. These outputs like the first output, $s = s_0$, are different weighted sum of inputs. The second output is the weighted sum of four inputs, the third output is the weighted sum of three outputs,

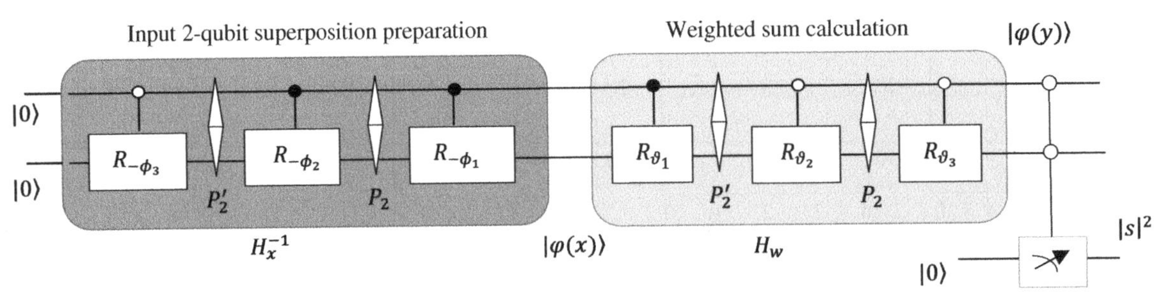

Fig. 13.15 The model of the quantum scheme of artificial neuron with two 2-qubit QsiHTs.

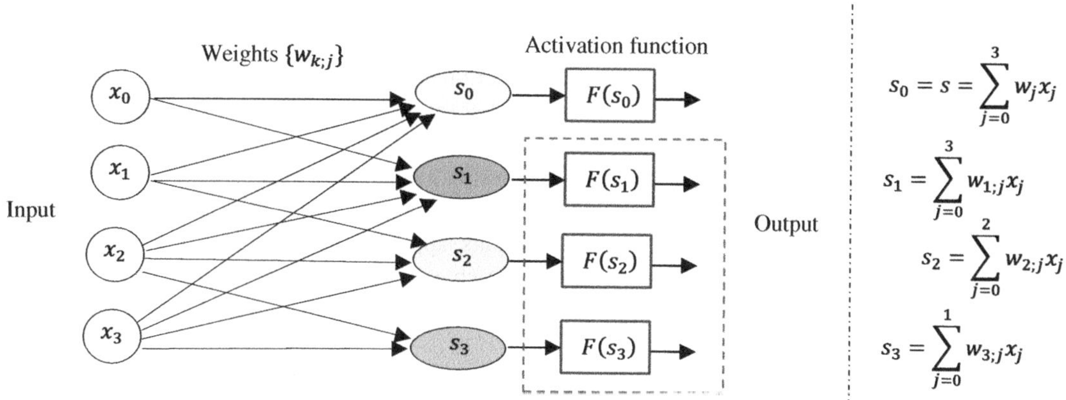

Fig. 13.16 The structure of four neurons, when using the 4-point DsiHT.

and the last output of the transform weights only two last outputs. This is the structure of the 4-point DsiHT with the strong carriage-wheel in general case. The classical model of the perceptron is described by only the first output of the DsiHT, that is, s_0. In this figure, we have a big picture of weighted sums of the inputs. The matrix $H_{\boldsymbol{w}}$ is the matrix of weights. Each column (row) in this matrix is the basis function of the DsiHT (inverse DsiHT). All four compose the basis in the space of 4-point discrete signals. The strong connection of these outputs is the $H_{\boldsymbol{w}}$ transformation itself. Therefore, we assume that these three outputs can also be used in CNN together with the classical perceptron.

This picture can be viewed as a network of four neurons with identical inputs $\boldsymbol{x}$. All weights for these four neurons are defined by the input weight vector $\boldsymbol{w}$.

Figure 13.17 shows the structure of these four neutrons in a more visual form. On the left part, the classical Rosenblatt perceptron is shown, which we call the main pyramid in this network. The three other perceptrons, or the chain of three other pyramids, are shown on the right side in this figure. It is considered that the same AF is used in all these pyramids.

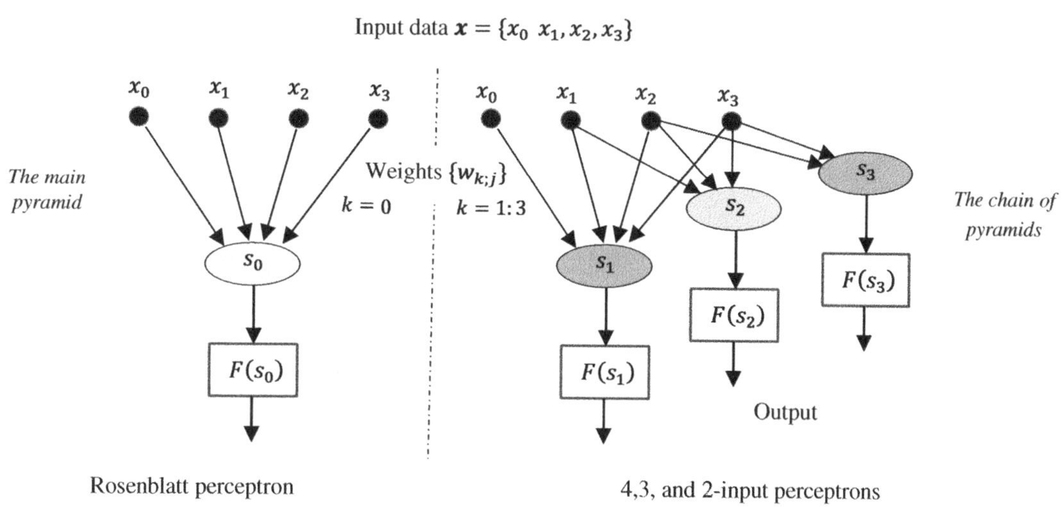

Fig. 13.17 The quantum 4-neuron structure of the 4-point discrete Heap transform.

In the general case with n inputs, $x = \{x_0, x_1, \ldots, x_{n-1}\}$, $n \geq 4$, such a pyramidal structure of the DsiHT network describes, together with first main perceptron, a chain of $(n - 1)$ perceptrons with a decreasing number of input data. This chain together with the first perceptron may give the additional feature maps that can be useful in practical application, such as future extraction and classification. Whether this chain of perceptrons should be thrown out of the network is a question that remains to be analyzed and resolved.

The main concept of QNNs is to emulate the information processing mechanisms of the classical NNs that are widely used in machine learning for the important task of pattern recognition. Classical NNs consist of artificial neurons, which are simple units that can receive one or more inputs, apply a function to them, and produce an output. The artificial neurons are connected by weighted links, which can modulate the signals passing through them. The weights of the links can be adjusted by a learning algorithm, which can optimize the performance of the network based on a given objective or criterion.

QNNs try to mimic classical NNs by using quantum computing principles and technologies, such as qubits, quantum gates, or quantum circuits. All gates correspond to unitary transforms with the same number of inputs and outputs. Figure 13.18 shows the generic QNN structure. QNNs use quantum neurons, which are quantum circuits that can receive one or more inputs, apply a quantum function to them, and produce an output. The quantum neurons are connected by quantum links, which can transfer quantum states or information between them. The quantum links can be adjusted by a quantum learning algorithm, which can optimize the performance of the network based on a quantum objective or criterion.

As one can see, these two perspectives are complementary and do not necessarily rely on strict definitions of concepts such as "quantum neuron" or what constitutes a QNN's "layer." Here are some key aspects of QNNs. The main components and operations of QNNs are:

A) **Input layer:** This is the layer that receives the input data and encodes it into quantum states using a quantum circuit or a classical pre-processing technique. The input data can be classical or quantum, depending on the type and application of the quantum network. For example, the input layer can receive a 4×1 classical data vector and encode it into quantum states using angle encoding. For example, if the data vector is [0.1, 0.2, 0.3, 0.4], then the corresponding angles of rotation are $[0.1\pi, 0.2\pi, 0.3\pi, 0.4\pi]$ in the range of $[0, \pi]$. Then, the angles can be used for rotation gates, such as a Pauli X, Y, or Z (see Table 1 in Section 2). Applying these quantum gates to four qubits, each initialized in the $|0\rangle$ state creates a quantum superposition of states representing the data vector.

B) **Hidden layer(s):** This is the layer that performs the main computation of the quantum network using a variational quantum circuit, which is a quantum circuit with trainable parameters. The variational circuit can implement various functions, such as convolution, pooling, activation, or output, depending on the type and structure of the quantum network. The hidden layer(s) can also include classical components, such as classical NNs or classical optimizers, to form a hybrid quantum-classical NN.

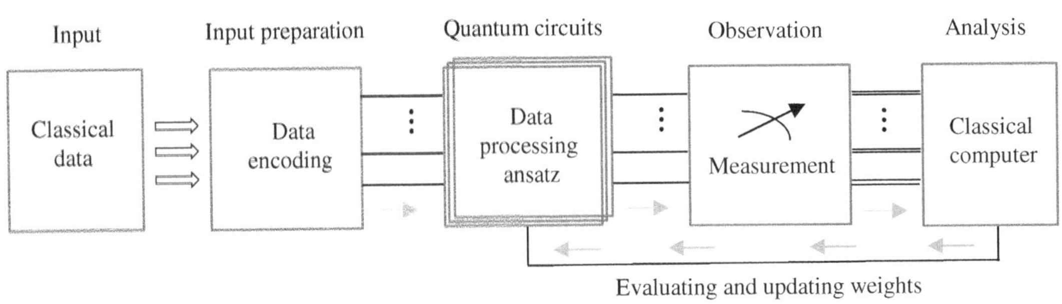

Fig. 13.18 Generic quantum neural network (QNN) structure with a classical computer.

C) **Output layer:** This is the layer that produces the output of the quantum network and sends it to the next layer or the destination. The output layer can involve quantum measurements, which are operations that extract classical data from quantum states, such as probabilities, expectations, or samples. The output layer can also involve classical post-processing, such as decoding, classification, or regression, to obtain the desired output format.

QNNs are anticipated to offer several advantages compared to classical NNs. These networks hold immense promise for surpassing classical NNs in various ways. Here are some key potential advantages:

a) **Efficiency with fewer neurons:** QNN may achieve superior performance with fewer hidden neurons than CNN. This stems from the ability of quantum gates to perform complex transformations using fewer parameters. In simpler terms, QNNs can learn intricate patterns with a more compact architecture.
b) **Exponential memory potential:** Unlike classical bits that are either 0 or 1, a single qubit in a QNN can simultaneously exist in a superposition of both states. This allows for exponential growth in memory capacity as the number of qubits increases (n qubits can store 2^n states) [12]. This vast memory potential could unlock entirely new applications for NNs.
c) **Faster learning through quantum parallelism** [17]: QNNs leverage the power of quantum parallelism, where computations can be performed simultaneously on multiple qubits. This has the potential to significantly accelerate the process of calculating gradients and cost functions, leading to faster learning compared to classical networks.
d) **Mitigating catastrophic forgetting:** Classical networks often suffer from "catastrophic forgetting," where learning new information erases previously learned knowledge. QNNs, due to the phenomenon of quantum interference, may be less susceptible to this problem, potentially retaining valuable past knowledge while acquiring new information.
e) **Tackling nonlinear problems with single layers:** Quantum entanglement, a unique property of quantum mechanics, allows qubits to be linked so that their fates are intertwined. This enables QNNs to solve problems that are linearly inseparable (not solvable by straight lines) using just a single layer of quantum neurons, unlike classical networks that typically require multiple layers for such tasks.

Additional Considerations:

a) **Noise and error sensitivity:** While generally more robust than classical networks, QNNs are still susceptible to noise and errors in the quantum hardware. Mitigating these errors remains an ongoing challenge.
b) **Scalability and practical applications:** Building and maintaining large-scale QNNs with many qubits is a significant technical hurdle. QNNs can operate efficiently at a small scale, with a density of 10^{11} neurons per cubic millimeter, as quantum devices can be miniaturized using nanotechnology [9].

Real-world applications of QNNs will likely emerge as hardware advancements overcome these scalability challenges. Overall, QNNs represent a fascinating frontier in AI. Their potential for superior performance, efficiency, and the ability to tackle complex problems make them a compelling area of research with the potential to revolutionize various fields.

13.7 The Main Difference Between QNNs and CNNs

The main difference between QNNs and CNNs is that QNNs use quantum computing principles and technologies, while CNNs use classical computing principles and technologies. Quantum computing is a potential solution to address the limitations of classical computing, such as the need for substantial time and data for training, the difficulty of dealing with noisy or incomplete datasets, and the lack of scalability and generalization. Quantum computing offers significant speedup over classical computers in various computational tasks, such as factorization, search, and optimization. Quantum computing also enables the representation and manipulation of high-dimensional and complex data, such as quantum states and operators.

QNNs use qubits, the fundamental units of quantum information, instead of bits, to represent and process data in a quantum superposition of states. QNNs use quantum circuits, composed of quantum gates and measurements, to implement the functions of NN layers and units. QNNs can leverage the advantages of quantum computing, such as quantum parallelism, interference, and entanglement, to enhance the performance and capabilities of machine learning algorithms. CNNs, on the other hand, use bits, the fundamental units of classical information, to represent and process data in a binary form. CNNs use matrices, composed of weights and biases, to implement the functions of NN layers and units. CNNs are specially designed for computer vision tasks, such as image classification, object detection, and face recognition. CNNs use convolutional filters, which are small matrices that slide over the input image and perform element-wise multiplication and summation to extract local features from the image. CNNs also use pooling layers, which are operations that reduce the size of the feature maps by applying a function, such as max or average, to a small region of the map to reduce the complexity and dimensionality of the model.

Some of the advantages of QNNs over CNNs are:

1) QNNs can exhibit exponential memory capacity, as a single qubit can store two states, and n qubits can store 2^n states, while a single bit can store only one state, and n bits can store only n states.
2) QNNs can achieve higher performance with fewer hidden neurons, as quantum gates can implement complex transformations with fewer parameters, while classical matrices require more parameters to implement the same transformations.
3) QNNs can facilitate faster learning processes, as quantum parallelism can speed up the computation of gradients and cost functions. At the same time, classical computers have to perform these computations sequentially or in parallel with limited resources. Quantum parallelism can enhance the performance and capabilities of machine learning algorithms, such as NNs, by speeding up the computation of gradients, cost functions, and feature extraction.
4) QNNs can eliminate catastrophic forgetting, as quantum interference can prevent the loss of previously learned information, while classical NNs tend to overwrite or forget the old information when learning new information.
5) QNNs can solve linearly inseparable problems with a single-layer network, as quantum entanglement can create nonlinear correlations between qubits, while classical NNs require multiple layers or nonlinear AFs to solve such problems.
6) QNNs can have different types and architectures depending on the structure and function of the quantum neurons and the quantum links.

Some of the common types and architectures of QNNs are the following:

- **Feedforward quantum networks:** These networks have a unidirectional flow of information from the input layer to the output layer without any feedback loops or cycles. Feedforward quantum networks are the most common and basic type of QNNs, and they can be used for supervised learning tasks, such as regression or classification. Feedforward quantum networks can have different architectures, such as fully connected quantum networks, quantum convolutional networks, or quantum radial basis function networks [18].
- **Recurrent quantum networks:** These have a bidirectional flow of information, with feedback loops or cycles that allow the network to have a memory of previous inputs or outputs. Recurrent quantum networks are more suitable for sequential or temporal data, such as time series or natural language. Recurrent quantum networks can be used for unsupervised or supervised learning tasks, such as clustering, anomaly detection, or sequence prediction. Recurrent quantum networks can have different architectures, such as simple recurrent quantum networks, quantum long short-term memory networks, or quantum-gated recurrent unit networks [19].
- **Self-organizing quantum networks:** These have a self-organizing or adaptive behavior, allowing the network to learn from unlabeled or unstructured data without external supervision or guidance. Self-organizing quantum

networks can be used for unsupervised learning tasks, such as dimensionality reduction, feature extraction, or data visualization. Self-organizing quantum networks can have different architectures, such as quantum Kohonen networks, quantum Hopfield networks, or quantum Boltzmann machines [20].

- **Quantum convolutional neural networks (QCNNs),** also known as quantum convolutional networks or quantum convnets, are a branch of QML that are inspired by the principles of quantum mechanics and the structure of classical CNNs. The QCNN proposed by Henderson et al. [4] uses quantum circuits to perform local transformations on the input data, like the convolutional filters used in classical CNNs. QCNNs are composed of quantum circuits that implement a CNN's convolution, pooling, and output layers, using quantum gates and measurements to perform various operations on the input data and produce the output data. QCNNs can learn from data and adapt to new situations, making them powerful tools for solving complex problems in various domains, such as computer vision, natural language processing, speech recognition, and robotics [3].

  The main concept of QCNNs is to emulate the information processing mechanisms of the classical CNNs, widely used in machine learning for feature extraction. Classical CNNs consist of multiple layers of artificial neurons, which can perform various operations on the input data, such as convolution, pooling, activation, or output. The convolution operation applies a set of filters (also called kernels) to the input data, producing a set of feature maps that capture the local patterns or features of the data. These feature maps are also known as convolution feature maps or activations maps. The pooling operation reduces the size or dimensionality of the feature maps, preserving the most important or relevant features. The activation operation introduces nonlinearity and complexity to the network, enhancing its expressive power and flexibility. The output operation produces the network's final output, such as a class label or score.

  However, classical CNNs face many limitations, such as the need for substantial time and data for training, the difficulty of dealing with noisy or incomplete datasets, and more scalability and generalization. Quantum computing, on the other hand, offers significant speedup over classical computers in various computational tasks, such as factorization, search, and optimization. Quantum computing also enables the representation and manipulation of high-dimensional and complex data, such as quantum states and operators. It should be noted that an open challenge for quantum computers is to efficiently access and read classical data.

  QCNNs try to mimic classical CNNs by using quantum computing principles and technologies, such as qubits, quantum gates, or quantum circuits. QCNNs use quantum circuits that can receive one or more inputs, apply a quantum function to them, and produce an output. A quantum circuit is a sequence of quantum gates that can implement a quantum algorithm or function. The quantum circuits can implement various functions, such as convolution, pooling, activation, or output, depending on the type and structure of the quantum network.

  One example of a quantum circuit-based algorithm for CNNs is the quanvolutional neural network, proposed by Henderson et al. [4]. QNNs use quantum circuits to perform local transformations on the input data, similar to the convolutional filters used in classical CNNs. QNNs can be trained in a variational manner using classical optimizers, or in a hybrid manner using both quantum and classical resources. QNNs have shown higher test set accuracy and faster training compared to classical CNNs on the MNIST dataset.

  Another example of a quantum circuit-based algorithm for CNNs is the quantum dilated convolutional neural network (QDCNN) proposed by Chen et al. [22]. QDCNNs extend the concept of dilated convolution, which has been widely applied in modern deep learning algorithms, to the context of hybrid NNs. Dilated convolution can increase the receptive field of the network without increasing the number of parameters or reducing the resolution. QDCNNs use quantum circuits to implement the dilated convolution operation, and they can be trained using backpropagation and gradient descent.

- **Hybrid quantum-classical networks:** The motivation behind QML is to integrate notions from quantum computing and classical machine learning to open the way for new and improved learning schemes. QNNs apply this generic principle by combining classical NNs and parametrized quantum circuits. Because they lie at an intersection between two fields, QNNs can be viewed from two perspectives: From a machine learning perspective, QNNs are, once again, algorithmic models that can be trained to find hidden patterns in data similar to their classical counterparts. These models can load classical data (inputs) into a quantum state and later process it with quantum gates parametrized by trainable weights. These are networks that combine quantum and classical computation to solve a problem [23, 24].

Typically, a hybrid network consists of the following steps: prepare the input data and encode it into quantum states using a quantum circuit or a classical pre-processing technique; apply a variational quantum circuit, which is a quantum circuit with trainable parameters, to the encoded quantum states; measure the output of the variational circuit and obtain classical data, such as probabilities, expectations, or samples; evaluate the performance of the variational circuit using a classical cost function, such as cross-entropy, mean squared error, or fidelity; update the parameters of the variational circuit using a classical optimizer; repeat the steps until the cost function reaches a minimum or a desired accuracy is achieved. Hybrid quantum-classical networks can leverage the advantages of both quantum and classical computation, such as quantum parallelism, interference, and entanglement, as well as classical data processing, optimization, and visualization.

An interesting hybrid quantum-classical convolutional and residual NNs based on classical architectures for phytoplankton classification is described in [25]. Phytoplankton are microscopic organisms that live in aquatic ecosystems and produce oxygen through photosynthesis. The taxonomic composition and abundance of phytoplankton directly impact marine ecosystem dynamics and global environment change and are listed as essential ocean variables. Phytoplankton classification is crucial for phytoplankton analysis, but it is very challenging due to its huge amount and tiny volume. Machine learning is the principal way of performing phytoplankton image classification automatically. However, it requires substantial training time and data and needs help dealing with noisy or incomplete datasets. Hybrid quantum-classical convolutional neural networks (HQCNNs) and residual NNs were developed. That combines quantum and classical computation to solve the problem of phytoplankton classification. Comparing the performance of our quantum models with their classical counterparts on a phytoplankton image dataset, it was shown that the proposed quantum models achieve faster convergence, higher classification accuracy, and lower accuracy fluctuation. The authors also analyze the effects of different quantum parameters, such as the number of qubits, the depth of thc quantum circuit, and the type of the quantum gate, on the performance of our quantum models. HQCNNs are based on the classical architectures of CNNs, widely used in machine learning for feature extraction and pattern recognition. HQCNNs use quantum circuits to implement a CNN's convolution, pooling, and output layers and use classical computers to perform the data encoding, measurement, and optimization steps. HQCNNs balance the limited function of current quantum devices and the large size of phytoplankton images, making it possible to classify phytoplankton on near-term quantum computers.

Figure 13.19 shows an example of a quantum circuit-based algorithm for CNNs [25]. The first three parts are implemented on the quantum device, while the optimization routine is executed on the classical computer, which then feeds the updated parameters back into the quantum device. The quantum circuit consists of four parts: data encoding, forward transformation performed by the ansatz, quantum measurement, and parameter optimization routine. The data encoding part encodes the input data into quantum states using a quantum circuit or a classical pre-processing technique. The forward transformation part applies a variational quantum circuit, which is a

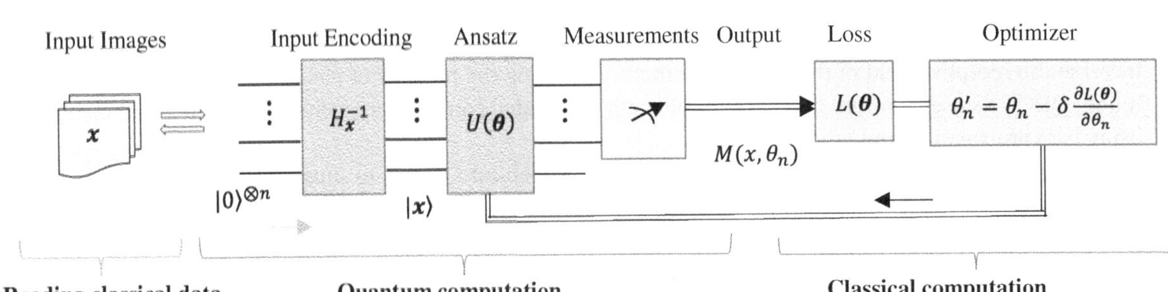

Fig. 13.19 Architecture of QNN model. QNN is a hybrid quantum-classical algorithm. The forward transformation is implemented by the quantum computer, while the optimization of parameters is done by the classical computer.

quantum circuit with trainable parameters, to the encoded quantum states. The quantum measurement part measures the output of the variational circuit and obtains classical data, such as probabilities, expectations, or samples. The parameter optimization part updates the parameters of the variational circuit using a classical optimizer, such as gradient descent, stochastic gradient descent, or its extension, Adam optimizer.

Figure 13.20 shows the QCCNN (QCCNN-1) that has a quantum convolutional layer and a classical convolutional layer, while in Fig. 13.21, the classical convolution layer is presented by two quantum convolutional layers (QCCNN-2) [25]. Note that to take full advantage of quantum feature map to process the raw data, QCCNN always takes the first convolutional layer as quantum one.

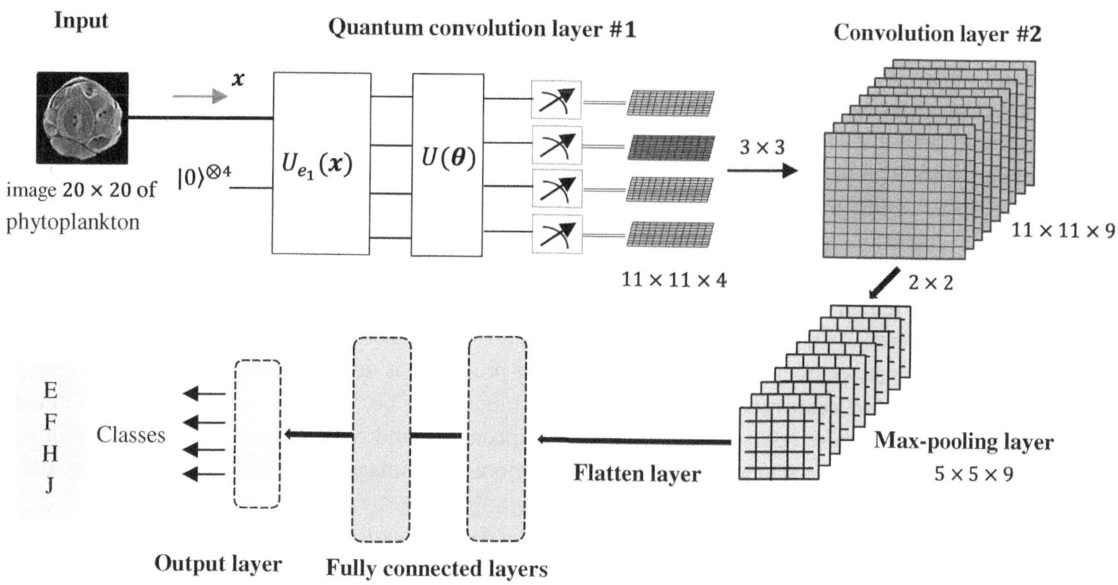

Fig. 13.20 An architecture of the QCCNN (Alfred-Wegener-Institut / Wikimedia Commons / CC BY 2.5).

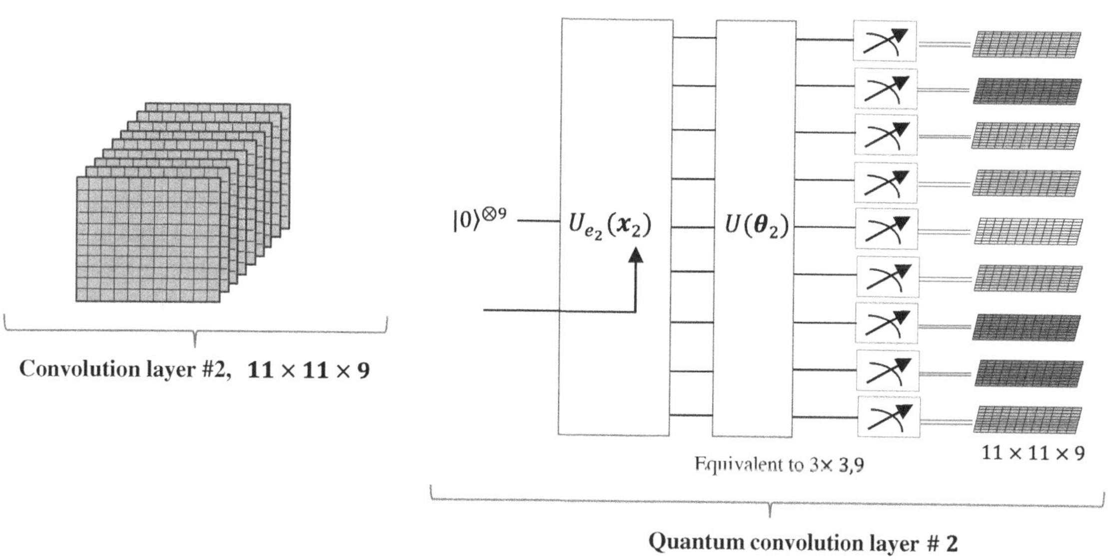

Fig. 13.21 A classical convolutional layer (named QCCNN-1) and two quantum convolutional layers (named QCCNN-2).

13.8 Applications of QNN in Image Processing

The potential applications of QNNs in image processing are vast and promising, mainly due to quantum computing's advantages in processing speed and complexity management. Below are some elaborated examples of how QNNs could revolutionize image processing:

1) **Image classification:** QNNs could outperform classical NNs in classifying images by leveraging quantum states to handle high-dimensional data. For example, a QNN might classify medical images, such as MRI scans or X-rays, more effectively, potentially detecting diseases like cancer at earlier stages by rapidly analyzing extensive datasets.
2) **Feature extraction:** Quantum circuits could expedite feature extraction by processing complex attributes of an image simultaneously through quantum superposition and entanglement. This advancement could significantly improve facial recognition systems and object detection in autonomous vehicles, enabling quicker and more accurate responses to visual stimuli.
3) **Pattern recognition:** The parallelism and high-dimensional vector spaces inherent in quantum computing could enhance QNNs' ability to recognize image patterns. This improvement is crucial for security systems that rely on pattern recognition to detect anomalies or suspicious activities within visual data. For instance, researchers from IBM used QNN to classify handwritten digits with an accuracy of 95% [26, 27]. In the numerical experiment conducted on the IBM Quantum simulator, they initially utilized the MNIST dataset [28] and a CNN with 6690 parameters with 13 qubits. The training and testing datasets comprise 60,000 and 10,000 samples, respectively. The test accuracy throughout the training process was depicted for 26 QNN layers.
4) **Image enhancement and restoration:** Quantum algorithms might be applied more efficiently to image enhancement tasks – like noise reduction, resolution enhancement, and color correction. Such capabilities would be invaluable in forensic science and astronomy, where clear images are essential for accurate data analysis.
5) **QCNNs:** QCNNs, akin to classical CNNs, could process visual data through quantum-based convolutions and pooling operations. This method could introduce novel image processing techniques, especially for noisy or complex environments like remote sensing or underwater imaging [29].

While the applications mentioned are based on current research, the field is rapidly evolving, and new developments may further expand the potential of QNNs in image processing [30].

Example 13.4 Image 2 × 4 Classifier

This example by Alexis Gomez explores the potential of QCNNs in processing pixelated images of size 2 × 4 for classifying whether a horizontal or vertical line is present in the image through the use of quantum circuits, featuring quantum convolutional and pooling layers. We dive into the use of parametrized quantum circuits for feature detection, dimensionality reduction, and efficient computation. We also detail the application of QCNNs in image classification tasks, particularly for identifying orientations such as horizontal and vertical lines in images.

QCNNs adapt the architecture of classical CNNs to leverage the unique properties of quantum mechanics. By encoding image data into quantum states and utilizing quantum operations, QCNNs aim to achieve high efficiency and performance in tasks like feature detection and image classification. In Fig. 13.22, the QCNN architecture for containing eight qubits for classifying horizontal and vertical lines in images is shown.

In QCNNs, encoding pixelated images into quantum circuits is a crucial step that bridges classical image data with quantum information processing. This process is facilitated by feature maps, which are specific functions or sets of operations designed to convert image pixels into quantum states. These maps apply a series of quantum gates to the initial state of qubits, resulting in a quantum state that encodes the information from the classical input. The "ZFeatureMap," provided by Qiskit's circuit library, is one of the most commonly used quantum feature

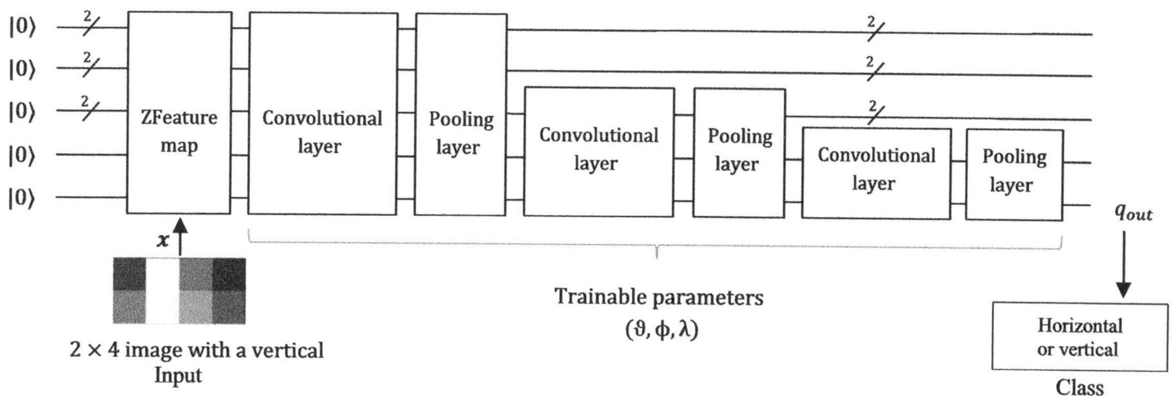

Fig. 13.22 8-qubit input QCNN for classifying whether a horizontal or vertical line is present in a 2 × 4 image.

maps. This feature map involves applying a Hadamard gate and single-qubit Z rotations to each qubit. This has the effect of not just rotating the phase of the state but also influencing how amplitudes interfere. The rotations are parameterized by the values of the image pixels, which provides a direct correlation between pixel values and rotation angles. This allows this map to encode the intensity or value of each pixel directly into the quantum state.

The "ZFeatureMap" employs single-qubit Z rotations, represented by the Pauli-Z gate matrix:

$$Z = \begin{bmatrix} 1 & 0 \\ 0 & -1 \end{bmatrix}$$

In "ZFeatureMap," these rotations are parameterized by the input features, applying a unitary operator $U_z(\theta)$ defined as:

$$U_z(\theta) = e^{-i\theta Z/2} = \begin{bmatrix} e^{-i\theta/2} & 0 \\ 0 & e^{i\theta/2} \end{bmatrix}.$$

In Qiskit, this unitary operator, $U_z(\theta)$, is defined as the "RZGate" and is referenced as $R_z(\theta)$.

Encoding classical data: The feature vector $\boldsymbol{x} = (x_1, x_2, \ldots, x_n)$ is considered, which represents classical data such as normalized image pixels. The feature map applies a rotation $U_z(\theta_r)$ to each qubit corresponding to each feature x_r, $r = 1{:}\,8$. The rotation angle θ_r for each qubit is a simple linear function of the corresponding feature such function could be $\theta_r = \pi x_r$ or a more complex function designed to represent more intricate relationships.

Applying the ZFeatureMap multiple times can effectively deepen the encoding process, creating a more complex quantum state that could capture higher-order correlations within the data. This can be particularly useful when the data has complex patterns that a single layer of the feature map might not sufficiently encode.

Linear encoding: For a normalized image with pixel intensity values in the range [0, 1], the image can be encoded onto a quantum state using a ZFeatureMap with the following approach:

1) **Initialization:** Start with all qubits in the state $|0\rangle$.
2) For each qubit, apply a Hadamard gate.
3) For each pixel x_r, apply $U_z(\pi x_r)$ to the corresponding qubit r, where

$$R_z(\vartheta_r = \pi x_r) = U_z(\pi x_r) = \begin{bmatrix} e^{-i\pi x_r/2} & 0 \\ 0 & e^{i\pi x_r/2} \end{bmatrix}.$$

4) Repeat Steps 2 and 3 (optionally): Apply the superposition and ZFeatureMap again.

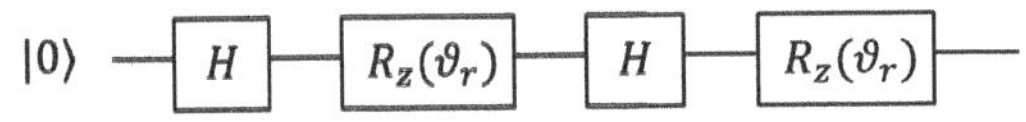

Fig. 13.23 Qiskit circuit for the feature map.

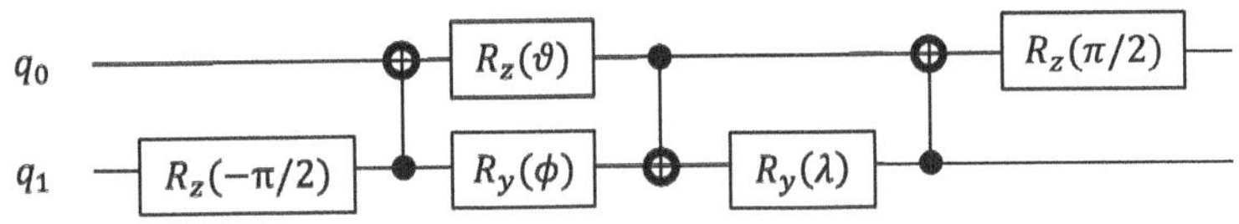

Fig. 13.24 The parameterized 2-qubit unitary circuit where ϑ, ϕ, and λ are the adjustable training parameters.

The resulting quantum state represents the encoded image, with each qubit state reflecting the intensity of the corresponding pixel through the phase factor introduced by the Z rotation, as shown in Fig. 13.23.

Parametrized unitary gate: At the core of a quantum convolutional layer is the parametrized unitary gate, a quantum circuit designed to act on pairs of qubits. This gate is akin to the kernel in classical convolutional networks, manipulating data through a series of quantum operations that depend on adjustable parameters. These parameters are optimized during the training process to minimize the loss function, much like the weights in a classical network are adjusted through backpropagation. The parametrized circuit used is based on the two-qubit unitary as proposed by Vatan and Williams [31] and is shown in Fig. 13.24.

This parameterized circuit is structured as follows:

1) **Initial phase setting:** A $R_z(-\pi/2)$ gate is applied to the second qubit. This gate adjusts the phase of the qubit, setting the stage for subsequent operations.
2) **Entanglement creation:** A controlled-NOT (CNOT) gate is used to entangle the second qubit (control) with the first qubit (target).
3) **Parameterized rotations:** The qubits undergo rotations through the $R_z(\vartheta)$, $R_y(\phi)$, and $R_y(\lambda)$ gates. These rotations are controlled by parameters ϑ, ϕ, and λ, which are analogous to the trainable weights in a kernel.
4) **Further processing and final adjustment:** A second CNOT and a final $R_z(\pi/2)$ on the first qubit are applied to complete the quantum state transformation as defined by Vatan and Williams.

Convolutional layer: This sequence of quantum operations forms a flexible and dynamic convolutional gate, capable of complex data transformations necessary for feature detection and extraction in quantum data. The convolutional layer in a QCNN is constructed by applying the two-qubit unitary gate across pairs of qubits within the quantum circuit in the following process [32]:

1) **Sequential application:** The convolutional unitary is first applied to all odd-indexed pairs of qubits, followed by its application to even-indexed pairs. This staggered approach ensures comprehensive coverage and interaction across all qubits, mimicking the sliding window operation of classical convolutional filters.
2) **Circular coupling:** To ensure that edge qubits also participate equivalently in the convolution process, the last qubit is paired with the first qubit, forming a circular topology (Fig. 13.25).

Pooling layer: Pooling layers play a critical role in CNNs by reducing the spatial dimensions of the feature maps. This helps to prevent overfitting and enhances the model's ability to capture essential features at a higher level of abstraction. In QCNNs, however, the concept of pooling takes on a different aspect due to the fundamental differences in quantum information processing. The primary goal of a pooling layer in a QCNN is to reduce the number of qubits in the circuit. Reducing the number of qubits directly decreases the computational cost and the complexity of the quantum state space that the network needs to manage. However, in quantum circuits, one cannot directly remove qubits to reduce dimensionality as is done in classical pooling.

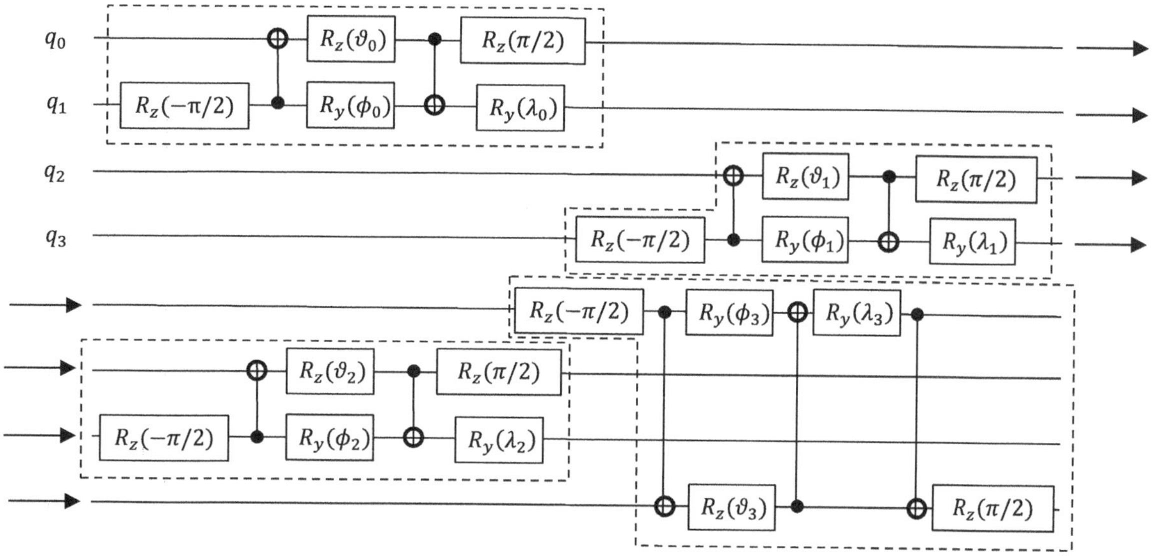

Fig. 13.25 Example of a convolutional layer for 4 qubits with sequential application and circular coupling where q_0, q_1, q_2, and q_3 are the output of the ZFeatureMap layer previously mentioned.

Fig. 13.26 Modified parameterized two-qubit unitary circuit for pooling, where q_0 and q_1 are the output of the feature map previously mentioned.

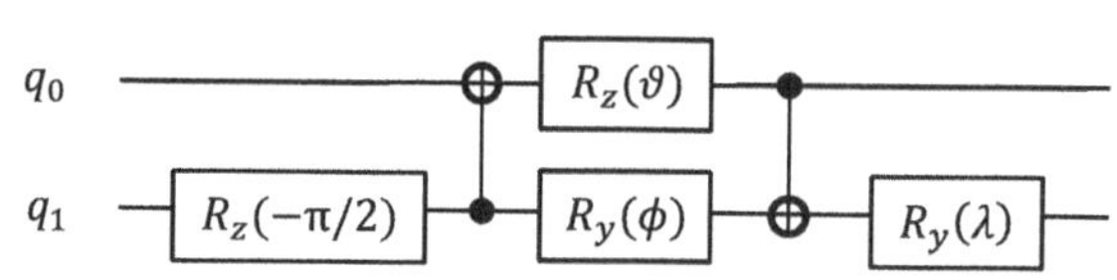

Instead, quantum pooling must be approached differently to ensure that the essential quantum information is not lost. To reduce the number of qubits in the circuit, pairs of the r qubits in the circuit are created. After pairing all the qubits, the generalized 2-qubit unitary circuit, demonstrated in Fig. 13.26, is applied to each pair. When this two-qubit unitary is applied, one qubit from each pair of qubits is ignored for the remainder of the circuit. This layer mimics the effect of combining the information of the 2 qubits into 1 qubit by applying the unitary circuit to encode the information from one qubit onto another qubit before ignoring one of the qubits for the remainder of the circuit and not performing any operations or measurements on it. For this QCNN, the first qubit is neglected in future layers where only the second qubit is used in the QCNN. This allows for the QCNN pooling layer to transform the quantum circuit from r qubits to $r/2$ qubits [32].

To implement this quantum pooling, the following steps are done:

1) **Qubit pairing:** Initially, the qubits are paired systematically within the circuit. This pairing is strategic, ensuring that each qubit is coupled with a neighbor to facilitate effective information compression.
2) **Two-qubit unitary application:** A generalized two-qubit unitary transformation, similar to those used in the convolutional layers, is applied to each pair of qubits. This transformation is crucial as it mixes the quantum states of the paired qubits, allowing for the entanglement and transfer of information between them.
3) **Selective qubit disregard:** After the application of the two-qubit unitary, one qubit (in this case, the first qubit) from each pair is selectively ignored or disregarded for the remainder of the processing within the network. This effectively reduces the number of active qubits in the circuit, achieving the pooling effect.

In the example in Fig. 13.27, the dimensionality of the four-qubit circuit is reduced to the last two qubits. These qubits would then be used in the next layer, while the first two are neglected for the remainder of the QCNN.

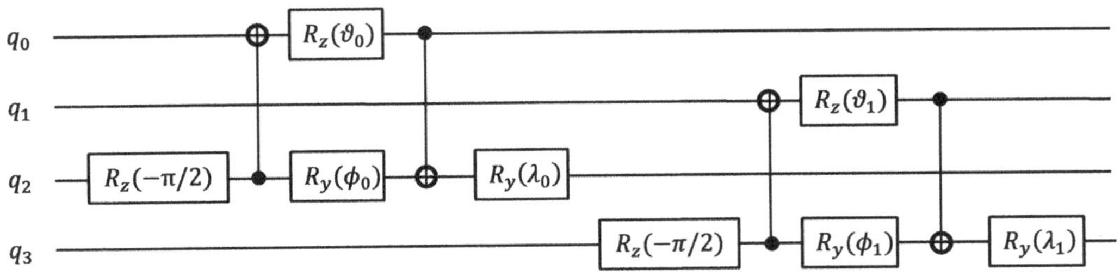

Fig. 13.27 Example of a pooling layer for four qubits where q_0, q_1, q_2, and q_3 are the output of the convolutional layer previously mentioned.

Now, we consider the process of initializing, training, and evaluating a QNN classifier using the COBYLA optimization algorithm.

Classifier initialization: Utilizing the Neural Network Classifier from Qiskit's machine learning library, the classifier used is the Constrained Optimization by Linear Approximations (COBYLA) optimizer, a robust method suited for handling nonlinear constraints without requiring derivatives of the objective function. This optimizer is favored for its proficiency in managing complex optimization landscapes that are typical in quantum computing scenarios. The initialization of the classifier with the COBYLA optimizer is important due to its parameter specifying the maximum number of iterations. This parameter balances the computational resource expenditure against the potential gains in model accuracy.

Training process: During the training phase, the classifier utilizes a callback function that acts as a monitoring tool. This function is invoked at each iteration of the optimization process, providing real-time updates on the progress of the model training. It typically captures metrics such as the value of the objective function at each step, facilitating an immediate visual assessment of the model's convergence behavior through plotted outputs. This continuous feedback allows for adjustments to be made promptly if the training does not progress as expected, enhancing the overall efficiency of the model development process. During this phase, the model learns to classify images by adjusting the parameters of the QNN based on the feedback received from the optimizer regarding the performance of the model, as measured by the callback function, where the expectation value of the Pauli Z operator of the final qubit is measured and based on the obtained value being $+1$ or -1, the input image can be classified as containing either a horizontal or a vertical line.

The classifier's training is executed by fitting it to labeled training data. This process adjusts the internal parameters of the QNN to minimize discrepancies between predicted and actual labels, measured by the objective function. The efficiency of this training is quantitatively evaluated by computing the accuracy of the classifier on the training data, providing a direct indicator of how well the classifier has learned to predict the correct labels based on the input features.

Model evaluation: After training, the model's performance is evaluated in two stages:

1) **Training set evaluation:** The classifier's accuracy on the training data is computed to gauge how well the model has learned to classify the data it was trained on. This is done using the "score" method, which returns the accuracy, the proportion of correctly classified instances in the training set.
2) **Testing set evaluation:** Similarly, the classifier's accuracy is also measured on a separate test dataset, which was not used during the training phase. This step is critical for assessing the model's generalization ability, that is, how well it performs on new, unseen data.

Both accuracy metrics are computed by comparing the predicted labels against the true labels of the respective datasets, and the results are expressed as percentages as shown in the code in Figure 13.28.

```
# TrainClassifier.py by Alexis Gomez
# Initializing the classifier with the QNN and the COBYLA optimizer
classifier = NeuralNetworkClassifier(
qnn,
optimizer = COBYLA(maxiter=1200), # Sets maximum iterations for the optimizer
callback=callback_function) # Callback function to for the training
# Convert training data to NumPy arrays for processing
train_images = np.asarray(train_images)
train_labels = np.asarray(train_labels)
# List to store values of the objective function after each iteration
objective_function_vals = []
# Train the quantum neural network classifier
classifier.fit(train_images, train_labels)
# Evaluate the classifier on the training data
train_accuracy = np.round(100 * classifier.score(train_images, train_labels), 2)
print(f"Accuracy from the train data: {train_accuracy}%")
# Evaluate the classifier on the test data
test_images = np.asarray(test_images)
test_labels = np.asarray(test_labels)
test_accuracy = np.round(100 * classifier.score(test_images, test_labels), 2)
print(f"Accuracy from the test data: {test_accuracy}%")
```

Fig. 13.28 Python code using Qiskit for training the classifier based the circuit in Fig. 13.22 and the subsequent circuits (feature map, convolutional, and pooling circuits) previously described.

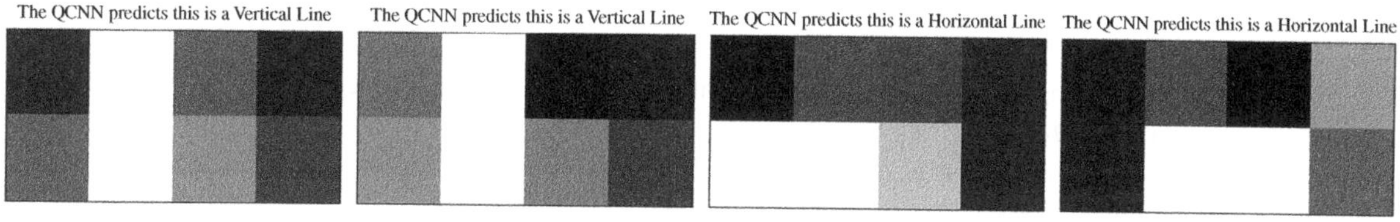

Fig. 13.29 QCNN visualization of some results with a train accuracy of 97.3% and a test accuracy of 96.0%.

A dataset with specified patterns of horizontal and vertical lines is generated where each image in the dataset is either a representation of a horizontal or a vertical line (see Fig. 13.29). These patterns are filled with a value of $\pi/2$ where random noise is added to blank spaces. This dataset is labelled and then fed to the classifier in the code given in Fig. 13.28 to yield the results as in Fig. 13.29.

13.9 The Current and Future Trends and Developments in Quantum Neural Networks

QNNs require quantum hardware and software, which are still in the early stages of development and not widely available or accessible. At the same time, CNNs can run on classical hardware and software, which are more mature and ubiquitous.

a) QNNs are subject to noise and decoherence, which degrades the quality and coherence of quantum states and operations, while CNNs are more robust and stable against noise and errors.

b) QNNs are challenging to interpret and explain, as quantum mechanics is based on probabilistic and counter-intuitive principles. At the same time, CNNs are more intuitive and understandable, as classical mechanics is based on deterministic and familiar principles.
c) QNNs can be difficult to interpret or explain, due to their black-box or opaque nature. The black-box or opaque nature of QNNs can make it hard to understand how the network works, how the network makes decisions, or what the network learns from the data. Note explainable QNNs can (i) increase the transparency and interpretability of the network, as well as the trust and confidence of the users or the stakeholders; (ii) facilitate the debugging and the improvement of the network, as well as the discovery of new insights or knowledge from the data; and (iii) pose problems for trust, accountability, or ethics, especially in sensitive or critical domains, such as health care, finance, or security.
d) QNNs can be difficult to scale up or down, due to their high computational or memory requirements. The high computational or memory requirements of QNNs can make it hard to train or test the network, especially on large-scale or high-dimensional data. This can pose problems for efficiency, performance, or accuracy, especially in resource-constrained or real-time scenarios, such as mobile devices, edge computing, or the IoT.
e) QNNs can be prone to overfitting or underfitting, due to their high flexibility or complexity. Overfitting occurs when the network learns too much from the training data and fails to generalize to new or unseen data. Underfitting occurs when the network learns too little from the training data and fails to capture the essential features or patterns of the data. This can pose problems for validity, reliability, or robustness, especially in noisy or incomplete data, such as missing values, outliers, or anomalies.

References

1 Yunseok K., Yun, W.J., Jung, S., and Kim, J. (2018). Quantum neural networks: concepts, applications, and challenges. 4, *arXiv:2108.01468*.

2 Caro, M.C., Huang, H.Y., Ezzell, N. et al. (2023). Out-of-distribution generalization for learning quantum dynamics. *Nature Communications* 14: 3751. https://doi.org/10.1038/s41467-023-39381-w.

3 Hur, T., Kim, L., and Park, D.K. (2022). Quantum convolutional neural network for classical data classification. *Quantum Machine Intelligence* 4 (3): https://doi.org/10.1007/s42484-021-00061-x.

4 Henderson, M., Shakya, S., Pradhan, S., and Cook, T. (2020). Quanvolutional neural networks: powering image recognition with quantum circuits. *Quantum Machine Intelligence* 2 (2): https://doi.org/10.1007/s42484-020-00012-y.

5 Nguyen, N.H. (2020). Quantum neural networks. Doctoral dissertation, Wichita State University.

6 Adriano B, Bennett, C.H., Cleve, R. et al. (1995) 'Elementary gates for quantum computation,' *Physical Review A*, 52 (2), pp. 3457–3467. doi: https://doi.org/10.1103/PhysRevA.52.3457.

7 Jeswal, S.K. and Chakraverty, S. (2019). Recent developments and applications in quantum neural network: a review. *Archives of Computational Methods in Engineering* 26: 793–807. https://doi.org/10.1007/s11831-018-9269-0.

8 Biamonte, J., Wittek, P., Pancotti, N. et al. (2017). Quantum machine learning. *Nature* 549 (7671): 195–202.

9 Huynh, L., Hong, J., Mian, A. et al. (2023). Quantum-inspired machine learning: a survey. *arXiv:2308.11269v2*.

10 Ventura, D. (1998). Artificial associative memory using quantum processes. *Proceedings of the International Conference on Computational Intelligence and Neuroscience* 2: 218–221. https://axon.cs.byu.edu/papers/ventura.jcis98.pdf.

11 Bokhan, D., Mastiukova, A.S., Boev, A.S. et al. (2022). Multiclass classification using quantum convolutional neural networks with hybrid quantum-classical learning. *Frontiers in Physics* 10: https://www.frontiersin.org/articles/10.3389/fphy.2022.1069985.

12 Ventura, D. and Martinez, T. (1998). Quantum associative memory with exponential capacity. *Proceedings of the International Joint Conference on Neural Networks*, 509–513. https://axon.cs.byu.edu/papers/ventura.ijcnn98a.pdf

13 Grigoryan, A.M. and Grigoryan, M.M. (2006). Nonlinear approach of construction of fast unitary transforms. *2006 40th Annual Conference on Information Sciences and Systems*, Princeton, NJ, USA, 1073–1078. https://ieeexplore.ieee.org/abstract/document/4067966.

14 Tacchino, F., Macchiavello, C., Gerace, D. et al. (2019). An artificial neuron implemented on an actual quantum processor. *npj Quantum Information* 5: 26. https://doi.org/10.1038/s41534-019-0140-4.

15 Grigoryan, A.M. and Grigoryan, M.M. (2009). *Brief Notes in Advanced DSP: Fourier Analysis with MATLAB.* CRC Press, Taylor and Francis Group.

16 Grigoryan, A.M. and Grigoryan, M.M. (2007). Discrete unitary transforms generated by moving waves. *Proceedings of the International Conference: Wavelets XII, SPIE: Optics + Photonics 2007*, 6701, article id. 670125. https://doi.org/10.1117/12.728383.

17 Menneer, T. and Narayanan, A. (1995). Quantum-inspired neural networks. *Technical report R329*. Department of Computer Science, University of Exeter, UK.

18 Changpeng Shao (2020). Data classification by quantum radial-basis-function networks, *Physical Review A*, 102 (4) 042418. doi: https://doi.org/10.1103/PhysRevA.102.042418.

19 Wan, K.H., Dahlsten, O., Kristjánsson, H. et al. (2017). Quantum generalization of feedforward neural networks. *NPJ Quantum Information* 3: 36. https://doi.org/10.1038/s41534-017-0032-4.

20 Choi, J. Oh, S., and Kim J. (2021). A Tutorial on quantum graph recurrent neural network (QGRNN). *2021 International Conference on Information Networking (ICOIN)*, 46–49. https://api.semanticscholar.org/CorpusID:231826284.

21 Alanis, D., Botsinis, P., Ng, S.X., and Hanzo, L. (2014). Quantum-assisted routing optimization for self-organizing networks. *IEEE Access* 2: 614–632. https://doi.org/10.1109/ACCESS.2014.2327596.

22 Chen, Y. (2022). Quantum dilated convolutional neural networks. *IEEE Access* 10: 20240–20246. https://doi.org/10.1109/ACCESS.2022.3152213.

23 Chalumuri, A., Kune, R., and Manoj, B.S. (2021). A hybrid classical-quantum approach for multi-class classification. *Quantum Information Processing* 20: 119. https://doi.org/10.1007/s11128-021-03029-9.

24 Hur, T., Kim, L., and Park, D.K. (2022). Quantum convolutional neural network for classical data classification. *Quantum Machine Intelligence* 4: 3. https://doi.org/10.1007/s42484-021-00061-x.

25 Shi, S., Wang, Z., Shang, R. et al. (2023). Hybrid quantum-classical convolutional neural network for phytoplankton classification. *Frontiers in Marine Science* 10: 1158548. https://doi.org/10.3389/fmars.2023.1158548.

26 Liu, C.Y., Kuo, E.J., Lin, C.H.A. et al. (2024). Training classical neural networks by quantum machine learning. 8. https://arxiv.org/html/2402.16465v1#bib.bib33

27 Abbas, A., Sutter, D., and Wörner, S. (2022). The power of quantum neural networks. *IBM Research Blog*, IBM, 3 August. https://research.ibm.com/blog/quantum-neural-network-power.

28 Deng, L. (2012). The MNIST database of handwritten digit images for machine learning research. *IEEE Signal Processing Magazine* 29: 141.

29 Chalumuri, Q., Kune, R., Kannan, S., and Manoj, B.S. (2021). Quantum-enhanced deep neural network architecture for image scene classification. *Quantum Information Processing* 20: 381. https://doi.org/10.1007/s11128-021-03314-7.

30 Weikang, L.Zhide, L., and Dong-Ling, D. (2022). Quantum neural network classifiers: a tutorial. *SciPost Physics Lecture Notes* 61. doi: https://doi.org/10.21468/SciPostPhysLectNotes.61

31 Vatan, F. and Williams, C. (2004). Optimal quantum circuits for general two-qubit gates. *Physical Review A* 69 (3): 032315.

32 Cong, I., Choi, S., and Lukin, M.D. (2019). Quantum convolutional neural networks. *Nature Physics* 15 (12): 1273–1278.

14

Conclusion and Opportunities and Challenges of Quantum Image Processing

We introduced essential quantum information concepts like qubits, superposition, entanglement, gates, and circuits, setting the stage for their creative usage in image analysis tasks. Core topics included various quantum image representations and associated operations, fast transforms, quaternion arithmetic, quantum neural networks, and more.

Several chapters presented original methods developed by the authors, such as novel quantum multiplication schemes, discrete signal-induced heap transforms, and color image and quaternion qubit image representation. These contributions address the limitations of existing techniques and expand the methodological toolkit available to quantum image processing researchers.

While still largely theoretical today, the field displays immense promise to exploit quantum advantages in areas like image compression, filtering, segmentation, and reconstruction. However, many open challenges remain, including demonstrating definitive quantum speedups, ensuring fault-tolerances and computational stability, addressing quantum measurement constraints, and ultimately realizing practical systems.

Combining quantum information science with image processing is a fascinating and rapidly evolving research frontier. The techniques covered in this book highlight the transformative potential of applying quantum principles to visualize, store, manipulate, and analyze high-dimensional visual data. Nonetheless, the fusion of quantum information science with image processing forms a fascinating and rapidly evolving frontier for both research and practical use. These preliminary concepts will seed further interdisciplinary innovations at the nexus of quantum computing and computer vision.

There remain abundant open problems to motivate future work [1–17], including the following:

1) **Benchmarking quantum image processing methods:** Creating meaningful benchmarks for quantum image processing is difficult because we still do not know exactly when quantum methods outperform classical ones. Simply running quantum algorithms on standard datasets is not enough. Instead, we must think carefully about what makes real-world imaging problems challenging and how to recreate these challenges in smaller, manageable test cases. Such carefully chosen data will ensure that benchmarks measure true quantum advantages, not just trivial gains. This search for the right data becomes a "chicken-and-egg" problem. We do not know which kinds of data highlight quantum advantages, so it is hard to pick suitable datasets. At the same time, without the right data, it is tough to prove that quantum methods can offer real benefits. Solving this issue will require experts in both classical machine learning and quantum computing to work together. By designing datasets that capture the unique strengths of quantum systems, we can create solid benchmarks that guide researchers toward developing genuinely superior quantum image processing solutions.
2) **Error-correction and robust quantum image representations:** Representing images as quantum states opens new ways to encode and process visual data, but these states are very delicate. Errors can arise from imperfect quantum gates, environmental interference, and the methods used to prepare the states.

Quantum Image Processing in Practice: A Mathematical Toolbox, First Edition. Artyom M. Grigoryan and Sos S. Agaian.

Companion website: www.wiley.com/go/grigoryan/quantumimageprocessing

Researchers are exploring specialized error-correcting codes, fault-tolerant qubits, and flexible encoding methods designed for image tasks to address this. By protecting quantum states from noise and building circuits that limit how errors spread, we can hope to process quantum images on a larger-scale reliably. A strong approach will likely blend ideas from quantum information science with classical image processing techniques. Some image features or transformations may handle errors better than others, allowing us to use quantum operations only where they provide the greatest benefit. In the end, the success of quantum image representations depends on controlling errors so that the advantages of quantum speedups and improved quality are not lost. As error-mitigation strategies improve, quantum image processing will become more practical and effective.

3) **Quantum image encoding, storage, and compression:** Quantum encoding and compression use superposition and entanglement to simultaneously represent and process many image states. Theoretically, this can greatly reduce storage and bandwidth compared to classical methods, making it attractive for medical imaging or satellite data analysis fields. However, current quantum hardware is too limited to realize these benefits. Decoherence, too few qubits, and high error rates make it hard to encode and compress images effectively. Achieving real improvements will require better quantum memory, fewer errors, and smarter compression strategies. Researchers must also compare quantum gains against classical methods to show that quantum compression is worthwhile.
4) **Quantum-assisted image encryption and geometric transformations:** As we rely more on digital images for communication and security, quantum computing brings both risks and opportunities. On the one hand, quantum algorithms might break current encryption, exposing private image data. On the other hand, new quantum-safe encryption methods and entanglement-based key sharing could offer stronger protection. Quantum algorithms may also perform geometric and color changes (like rotation or scaling) more efficiently than classical methods. However, current hardware limits still prevent practical demonstrations. Making quantum-based transformations useful will require careful algorithm design, reliable circuits, and easy ways to add them to existing image-processing tools.
5) **Quantum image retrieval:** Image retrieval finds the right images in a large dataset based on similarity, features, or content. Quantum approaches could speed this up using quantum parallelism and special similarity measures. Quantum search algorithms like Grover's might help find relevant images faster than classical methods. However, storing, preparing, and handling images in a quantum format without errors are tricky. Converting classical image features into quantum states and getting classical results back is also challenging. Researchers must solve these issues and create fair tests to prove that quantum retrieval can work well in real-world settings.
6) **Quantum pattern recognition and feature extraction:** Pattern recognition is central to many image tasks, like detecting objects or faces. Quantum algorithms might find features and patterns faster by handling image data in superposition. Entanglement could help detect geometric shapes under different sizes and angles more efficiently than classical approaches. Still, turning theory into practice is hard. You cannot clone quantum states, preparing them from classical images is costly. Also, most existing feature extraction methods were designed for classical images and need to be adapted to quantum formats. Only by solving these problems can quantum pattern recognition reach its full potential.
7) **Algorithmic design and implementation of quantum image circuits:** Building efficient quantum circuits for classification, denoising, or segmentation is challenging. While some early tests show that quantum methods can classify images, it is unclear whether they can beat classical techniques. Researchers must simplify circuits, use fewer gates, and design structures that fully leverage quantum properties. Combining knowledge from computer vision, quantum algorithms, and hardware engineering is key. Hybrid workflows, where some parts run on classical computers before passing the data to a quantum device, may help. By comparing quantum and classical results, we can learn which tasks benefit most from quantum computing.
8) **Security and robustness against attacks:** Quantum image processing introduces new security concerns. Attackers might try to read or change quantum states or poison the training data. They could insert hidden

backdoors into quantum circuits. Defending against these threats requires quantum-safe protocols, secure key distribution, and strong protective measures. Another challenge is scaling up to large, complex images without losing stability. Bigger images need more qubits and deeper circuits, which increases the chance of errors. Finding ways to handle noise, reduce errors, and make algorithms more stable will be critical for using quantum image processing in sensitive applications.

9) **Identifying quantum advantages and hybrid approaches:** A major question is when quantum methods actually outperform classical ones. Quantum learning theory tries to find problems that quantum computers can solve more efficiently. Can quantum methods do feature extraction or compression better for image processing than classical techniques? Hybrid models, where some steps happen on a quantum device and others on a classical computer, might help us get benefits sooner. Showing clear gains is crucial. If we can prove that certain image tasks are faster or more accurate with quantum help, it will guide the future of quantum computing in imaging.
10) **Fault-tolerance and quantum error correction strategies:** Fault-tolerant quantum computing aims to operate reliably even if individual qubits fail or encounter errors. Image data, which is inherently noisy, poses additional challenges, as quantum states are susceptible to errors from imperfect gates and environmental interference. This makes robust error correction methods and stable state preparation techniques essential for accurate, real-world quantum image processing. These strategies pave the way for practical quantum image representations by ensuring that error-corrected qubits outperform their noisy counterparts. Advances in topological quantum computing and other resilient architectures may significantly reduce or even eliminate quantum noise. Interestingly, some noisy quantum systems can tackle problems that remain too difficult for equally noisy classical systems. Whether these theoretical advantages translate into real imaging benefits depends on balancing the cost of error correction with the speedups promised by quantum computing. As error correction techniques improve, quantum image processing will become more stable, scalable, and ready for practical use.
11) **Linear quantum mechanics vs. nonlinear classical techniques:** Quantum mechanics is linear, but many successful image processing methods rely on nonlinear operations like the sigmoid function in deep learning. Introducing nonlinearity into quantum algorithms is tricky. One option is to use measurements or classical post-processing, but this might lose some quantum advantages. Another idea is to design special quantum kernels or hybrid algorithms where classical parts handle nonlinear steps. The goal is to find the right mix so that quantum image processing can offer at least the same flexibility and power as classical techniques, if not more.
12) **Quantum image quality assessment:** Current image quality measures, such as Peak signal-to-noise ratio (PSNR), were made for classical images and may not work well for quantum images. We need new metrics that consider quantum features like entanglement and how we measure quantum states. These new metrics will help us check if quantum methods improve image clarity, keep essential details, or create hard-to-spot artifacts. It can help benchmark and optimize algorithm parameters. By defining clear standards, researchers can tune their algorithms, improve hardware, and fairly compare quantum and classical approaches. As the field matures, these metrics will be key tools for ensuring that quantum processing helps with image analysis.
13) **Quantum data hiding and steganography:** Quantum data hiding uses quantum states to embed information invisibly, supporting both copyright protection and covert communication. Quantum watermarking adds ownership marks to quantum multimedia, while quantum steganography conceals secret messages. Scaling these techniques to large images is challenging since more qubits and operations are needed. Researchers must develop more qubit-efficient encodings, better compression, and simpler circuits. Although some progress has been made, quantum data hiding is still in its early stages. Reliable data extraction, defense against quantum-based attacks, and integration with quantum communication networks remain open problems. Stronger error correction, well-defined evaluation metrics, and clear demonstrations of quantum advantage will be crucial to proving that quantum data hiding can securely protect multimedia in the future.

14) **Incorporating color into quantum image processing:** Color contains richer information than grayscale, aiding object identification, segmentation, and interpretation. However, treating each color channel separately can break essential relationships between them, complicating analysis. Additionally, classical nonlinear operations can collapse quantum states, erasing potential quantum benefits. Using quaternion-based math to treat all primary color components as a single unit may preserve these relationships and enable nonlinear transformations without losing quantum properties. This strategy, possibly enhanced by quaternion Fourier transforms on multi-qubit states, could lead to more efficient and accurate quantum-based color image processing.
15) **The intersection of machine learning and quantum computing:** Combining machine learning with quantum computing promises faster learning and the ability to handle complex data structures that challenge classical methods. Quantum neural networks (QNNs), which apply neural network principles to quantum states, could potentially outperform classical deep learning. However, these systems face significant challenges: current quantum hardware and Noisy Intermediate-Scale Quantum (NISQ) devices offer limited qubits and are prone to errors from noise and decoherence. Scaling QNNs to handle large datasets is difficult since the required number of qubits grows quickly with problem size. Moreover, QNNs suffer from the vanishing gradient problem, a known issue in classical deep learning. Classical solutions rely on nonlinear activation functions, which do not translate directly to QNNs, necessitating new training strategies. Designing QNN algorithms that benefit from quantum speedups while tolerating noise and errors is challenging. Classical neural network techniques often cannot be applied as-is to the quantum domain, requiring innovative approaches. Training QNNs involves high-dimensional optimizations and balancing both quantum and classical computational resources, a task made even more complex by the lack of mature quantum optimization algorithms. Finally, developing QNNs demands expertise in both quantum mechanics and machine learning, a combination not yet widespread. Despite these challenges, the potential for QNNs to surpass classical networks in speed, capacity, and problem complexity makes this a promising, if still nascent, field of research.

To gain an in-depth understanding of some current research frontiers and limitations in quantum image processing, we refer the reader to the following papers [1, 2, 4, 6], which provide comprehensive reviews, surveys, and analyses of the state-of-the-art methods and challenges in this field.

Quantum image processing is a fascinating and promising field that offers many opportunities and challenges for innovation and discovery. We invite you to join us in exploring the potential and limitations of quantum technologies for visual data processing.

Contact us if you have any questions, comments, or feedback. We would love to hear from you and learn from your perspectives.

References

1 Ruan, Y., Xue, X., and Shen, Y. (2021). Quantum image processing: opportunities and challenges. *Mathematical Problems in Engineering* 2021 (1): 6671613.

2 Chakraborty, S., Mandal, S.B., and Shaikh, S.H. (2022). Quantum image processing: challenges and future research issues. *International Journal of Information Technology* 14: 475–489. https://doi.org/10.1007/s41870-018-0227-8.

3 Yan, F., Venegas-Andraca, S.E., and Hirota, K. (2023). Toward implementing efficient image processing algorithms on quantum computers. *Soft Computing* 27: 13115–13127. https://doi.org/10.1007/s00500-021-06669-2.

4 Holter, C.T., Inglesant, P., and Jirotka, M. (2023). Reading the road: challenges and opportunities on the path to responsible innovation in quantum computing. *Technology Analysis & Strategic Management* 35 (7): 844–856. https://doi.org/10.1080/09537325.2021.1988070.

5 Singh, J. and Bhangu, K.S. (2022). Contemporary quantum computing use cases: taxonomy, review, and challenges. *Archives of Computational Methods in Engineering* 30 (1): 615–638.

6 Zhang, L. and Zhang, L. (2022). Artificial intelligence for remote sensing data analysis: a review of challenges and opportunities. *IEEE Geoscience and Remote Sensing Magazine* 10 (2): 270–294. https://doi.org/10.1109/MGRS.2022.3145854.

7 Peral-García, D., Cruz-Benito, J., and García-Peñalvo, F. (2024). Systematic literature review: Quantum machine learning and its applications, *Computer Science Review*, Volume 51, 100619, ISSN 1574–0137, https://doi.org/10.1016/j.cosrev.2024.100619.

8 Wang, Z., Xu, M., and Zhang, Y. (2022). Review of quantum image processing. *Archives of Computational Methods in Engineering* 29: 737–761.

9 Grilo, B., Kerenidis, I., and Zijlstra, T. (2019). Learning-with-errors problem is easy with quantum samples. *Physical Review A* 99 (3): 032314, 2019.

10 Grigoryan, A. Agaian, S. (2024) "Quaternion-based arithmetic in quantum information processing: a promising approach for efficient color quantum imaging [Hypercomplex Signal and Image Processing]," *IEEE Signal Processing Magazine*, vol. 41, no. 2, pp. 64–74, doi: https://doi.org/10.1109/MSP.2023.3327627.

11 Grigoryan, A. and Agaian, S. (2022). Quantum representation of 1-D signals on the unit circle. *Quantum Information Processing* 21: 24. https://doi.org/10.1007/s11128-021-03237-3.

12 Grigoryan, A. and Agaian, S. (2020). New look on quantum representation of images: Fourier transform representation. *Quantum Information Processing* 19: 148. https://doi.org/10.1007/s11128-020-02643-3.

13 Collins, J. and Agaian, S. (2016). Trends toward real-time network data steganography. *arXiv preprint arXiv*:1604.02778.

14 Collins, J. and Agaian, S. (2016). High-capacity image steganography using adjunctive numerical representations with multiple bit-plane decomposition methods. *arXiv preprint arXiv*:1606.02312.

15 Collins J, Agaian S. (2014). Taxonomy for spatial domain LSB steganography techniques, *Proceedings of the SPIE*, Volume 9120, id. 912006 15 pp.

16 Agaian, S. (2008). *Steganography & Steganalysis, An Overview of Research & Challenges*, Aspects of Network Security and Information Security, NATO Science for Peace and Security Series D, Information and Communication Security, vol. 17, 179–210.

17 Cerezo, M., Verdon, G., Huang, H.Y. et al. (2022). Challenges and opportunities in quantum machine learning. *Nature Computational Science* 2: 567–576. https://doi.org/10.1038/s43588-022-00311-3.

Index

Printed and bound by CPI Group (UK) Ltd, Croydon, CR0 4YY
21/09/2026
14965807-0001